Dermatological and Transdermal Formulations

DRUGS AND THE PHARMACEUTICAL SCIENCES

A Series of Textbooks and Monographs

1. Pharmacokinetics, *Milo Gibaldi and Donald Perrier*
2. Good Manufacturing Practices for Pharmaceuticals: A Plan for Total Quality Control, *Sidney H. Willig, Murray M. Tuckerman, and William S. Hitchings IV*
3. Microencapsulation, *edited by J. R. Nixon*
4. Drug Metabolism: Chemical and Biochemical Aspects, *Bernard Testa and Peter Jenner*
5. New Drugs: Discovery and Development, *edited by Alan A. Rubin*
6. Sustained and Controlled Release Drug Delivery Systems, *edited by Joseph R. Robinson*
7. Modern Pharmaceutics, *edited by Gilbert S. Banker and Christopher T. Rhodes*
8. Prescription Drugs in Short Supply: Case Histories, *Michael A. Schwartz*
9. Activated Charcoal: Antidotal and Other Medical Uses, *David O. Cooney*
10. Concepts in Drug Metabolism (in two parts), *edited by Peter Jenner and Bernard Testa*
11. Pharmaceutical Analysis: Modern Methods (in two parts), *edited by James W. Munson*
12. Techniques of Solubilization of Drugs, *edited by Samuel H. Yalkowsky*
13. Orphan Drugs, *edited by Fred E. Karch*
14. Novel Drug Delivery Systems: Fundamentals, Developmental Concepts, Biomedical Assessments, *Yie W. Chien*
15. Pharmacokinetics: Second Edition, Revised and Expanded, *Milo Gibaldi and Donald Perrier*
16. Good Manufacturing Practices for Pharmaceuticals: A Plan for Total Quality Control, Second Edition, Revised and Expanded, *Sidney H. Willig, Murray M. Tuckerman, and William S. Hitchings IV*
17. Formulation of Veterinary Dosage Forms, *edited by Jack Blodinger*
18. Dermatological Formulations: Percutaneous Absorption, *Brian W. Barry*
19. The Clinical Research Process in the Pharmaceutical Industry, *edited by Gary M. Matoren*
20. Microencapsulation and Related Drug Processes, *Patrick B. Deasy*
21. Drugs and Nutrients: The Interactive Effects, *edited by Daphne A. Roe and T. Colin Campbell*
22. Biotechnology of Industrial Antibiotics, *Erick J. Vandamme*
23. Pharmaceutical Process Validation, *edited by Bernard T. Loftus and Robert A. Nash*
24. Anticancer and Interferon Agents: Synthesis and Properties, *edited by Raphael M. Ottenbrite and George B. Butler*
25. Pharmaceutical Statistics: Practical and Clinical Applications, *Sanford Bolton*
26. Drug Dynamics for Analytical, Clinical, and Biological Chemists, *Benjamin J. Gudzinowicz, Burrows T. Younkin, Jr., and Michael J. Gudzinowicz*
27. Modern Analysis of Antibiotics, *edited by Adjoran Aszalos*
28. Solubility and Related Properties, *Kenneth C. James*
29. Controlled Drug Delivery: Fundamentals and Applications, Second Edition, Revised and Expanded, *edited by Joseph R. Robinson and Vincent H. Lee*
30. New Drug Approval Process: Clinical and Regulatory Management, *edited by Richard A. Guarino*

ADDITIONAL VOLUMES IN PREPARATION

Dermatological and Transdermal Formulations

edited by

Kenneth A. Walters
An-eX Analytical Services Ltd.
Cardiff, Wales

CRC Press
Taylor & Francis Group
Boca Raton London New York

CRC Press is an imprint of the
Taylor & Francis Group, an **informa** business

CRC Press
Taylor & Francis Group
6000 Broken Sound Parkway NW, Suite 300
Boca Raton, FL 33487-2742

First issued in paperback 2019

© 2007 by Taylor & Francis Group, LLC
CRC Press is an imprint of Taylor & Francis Group, an Informa business

No claim to original U.S. Government works

ISBN-13: 978-0-8247-9889-5 (hbk)
ISBN-13: 978-0-367-39634-3 (pbk)

Visit the Taylor & Francis Web site at
http://www.taylorandfrancis.com

and the CRC Press Web site at
http://www.crcpress.com

Preface

The past two decades have witnessed brilliant discoveries regarding the structure and functions of the stratum corneum.

—*Albert Kligman, 2000*

An immense amount of research has been carried out over the past two decades on the micromorphology of the skin, in particular of the stratum corneum, and the important role that this organ plays in the maintenance of human life. It has also been nearly two decades since the publication of Brian Barry's book *Dermatological Formulations—Percutaneous Absorption*. This book remains one of the most widely and frequently cited references in the field of skin transport and also has been used extensively as an introduction to the complexities surrounding the theory and development of topical pharmaceutical products.

The introduction and subsequent success of transdermal therapeutic systems have advanced our understanding of the structure of the skin and the mechanisms of transport through the barrier membrane. In addition, technological developments in molecular biology and pharmacology have led to an increased understanding of the biochemistry of skin diseases. The result is the introduction of new therapeutic strategies that use both existing and new chemical entities to treat skin diseases. This volume serves as a useful addition to the literature in the dermatopharmaceutics field.

The rational treatment of skin diseases, based on the biochemical mechanisms underlying the pathology, is discussed in Chapter 2. For example, vitamin D_3 derivatives, such as calcipotriol and calcitriol, have recently been introduced as topical therapeutic modalities for the treatment of psoriasis. This is a result of the finding that some compounds of this type possess a high binding affinity to specific cellular receptors and are potent regulators of cell differentiation and inhibitors of cell proliferation in human keratinocytes. Cosmetic scientists have long known the epidermal advantages of another vitamin, retinol (vitamin A). Deficiency of this vitamin has been implicated in squamous metaplasia and keratinization of epithelial tissue, and several derivatives have been synthesized and evaluated for their effects in such diseases as acne, psoriasis, and hyperkeratosis. The results have been somewhat variable; however, the recent identification of several receptor proteins for retinoic acid should lead to the development of more potent analogs with fewer side effects.

The ability to enhance skin penetration and permeation has been the subject of considerable research over the past two decades and is reviewed in Chapter 6. The science of penetration enhancement has expanded considerably over the past few years, and it is now possible to increase drug delivery across the skin using both chemical and physical means. Various synthetic (e.g., SEPA® and Azone®) and natural (e.g., terpenes) compounds have proved useful in this respect. Moreover, there is evidence that the skin penetration of large molecules such as insulin can be increased using physical methods of enhancement, such as iontophoresis.

The use of the skin as a drug delivery route for both topical and systemic therapy is covered in Chapter 7. Transdermal drug delivery using patches or semisolid formulations is now a reality with products available for travel sickness, hypertension, angina, postmenopausal symptoms, male hypogonadism, pain, inflammation, and smoking cessation. Problems of irritation are being overcome with the development of skin-compatible materials, such as some of the newer pressure-sensitive adhesives. The success of such systems has been achieved only by means of a greater understanding of the physical and biochemical nature of the permeation routes through the skin, especially in relation to the intercellular lipid lamellae of the stratum corneum (as discussed in Chapters 1, 3, and 4). In addition, the methods of studying percutaneous absorption, both in vivo and in vitro, have become more standardized thanks to the efforts of the American Association of Pharmaceutical Scientists (AAPS), the U.S. Food and Drug Administration (FDA), and other industrial and regulatory bodies. Chapter 5 provides a complete description of the AAPS/FDA guidelines for such experimentation, together with a full evaluation of the in vivo tape-stripping procedure. This chapter also gives a description of the use of cultured skin membranes for the study of irritation and other toxic responses to materials applied to the skin.

The formulation of dermatological vehicles has become more innovative with the introduction of many new excipient materials and the development of delivery systems made up of vesicles such as liposomes and niosomes. These are discussed in Chapters 6 and 7. As an example, the feasibility of using supersaturated solutions as vehicles for improving dermal drug delivery has been established. This type of strategy will undoubtedly reduce the amount of drug necessary for a therapeutic effect, which should result in fewer local side effects and a lower incidence of unwanted systemic effects. The use of liposomes and niosomes in cosmetic formulation is reputed to impart beneficial properties to such products as moisturizers. Although these types of vesicles can be useful in the targeting of pharmaceutical agents to epidermal sites of action, their usefulness in transporting drugs across the skin to systemic sites has yet to be fully established. The usefulness of multivariate optimization in the scale-up of dermatological dosage forms is discussed in Chapter

9. Chapter 8 covers bioequivalence of dermal and transdermal systems. Safety considerations are outlined in Chapters 10 and 11.

This book will be useful to pharmacy students and practitioners and cosmetic and veterinary scientists. It may also prove useful to toxicologists working in the field of risk assessment and dermatologists requiring a deeper understanding of the mechanisms of drug transport through the skin. The chapters have been authored by international experts in their fields and provide a comprehensive review of current dermatopharmaceutics.

I acknowledge, and am extremely grateful for, the hard work and infinite patience shown by the contributors to this volume. I would also like to acknowledge the unreserved help provided by my colleague Dr. Keith Brain. Finally, since the written word cannot fully express my love and gratitude to Peggy for her support and encouragement during many hours of word processing and reference hunting, this absentee husband will find other ways.

Kenneth A. Walters

Contents

Contributors

Keith R. Brain, Ph.D. An-eX Analytical Services Ltd., Cardiff, Wales

Sheree Elizabeth Cross, Ph.D. Department of Medicine, University of Queensland, Princess Alexandra Hospital, Brisbane, Queensland, Australia

Adrian F. Davis, Ph.D. GlaxoSmithKline Consumer Healthcare, Weybridge, Kent, England

Peter J. Dykes, B.Sc., Ph.D. Cutest (Skin Toxicity Testing Company), Cardiff, Wales

Bruce A. Firestone, Ph.D. Allergan, Inc., Irvine, California

Robert J. Gyurik MacroChem Corporation, Lexington, Massachusetts

Jonathan Hadgraft, D.Sc., F.R.S.C. Medway Sciences, University of Greenwich, Chatham Maritime, Kent, England

C. Colin Long, F.R.C.P. Department of Dermatology, Cardiff and Vale NHS Trust, Cardiff, Wales

Howard I. Maibach, M.D. Department of Dermatology, University of California School of Medicine, San Francisco, California

Orest Olejnik, Ph.D., M.R.Pharm.S.   Allergan, Inc., Irvine, California

Anthony D. Pearse, M.Sc., M.I.Biol., C.Biol., F.I.Sc.T.   Department of Dermatology, University of Wales College of Medicine, and Cutest (Skin Toxicity Testing Company), Cardiff, Wales

Mark A. Pellett, Ph.D., M.R.Pharm.S.   Whitehall International, Havant, England

Michael S. Roberts, Ph.D., D.Sc.   Department of Medicine, University of Queensland, Princess Alexandra Hospital, Brisbane, Queensland, Australia

Jagdish Singh   Department of Pharmaceutical Sciences, College of Pharmacy, North Dakota State University, Fargo, North Dakota

Christian Surber, Ph.D., Priv.-Doz. Dr.   Institute of Hospital-Pharmacy, University Clinics, Kantonsspital Basel, Basel, Switzerland

Kenneth A. Walters, F.I.Biol., Ph.D.   An-eX Analytical Services Ltd., Cardiff, Wales

Adam C. Watkinson, Ph.D., M.B.A.* An-eX Analytical Services Ltd., Cardiff, Wales

---

*Current affiliation: Strakan Pharmaceuticals Ltd., Galashiels, Scotland.

Dermatological
and Transdermal
Formulations

1

The Structure and Function of Skin

KENNETH A. WALTERS

An-eX Analytical Services Ltd., Cardiff, Wales

MICHAEL S. ROBERTS

University of Queensland, Princess Alexandra Hospital, Brisbane, Queensland, Australia

I. INTRODUCTION

The skin is the largest organ of the body, accounting for more than 10% of body mass, and the one that enables the body to interact most intimately with its environment. Figure 1 shows a diagrammatic illustration of the skin. In essence, the skin consists of four layers: the stratum corneum (nonviable epidermis), the remaining layers of the epidermis (viable epidermis), dermis, and subcutaneous tissues. There are also several associated appendages: hair follicles, sweat ducts, apocrine glands, and nails. Many of the functions of the skin can be classified as essential to survival of the body bulk of mammals and humans in a relatively hostile environment. In a general context, these functions may be classified as protective, maintaining homeostasis, or sensing. The importance of the protective and homeostatic role of the skin is illustrated in one context by its barrier property. This allows the survival of humans in an environment of variable temperature; water content (humidity and bathing); and the presence of environmental dangers, such as chemicals, bacteria, allergens, fungi, and radiation. In a second context, the skin is a major organ for maintaining the homeostasis of the body, especially in terms of its composition, heat regulation, blood pressure control, and excretory roles. It has been argued that the basal metabolic rate of animals differing in size should be scaled to the surface area of the body to maintain a constant temperature through the skin's thermoregulatory control (1). Third, the skin is a major sensory organ in terms of sensing environmental influences, such as heat, pressure, pain, allergen, and microorganism entry. Finally, the skin is an organ that is in a continual state of regeneration and repair. To fulfill

1

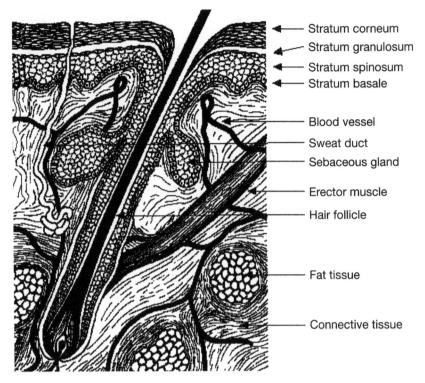

Stratum corneum
Stratum granulosum
Stratum spinosum
Stratum basale

Blood vessel
Sweat duct
Sebaceous gland

Erector muscle
Hair follicle

Fat tissue

Connective tissue

Figure 1 Components of the epidermis and dermis of human skin.

each of these functions, the skin must be tough, robust, and flexible, with effective communication between each of its intrinsic components.

Many agents are applied to the skin either deliberately or accidentally, with either beneficial or deleterious outcomes. The use of topical products was evident in ancient times, and there are reports of systemic benefits of topical anti-infective and hormonal agents in the 1940s. Modern transdermal patch technology was introduced in the late 1970s. The main interests in dermal absorption assessment are in the application of compounds to the skin (a) for local effects in dermatology (e.g., corticosteroids for dermatitis); (b) for transport through the skin for systemic effects (e.g., nicotine patches for smoking cessation); (c) for surface effects (e.g., sunscreens, cosmetics, and anti-infectives); (d) to target deeper tissues (e.g., nonsteroidal anti-inflammatory agents [NSAIDs] for muscle inflammation); and (e) unwanted absorption (e.g., solvents in the workplace, agricultural chemicals, or allergens). Figure 2 summarizes these processes and sites of effect of compounds applied to the skin. The skin became popular as a potential site for systemic drug delivery because it

$\longrightarrow$

Figure 2 (A) Structure of the skin and processes of percutaneous absorption and transdermal delivery. Absorption can occur through sweat ducts (1), intercellular regions of the stratum corneum (2), and through the hair follicles (3). (B) Dermal absorption, sites of action and toxicity.

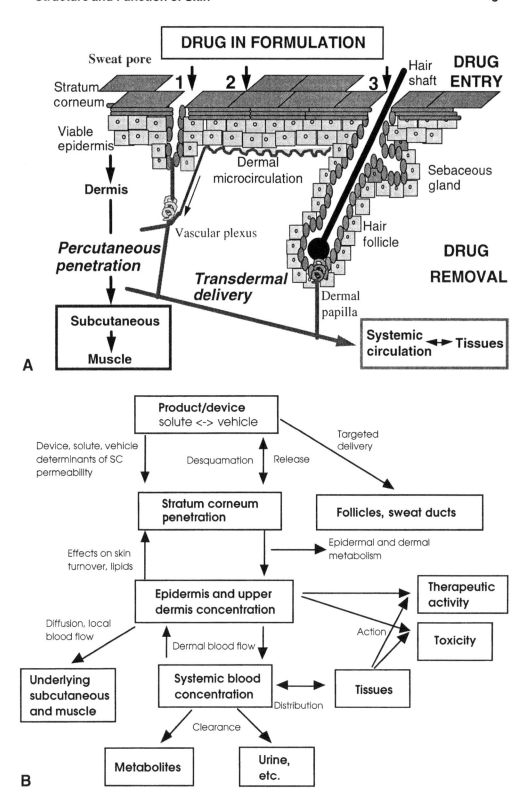

was thought to (a) avoid the problems of stomach emptying, pH effects, and enzyme deactivation associated with gastrointestinal passage; (b) to avoid hepatic first-pass metabolism; and (c) to enable control of input, as exemplified by termination of delivery through removal of the device. In practice, as discussed later in this book, delivery of solutes through the skin is associated with various difficulties, such as (a) the variability in percutaneous absorption owing to site, disease, age, and species differences; (b) the skin's "first-pass" metabolic effect; (c) the reservoir capacity of the skin; (d) irritation and other toxicity caused by topical products; (e) heterogeneity and inducibility of the skin in both turnover and metabolism; (f) inadequate definition of bioequivalence criteria; and (g) an incomplete understanding of the technologies that may be used to facilitate or retard percutaneous absorption. However, the controlled delivery of solutes through the skin continues to be of interest, with the further development of technologies, such as chemical penetration enhancement, sonophoresis, transferosomes, and electroporation. The extent to which these are translated into practice will be defined by time.

II. GROSS STRUCTURE AND FUNCTION OF THE SKIN

Whereas Figure 1 provides an overview of the gross structure of the skin, Figure 3 represents the skin components in terms of the various functions they perform. It needs to be emphasized that the protection, homeostatic, and sensing functions of the skin are both overlapping and integrated. For instance, barrier properties to a chemical entity involves resistance to its entry (barrier provided by stratum corneum), metabolism for that proportion of entity bypassing the stratum corneum (in viable epidermis), sensing of and attention to damage caused by entry (inflammatory mediator release in epidermis, with involvement of dermis), and removal of entity from site by dermal blood supply and distribution into those body organs specifically responsible for elimination of the entity by metabolism (liver) and excretion (kidney). Heat regulation occurs through the use of the subcutaneous fat pad, physiological regulation of blood flow to effect, for instance, heat loss by vasodilation, and cooling by perspiration. We now consider the structure and functions provided by each skin component in some detail.

A. The Epidermis

The epidermis performs a number of functions, as shown in Figure 3, one of the most important being the generation of the stratum corneum, as described later. The stratum corneum is the heterogeneous outermost layer of the epidermis and is approximately 10–20 μm thick. It is nonviable epidermis and consists, in a given cross-section, of 15–25 flattened, stacked, hexagonal, and cornified cells embedded in a mortar of intercellular lipid. Each cell is approximately 40 μm in diameter and 0.5 μm thick. The thickness varies, however, and may be a magnitude of order larger in areas such as the palms of the hand and soles of the feet, areas of the body associated with frequent direct and substantial physical interaction with the physical environment. Not surprisingly, the absorption of solutes, such as methyl salicylate, is slower through these regions than through the skin of other parts of the body. The stratum corneum barrier properties may be partly related to its very high density (1.4 g/cm^3 in the dry state), its low hydration of 15–20%, compared with the usual 70%

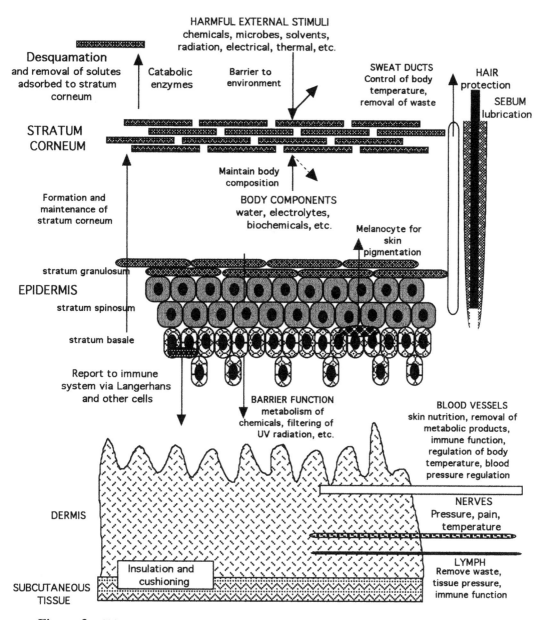

Figure 3 Skin components and their function.

for the body, and its low surface area for solute transport (it is now recognized that most solutes enter the body through the less than $0.1\text{-}\mu\text{m}$–wide intercellular regions of the stratum corneum). Each stratum corneum cell is composed mainly of insoluble bundled keratins (~70%) and lipid (~20%) encased in a cell envelope, accounting for about 5% of the stratum corneum weight. The intercellular region consists mainly of lipids and desmosomes for corneocyte cohesion, as described later. The barrier function is further facilitated by the continuous desquamation of this horny layer with a total turnover of the stratum corneum occurring once every 2–3 weeks. Accordingly, very lipophilic agents, such as sunscreens and substances binding to the horny layer (e.g., hexachlorophane), may be less well absorbed into the body than would be indicated by the initial partitioning of the agents into the horny layer from an applied vehicle. The stratum corneum also functions as a barrier to prevent the loss of internal body components, particularly water, to the external environment. It is estimated that the efficiency of this barrier is such that water loss from "insensible perspiration" is restricted to 0.5 $\mu L/cm^2/h^{-1}$, or 250 mL of water per day for a normal adult. Disorders of epithelization, such as psoriasis, lead to a faster skin turnover, sometimes being reduced to 2–4 days, with improper stratum corneum barrier function formation.

The cells of the stratum corneum originate in the viable epidermis and undergo many morphological changes before desquamation. Thus the epidermis consists of several cell strata at varying levels of differentiation (Fig. 4). The origins of the cells of the epidermis lie in the basal lamina between the dermis and viable epidermis. In this layer there are melanocytes, Langerhans cells, Merkel cells, and two major keratinic cell types: the first functioning as stem cells having the capacity to divide and produce new cells; the second serving to anchor the epidermis to the basement

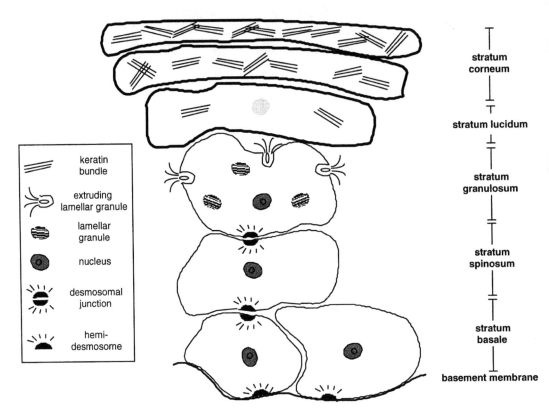

Figure 4   Epidermal differentiation: major events include extrusion of lamellar bodies, loss of nucleus, and increasing amount of keratin in the stratum corneum. The diagram is not to scale and only a few cells are shown for clarity.

membrane (2). The basement membrane is 50–70 nm thick and consists of two layers—the lamina densa and lamina lucida—which comprise mainly proteins, such as type IV collagen, laminin, nidogen, and fibronectin. Type IV collagen is responsible for the mechanical stability of the basement membrane, whereas laminin and fibronectin are involved with the attachment between the basement membrane and the basal keratinocytes.

The cells of the basal lamina are attached to the basement membrane by hemidesmosomes, which are found on the ventral surface of basal keratinocytes (3). Hemidesmosomes appear to comprise three distinct protein groups: two of which are bullous pemphigoid antigens (BPAG1 and BPAG2), and the other epithelial cell-specific integrins (4–6). BPAG1 is associated with the organization of the cytoskeletal structure and forms a link between the hemidesmosome structure and the keratin intermediate filaments. The integrins are transmembrane receptors that mediate attachment between the cell and the extracellular matrix. Human epidermal basal cells contain integrins $\alpha_2\beta_1$, $\alpha_3\beta_1$, and $\alpha_6\beta_4$. Integrin $\alpha_6\beta_4$ and BPAG2 appear to be the major hemidesmosomal protein contributors to the anchoring of the keratinocyte, spanning from the keratin intermediate filament, through the lamina lucida, to the lamina densa of the basement membrane (7). In the lamina densa, these membrane-spanning proteins interact with the protein laminin-5 which, in turn, is linked to collagen VII, the major constituent of the anchoring fibrils within the dermal matrix. It has also been suggested that both BPAG2 and integrin $\alpha_6\beta_4$ mediate in the signal transductions required for hemidesmosome formation (8) and cell differentiation and proliferation. Integrin $\alpha_3\beta_1$ is associated with actin and may be linked with laminin-5. Epidermal wounding results in an up-regulation of these proteins that appears to be involved with cell motility and spreading. The importance of maintaining a secure link between the basal lamina cells and the basement membrane is obvious, and the absence of this connection results in chronic blistering diseases such as pemphigus and epidermolysis bullosa.

In addition to hemidesmosome cell–matrix binding, another site for adhesion of the cells of the epidermal basal layer and the basal membrane is the adherens junction (9). The adherens junction expresses a protein profile different from desmosomes and hemidesmosomes (10,11) and contains talin, vinculin, and cadherins, and with the possible participation of type XIII collagen (12). Whereas the desmosomes and hemidesmosomes are linked to cytoplasmic keratin, the adherens junctions are linked to cytoplasmic actin microfilaments.

The cohesiveness of, and the communication between, the viable epidermal cells, the cell–cell interaction, is maintained in a fashion similar to the cell–matrix connection, except that desmosomes replace hemidesmosomes. Adherens junctions are also located at keratinocyte–keratinocyte borders (13,14). At the desmosomal junction there are two transmembrane glycoprotein types: desmogleins and desmocollins (each of three subtype desmosome-specific cadherins), which are associated with the cytoplasmic plaque proteins, desmoplakin, plectin, periplakin, envoplakin, and plakoglobin (Fig. 5), and provide a linkage to keratin intermediate filaments. On the other hand, in the adherens junction, classic cadherins act as transmembrane glycoproteins and these are linked to the actin filaments by α-, β-, and γ-catenins (15,16). Thus, in the epidermis, the desmosomes are responsible for interconnecting individual cell keratin cytoskeletal structures, thereby creating a tissue very resistant to shearing forces.

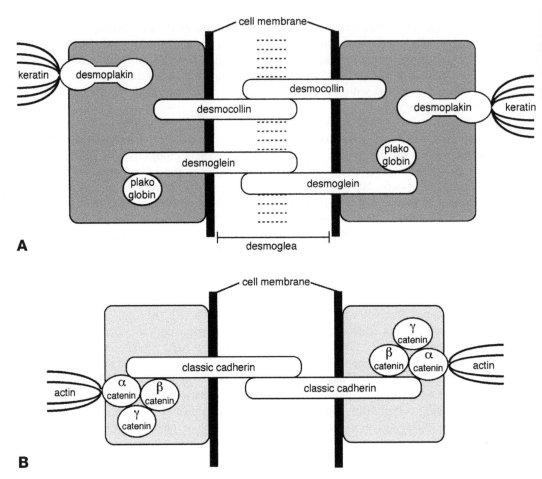

Figure 5 Epidermal cell cohesion and communication is provided by desmosomes and adherens junctions. Different functions are attributed to these junctions. See text for details.

The importance of the calcium ion in cell regulation and intercellular communication has been known for some time (17). It is not surprising, therefore, that the formation of desmosomes and hemidesmosomes appears to be induced by Ca^{2+} and mediated by protein kinase C (PKC) activation (18). The presence of Ca^{2+} activates the metabolism of inositol phospholipids, resulting in the generation of diacylglycerol and inositol-1,4,5-triphosphate. The diacyglycerol subsequently activates protein kinase C, which plays an important role in keratinocyte differentiation (19), and the inositol-1,4,5-triphosphate generates further Ca^{2+} influx into the cytoplasm. The epidermal distribution and function of Ca^{2+}, and several other physiological elements, has been recently reviewed (20). As discussed later, Ca^{2+} also plays a role in proteolysis and desquamation.

Another cell type found in the epidermal basal layer is the Langerhans cell. Although it has been proposed that these dendritic cells plays a role in the control of the proliferation of keratinocytes (21), they have since become recognized as the prominent antigen-presenting cells of the skin immune system (22,23). As such, their main function appears to be to pick up contact allergens in the skin and present these agents to T lymphocytes in the skin-draining lymph nodes; thus, they play an important role in contact sensitization. Cell surface moieties on the Langerhans cells are modified and the cells increase in size following topical application of hapten. The activated cells migrate from the epidermis to the dermis and from there to the regional lymph nodes where they sensitize T cells. The ability of Langerhans cells to migrate from bone marrow, localize in a specific region of the epidermis, and further migrate when activated, suggests that there is some mechanism for accumulation in the epidermis, adhesion to keratinocytes and the basement membrane, and for disruption of the adhesive bond. Migration into the epidermis may be mediated by granulocyte–macrophage colony-stimulating factor (GM-CSF), tumor necrosis factor-α (TNF-α), interleukin-6 (IL-6), transforming growth factor-β (TGF-β), chemotactic cytokines, such as monocyte chemotactic protein (MCP), and cutaneous lymphocyte-associated antigen (CLAA) (24). The adhesive bonds within the epidermis appear to be formed by interaction of Langerhans cells with extracellular matrix proteins, such as fibronectin and laminin, through β_1-integrins (25). Detachment of Langerhans cells from keratinocytes and the basement membrane following skin sensitization may be mediated by epidermal cytokines, including GM-CSF and TNF-α (26), whereas cell maturation, which occurs during transit to the local lymph nodes, may be mediated by GM-CSF. Recently, it was shown that IL-1β and TNF-α acted directly on Langerhans cells to reduce adhesion to keratinocytes and the basement membrane by down-regulating the binding protein E-cadherin (27).

Melanocytes are a further functional cell type of the epidermal basal layer and are also present in hair and eyes. The main function of these cells is to produce melanins, high molecular weight polymers of indole quinone, which affect pigmentation of the skin, hair, and eyes (28,29). Melanin is produced in the melanosomes: membrane-bound organelles that are transferred to keratinocytes, probably by a process involving phagocytosis (30), to provide a uniform distribution of pigmentation. Intracellular movement of melanosomes is possibly mediated by actin and microtubules (31). Visible pigmentation is dependent not only on the number, shape, and size of the melanosomes, but also on the chemical nature of the melanin. Hair color is governed by melanocytes which reside in the hair bulbs within the dermis (32). Melanosomes are transferred to the growing hair shaft. The major function of skin pigmentation is to provide protection against harmful environmental effects, such as ultraviolet (UV) radiation, especially in the proliferating basal layers where the mutagenic effects of this type of insult have particularly serious implications. Melanocytes remain attached to the basal layer and are thought to exist in a nonproliferative state when in contact with undifferentiated keratinocytes. Recent studies have indicated, however, that melanocytes can proliferate if they are separated from the basal layer and surrounded by differentiated keratinocytes (33).

Regulation of melanogenesis is a complex process (34) involving some 80 genetic loci. It is mutations of these genes that lead to pathological states, such as albinism, vitiligo, and Waardenberg's syndrome. The initial substrate for melanin is tyrosine, which is hydroxylated to dihydroxyphenylalanine (DOPA), and from there may be processed through several routes to produce either the eumelanins (5,6-dihydroxyindole melanin and 5,6-dihydroxyindole-2-carboxylic acid melanin) or pheomelanins (Fig. 6). Eumelanins are brown–black, whereas pheomelanins are yellow–red. It is thought that interactions between the tridecapeptide α-melanocyte-

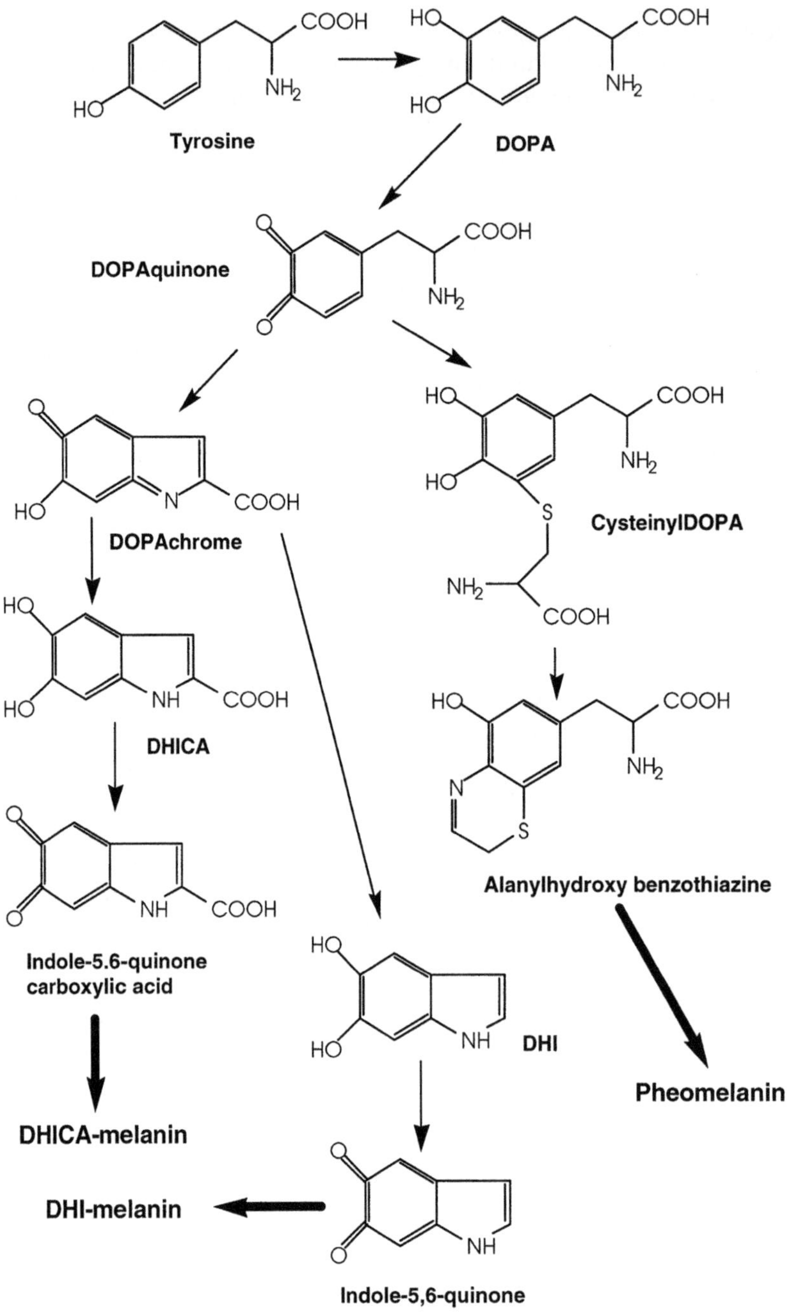

Figure 6 Routes of synthesis of the melanins: tyrosine is converted to dihydroxyphenyl-alanine (DOPA) by tyrosine hydroxylase and is subsequently converted to DOPAquinone by DOPA oxidase. Subsequent processing is by several different pathways to yield either dihydroxyindole melanin, dihydroxyindole-2-carboxylic acid melanin or pheomelanin. (From Ref. 34.)

stimulating hormone (α-MSH) and agouti signal protein are responsible for governing the type of melanogenesis pathway followed and that, in conditions in which α-MSH dominates, eumelanins are produced. UV radiation appears to increase production of the precursor hormone proopiomelanocortin, which increases α-MSH production, resulting in increased levels of eumelanins (35,36).

The final type of cell found in the basal layer of the stratum corneum is the Merkel cell. These cells can be distinguished from the keratinocytes by their clear cytoplasm and lack of tonofilaments. The cells are closely associated with nerve endings, present on the other side of the basement membrane, which suggests they function as sensory receptors of the nervous system. Although histochemical evidence demonstrating the presence of acetylcholinesterase suggests a sensory role for Merkel cells, there has been no direct evidence for the release of neurotransmitters. Indeed, acetylcholinesterases have been found in keratinocytes (37). Despite this lack of confirmation, most researchers in the field agree that Merkel cells play a role (a) in the mechanosensory system; (b) in trophic action on peripheral nerve fibers; (c) in stimulating and maintaining proliferation and keratinocytes; and (d) in release of bioactive substances to subepidermal structures (38,39).

B. The Dermis

The dermis, a critical component of the body, not only provides the nutrative, immune, and other support systems for the epidermis, through a thin papillary layer adjacent to the epidermis, but also plays a role in temperature, pressure, and pain regulation. The main structural component of the dermis is referred to as a coarse reticular layer. The dermis is about 0.1–0.5 cm thick and consists of collagenous fibers (70%), providing a scaffold of support and cushioning, and elastic connective tissue, providing elasticity, in a semigel matrix of mucopolysaccharides. In general, the dermis has a sparse cell population. The main cells present are the fibroblasts, which produce the connective tissue components of collagen, laminin, fibronectin, and vitronectin; mast cells, which are involved in the immune and inflammatory responses; and melanocytes involved in the production of the pigment melanin.

Contained within the dermis is an extensive vascular network providing for the skin nutrition, repair, and immune responses and, for the rest of the body, heat exchange, immune response, and thermal regulation. The blood flow rate to the skin is about 0.05 mL min^{-1} cc^{-3} of skin, providing a vascular exchange area equivalent to that of the skin surface area. Skin blood vessels derive from those in the subcutaneous tissues, with an arterial network supplying the papillary layer, the hair follicles, the sweat and apocrine glands, the subcutaneous area, as well as the dermis itself. These arteries feed into arterioles, capillaries, venules, and, thence, into veins. Of particular importance in this vascular network is the presence of arteriovenous anastomoses at all levels in the skin. These arteriovenous anastomoses, which allow a direct shunting of up to 60% of the skin blood flow between the arteries and veins, thereby avoiding the fine capillary network, are critical to the skin's functions of heat regulation and blood vessel control. Blood flow changes are most evident in the skin in relation to various physiological responses and include psychological effects, such as shock ("draining of color from the skin") and embarrassment ("blushing"); temperature effects; and physiological responses to exercise, hemorrhage, and alcohol consumption.

The lymphatic system is an important component of the skin in regulating its interstitial pressure, mobilization of defense mechanisms, and in waste removal. It exists as a dense, flat meshwork in the papillary layers of the dermis and extends into the deeper regions of the dermis. Cross and Roberts (40) have shown that whereas blood flow determines the clearance of small solutes, such as water and lidocaine, lymphatic flow is an important determinant in the dermal removal of larger solutes, such as interferon. Also present in the dermis are a number of different types of nerve fibers supplying the skin, including those for pressure, pain, and temperature.

C. The Subcutis

The deepest layer of the skin is the subcutaneous tissue or hypodermis. The hypodermis acts as a heat insulator, a shock absorber, and an energy storage region. This layer is a network of fat cells arranged in lobules and linked to the dermis by interconnecting collagen and elastin fibers. As well as fat cells (possibly 50% of the body's fat), the other main cells in the hypodermis are fibroblasts and macrophages. One of the major roles of the hypodermis is to carry the vascular and neural systems for the skin. It also anchors the skin to underlying muscle. Fibroblasts and adipocytes can be stimulated by the accumulation of interstitial and lymphatic fluid within the skin and subcutaneous tissue (41).

D. Skin Appendages

There are four skin appendages: the hair follicles with their associated sebaceous glands, eccrine sweat glands, apocrine sweat glands, and the nails. Each appendage has a different function as outlined in Table 1. The hair follicles are distributed across the entire skin surface with the exception of the soles of the feet, the palms of the hand and the lips. A smooth muscle, the erector pilorum, attaches the follicle to the dermal tissue and enables hair to stand up in response to fear. Each follicle is associated with a sebaceous gland that varies in size from 200 to 2000 μm in diameter. The sebum secreted by this gland (Table 2), consisting of triglycerides, free fatty acids, and waxes, protects and lubricates the skin as well as maintaining a pH of about 5. The fractional area for these is slightly more than 1/1000 of the total skin surface (see Table 1). Also described in Table 1 are the eccrine or sweat glands and apocrine glands, accounting for about two-thirds and one-third of all glands, respectively. The eccrine glands are epidermal structures that are simple, coiled tubes arising from a coiled ball, of approximately 100 μm in diameter, located in the lower dermis. It secretes a dilute salt solution with a pH of about 5, this secretion being stimulated by temperature-controlling determinants, such as exercise and high environmental temperature, as well as emotional stress through the autonomic (sympathetic) nervous system (see Table 1). These glands have a total surface area of about 1/10,000 of the total body surface. The apocrine glands are limited to specific body regions and are also coiled tubes. These glands are about ten times the size of the eccrine ducts, extend as low as the subcutaneous tissues and are paired with hair follicles.

In many respects the nail may be considered as vestigial in humans. However, some manipulative and protection function can be ascribed. Certainly nail plate composition, layers of flattened keratinized cells fused into a dense, but somewhat elastic

Table 1 Appendages Associated with the Skin

Parameter	Appendage			
	Hair follicle and sebaceous gland	Eccrine gland	Apocrine gland	Nails
Function	Protection (hair) and lubrication (sebum)	Cooling	Vestigal secondary sex gland?	Protection
Distribution	Most of the body	Most of the body	Axillae, nipples, anogenital	Ends of fingers and toes
Average/cm^2	57–100	100–200	Variable	—
Fractional area	2.7×10^3	10^{-4}	Variable	—
Secretions	Sebum	Sweat (dilute saline)	"Milk" protein, lipoproteins, lipid	Nil
Secretions stimulated by	Heat (minor)	Heat, cholinergic	Heat	—
Biochemical innervation of gland response	—	Cholinergic	Cholinergic (?)	—
Control	Hormonal	Sympathetic nerves	Sympathetic nerves	—

Table 2 Lipid Composition of Human Sebum

Lipid	Lumen of gland[a]	Skin surface[a]
Squalene	15	15
Wax esters	25	25
Cholesteryl esters	2	2
Triglycerides	57	42
Fatty acids	0	15[b]
Cholesterol	1	1

[a]Expressed as weight%.
[b]Most abundant fatty acid in human sebum is $C_{16:1}$.
Source: Courtesy of P. Wertz.

mass, will afford some protection to the highly sensitive terminal phalanx. The cells of the nail plate originate in the nail matrix and grow distally at a rate of about 0.1 mm/day. In the keratinization process the cells undergo shape and other changes, similar to those experienced by the epidermal cells forming the stratum corneum. This is not surprising because the nail matrix basement membrane shows many biochemical similarities to the epidermal basement membrane (42). Thus, the expression of integrins $\alpha_2\beta_1$ and $\alpha_3\beta_1$ within the nail matrix basement membrane zone is indicative of a highly proliferative tissue. The structure of the keratinized layers is very tightly knit but, unlike the stratum corneum, no exfoliation of cells occurs. Given that it is a cornified epithelial structure, the chemical composition of the nail plate is not remarkable, and there are many similarities to that of the hair (Table 3) (43,44). Thus, the major components are highly folded keratin proteins (containing many disulfide linkages) with small amounts (0.1–1.0%) of lipid, the latter presumably located in the intercellular spaces. The principal plasticizer of the nail plate is water, which is normally present at a concentration of 7–12%.

The nail plate comprises two major layers (the dorsal and intermediate layer) with, possibly, a third layer adjacent to the nail bed (45,46). The dorsal nail plate is harder and thinner than the intermediate plate, suggesting that there are differences in the chemical composition of the two layers, which further suggests that applied drugs may possess differing partitioning tendencies between the layers. The latter is a particularly important consideration for the topical treatment of fungal infections of the nail (onychomycoses) (47), and mechanisms for enhancing solubility and diffusivity of drugs within these layers have been suggested (48).

III. DEVELOPMENT OF THE STRATUM CORNEUM

It is the structure of the stratum corneum that enables terrestrial animals to exist in a nonaquatic environment without desiccation. The ability to control both the loss of water and the influx of potentially harmful chemicals and microorganisms is the result of the evolution of a unique mixture of protein and lipid materials that collectively form this coherent membrane composed of morphologically distinct domains. These domains are principally proteinaceous, the keratinocytes; or lipophilic, the intercellular spaces. Because of the lack of nuclei in the keratinocytes, the stratum corneum is often considered to be a dead tissue, but this is certainly a simplistic

Table 3 Amino Acid Composition of Nail, Hair, and Stratum Corneum

Amino acid	Nail	Hair	Stratum corneum
Lysine	3.1[a]	2.5[a]	4.2[a]
Histidine	1.0	0.9	1.5
Arginine	6.4	6.5	3.9
Aspartic acid	7.0	5.4	7.9
Threonine	6.1	7.6	3.0
Serine	11.3	12.2	13.6
Glutamic acid	13.6	12.2	12.6
Proline	5.9	8.4	3.0
Glycine	7.9	5.8	24.5
Alanine	5.5	4.3	4.4
Valine	4.2	5.5	3.0
Methionine	0.7	0.5	1.1
Isoleucine	2.7	2.3	2.7
Leucine	8.3	6.1	6.9
Tyrosine	3.2	2.2	3.4
Phenylalanine	2.5	1.7	3.2
Half-cysteine	10.6	15.9	1.2
Sulfur	3.2%[b]	4.5%[b]	1.4%[b]

[a]Expressed as residues per 100 residues.
[b]Expressed as % dry weight.
Source: Ref. 43.

description and many events occur that indicate that "dead" does not necessarily indicate a lack of response.

A. Epidermal Differentiation

The development of the stratum corneum from the keratinocytes of the basal layer involves several steps of cell differentiation, which has resulted in a structure-based classification of the layers above the basal layer (the stratum basale). Thus the cells progress from the stratum basale through the stratum spinosum, the stratum granulosum, the stratum lucidium, to the stratum corneum (49). Cell turnover, from stratum basale to stratum corneum, is estimated to be on the order of 21 days.

The exact mechanism whereby keratinocytes in the basal layer are stimulated to initiate the differentiation process is not fully understood. Protein kinase C and several keratinocyte-derived cytokines may play a regulatory role in the differentiation process (50,51). Thus, for example, interleukin-1 (IL-1) stimulates the production of other cytokines, in both autocrine and paracrine fashion, which act to induce proliferation and chemotaxis (Fig. 7). These inducing cytokines include GM-GSF, transforming growth factor-α (TGF-α), nerve growth factor, amphiregulin, IL-6, and IL-8. On the other hand, transforming growth factor-β (TGF-β), the production of which is also initiated by IL-1, suppresses keratinocyte growth, but stimulates keratinocyte migration, the latter possibly as a result of modulation of hemidesmosomal and desmosomal integrins. It has also been suggested that urokinase-type plasminogen activator (uPA) may activate growth factors and stimulate epidermal prolifer-

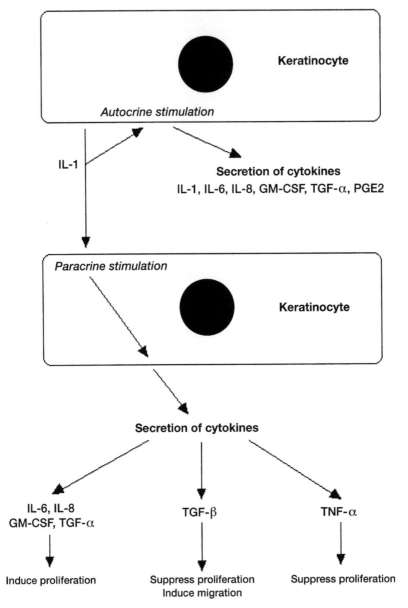

Figure 7 Cytokine involvement in keratinocyte proliferation. Secretion of interleukin-1 (IL-1) leads to the autocrine and paracrine stimulation of the release of other cytokines, including interleukins-6 and -8 (IL-6 and IL-8), granulocyte–macrophage colony-stimulating factor (GM-CSF) and transforming growth factor-α (TGF-α), all of which induce epidermal proliferation. Transforming growth factor-β (TGF-β) and tumor necrosis factor-α (TNF-α), both of which suppress proliferation, are also released. TGF-β can induce keratinocyte migration. (From Ref. 51.)

ation (52). Generation and activation of the serine protease, plasmin, from plasminogen is induced by uPA and the activated plasmin instigates localized extracellular proteolysis of cell surface adherent proteins and eventual disruption of the hemidesmosomes. The increasing understanding of the role of cytokines in the maintenance of epidermal homeostasis has stimulated research into the possibility of using such compounds as active principles in cosmetic products (53).

The stratum spinosum (prickle cell layer), which lies immediately above the basal layer, consists of several layers of cells that are connected to each other and to the stratum basale cells by desmosomes and contain prominent keratin tonofilaments. The cells of the stratum spinosum have a larger cytoplasm than those of the stratum basale. Within the cytoplasm are numerous organelles and filaments. It is clear that the α-keratins of the stratum spinosum are somewhat different from those found in the stratum basale (54), indicating that, although mitosis has ceased and a phase of terminal differentiation has been initiated, the cell still maintains a capacity to alter the transcriptional expression of its genes. In the outer cell layers of the stratum spinosum, intracellular membrane-coating granules (100–300 nm in diameter) appear within the cytosol, marking the transition between the stratum spinosum and stratum granulosum.

Although further keratin differentiation occurs in the stratum granulosum (55,56), new keratin synthesis stops. The most characteristic features of this layer are the presence of many intracellular keratohyalin granules and membrane-coating granules, the assembly of the latter appearing to take place in the endoplasmic reticulum and Golgi regions (57,58). Within these granules lamellar subunits arranged in parallel stacks are observed. These are believed to be the precursors of the intercellular lipid lamellae of the stratum corneum (57,59). Also present in the lamellar granules are hydrolytic enzymes, the most important of which is stratum corneum chymotryptic enzyme (SCCE). SCCE is a serine protease that, because of its ability to locate at desmosomal regions in the intercellular space, has been implicated in the desquamation process (60–62). In the outermost layers of the stratum granulosum the lamellar granules migrate to the apical plasma membrane where they fuse and eventually extrude their contents into the intercellular space (57). At this stage in the differentiation process, as a result of the release of selective lysing enzymes, the keratinocytes lose their nuclei and other cytoplasmic organelles, become flattened and compacted to form the stratum lucidum, which eventually forms the stratum corneum. The extrusion of the contents of lamellar granules is a fundamental requirement for the formation of the epidermal permeability barrier (63,64), and disturbances in this process have been implicated in various dermatological disorders (65).

The entire process of epidermal terminal differentiation is geared toward the generation of the specific chemical morphology of the stratum corneum. Thus, the end products of this process are the intracellular protein matrix and the intercellular lipid lamellae.

B.   Cornified Cell Envelope

The cornified cell envelope (about 15 nm thick and 10% mass of the stratum corneum) is the outermost layer of a corneocyte, which consists mainly of tightly bundled keratin filaments aligned parallel to the main face of the corneocyte. The envelope consists of both protein and lipid components. The protein envelope (~10 nm thick) is a covalent, cross-linking of several proteins as a result of actions by sulfhydryl oxidases and transglutaminases; whereas the lipid envelope (~5 nm thick) are lipid attached covalently to the protein envelope. Sulfhydryl oxidases and transglutaminases lead to the formation of disulfide and isopeptide bonds, respectively (66). It has been suggested that cross-linking by the creation of N-(γ-glutamyl) lysine isodipeptide bonds formed by epidermal transglutaminases is a reaction possibly mediated by cholesterol sulfate (67). The envelope lies adjacent to the interior surface of the plasma membrane. In addition to the predominant protein loricrin, several other envelope precursor proteins have been identified including cystatin-α (68), cornifin-α (69), elafin, and filaggrin (70). The predominance of the structural proteins in the cornified envelope are as follows: involucrin (65 kDa; 2–5%), loricin (26 kDa; 80%), small proline-rich proteins (a family of 11–14 closely related proteins, including cornifins and pancornulins, 6–26 kDa; 3–5%), and cystalin A or keratolinin (12 kDa; 2–5%). There are also a range of proteins with an expression of less than 1%, including elafin, profilaggrin, keratin intermediate filaments, desmoplakin I and II, S100 proteins, and annexin I (also called lipocortin I) (66).

Formation of the envelope is believed to occur in two stages. In the first stage soluble proteins, such as involucrin and cystatin-α, form a scaffold to which other insoluble precursors, including loricrin, are added in the latter stage. Thus, the cornified envelope is formed by the sequential deposition of consecutively expressed proteins starting with the fixation of involucrin as a scaffold on the intracellular surface of the plasma membrane in a calcium- and phospholipid-dependent manner. It is cross-linked to desmoplakin and envoplakin and also covalently bound to ω-hydroxyceramides. Other proteins then reinforce the envelope by attaching, including loricin and small proline-rich proteins (66). The cross-linked protein complex of the corneocyte envelope is very insoluble and chemically resistant. Cornified cell envelopes are also present in the hair follicle and nail matrix but, although morphologically similar, the pattern and types of precursor are slightly different from those of the epidermis (71).

It is currently proposed that the corneocyte protein envelope plays an important role in the structural assembly of the intercellular lipid lamellae of the stratum corneum. The work of Downing and colleagues (72–75) has demonstrated that the corneocyte possesses a chemically bound lipid envelope comprised of N-ω-hydroxyceramides that are ester-linked to the numerous glutamate side chains provided by the β-sheet conformation of involucrin in the envelope protein matrix (74) (Fig. 8). Recent molecular modeling of the human involucrin molecule has suggested that a conventional α-helical conformation could also provide the requisite number of glutamate side chains for ester linkage (76). The lipids of the cornified cell envelope are resistant to extraction by chloroform–methanol mixtures, but can be extracted following alkaline hydrolysis. This lipid envelope may provide the framework for the generation of the intercellular lipid lamellae. Inhibition of the formation of N-ω-hydroxyceramides may be achieved using aminobenzotriazole, an inhibitor of type 4 cytochrome P450. Behne et al. (77) demonstrated that, in the absence of N-ω-hydroxyceramides, the stratum corneum intercellular lipid lamellae were abnormal and permeability barrier function was disrupted. These data provide direct evidence supporting the important roles of N-ω-hydroxyceramides in epidermal barrier homeostasis and corneocyte lipid envelope formation.

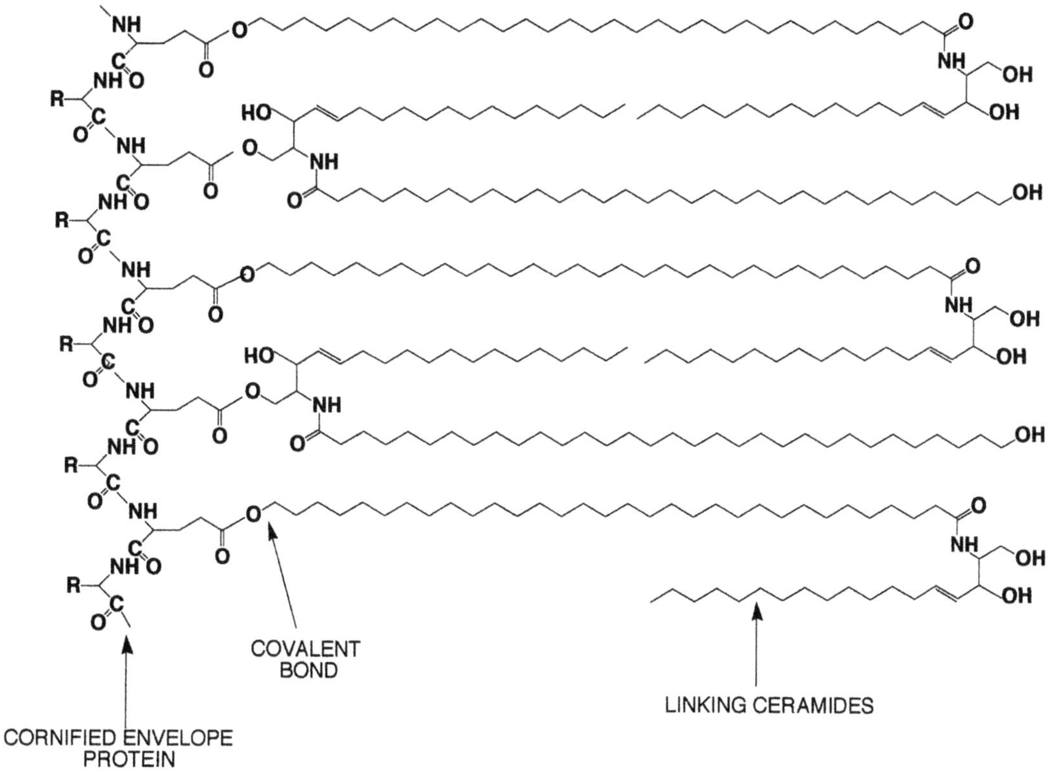

Figure 8 The corneocyte protein envelope plays an important role in the structural assembly of the intercellular lipid lamellae of the stratum corneum. The corneocyte possesses a chemically bound lipid envelope comprised of N-ω-hydroxyceramides that are ester-linked to the numerous glutamate side chains provided by the β-sheet conformation of involucrin in the envelope protein matrix. (From Ref. 74.)

C. Stratum Corneum Proteins

It has been suggested that the keratinization process contributes to the barrier function of the stratum corneum by making the corneocytes practically insoluble in most diffusing solutes (78). The poor stratum corneum protein solubility partly results from the extensive cross-linking of both the cell envelope and intracellular proteins. The majority of the intracellular protein in the stratum corneum is composed of keratin filaments—which are cross-linked by intermolecular disulfide bridges (79,80) —and the components of the cornified cell envelope. In the terminal stages of differentiation, the keratinocytes contain keratin intermediate filaments (keratins 1 and 10, which are derived from keratins 5 and 14 present in basal keratinocytes) together with several other proteins, including involucrin, loricrin, and profilaggrin (81–83). Loricrin and involucrin are major components of the cornified cell envelope; whereas profilaggrin is implicated in both the alignment of the keratin filaments (84) and epidermal flexibility; the latter based on the water-holding capacity of the constituent amino acids (85). Profilaggrin is a large, highly phosphorylated protein, consisting of multiple (10–12) filaggrin units with N- and C-terminal domains, which first appears in keratohyalin granules in the stratum granulosum. The profilaggrin molecule is processed in a calcium-dependent manner by dephosphorylation and proteolysis into individual filaggrin molecules that serve to aggregate keratin filaments. The term filaggrin (*fil*ament *aggr*egating prote*in*) was used to name the keratin matrix proteins. The interaction between filaggrin and keratin is believed to be ionic (86). Evidence of a role for profilaggrin as a calcium binder in epidermal differentiation has also been presented (87). It is important to recognize that filaggrin does not exist beyond the lower layers of the stratum corneum. It is completely proteolyzed into constituent amino acids about 2–3 days after its formation from profilaggrin (88).

D. The Intercellular Lipids

The composition of the stratum corneum intercellular lipids is unique in biological systems (Table 4). These lipids exist as a continuous lipid phase; occupying about 20% of the stratum corneum volume, and arranged in multiple lamellar structures. A remarkable feature is the lack of phospholipids and the preponderance of cholesterol (27%) and ceramides (41%), together with free fatty acids (9%), cholesteryl esters (10%), and cholesteryl sulfate (2%) (89). This composition varies with body site. There are distinct alterations in the distribution of lipid type during the course of epidermal differentiation (Fig. 9). Phospholipids, which dominate in the basal layer, are converted to glucosylceramides and, subsequently, to ceramides and free fatty acids, and are virtually absent in the outer layers of the stratum corneum. Levels of sphingolipids and free fatty acids increase in the latter stages of terminal differentiation.

   Eight classes of ceramides (designated ceramides 1–8) have been isolated and identified in human stratum corneum (90,91). Ceramide classification is arbitrarily based on polarity, with ceramide 1 being the least polar. Their structures are based on sphingolipids (Fig. 10), and they contain long-chain bases, such as sphinganine

Table 4 Lipid Content of the Stratum Corneum
Intercellular Space

Lipid	% (w/w)	mol %
Cholesterol esters	10.0	7.5[a]
Cholesterol	26.9	33.4
Cholesterol sulfate	1.9	2.0
Total cholesterol derivatives	38.8	42.9
Ceramide 1	3.2	1.6
Ceramide 2	8.9	6.6
Ceramide 3	4.9	3.5
Ceramide 4	6.1	4.2
Ceramide 5	5.7	5.0
Ceramide 6	12.3	8.6
Total ceramides	41.1	29.5
Fatty acids	9.1	17.0[a]
Others	11.1	10.6[b]

[a]Based on C_{16} alkyl chain.
[b]Based on MW of 500.

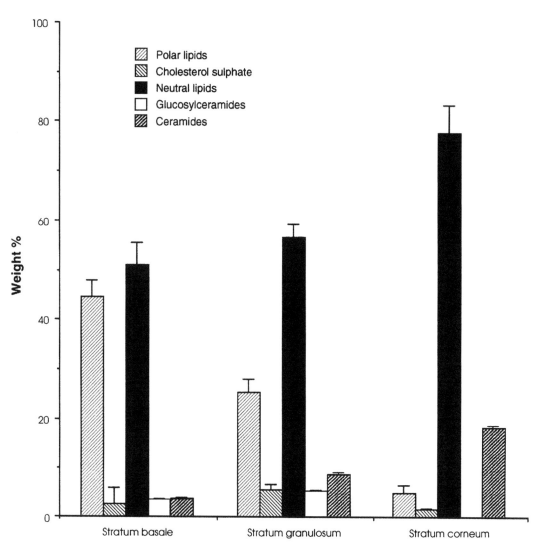

Figure 9 There are distinct alterations in the distribution of lipid type during the course of epidermal differentiation. Polar phospholipids, which dominate in the basal layer, are virtually absent in the outer layers of the stratum corneum. Levels of ceramides and neutral lipids increase in the latter stages of terminal differentiation.

and 4-hydroxysphinganine, N-acetylated by different fatty acids. Because of its unique structure, ceramide 1, an acylceramide, may function as a stabilizer of the intercellular lipid lamellae. This molecule consists of a sphingosine base, with an amide-linked ω-hydroxyacid and a nonhydroxyacid ester linked to the ω-hydroxyl group (75). It is possible that ceramide 1 acts as a "molecular rivet" in the intercellular lipid lamellae of the stratum corneum. There is strong evidence to indicate that the intercellular lipid lamellae are further stabilized by the chemical links between the long-chain ceramides and glutamate residues on the corneocyte protein envelope (74). Extracellular calcium may also be involved in the formation and

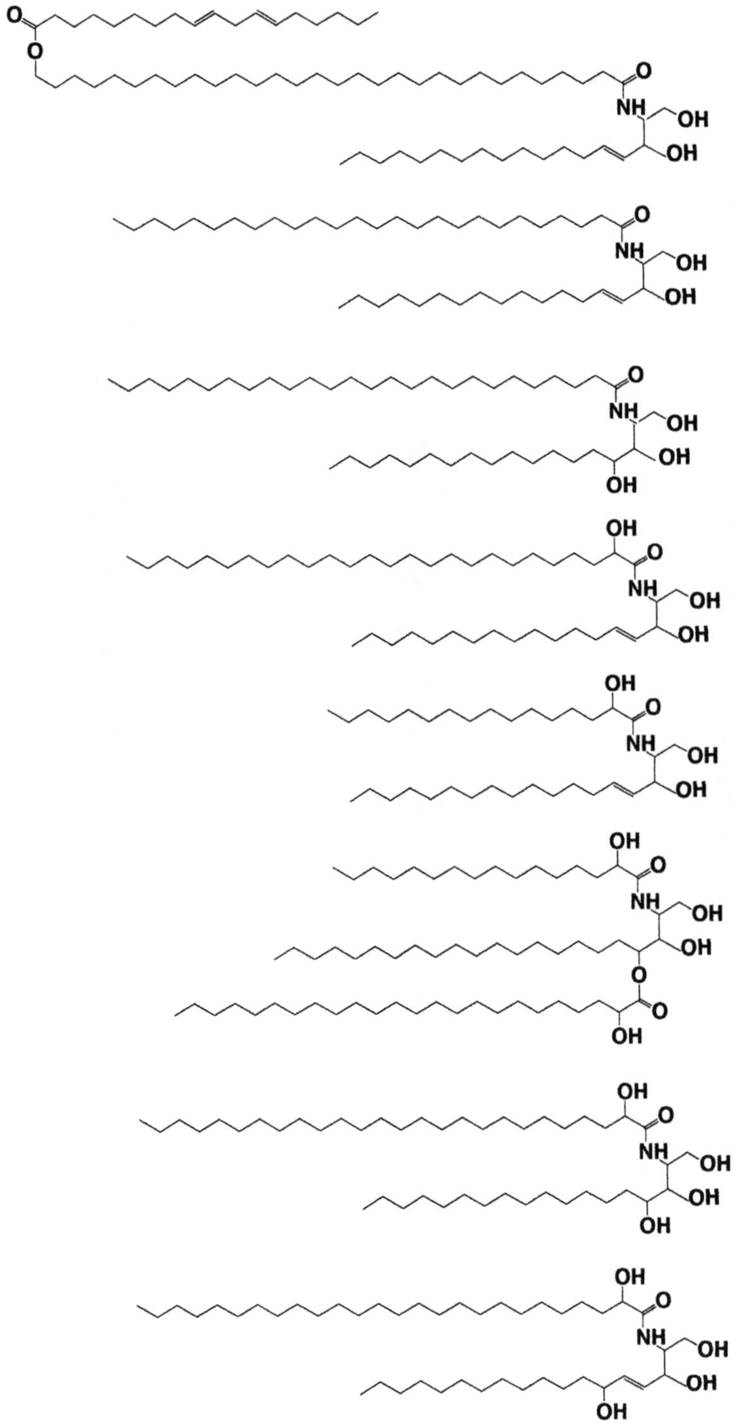

Ceramide 1

Ceramide 2

Ceramide 3

Ceramide 4

Ceramide 5

Ceramide 6.1

Ceramide 6.2

Ceramide 8

Figure 10 Ceramides of the human stratum corneum intercellular space.

stabilization of the lamellae (92). Wertz (93) and Nemes and Steinhardt (66) have suggested that the long-chain ceramides constituting the lipid envelope and attached covalently to the protein envelope function in a "Velcro-like" fashion by interdigitating with the intercellular lipids, allowing the structural integrity of the lipid lamellae to be maintained.

The functions of the individual ceramide type are not fully understood, and the knowledge that has been accumulated is based mainly on examination of barrier function, lipid content, and lipid distribution in diseased skin. For example, acylceramides, isolated from acne comedones and the skin of patients with acne, contained higher proportions of saturated and monounsaturated C_{16} and C_{18} fatty acids, and less linoleate than those isolated from control subjects (94). Distribution of free fatty acids showed a similar pattern. Reduction in dietary linoleate in experimental animals results in epidermal hyperproliferation and impaired skin barrier function (95). In patients with acne, epidermal hyperproliferation produces a keratinous follicular plug that results in the formation of a comedone. These observations suggested a potential role of ceramide 1 as an essential constituent of the skin barrier and, possibly, as a mediator of epidermal proliferation (96). Also, the distribution and amount of ceramide types in psoriatic scales is different from that in normal skin (97), but the significance of this anomaly is unknown. Similarly, ceramide content was reduced in the stratum corneum of patients with atopic dermatitis (98).

In many biological membranes cholesterol acts as a stabilizer and reduces the mobility of the alkyl chains. The exact function of cholesterol and cholesterol esters in the stratum corneum intercellular lamellae are unknown, although it is likely that cholesterol acts to reduce fluidity of the ceramide alkyl chains. Cholesterol and ceramides are present at almost equimolar proportions throughout the stratum corneum. Norlén et al. (99) obtained values of 37%mol for ceramides and 32%mol for cholesterol using human forearm skin (interestingly, the molar distribution of cholesterol esters and free fatty acids was also similar at 15%mol and 16%mol, respectively). This finding supports the suggestion that cholesterol and ceramide may interact on a molecular one-to-one basis in the stratum corneum intercellular lamellae (100). There is strong evidence that cholesterol interacts with phospholipids to form one-to-one molar complexes involving hydrogen bonding of the 3-β-hydroxyl of cholesterol with the glyceryl oxygen at the 2 position of the phospholipid (101). It is possible that a similar type of binding occurs between ceramides and cholesterol within skin lipids.

The exact functions of cholesterol esters within the stratum corneum lamellae are also elusive. It is theoretically possible that cholesterol esters may span adjacent bilayers and serve as additional stabilizing moieties. Similarly, the role of fatty acids is unclear. The recent work of Norlén and colleagues (102) has indicated that the free fatty acids of the stratum corneum are composed entirely of saturated long-chain acids, the majority of which are lignoceric acid (C_{24}, 39%mol) and hexacosanoic acid (C_{26}, 23%mol). The authors extracted lipid from the deeper layers of the stratum corneum and concluded that the sometimes reported presence of shorter-chain saturated and unsaturated fatty acids in the outer layers of the stratum corneum is the result of contamination from sebaceous gland lipid and the environment.

Overall, the intercellular lipid lamellae appear to be highly structured, very stable, and constitute a highly effective barrier to chemical penetration and permeation. However, the exact structure and physical state of the stratum corneum intercellular lipid lamellae are not known. Forslind and colleagues (103,104) proposed a domain mosaic model in which the long-chain ceramides are in a crystalline state, whereas short-chain and unsaturated free fatty acids are in the liquid state. The model proposes that large crystalline domains are surrounded by thin liquid crystalline channels and suggests that any water present in the region is associated with the liquid crystalline phase or the corneocytes. Considerable information on lipid structure within the stratum corneum has been generated by Bouwstra and colleagues (105–108) using small-angle X-ray diffraction and transmission electron microscopic techniques. These and earlier studies have shown that the lipid lamellae of the stratum corneum are orientated parallel to the corneocyte surface and have repeat distances of approximately 6.0–6.4 and 13.2–13.4 nm. Bouwstra et al. (107) have proposed that the broad band represents regions where ceramide moieties are partly interdigitating, and the narrow band represents regions of full interdigitation.

In a more recent study on lipid packing (109), the Leiden group have evaluated lipid organization of the stratum corneum using electron diffraction. Whereas wide-angle X-ray diffraction techniques were able to demonstrate lattice spacings that were consistent with orthorhombic (crystalline) packing of the lipids (reflections at 0.415 and 0.375 nm), they cannot confirm the presence or absence of hexagonal (gel) packing, where only the 0.415 nm reflection occurs. On the other hand, electron diffraction technology can distinguish between orthorhombic and hexagonal packing. In this elegant study the authors found that, although the majority of lipids in the intercellular space were present in the crystalline state, there were some lipids existing in the gel state that has a slightly looser hexagonal packing arrangement in the outer layers of the stratum corneum. It was suggested that the existence of the gel phase represents the influence of contaminating sebaceous lipid in this region, but it is tempting to speculate that the alteration in lipid states in these outer layers is somehow related to the process of desquamation. Fenske et al. (110) showed a similar lateral packing in model membrane systems made of stratum corneum lipids.

E. Desquamation

The mechanisms underlying the desquamation of stratum corneum cells are not fully understood. Suzuki et al. (111) suggested that, through the action of two types of serine protease, the degradation of desmosomes leads to desquamation. Certainly there has to be proteolysis of any intercellular adhesive structures between the terminal keratinocytes. Egelrud's group have suggested that desquamation may be regulated by the extent of activation of protease precursors and changes in the pH of the stratum corneum intercellular space (112–114). Tape strips of the outer layers of human stratum corneum contained precursors and active forms of both stratum corneum chymotryptic enzyme and stratum corneum tryptic enzyme (113). Although both enzymes possessed maximum activity at pH 8.0, considerable activity was retained at pH 5.5 (the pH of the skin surface).

Other proteins that may play a role in desquamation include cathepsin D, a protease active in the acid range (115), desquamin (116), and stratum corneum gelatinase (117).

Relevant to the discussion on desquamation is the role of corneodesmosomes or corneosomes, a description for homogeneously electron-dense desmosomes in the intercellular region. Much emphasis has been placed on the protein corneodesomosin, which is located in the extracellular part of the desmosomes and adjacent parts of the cornified cell envelope. It has been suggested that this protein is continuously degraded, providing an explanation for the gradient of increased corneocyte cohesiveness from the skin surface toward deeper layers (118). It has been postulated that cell cohesion is lost through proteolytic degradation, which may be inhibited by calcium ions.

Scaly skin diseases may sometimes be a consequence of a disrupted desquamation process. Desquamation is associated with a conversion of cholesterol sulfate to cholesterol (119). Interestingly, X-linked ichthyosis, a scaly disease characterized by a disrupted desquamation process, is identified with a lack of the enzyme cholesterol sulfatase (120). More recent work (121) has shown that hyperkeratosis attributable to desmosomes is associated with an increased content of cholesterol sulfate in patients with X-linked ichthyosis. It is apparent that cholesterol sulfate retards desquamation by acting as a serine protease inhibitor.

IV. EPIDERMAL REPAIR MECHANISMS

A. The Effects of Hydration

Hydration of the stratum corneum can lead to profound changes in its barrier properties (122). The mechanisms involved in the hydration response are not full defined, although it is likely that it is the result of a combination of water-induced swelling of the corneocytes and some form of water-induced expansion of the intercellular lipid lamellae. In the normal state, the stratum corneum holds between 15 and 20% (dry weight) water, most of which appears to be associated with intracellular keratin (123). Stratum corneum water can be increased up to about 400% (dry weight) following excessive soaking. Swelling of corneocytes is possibly due to increased uptake of water, which then interacts with keratin to expand the spatial orientation of the protein. The observation that the corneocytes of the nail plate and hair do not swell to the same extent as those of the stratum corneum following excessive hydration indicates that the degree of interaction between water and keratin is a function of the positioning and stability of disulfide bonds in the peptide (44). Thus, where the α-helix keratin filaments are loosely packed and more flexible, as in the stratum corneum keratinocytes, there is a greater ability to alter conformation to accommodate water.

The site of interaction of water with intercellular lipid lamellae is less clear. Wide-angle X-ray diffraction studies indicated that no bilayer swelling occurred with hydration (124). This suggests that water molecules are not absorbed between the lamellar regions. Bouwstra's group has used freeze–fracture electron microscopy and other techniques to examine fully hydrated human stratum corneum (125,126). By using an elegant technique of sample preparation (126), which involved multiple folding of the sample, visualization of the interface between the stratum corneum and the hydration medium was possible. As would be expected for fully hydrated stratum corneum the observed corneocytes were swollen, with pools of water apparently displacing and separating keratin filaments. Distinct water domains (pools) were observed in the intercellular spaces of hydrated stratum corneum. Also observed in the intercellular spaces were vesicle-like lipid structures, which suggested that lipids were extracted, presumably from the lamellae.

It has been suggested that the vesicle-like structures observed by van Hal et al. (126) may depict the lacunae that result from desmosomal degradation (127). Lacunae are discontinuous microdomains located in the extracellular space in the middle to outer layers of the stratum corneum (128). During hydration the lacunae provide an obvious site for water pooling and, during prolonged exposure to water, lateral expansion of the lacunae occurs through polar head regions of the intercellular lipids (Fig. 11) (127). Although the expansion of the individual lacunae may lead to a continuous lacunar system, this process does not appear to disrupt the lipid lamellae. Menon and Elias (127) have proposed that the continuous lacunal system may represent a putative "poor" pathway through the stratum corneum.

It is well recognized that natural moisturizing factor (NMF) can make up to 10% of the corneocyte dry weight and, as humectants, these materials can sorb water extensively. There appears to be an absence of NMF in severe, dry flaking skin in both psoriasis and ichthyosis vulgaris. Rawlings et al. (88) have pointed out that the amino acids to which filaggrin is proteolyzed are themselves precursors for the natural moisturizing factor. Glutamine is converted to the potent humectant, pyrrolidone carboxylic acid, a major component of NMF, whereas histidine is converted to urocanic acid. Interestingly, filaggrin is converted to NMF only when the water activity is between 0.70 and 0.95, filaggrin being stable at higher water activities and proteolysis being impeded by low water activity. Hence, under occlusive conditions the stratum corneum NMF level decreases to close to zero, and all corneocytes contain filaggrin. The result of this homeostatic mechanism is that the skin has prevented itself from being "overhydrated."

In conclusion, the current observations suggest that stratum corneum hydration does not lead to an overall decrease in intercellular lipid order and only small amounts of water are present in the intercellular polar head group regions (89). Therefore, it is tempting to revisit a possible mechanism by which hydration promotes percutaneous absorption, which has been raised in an earlier review (122). In that model, swelling of the keratin is akin to the "bricks" becoming swollen in the "bricks-and-mortar" model of the stratum corneum, with a loosening of the intercellular lipid "mortar." The overall effect should be an increase in the mobility of the chains and in permeability, without an effect on the lipid ordering.

B. Chemical Damage

When the stratum corneum is perturbed, several localized biochemical events occur that result in rapid reconstitution of barrier function (64,129–137). Thus, in extreme cases of stratum corneum damage, such as acetone-induced delipidation (129–131) or tape-stripping (137), there appears to be a biphasic pattern of recovery: a rapid phase of repair, followed by a slower phase of normalization. The initial rapid phase of barrier recovery involves the expeditious secretion of preformed lamellar bodies from the granular cells into the intercellular space (64), an increase in epidermal cholesterol and fatty acid synthesis (134,135), and accelerated production and secretion, into the intercellular space, of new lamellar bodies. The subsequent and slower phase of barrier repair involves an increase in ceramide synthesis (135) and an increase in DNA synthesis (136) leading to epidermal hyperplasia. A similar response to barrier perturbation occurs following treatment of the skin with sodium dodecyl sulfate (SDS) (137), but the magnitude of the response depends on the severity of

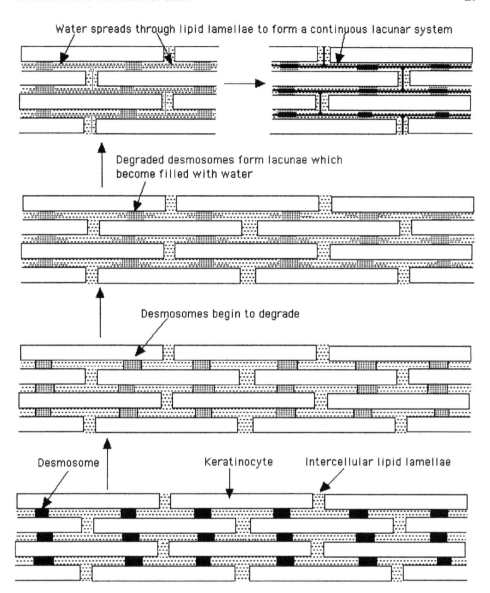

Figure 11 During hydration the lacunae formed by degrading desmosomes provide an obvious site for water pooling and, during prolonged exposure to water, lateral expansion of the lacunae occurs through polar-head regions of the intercellular lipids. Although the expansion of the individual lacunae may lead to a continuous lacunar system, this process does not appear to disrupt the lipid lamellae. Menon and Elias (127) have proposed that the continuous lacunal system may represent a putative aqueous "pore" pathway through the stratum corneum.

the induced perturbation. It is remarkable that the initial perturbation, which occurs in the outermost layers of the stratum corneum, can rapidly stimulate biochemical events in the stratum granulosum and lower levels of the epidermis.

Although the exact mechanisms stimulating these events are unknown, there is some indication that a change in the rate of transepidermal water loss (TEWL) induced by barrier alterations, may play a role (131). This increase in TEWL may lead to focal changes in the concentration of certain ions in the outer epidermis. In the normal state, the epidermis possesses a Ca^{2+} ion gradient such that there is more Ca^{2+} in the outer layers than the inner (138). Following barrier disruption the Ca^{2+} gradient is lost. The presence of higher levels of intracellular Ca^{2+} in the outer epidermis is believed to block lamellar body secretion (139,140), and reduced levels will stimulate secretion. In addition, K^+ may play a role in this homeostatic mechanism and may also influence barrier repair independently of Ca^{2+} (141). Thus, although there are still many uncertainties concerning the biochemistry of barrier repair, there is much evidence that suggests the role of ion concentration and the induction of lipid-producing enzymes; such as 3-hydroxy-3-methylglutaryl coenzyme A and serine palmitoyl transferase (142).

Perturbation of barrier function sometimes, but not always, also induces an inflammatory response that results in irritation. It is important to appreciate that *irritation* is used to describe skin reactions that can range from a mild and transient erythema or itch, to serious vesiculation (see Chaps. 12 and 13). Whereas the insults of solvent delipidation and tape-stripping of the stratum corneum result in barrier repair and epidermal hyperplasia, they do not necessarily lead to an irritant reaction. On the other hand, application of SDS almost always results in an irritant response (143,144). Although solvent delipidation and tape-stripping of the stratum corneum both physically remove the intercellular lipid lamellae, which results in considerable increases in TEWL, SDS intercalates with the lamellae and increases fluidity in this region (145), resulting in an increase in TEWL. Furthermore, although other surface-active agents, such as sodium laurate and polysorbates, can increase TEWL to levels similar to SDS, the resultant irritation is much less and, in some cases, not significantly different from untreated skin (146). It follows that irritation subsequent to exposure to SDS must be a result of factors other than an increase in water transport and the stimulation of lipogenesis.

That surface-active agents can cause skin irritation is well established and has been so for many years (147). Also, whereas ionic surfactants can cause severe irritation, nonionic surfactants are considered virtually nonirritant in normal use (148,149). Thus, much of the research on surfactant-induced skin irritation has involved studies on SDS. The collective data suggest that SDS can interact with both lipid and protein structures in the stratum corneum. Interaction with lipids will increase lipid fluidity and thereby enhance skin permeability. This alone, however, apart from increasing its own permeation, will not account for the irritation caused by SDS. Although SDS can penetrate into the corneocyte and interact with the protein structure such that α-keratin is uncoiled (150), it is difficult to relate this aspect to an irritant response. A more likely explanation for the irritation induced by SDS is its capacity to stimulate keratinocyte production of inflammatory mediators such as IL-1 and PGE_2 (151). Whether this induction is secondary to some interaction between SDS and the corneocyte cell membrane is unknown.

There is a range of mechanisms by which solvents may affect skin permeability, as proposed by Menon et al. (152). Suhonen et al. (89) recently reviewed chemical enhancement in terms of stratum corneum alterations. They suggested that both hydration and temperature effects by transitions were involving the hydrocarbon chains of the stratum corneum lipid components. They suggested that enhancer actions could be located either in the lipid region near the polar head group or between the hydrophobic tails. Extraction of lipids would also lead to an increase in disorder because there is more space (free volume) in which the hydrocarbon chains can move (Fig. 12). Suhonen et al. (89) suggest that ethanol may act by displacing bound water molecules at the lipid headgroup–membrane interface region with a resulting increased interdigitation of the hydrocarbon chains. Ethanol extracts lipids from the stratum corneum only at high concentrations, with a resulting greater free volume for the lipid chains, as shown by infrared spectroscopy.

Keratolytics are becoming increasing used to promote permeability of the human epidermis, especially the nail plate, to topical agents. Walters et al. (153) have previously reported that agents that act as accelerants on the epidermis may not necessarily do so with the nail plate. Many enhancers have an effect on intercellular lipids, which constitute less than 1% of the nail weight. Quintanar-Guerrero et al. (154) suggest that keratolytic agents may facilitate antimycotic penetration through the nail plate by pore formation. This effect, observed using scanning electron microscopy, was most pronounced for papain, followed by salicylic acid, and then urea. Effects on deeper regions of the nail required a combination of papain and salicylic acid. In general, keratolytics have a limited effect on skin permeability, emphasizing the role of the intercellular lipids as a barrier to skin penetration by solutes. It has been suggested that other agents that affect stratum corneum protein structures (e.g., propylene glycol, ethanol, and dimethyl sulfoxide), create a reversible conformation change in the keratin protein from an α-helix to a β-sheet as a consequence of a replacement of water that is bound to polar protein side chains. Dithiothreitol, a disulfide-reducing agent, has also been suggested to enhance hydrophilic solute penetration by an altered protein conformation to the β-sheet as a consequence of the appearance of free thiols.

C. Biochemical Abnormalities

There are a large number of diseases that can affect epidermal barrier function, and it is beyond the scope of this chapter to consider any of these in great depth. Some of the diseases affect the formation of the corneocyte ("broken brick syndrome") whereas other affect the intercellular lipid ("weak mortar syndrome"). For instance, Nemes and Steinhardt (66) refer to more than ten different diseases involving genes that encode keratin intermediate filaments, including Unna–Thost disease and tylosis. Other genetic diseases are related to defects in the genes associated with the structural proteins of the cornified envelope or transglutaminases. For instance, a genetic defect in *TGM1*, the gene that encodes the transglutaminase I enzyme, leads to a life-threatening disease: lamellar ichthyosis (66).

A reduction of the effective intercellular lipid barrier properties can lead to deficiencies ranging from dry skin (depletion of lipids owing to excessive use of detergents), to hyperproliferation and abnormal scaling. Causes include essential fatty acid deficiency, abnormal intercellular deposition of various lipids, accumulation of

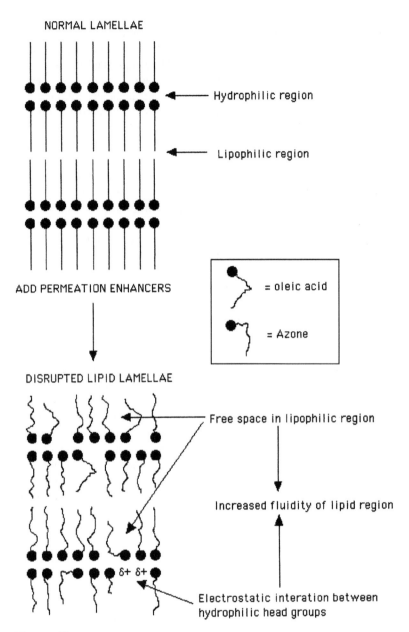

Figure 12 Creation of free space (free volume) in the intercellular lipid lamellae of the stratum corneum allows a greater mobility of the hydrocarbon chains that may result in enhanced diffusivity. This could be induced by enhancer shape [e.g., oleic acid and laurocapram (Azone)] or by electrostatic headgroup interactions.

cholesterol sulfate in X-linked ichthyosis, genetic defects of lipid metabolism (e.g., Refsum's disease and Sjogren–Larsson syndrome resulting from phytanoyl-CoA hydroxylase and fatty aldehyde deficiencies, respectively).

V. CONCLUDING REMARKS

The aim of this chapter has been to introduce the reader to the basic morphology and function of skin, to outline the stages in the development of the barrier layer, to define the chemical makeup of the stratum corneum, and to illustrate repair mechanisms following barrier disruption. What is evident is that the skin is more than another simple biological barrier membrane, into and through which therapeutic agents can be delivered. Rather, the skin should be viewed as an extremely selective semipermeable membrane overlying a powerful immune system ready to react to any given insult. Although it is difficult to resolve the latter, without pharmacological intervention, the problem of membrane permeability may be approached by several diverse strategies. It is important to fully understand the mechanisms of percutaneous absorption and the means by which skin permeation is quantified. The remaining chapters in this book develop the concept of the skin, both as a therapeutic target and as a portal for drug delivery to the systemic sites. Formulation development and scale-up are addressed together with a comprehensive evaluation of bioequivalence for dermatological and transdermal dosage forms. Finally, consideration is given to adverse cutaneous reactions and safety aspects of formulations applied to the skin.

REFERENCES

1. Nevill AM. (1994) The need to scale for differences in body size and mass: and explanation of Klieber's 0.75 mass exponent. Am Physiol Soc pp 2870–2873.
2. Lavker RM, Sun T. (1982) Heterogeneity in epidermal basal keratinocytes: morphological and functional correlations. Science 215:1239–1241.
3. Borradori L, Sonnenberg A. (1999) Structure and function of hemidesmosomes: more than simple adhesion complexes. J Invest Dermatol 112:411–418.
4. Sawamura D, Li K, Chu M–L, Uitto J. (1991) Human bullous pemphigoid antigen (BPAG1): amino acid sequences deduced from cloned cDNAs predict biologically important peptide segments and protein domains. J Biol Chem 266:17784–17790.
5. Li K, Tamai K, Tan EML, Uitto J. (1993) Cloning of type XVII collagen. Complementary and genomic DNA sequences of mouse 180-kDa bullous pemphigoid antigen (BPAG2) predict an interrupted collagenous domain, a transmembrane segment, and unusual features in the 5′-end of the gene and the 3′-untranslated region of mRNA. J Biol Chem 268:8825–8834.
6. Stepp MA, Spurr–Michaud S, Tisdale A, Elwell J, Gipson IK. (1990) $\alpha_6\beta_4$ Integrin heterodimer is a component of hemidesmosomes. Proc Natl Acad Sci USA 87:8970–8974.
7. Burgeson RE, Christiano AM. (1997) The dermal–epidermal junction. Curr Opin Cell Biol 9:651–658.
8. Dowling J, Yu QC, Fuchs E. (1996) β_4 Integrin is required for hemidesmosome formation, cell adhesion and cell survival. J Cell Biol 134:559–572.
9. Kaiser HW, Ness W, Jungblut I, Briggaman RA, Kreysel HW, O'Keefe EJ. (1993) Adherens junctions: demonstration in human epidermis. J Invest Dermatol 100:180–185.

10. Kaiser HW, Ness W, Offers M, O'Keefe EJ, Kreysel HW. (1993) Talin: adherens junction protein is localized at the epidermal–dermal interface in skin. J Invest Dermatol 101:789–793.

11. Amagai M. (1995) Adhesion molecules. I: Keratinocyte–keratinocyte interactions; cadherins and pemphigus. J Invest Dermatol 104:146–152.

12. Peltonen S, Hentula M, Hägg P, Ylä–Outinen H, Tuukkanen J, Lakkakorpi J, Rehn M, Pihlajaniemi T, Peltonen J. (1999) A novel component of epidermal cell–matrix and cell–cell contacts: transmembrane protein type XII collagen. J Invest Dermatol 113: 635–642.

13. O'Keefe E, Briggaman RA, Herman B. (1987) Calcium-induced assembly of adherens junctions in keratinocytes. J Cell Biol 105:807–817.

14. Haftek M, Hansen MU, Kaiser HW, Kreysel HW, Schmitt D. (1996) Interkeratinocyte adherens junctions: immunocytochemical visualization of cell–cell junctional structures, distinct from desmosomes, in human epidermis. J Invest Dermatol 106:498–504.

15. Buxton RS, Magee AI. (1992) Structure and interactions of desmosomal and other cadherins. Semin Cell Biol 3:157–167.

16. Tsukita S, Tsukita S, Nagafuchi A. (1990) The undercoat of adherence junctions: a key specialized structure in organogenesis and carcinogenesis. Cell Struct Funct 15:7–12.

17. Petersen OH. (1999) Waves of excitement: calcium signals inside cells. Biologist 46: 227–230.

18. Kitajima Y, Aoyama Y, Seishima M. (1999) Transmembrane signaling for adhesive regulation of desmosomes and hemidesmosomes, and for cell–cell detachment induced by pemphigus IgG in cultured keratinocytes: involvement of protein kinase C. J Invest Dermatol Symp Proc 4:137–144.

19. Yuspa SH. (1994) The pathogenesis of squamous cell cancer: lessons learned from studies of skin carcinogenesis. Cancer Res 54:1178–1189.

20. Warner RR. (2000) The distribution and function of physiological elements in skin. In: Lodén M, Maibach HI, eds. Dry Skin and Moisturizers, Boca Raton: FL, CRC Press, pp 89–108.

21. Potten CS, Allen TD. (1976) A model implicating the Langerhans cell in keratinocyte proliferation control. Differentiation 5:43–47.

22. Aiba S, Katz SI. (1990) Phenotypic and functional characteristics of in vivo activated Langerhans cells. J Immunol 145:2791–2796.

23. Teunissen MBM. (1992) Dynamic nature and function of epidermal Langerhans cells in vivo and in vitro: a review, with emphasis on human Langerhans cells. Histochem J 24:697–716.

24. Nakamura K, Saitoh A, Yasaka N, Furue M, Tamaki K. (1999) Molecular mechanisms involved in the migration of epidermal dendritic cells in the skin. J Invest Dermatol Symp Proc 4:169–172.

25. Staquet M–J, Levarlet B, Dezutter–Dambuyant C, Schmitt D. (1992) Human epidermal Langerhans cells express β_1 integrins that mediate their adhesion to laminin and fibronectin. J Invest Dermatol 99:12S–14S.

26. Kimber I, Cumberbatch M. (1992) Stimulation of Langerhans cell migration by tumour necrosis factor α (TNF-α). J Invest Dermatol 99:48S–50S.

27. Jakob T, Udey MC. (1998) Regulation of E–cadherin-mediated adhesion in Langerhans cell-like dendritic cells by inflammatory mediators that mobilize Langerhans cells in vivo. J Immunol 160:4067–4073.

28. Jimbow K, Fitzpatrick TB, Quevedo WC. (1986) Formation, chemical composition and functions of melanin pigments in mammals. In: Matolsky A, ed. Biology of the Integument. Vol. 2. New York: Springer-Verlag, pp 278–296.

29. Jimbow K, Lee SK, King MG, Hara H, Chen H, Dakour J, Marusyk H. (1993) Melanin pigments and melanosomal proteins as differentiation markers unique to normal and neoplastic melanocytes. J Invest Dermatol 100:259S–268S.
30. Prota G. (1996) Melanins and melanogenesis. Cosmet Toiletr 111:43–51.
31. Tuma MC, Gelfand VI. (1999) Molecular mechanisms of pigment transport in melanophores. Pigment Cell Res 12:283–294.
32. Ortonne J–P, Prota G. (1993) Hair melanins and hair color: ultrastructural and biochemical aspects. J Invest Dermatol 101:82S–89S.
33. Valyi–Nagy IT, Hirka G, Jensen PJ, Shih I–M, Juhasz I, Herlyn M. (1993) Undifferentiated keratinocytes control growth, morphology, and antigen expression of normal melanocytes through cell–cell contact. Lab Invest 69:152–159.
34. Hearing VJ (1999). Biochemical control of melanogenesis and melanosomal organization. J Invest Dermatol Symp Proc 4:24–28.
35. Chakraborty A, Slominski A, Ermak G, Hwang J, Pawelek J. (1995) Ultraviolet B and melanocyte-stimulating hormone (MSH) stimulate mRNA production for αMSH receptors and proopiomelanocortin-derived peptides in mouse melanoma cells and transformed keratinocytes. J Invest Dermatol 105:655–659.
36. Suzuki I, Tada A, Ollmann MM, Barsh GS, Im S, Lamoreux ML, Hearing VJ, Nordlund JJ, Abdel–Malek ZA. (1997) Agouti signaling protein inhibits melanogenesis and the response of human melanocytes to α-melanotropin. J Invest Dermatol 108:838–842.
37. Grando SA, Kist DA, Qi M, Dahl MV. (1993) Human keratinocytes synthesize, secrete, and degrade acetylcholine. J Invest Dermatol 101:32–36.
38. Tachibana T. (1995) The Merkel cell: recent findings and unresolved problems. Arch Histol Cytol 58:379–396.
39. Halata Z. (1993) Sensory innervation of the hairy skin (light- and electronmicroscope study). J Invest Dermatol 101:75S–81S.
40. Cross SE, Roberts MS. (1993) Subcutaneous absorption kinetics of interferon and other solutes. J Pharm Pharmacol 45:606–609.
41. Szuba A, Rockson SG. (1997) Lymphedema: anatomy, physiology and pathogenesis. Vasc Med 2:321–326.
42. Cameli N, Picardo M, Perrin C. (1994) Expression of integrins in human nail matrix. Br J Dermatol 130:583–588.
43. Baden HP, Goldsmith LA, Fleming B. (1973) A comparative study of the physicochemical properties of human keratinized tissues. Biochim Biophys Acta 322:269–278.
44. Gniadecka M, Nielsen OF, Christensen DH, Wulf HC. (1998) Structure of water, proteins, and lipids in intact human skin, hair, and nail. J Invest Dermatol 110:393–398.
45. Dawber RPR. (1980) The ultrastructure and growth of human nails. Arch Dermatol Res 269:197–204.
46. Jemec GBE, Serup J. (1989) Ultrasound structure of the human nail plate. Arch Dermatol 125:643–646.
47. Kobayashi Y, Miyamoto M, Sugibayashi K, Morimoto Y. (1999) Drug permeation through the three layers of the human nail plate. J Pharm Pharmacol 51:271–278.
48. Wang JCT, Sun Y. (1999) Human nail and its topical treatment: brief review of current research and development of topical antifungal drug delivery for onychomycosis treatment. J Cosmet Sci 50:71–75.
49. Eckert EL. (1989) Structure, function, and differentiation of the keratinocyte. Physiol Rev 69:1316–1346.
50. Dinarello CA. (2000) Interleukin-1 receptors and signal transduction. In: Kydonieus AF, Wille JJ, eds. Biochemical Modulation of Skin Reactions. Boca Raton, FL: CRC Press, pp 173–187.

51. Kondo S. (1999) The roles of keratinocyte-derived cytokines in the epidermis and their possible responses to UVA-irradiation. J Invest Dermatol Symp Proc 4:177–183.
52. Jensen PJ, Lavker RM. (1999) Urokinase is a positive regulator of epidermal proliferation in vivo. J Invest Dermatol 112:240–244.
53. Petersen R–D, Reinhold W, Tyborczyk J. (1997) Cytokines in cosmetology. Cosmet Toiletr 112:65–70.
54. Skerrow D, Skerrow CJ. (1983) Tonofilament differentiation in human epidermis: isolation and polypeptide chain composition of keratinocyte subpopulations. Exp Cell Res 143:27–35.
55. Baden HP. (1979) Keratinization in the epidermis. Pharm Ther 7:393–411.
56. Hunter I, Skerrow D. (1989) Relationship of the final α-keratin compositions in dyskeratoses to those along the normal keratinization pathway: aetiological and diagnostic significance. Br J Dermatol 120:363–369.
57. Landmann L. (1988) The epidermal permeability barrier. Anat Embryol 178:1–13.
58. Madison KC, Howard EJ. (1996) Ceramides are transported through the Golgi apparatus in human keratinocytes in vitro. J Invest Dermatol 106:1030–1035.
59. Wertz PW, Downing DT. (1982) Glycolipids in mammalian epidermis: structure and function in the water barrier. Science 217:1261–1262.
60. Egelrud T, Lundström A. (1991) A chymotrypsin-like proteinase that may be involved in desquamation in plantar stratum corneum. Arch Dermatol Res 283:108–112.
61. Egelrud T, Régnier M, Sondell B, Shroot B, Schmidt R. (1993) Expression of stratum corneum chymotryptic enzyme in reconstructed human epidermis and its suppression by retinoic acid. Acta Derm Venereol 73:181–184.
62. Sondell B, Thornell L–E, Egelrud T. (1995) Evidence that stratum corneum chymotryptic enzyme is transported to the stratum corneum extracellular space via lamellar bodies. J Invest Dermatol 104:819–823.
63. Wertz PW, Downing DT, Freinkel RK, Traczyk TN. (1984) Sphingolipids of the stratum corneum and lamellar granules of fetal rat epidermis. J Invest Dermatol 83:193–195.
64. Menon GK, Feingold KR, Elias PM. (1992) Lamellar body secretory response to barrier disruption. J Invest Dermatol 98:279–289.
65. Fartasch M, Bassukas ID, Diepgen TL. (1992) Disturbed extruding mechanism of lamellar bodies in dry non-eczematous skin of atopics. Br J Dermatol 127:221–227.
66. Nemes Z, Steinert PM. (1999) Bricks and mortar of the epidermal barrier. Exp Mol Med 31:5–19.
67. Kawabe S, Ikuta T, Ohba M, Chida K, Ueda E, Yamanishi K, Kuroki T. (1998) Cholesterol sulfate activates transcription of transglutaminase 1 gene in normal human keratinocytes. J Invest Dermatol 111:1098–1102.
68. Takahashi M, Tezuka T, Katunuma N. (1992) Phosphorylated cystatin alpha is a natural substrate of epidermal transglutaminase for formation of skin cornified envelope. FEBS Lett 308:79–82.
69. Fujimoto W, Nakanishi G, Arata J, Jetten AM. (1997) Differential expression of human cornifin α and β in squamous differentiating epithelial tissues and several skin lesions. J Invest Dermatol 108:200–204.
70. Steinert PM, Marekov LN. (1995) The proteins elafin, filaggrin, keratin intermediate filaments, loricrin, and small proline-rich proteins 1 and 2 are isodipeptide cross-linked components of the human epidermal cornified cell envelope. J Biol Chem 270:17702–17711.
71. Baden HP, Kvedar JC. (1993) Epithelial cornified envelope precursors are in the hair follicle and nail. J Invest Dermatol 101:72S–74S.
72. Swartzendruber DC, Wertz PW, Madison KC, Downing DT. (1987) Evidence that the corneocyte has a chemically bound lipid envelope. J Invest Dermatol 88:709–713.

73. Downing DT. (1992) Lipid and protein structures in the permeability barrier of mammalian epidermis. J Lipid Res 33:301–313.

74. Lazo ND, Meine JG, Downing DT. (1995) Lipids are covalently attached to rigid corneocyte protein envelopes existing predominantly as β-sheets: a solid-state nuclear magnetic resonance study. J Invest Dermatol 105:296–300.

75. Wertz PW, Downing DT. (1989) Stratum corneum: biological and biochemical considerations. In: Hadgraft J, Guy RH, eds. Transdermal Drug Delivery. New York: Marcel Dekker, pp 1–22.

76. Downing DT, Lazo ND. (2000) Lipid and protein structures in the permeability barrier. In: Lodén M, Maibach HI, eds. Dry Skin and Moisturizers. Boca Raton, FL: CRC Press, pp 39–44.

77. Behne M, Uchida Y, Seki T, Ortiz de Montellano P, Elias PM, Holleran WM. (2000) omega-Hydroxyceramides are required for corneocyte lipid envelope (CLE) formation and normal epidermal permeability barrier function. J Invest Dermatol 114:185–192.

78. Kurihara–Bergstrom T, Good WR. (1987) Skin development and permeability. J Controlled Release 6:51–58.

79. Sun T–T, Green H. (1978) Keratin filaments of cultured human epidermal cells. J Biol Chem 253:2053–2060.

80. Steinert PM, North ACT, Parry DAD. (1994) Structural features of keratin intermediate filaments. J Invest Dermatol 103:19S–24S.

81. Steven AC, Bisher ME, Roop DR, Steinert PM. (1990) Biosynthetic pathways of filaggrin and loricrin—two major proteins expressed by terminally differentiated mouse keratinocytes. J Struct Biol 104:150–162.

82. Yaffe MB, Murthy S, Eckert RL. (1993) Evidence that involucrin is a covalently linked constituent of the highly purified cultured keratinocyte cornified envelopes. J Invest Dermatol 100:3–9.

83. Robinson NA, LaCelle PT, Eckert RL. (1996) Involucrin is a covalently crosslinked constituent of highy purified epidermal corneocytes: evidence for a common pattern of involucrin crosslinking in vivo and in vitro. J Invest Dermatol 107:101–107.

84. Dale BA, Holbrook KA, Steinert PM. (1978) Assembly of stratum corneum basic protein and keratin filaments in macrofibrils. Nature 276:729–731.

85. Scott IR, Harding CR, Barrett JG. (1982) Histidine-rich protein of the keratohyalin granules: source of free amino acids, urocanic acid and pyrolidone carboxylic acid in the stratum corneum. Biochim Biophys Acta 719:110–117.

86. Mack JW, Stevens AC, Steinert PM. (1993) The mechanism of interaction of filaggrin with intermediate filaments. The ionic zipper hypothesis. J Mol Biol 232:50–66.

87. Markova NG, Marekov LN, Chipev CC, Gan S–Q, Idler WW, Steinert PM. (1993) Profilaggrin is a major epidermal calcium-binding protein. Mol Cell Biol 13:613–625.

88. Rawlings AV, Scott IR, Harding CR, Bowser PA. (1994) Stratum corneum moisturisation at the molecular level. J Invest Dermatol 103:731–740.

89. Suhonen TM, Bouwstra JA, Urti A. (1999) Chemical enhancement of percutaneous absorption in relation to stratum corneum structural alterations. J Controlled Release 59:149–161.

90. Wertz WP, Miethke MC, Long SA, Strauss JS, Downing DT. (1985) The composition of the ceramides from human stratum corneum and from comedones. J Invest Dermatol 84:410–412.

91. Stewart ME, Downing DT. (1999) A new 6-hydroxy-4-sphingenine-containing ceramide in human skin. J Lipid Res 40:1434.

92. Abraham W, Wertz PW, Landmann L, Downing DT. (1987) Stratum corneum lipid liposomes: calcium-induced transformation into lamellar sheets. J Invest Dermatol 88:212–214.

93. Wertz P. (1997) Integral lipids of hair and stratum corneum. Experientia 78:227–237.

94. Perisho K, Wertz PW, Madison KC, Stewart ME, Downing DT. (1988) Fatty acids of acylceramides from comedones and from the skin surface of acne patients and control subjects. J Invest Dermatol 90:350–353.
95. Holman RT. (1968) Essential fatty acid deficiency. Prog Chem Fats Other Lipids 9: 275–348.
96. Jung E, Griner RD, Mann–Blakeney R, Bollag WB. (1998) A potential role for ceramide in the regulation of mouse epidermal keratinocyte proliferation and differentiation. J Invest Dermatol 110:318–323.
97. Motta S, Monti M, Sesana S, Caputo R, Carelli S, Ghidoni R. (1993) Ceramide composition of the psoriatic scale. Biochim Biophys Acta 1182:147–151.
98. Imokawa G, Kuno H, Kawai M. (1991) Stratum corneum lipids serve as a bound-water modulator. J Invest Dermatol 96:845–851.
99. Norlén L, Nicander I, Rozell BL, Ollmar S, Forslind B. (1999) Inter- and intra-individual differences in human stratum corneum lipid content related to physical parameters of skin barrier function in vivo. J Invest Dermatol 112:72–77.
100. Brain KR, Walters KA. (1993) Molecular modeling of skin permeation enhancement by chemical agents. In: Walters KA, Hadgraft J, eds. Pharmaceutical Skin Penetration Enhancement. New York: Marcel Dekker, pp 389–416.
101. Finean JB. (1990) Interaction between cholesterol and phospholipid in hydrated bilayers. Chem Phys Lipids 54:147–156.
102. Norlén L, Nicander I, Lundsjö A, Cronholm T, Forslind B. (1998) A new HPLC-based method for the quantitative analysis of inner stratum corneum lipids with special reference to the free fatty acid fraction. Arch Dermatol Res 290:508–516.
103. Forslind B. (1994) A domain mosaic model of the skin barrier. Acta Derm Venereol (Stockh) 74:1–6.
104. Engström S, Forslind B, Engblom J. (1996) Lipid polymorphism—a key to understanding of skin penetration. In: Brain KR, James VJ, Walters KA, eds. Prediction of Percutaneous Penetration. Vol. 4b. Cardiff: STS Publishing, pp 163–166.
105. Bouwstra JA, Cheng K, Gooris GS, Weerheim A, Ponec M. (1996) The role of ceramides 1 and 2 in the stratum corneum lipid organisation. Biochim Biophys Acta 1300: 177–186.
106. Bouwstra JA, Gooris GS, Cheng K, Weerheim A, Bras W, Ponec M. (1996) Phase behavior of isolated skin lipids. J Lipid Res 37:999–1011.
107. Bouwstra JA, Gooris GS, Dubbelaar FER, Weerheim A, IJzerman AP, Ponec M. (1998) Role of ceramide 1 in the molecular organization of the stratum corneum lipids. J Lipid Res 39:186–196.
108. van der Meulen J, van den Bergh BAI, Mulder AA, Mommaas AM, Bouwstra JA, Koerten HK. (1996) The use of vibratome sections for the ruthenium tetroxide protocol: a key for optimal visualization of epidermal lipid bilayers of the entire human stratum corneum in transmission electron microscopy. J Microstruct 184:67–70.
109. Pilgram GSK, Engelsma–van Pelt AM, Bouwstra JA, Koerten HK. (1999) Electron diffraction provides new information on human stratum corneum lipid organization studied in relation to depth and temperature. J Invest Dermatol 113:403–409.
110. Fensky DB, Thewalt JL, Bloom N, Kitson N. (1994) Models of stratum corneum intercellular membranes: ^{3}H NMR of macroscopically oriented multilayers. Biophys J 67:1562–1573.
111. Suzuki Y, Nomura J, Koyama J, Horii I. (1984) The role of proteases in stratum corneum: involvement in stratum corneum desquamation. Arch Dermatol Res 286:249–253.
112. Brattsand M, Egelrud T. (1999) Purification, molecular cloning, and expression of a human stratum corneum trypsin-like serine protease with possible function in desquamation. J Biol Chem 274:30033–30040.

113. Ekholm IE, Brattsand M, Egelrud T. (2000) Stratum corneum tryptic enzyme in normal epidermis: a missing link in the desquamation process? J Invest Dermatol 114:56–63.

114. Egelrud T. (2000) Desquamation. In: Lodén M, Maibach HI, eds. Dry Skin and Moisturizers. Boca Raton, FL: CRC Press, pp 109–117.

115. Horikoshi T, Chen S–H, Rajaraman S, Brysk H, Brysk MM. (1998) Involvement of cathepsin D in the desquamation of human stratum corneum. J Invest Dermatol 110: 547.

116. Brysk MM, Bell T, Brysk H, Selvanayagum P, Rajaraman S. (1994) Enzymic activity of desquamin. Exp Cell Res 214:22.

117. Watkinson A. (1998) Stratum corneum gelatinase: a novel late differentiation, epidermal cystein protease. J Invest Dermatol 110:539.

118. Lundstrom A, Serre G, Haftek M, Egelrud T. (1994) Evidence for a role of corneodesmosin, a protein which may serve to modify desmosomes during cornification, in stratum corneum cell cohesion and desquamation. Arch Dermatol Res 286:369–375.

119. Long SA, Wertz PW, Strauss JS, Downing DT. (1985) Human stratum corneum polar lipids and desquamation. Arch Dermatol Res 277:284–287.

120. Koppe JG, Marinkovic–Ilsen A, Rijken Y, DeGroot WP, Jobsis AC. (1978) X-linked ichthyosis. A sulfatase deficiency. Arch Dis Child 53:803–806.

121. Sato J, Denda M, Nakanishi J, Nomura J, Koyama J. (1998) Cholesterol sulfate inhibits proteases that are involved in desquamation of stratum corneum. J Invest Dermatol 111:189–193.

122. Roberts MS, Walker M. (1993) Water—the most natural penetration enhancer. In: Walters KA, Hadgraft J, eds. Pharmaceutical Skin Penetration Enhancement. New York: Marcel Dekker, pp 1–30.

123. Vavosour I, Kitson N, MacKay A. (1998) What's water got to do with it? A nuclear magnetic resonance study of molecular motion in pig stratum corneum. J Invest Dermatol Symp Proc 3:101–104.

124. Bouwstra JA, Gooris GS, Salomons–de Vries, van der Spek JA, Bras W. (1992) Structure of human stratum corneum as a function of temperature and hydration: a wide-angle X-ray diffraction study. Int J Pharm 84:205–216.

125. Hofland HEJ, Bouwstra JA, Boddé HE, Spies F, Nagelkerke JF, Cullander C, Junginger HE. (1995) Interactions between nonionic surfactant vesicles and human skin in vitro: freeze fracture electron microscopy and confocal laser scanning microscopy. J Liposome Res 5:241–263.

126. van Hal DA, Jeremiasse E, Junginger HE, Spies F, Bouwstra JA. (1996) Structure of fully hydrated human stratum corneum: a freeze–fracture electron microscopy study. J Invest Dermatol 106:89–95.

127. Menon GK, Elias PM. (1997) Morphologic basis for a pore-pathway in mammalian stratum corneum. Skin Pharmacol 10:235–246.

128. Hou SYE, Mitra AK, White SH, Menon GK, Ghadially R, Elias PM. (1991) Membrane structures in normal and essential fatty acid-deficient stratum corneum: characterization by ruthenium tetroxide staining and X-ray diffraction. J Invest Dermatol 96:215–223.

129. Feingold KR. (1991) The regulation and role of epidermal lipid synthesis. Adv Lipid Res 24:57–82.

130. Proksch E, Holleran WM, Menon GK, Elias PM, Feingold KR. (1993) Barrier function regulates epidermal lipid and DNA synthesis. Br J Dermatol 128:473–482.

131. Grubauer G, Elias PM, Feingold KR. (1989). Transepidermal water loss: the signal for recovery of barrier structure and function. J Lipid Res 30:323–334.

132. Ghadially R, Brown BE, Sequeira–Martin SM, Feingold KR, Elias PM. (1995) The aged epidermal permeability barrier: structural, functional, and lipid biochemical abnormalities in humans and a senescent murine model. J Clin Invest 95:2281–2290.

133. Menon GK, Feingold KR, Moser AH, Brown BE, Elias PM. (1985) De novo sterologenesis in the skin: fate and function of newly synthesized lipids. J Lipid Res 26: 418–427.

134. Grubauer G, Feingold KR, Elias PM. (1987) The relationship of epidermal lipogenesis to cutaneous barrier function. J Lipid Res 28:746–752.

135. Holleran WM, Feingold KR, Mao–Qiang M, Gao WN, Lee JM, Elias PM. (1991) Regulation of epidermal sphingolipid synthesis by permeability barrier function. J Lipid Res 32:1151–1158.

136. Proksch E, Feingold KR, Mao–Qiang M, Elias PM. (1991) Barrier function regulates epidermal DNA synthesis. J Clin Invest 87:1668–1673.

137. Taljebini M, Warren R, Mao–Qiang M, Lane E, Elias PM, Feingold KR. (1996) Cutaneous permeability barrier repair following various types of insults: kinetics and effects of occlusion. Skin Pharmacol 9:111–119.

138. Menon GK, Grayson S, Elias PM. (1985) Ionic calcium reservoirs in mammalian epidermis: ultrastructural localization by ion-capture cytochemistry. J Invest Dermatol 84: 508–512.

139. Menon GK, Elias PM, Lee SH, Feingold KR. (1992) Localization of calcium in murine epidermis following disruption and repair of the permeability barrier. Cell Tissue Res 270:503–512.

140. Lee SH, Elias PM, Feingold KR, Mauro T. (1994) A role for ions in barrier recovery after acute perturbation. J Invest Dermatol 102:976–979.

141. Lee SH, Elias PM, Proksch E, Menon GK, Mao–Qiang M, Feingold KR. (1992) Calcium and potassium are important regulators of barrier homeostasis in murine epidermis. J Clin Invest 89:530–538.

142. Elias PM, Feingold KR. (1992) Lipids and the epidermal water barrier: metabolism, regulation, and pathophysiology. Semin Dermatol 11:176–182.

143. Imokawa G. (1980) Comparative study on the mechanism of irritation by sulfate and phosphate type of anionic surfactants. J Soc Cosmet Chem 31:45–66.

144. Patil SM, Singh P, Maibach HI. (1994) Cumulative irritancy in man to sodium lauryl sulfate: the overlap phenomenon. Int J Pharm 110:147–154.

145. Downing DT, Abraham W, Wegner BK, Willman KW, Marshall JL. (1993) Partition of sodium dodecyl sulfate into stratum corneum lipid liposomes. Arch Dermatol Res 285:151–157.

146. van der Valk PGM, Nater JP, Bleumink E. (1984) Skin irritancy of surfactants as assessed by water vapor loss measurements. J Invest Dermatol 82:291–293.

147. Prottey C. (1978) The molecular basis of skin irritation. In: Breuer MM, ed. Cosmetic Science. Vol. 1. New York: Academic Press, pp 275–349.

148. Singer EJ, Pittz EP. (1985) Interaction of surfactants with epidermal tissues: biochemical and toxicological aspects. In: Rieger MM, ed. Surfactants in Cosmetics. New York: Marcel Dekker, pp 133–194.

149. Zhou J, Mark R, Stoudemayer T, Sakr A, Lichtin JL, Gabriel KL. (1991) The value of multiple instrumental and clinical methods, repeated patch applications, and daily evaluations for assessing stratum corneum changes induced by surfactants. J Soc Cosmet Chem 42:105–128.

150. Imokawa G, Sumura K, Katsumi M. (1975) Study on skin roughness caused by surfactants. II Correlation between protein denaturation and skin roughness. J Am Oil Chem Soc 52:484–489.

151. Cohen C, Dossou G, Rougier A, Roguet R. (1991) Measurement of inflammatory mediators produced by human keratinocytes in vitro: a predictive assessment of cutaneous irritation. Toxicol In Vitro 5:407–410.

152. Menon GK, Lee SH, Roberts MS. (1998) Ultrastructural effects of some solvents and vehicles on the stratum corneum and other skin components: evidence for an "extended

mosaic-partitioning model of the skin barrier." In: Roberts MS, Walters KA, eds. Dermal Absorption and Toxicity Assessment. New York: Marcel Dekker, pp 727–751.

153. Walters KA, Flynn GL, Marvel JR. (1985) Physicochemical characterization of the human nail: solvent effects on the permeation of homologous alcohols. J Pharm Pharmacol 37:771–775.

154. Quintanar–Guerrero D, Ganem–Quintanar A, Tapia–Olguin P, Kalia YN, Buri P. (1998) The effect of keratolytic agents on the permeability of three imidazole antimycotic drugs through the human nail. Drug Dev Ind Pharm 24:685–690.

2

Common Skin Disorders and Their Topical Treatment

C. COLIN LONG

Cardiff and Vale NHS Trust, Cardiff, Wales

I. INTRODUCTION

Dermatologists have the advantage over other clinicians in that their patients present with diseases that are usually visible and accessible. This means that the majority of skin diseases may be treated topically with treatment delivered directly to the desired site of action, thereby avoiding, or at least we hope attenuating, the potential for systemic side effects. Systemic treatment may be needed if the skin disease is severe, recalcitrant, or fails to respond to topical therapy. The aim of this chapter is to provide a brief synopsis of common skin disorders and their current topical treatment for nonspecialists. A glossary of common dermatological terms has been included at the end of this chapter (see Sec. XII).

II. PSORIASIS

A. Introduction

Psoriasis is a chronic inflammatory skin disease of unknown etiology that affects between 1 and 3% of the population. There is increased proliferation of the epidermis with infiltration of inflammatory cells within the dermis and epidermis, coupled with dilation of the upper dermal capillaries. Psoriasis tends to run in families and may be associated with certain HLA phenotypes; individuals with first-degree relatives with psoriasis are more likely to develop the disease themselves. Abnormalities of arachidonic acid metabolism have been demonstrated within psoriatic plaques, and it is possible that arachidonic acid and its metabolites may be intimately involved in the psoriatic process. An increase in the levels of prostaglandins (PGs) may cause vasodilation and erythema, and leukotrienes, such as 5,12-dihydroxy-6,14-*cis*-8,10-*trans*-eicosatetraenoic acid (LT-B$_4$) and 12-hydroxy-5,8,14-*cis*-10-*trans*-eicosatetraenoic acid (12-HETE), as well as interleukin (IL)-8, and the complement product C5a *des arg* may cause neutrophil accumulation (1–4). Raised levels of calmodulin, a cellular receptor protein for calcium, have been demonstrated in psoriatic lesions. The calcium–calmodulin complex may influence cell proliferation in psoriasis by modulating the activities of phospholipase-A$_2$ (which releases arachidonic acid from cell membranes) and phosphodiesterase. Drugs such as anthralin (dithranol) and cyclosporine, which are beneficial in psoriasis, are calmodulin antagonists. Intracellular cAMP levels are decreased within psoriatic lesions and drugs that decrease cAMP levels, such as β-blockers or lithium, may worsen psoriasis. Conversely, drugs such as benoxaprofen (now withdrawn) which elevate cAMP levels may improve psoriasis.

The role of T lymphocytes in psoriasis is unclear. However, T-helper lymphocytes form a major part of the dermal inflammatory cell infiltrate, and thus the effect of cyclosporine in psoriasis may be partly due to its anti–T-helper cell activity.

B. Clinical Patterns

Several clinical patterns in psoriasis are recognized. The most common is chronic plaque psoriasis in which there are erythematous plaques of psoriasis with an overlying silvery scale usually affecting the elbows, knees, and at times, the scalp and lower back (Fig. 1). Guttate psoriasis, which may be precipitated by a streptococcal infection of the throat, is characterized by numerous small, scaling erythematous plaques on the trunk and limbs. Psoriasis may also affect the flexures and may cause a glazed erythematous appearance similar to that seen in seborrheic eczema.

Erythrodermic psoriasis is characterized by severe erythema affecting the whole of the patient's skin. This may develop following deterioration of the patient's psoriasis or be precipitated by use of potent topical or systemic steroids. There may be associated systemic symptoms, and the patient is at risk from hypothermia owing to excessive heat loss, dehydration, and cardiac failure.

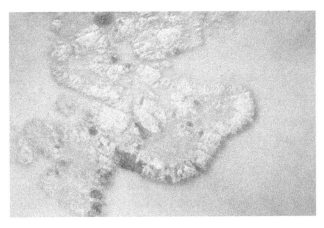

Figure 1   A large plaque of psoriasis on the trunk: There is erythema with overlying silvery scale.

Generalized pustular psoriasis is another rare, but extremely serious, type of psoriasis in which the patient has widespread areas of erythema with overlying sheets of sterile pustules. Localized pustular psoriasis may also affect the palms and soles.

The nails may be involved in psoriasis, and patients may develop pitting or lifting of the nail plate from the nail bed (onycholysis). Patients may also present with gross thickening of the nails owing to subungual hyperkeratosis.

C. Topical Therapy

1. Emollients

Emollients act by blocking transepidermal water loss and help to soften and soothe the skin. In psoriasis, they help reduce scaling and may make the skin more comfortable. Patients should be encouraged to use an emollient bath oil or shower gel when bathing and to apply emollients when other treatments (see later discussion) have been washed off. There are numerous emollients available, and it is important that the patient tries several until they find one that suits them best. Emollients are particularly beneficial in patients with erythrodermic or pustular psoriasis who are unable to tolerate other more "active" forms of topical therapy.

2. Coal Tar

Many preparations of coal tar are available, and there are differences between the composition of coal tar obtained from different sources. Coal tar may be formulated in different vehicles to produce creams, gels, and pastes. The mode of action is unclear, but it is believed to have an antimitotic effect on proliferative cells (5). Tar is also a photosensitizer and may be used in combination with ultraviolet light (UV)B radiation therapy in the Goeckerman regimen. Crude coal tar preparations have the disadvantages of mess and smell, but do not cause systemic toxicity. Tar preparations may be applied to normal skin without ill-effect, but may be irritant to the skin of the face, flexures, and genitals, and can cause folliculitis. It is doubtful whether concentrations of crude coal tar greater than 10% are of any additional benefit. Coal tar solutions (in alcohol) are cleaner and more cosmetically acceptable, but they are less potent and less effective.

3. Anthralin (Dithranol)

Anthralin [known abroad as dithranol], a synthetic derivative of chrysarobin (an extract of tree bark) is the most effective topical antipsoriatic treatment. Similar to tar, its mechanism of action is unknown, but anthralin is known to act as an antimitotic agent by reducing DNA synthesis (6), and it also inhibits the enzyme glucose-6-phosphate dehydrogenase. Anthralin is active only in its reduced form and thus must be combined with an antioxidant. In Lassar's paste, anthralin is combined with salicylic acid, which acts as a kerotolytic agent (and will help remove scale), and in Dithrocream, ascorbic acid is present as an antioxidant. Anthralin may cause irritation, burning, and staining of the skin, and may also stain clothing. It is important to initiate therapy with a weak concentration (such as 0.1%) and then increase the concentration as the patient's tolerance increases. In Ingram's method of applying anthralin the patient soaks in a warm bath containing coal tar solution (1:800) and, after drying, is exposed to UVB radiation. A paste of the agent is then applied to the lesions. This procedure is repeated daily. In the hospital anthralin is frequently left on for several hours, but short-contact therapy, in which it is washed off 30–60 min after application, may be more convenient for outpatients.

4. Calcipotriol [Calcipotriene]

Calcipotriol [known in the United States as calcipotriene] is a vitamin D analogue that has benefit in psoriasis. Vitamin D analogues reverse the increased proliferation and other changes seen in psoriatic skin, and this may be through intracellular vitamin D receptors known to be present in epidermal keratinocytes, Langerhans cells, T lymphocytes, and macrophages (7). Vitamin D analogues may also affect the inflammatory cell infiltrate. Calcipotriene (ointment or cream) has the advantage of being cosmetically more acceptable to patients than tar or anthralin (dithranol) preparations, but may cause irritant dermatitis in some patients. Patients are limited to a maximum of 100 g/week because there is a potential risk of hypercalcemia and hypercalciuria.

5. Topical Steroids

Topical steroids have anti-inflammatory, immunosuppressive, antimitotic, and vasoconstrictive effects on the skin. Topical steroids are effective in reducing the inflammatory changes seen in psoriasis, but there is a risk of precipitating widespread erythroderma, particularly if potent preparations are used, or if topical steroids are suddenly withdrawn. For these reasons, topical steroids are usually avoided in the general treatment of psoriasis. However, weak topical steroids are useful in treating areas such as the scalp, face, and flexures, where other treatments such as tar, anthralin, or calcipotriene are likely to cause irritation.

6. Treatment of Scalp Psoriasis

Scalp psoriasis, similar to psoriasis elsewhere, will respond to a similar range of topical therapy although the presence of hair makes treatment more difficult. Application is messy. For thick plaques on the scalp topical therapies include oil of cade, "ung. cocois," and combinations of coal tar and salicylic acid (such as 6% coal tar and 3% salicylic acid), all of which are effective at removing scale and settling inflammation. These preparations may be applied to the scalp and left on overnight before being washed out in the morning with a tar-based shampoo. Patients should be warned to use old pillowcases or towels, or to wear a showercap to protect bedding. Other more cosmetically acceptable preparations include calcipotriene scalp solution (Dovonex), 0.025% fluocinolone acetonide gel (Synalar gel), and 0.1% betamethasone valerate (Betnovate scalp application). Various tar-based shampoos can be used to reduce mild inflammation and scaling.

III. ECZEMA

A. Classification of Eczema

Eczema may be considered as either endogenous or exogenous. The terms eczema and dermatitis are synonomous, although dermatitis is sometimes used to imply that the eczema has been caused by an external agent (exogenous). Endogenous eczemas include atopic eczema, seborrheic eczema, discoid eczema, pompholyx, and varicose eczema. Exogenous eczemas include both irritant and allergic contact dermatitis as well as photodermatitis (caused by the interaction of light and chemicals absorbed by the skin).

Acute eczema presents as a pruritic erythematous confluent papular rash with an ill-defined border. There may be vesicles present, and if these rupture there may be exudation and "weeping." Chronic eczema tends to be erythematous, scaly, and is less likely to be vesicular. There may also be some degree of lichenification and fissuring.

B. Atopic Eczema (Dermatitis)

The etiology of atopic eczema is unknown. Patients have increased levels of serum IgE and some have precipitating antibodies to environmental allergens, including foods and inhaled materials. Many patients will have a positive response to intracutaneous challenge with pollen, house dust mite, cat fur, and fish antigens. However, the significance of these positive reactions is unclear. Patients with atopic eczema have reduced numbers of circulating T-suppressor cells which are responsible for modulating immunoglobulin-producing B lymphocytes. Low levels of the unsaturated fatty acids γ-linoleic and dihomo-γ-linolenic acid have been reported (8).

In most patients there is a family history of eczema or of other atopic diseases, such as asthma or allergic rhinitis. Atopic eczema usually presents during infancy and, often, may resolve during childhood, whereas in others it may persist into adult life. Atopic eczema usually affects the face, wrists, and the flexural aspects of the elbows and knees (Fig. 2). There may be some involvement of the trunk, and the rash may become generalized. The eczema may be complicated by bacterial infection, and there is evidence to suggest that many exacerbations of atopic eczema may be due to occult infection with *Staphylococcus aureus*. Eczematous skin is also more prone to infections with wart viruses, molluscum contagiosum, and herpesviruses. Patients with atopic dermatitis may develop a widespread and potentially fatal rash, eczema herpeticum, following the development of herpes simplex or following contact with individuals affected with herpes simplex.

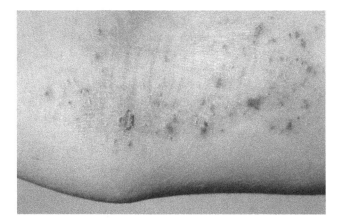

Figure 2 Excoriated atopic eczema in the flexural aspects of the elbow: There is lichenification with increased prominence of the skin markings.

C. Contact Dermatitis

Irritant contact dermatitis is caused by (usually repeated) exposure to chemical or mechanical trauma. Some individuals seem more prone than others. In allergic contact dermatitis the sensitizing agent (antigen) crosses the stratum corneum to reach the epidermal Langerhans cells. The antigen is processed by the Langerhans cells and presented to circulating T lymphocytes with subsequent development of a clone of T lymphocytes with a specific memory for that particular antigen. In an individual sensitized to a particular antigen, repeat exposure to that antigen will result in migration of the sensitized T lymphocytes to the site of exposure, with initiation of the inflammatory eczematous process. Both irritant and allergic contact dermatitis usually start at the site(s) of initial or more frequent contact, but may spread to involve other areas.

D. Treatment of Eczema

In exogenous irritant or allergic contact dermatitis the mainstay of treatment is to identify the precipitating agent and to avoid it if at all possible. Otherwise the same general principles apply to the treatment of all forms of eczema. If there is a possibility that the eczema may be infected, skin swabs should be submitted for bacterial culture and sensitivity and, if appropriate, an antibiotic such as flucloxacillin should be prescribed. For acute, particularly wet and weeping eczema, astringent solutions such as potassium permanganate (1:10,000) are indicated. The involved area, such as hands or feet, can be placed in a bowl of the solution or, alternatively, wet gauze swabs may be applied directly to the skin. The majority of patients with eczema have a chronic, dry scaling, rash. These patients should be advised to use emollients frequently. They should also avoid soap and use soap substitutes, such as emulsifying ointment or Diprobase, wherever possible. Emollient bath oil and gels should be used when bathing, and ointments and creams applied to the skin after bathing. If these simple measures fail to settle the eczema, a topical corticosteroid may be necessary. Topical corticosteroids have an anti-inflammatory, immunosuppressive, antimitotic, and vasoconstrictive action on the skin. These actions are mediated by a nuclear receptor for hydrocortisone to which other steroids also bind (9). The stronger the affinity of the steroid for the receptor the more potent the steroid. The anti-inflammatory action of the corticosteroids depends on the induction of peptides, known as lipocortins, which antagonize the actions of phospholipase-A_2 which acts to release arachidonic acid from membrance phospholipids. Other effects of corticosteroids include lysosomal and cellular membrane stabilization, a reduction in the number of epidermal Langerhans cells, and modulation of the migration of inflammatory cells.

Topical corticosteroids are classified as being mild (e.g., hydrocortisone 1%), moderately potent (e.g., clobetasone butyrate 0.05%), potent (e.g., betamethasone valerate 0.1% or hydrocortisone butyrate 0.1%), or as very potent (e.g., clobetasol propionate 0.05%).

Creams are suitable for moist or weeping areas of eczema, whereas ointments should be used for dry, scaly, or lichenified areas. Local side effects of topical steroids include masking or worsening of infection (especially fungal infections), thinning of the skin, induction of striae, bruising and telangectasia, on aggravation of rosacea.

Use of more potent topical steroids may result in pituitary–adrenal axis suppression, iatrogenic Cushing's syndrome, and stunted growth.

In general, one should aim to use a steroid of sufficient potency to control the eczema, and then aim to reduce the potency of the topical steroid preparation as the eczematous rash improves. There is a tendency for the eczematous rash to rebound when treatment is stopped or the potency of the topical steroid is decreased. The aim should be for the patient to use the least potent topical corticosteroid that will control the rash and, preferably, to use simple emollients only. It is important that patients are prescribed adequate quantities of topical therapy and that these are applied regularly. There is doubt as to whether topical steroids need to be applied twice daily and some topical steroids such as fluticasone propionate 0.05% (Cutivate) or mometasone furoate 0.1% (Elocon) are claimed to be effective when applied once daily. Patients may apply moisturizers ad lib in between applications of corticosteroids.

Coal tar preparations may be helpful with reducing the pruritis of eczema and may be particularly valuable in lichenified and localized eczema, such as lichen simplex. Occlusive bandaging, such as Viscopaste, Coltapaste, or Ichthopaste bandages, may be of benefit, particularly if the eczema is excoriated. The bandages are soothing and prevent further excoriation. Some patients, however, may develop sensitivity to the preservatives found in these bandages.

Infection with *S. aureus* can frequently exacerbate eczema and patients may benefit from the use of a combined topical steroid antibiotic preparation such as Fucibet. However, topical antibiotics may cause sensitization, and, in recent years, the problem of sensitization to topically applied corticosteroids themselves has also been recognized, and should be considered if the eczema fails to settle.

E. Seborrheic Eczema

In infants, seborrheic eczema may present as greasy adherent scale on the scalp (cradle cap). Seborrheic eczema in adults principally affects the greasier areas of the body, including the scalp, eyebrows, eyelids, nasal–labial areas, and chin. In young men, it may also affect the presternal area and upper back and, in the elderly, may involve the flexures and may become generalized. Individuals who are immunosuppressed (including those with human immunodeficiency virus; HIV) are more prone to develop seborrheic eczema. It is possible that, in affected individuals, the commensal yeast-like microorganism *Pityrosporum ovale* has become pathogenic and provokes an inflammatory response (10). Cradle cap in infants may be treated with olive oil or arachis (peanut) oil. In adults, weak topical steroids, with or without azoles such as miconazole, clotrimazole, or ketoconazole, may be used. Ketoconazole shampoo, used two or three times weekly as a liquid soap to wash the affected areas, can also be helpful. Topical 8% lithium succinate cream has also been of use.

IV. ACNE

A. Introduction

Acne is one of the most common and distressing of skin diseases commonly present during adolescence and usually (but not always) resolves in early adult life. Seventy percent of the population develop acne, but only a relatively small proportion seek

Figure 3 Nodular acne of the face and neck: There are pustules, papules, nodules, and cysts present. There is a high risk of scarring without treatment.

medical attention. Several variants of acne are recognized, including infantile acne, which occurs on the face during the first few months and usually settles spontaneously, and occupational acne, resulting from exposure to oil, coal tar, chlorinated hydrocarbons, or insecticides. Acne may be precipitated or exacerbated by certain combined oral contraceptive pills or by androgenic hormones.

Acne vulgaris commonly affects the face, chest, and upper back, and usually presents during puberty. The clinical features include an increased rate of sebum secretion, comedones, papules, and pustules (Fig. 3). Severe acne may be complicated by atrophic or nodular keloid-type scars or by the formation of chronic nodules and cysts (Fig. 4).

Patients with acne tend to have a higher sebum excretion rate than others, and there is a degree of correlation between the sebum secretion rate and the severity of the acne (11,12). Circulating androgens stimulate the sebaceous glands with resulting hypertrophy and increased sebum secretion. Furthermore, there is abnormal keratin-

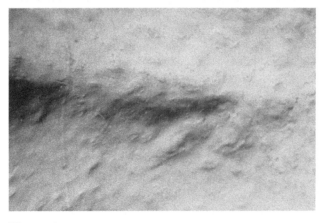

Figure 4 Pitted scars on the chest following acne.

ization of the epithelium lining the hair follicle, which may lead to obstruction of the follicle with resulting comedone (blackhead) formation. *Propionibacterium acnes*, a gram-positive commensal bacterium, proliferates within the obstructed hair follicle, and may break down the lipid esters of sebum to liberate potentially irritating fatty acids (13). Eventual rupture of the wall of the obstructed follicle and the release of fatty acids into the surrounding dermis will result in an inflammatory response.

B. Treatment of Acne

The aims of treatment are to reduce the bacterial population of the hair follicles; to encourage the shedding of comedones; to reduce the rate of sebum production; and to reduce the degree of inflammation. Topical therapy is appropriate for mild-to-moderate acne, but more severe forms of acne, in which there is a risk of scarring, will require systemic therapy. Skin cleansers such as Phisomed or Hibiscrub are of some value. Benzoyl peroxide reduces comedone formation, as well as reducing the population of *P. acnes*, and may also have an anti-inflammatory effect. Benzoyl peroxide cream may be applied twice daily at an initial concentration of 2.5% and increased to 5 or 10% as tolerated. Benzoyl peroxide can have an irritant effect and may also bleach both hair and clothing.

1. Topical Antibiotics

Topical erythromycin, clindamycin, and tetracycline are all effective in acne (14–17). These antibiotics reduce the population of *P. acnes* and *Staph. epidermidis*, and may have a separate anti-inflammatory action. The advantage of topical antibiotics is the reduction in the risk of potential systemic side effects, and this is particularly true with topical clindamycin. Topical tetracyclines may cause some yellow staining of clothing and fluoresce under ultraviolet radiation. It is also possible that they may exacerbate the problem of bacterial antibiotic resistance.

2. Topical Retinoids

Topical retinoids, including tretinoin and isotretinoin, act by decreasing epidermal proliferation and reducing the abnormal keratinization process in the hair follicle. This prevents new comedones forming and softens and removes existing comedones. There is also a reduction in the level of *P. acnes* within the hair follicle. Although topical retinoids are of particular value in severe acne, when there are numerous comedones present, they are also effective in other forms of mild to moderate acne. Initially, there may be some increased irritation and pain, but this usually settles with use. Tretinoin cream is preferable to tretinoin gel for those with dry or fair skin.

3. Azelaic Acid

Azelaic acid, a dicarboxylic acid, produced by the yeast *Pityrosporum*, has been of benefit in mild-to-moderate acne (18). The mode of action is unknown, but similar to other acne treatments, may normalize follicular keratinization (possibly by reducing filagrin formation) and may reduce the population of *P. acnes*. Twenty percent azelaic acid cream is well tolerated but may cause some local irritation.

V. ROSACEA

Rosacea is a chronic inflammatory skin disorder, affecting the face, which causes persistent erythema associated with telangectasia and papules. It most commonly

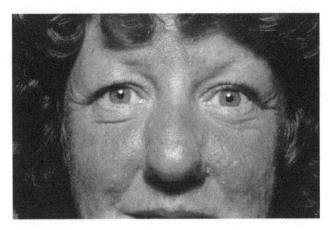

Figure 5 Rosacea with erythema, telangiectasia, and papules: There is also a bulbous enlargement of the nose (rhinophyma).

affects the forehead, nose, cheeks, and chin. Characteristically small pustules and papules arise on a background of erythema and telangectasia (Fig. 5). The patients may also complain of flushing in response to trivial stimuli. The rash may resemble that seen in acne, but rosacea usually affects an older-aged group and is not characterized by comedones. Persistent inflammation of the nose may result in rhinophyma (an irregular bulbous enlargement of the nose characterized by prominent hair follicles). Over 30% of patients with rosacea may also suffer with conjunctivitis and blepharitis. The etiology of rosacea is unknown, although the *Demodex folliculorum* mite is present in increased numbers (19).

It is possible that rosacea is the result of repeated environmental trauma (cold wind, ultraviolet radiation, and heat) which damage the upper dermal collagen and vasculature. Rosacea-like symptoms may be precipitated by the prolonged use of even moderately potent topical steroids on the face. Topical metronidazole gel (0.75%) applied twice daily is effective therapy for most individuals with rosacea.

VI. LICHEN PLANUS

Lichen planus is an inflammatory skin disorder, of unknown etiology, characterized by the presence of pruritic violaceous papules. Common sites are the flexural aspects of the wrists and forearms, but the rash may also affect the trunk and limbs (Fig. 6). In approximately 30% of patients there is mucosal involvement with a reticulate rash affecting the inside of the mouth. The nails may be affected in 10% of patients.

Lichen planus may be associated with other autoimmune diseases, including vitiligo and myasthenia gravis. The etiology is unknown but the deposition of IgM at the dermoepidermal junction, coupled with a dense inflammation in the upper dermis, suggest an autoimmune process. Lichen planus-like rashes may be precipitated by various drugs, including thiazide diuretics, gold, tetracyclines, and *para*-aminosalicylic acid (PAS). The disease is usually self-limiting, but topical corticosteroids may be helpful.

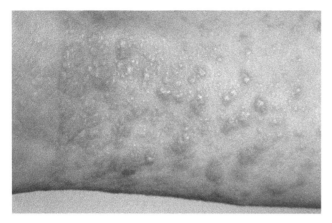

Figure 6 Multiple flat-topped violaceous papules of lichen planus on the flexural aspects of the wrist.

VII. PITYRIASIS ROSEA

Pityriasis rosea is an erythematous scaling rash, of unknown etiology, thought to be secondary to infection with an, as yet unidentified, virus. Patients develop a solitary erythematous scaling patch some 2–3 cm in diameter. After a few days, other smaller plaques develop on the trunk. The individual lesions tend to be oval and their longitudinal axis run parallel to the lines of the ribs. The rash may be associated with mild pruritus and can last for several weeks before resolving spontaneously. Topical steroids may be of benefit.

VIII. SOLAR KERATOSES

Solar (actinic) keratoses present as scaling hyperkeratotic plaques or papules on skin exposed to light. They are most commonly seen on elderly subjects with fair skin who have had high levels of ultraviolet exposure over many years. They may be associated with other signs of photodamage such as yellowing, coarsening, and wrinkling of the skin. Individuals with large numbers of solar keratoses are at increased risk of developing nonmelanoma skin cancer. Histologically, the lesions show epidermal thickening, with abnormal epidermal differentiation and scaling. Solar keratoses are common, and up to 20% of individuals over the age of 60 are affected. Some solar keratoses may resolve spontaneously, but a small proportion may develop into squamous cell carcinoma. Solar keratoses may respond to topical 5% 5-fluorouracil cream (Efudix) an antimetabolite that inhibits DNA synthesis. This should be applied once or twice daily for 10–14 days, and may cause the lesions and the surrounding skin to become sore and inflamed. Alternatively, the 5% 5-fluorouracil cream can be applied twice daily for 2 days each week and the frequency increased until the level of the patient's tolerance is reached, but it is probable that some degree of irritation is necessary for the treatment to be effective. The accompanying irritation and soreness may be treated with topical corticosteroids. Topical 5-fluorouracil has also been claimed to be effective for superficial basal cell carcinoma and for Bowen's disease. Topical tretinoin 0.05% and isotretinoin 0.1% have also been used to treat multiple solar keratoses.

IX. SUNSCREENS

Sunscreens are preparations that filter out or reflect harmful ultraviolet radiation. Various disorders may be precipitated or aggravated by exposure to ultraviolet light. These include polymorphic light eruption, Hutchinson's summer prurigo, cutaneous porphyria, rosacea, and lupus erythematosus. Exposure to ultraviolet radiation is also a major risk factor for both malignant melanoma and nonmelanoma skin cancer as well as photoaging. Recurrent attacks of herpes simplex may also be precipitated by ultraviolet exposure. Sunscreens differ greatly in their protective capacity and some of the newer sunscreens also offer some degree of protection against ultraviolet A.

X. ULCERS

Ulcers may be seen at any body site, but are most commonly seen on the legs, probably caused by a combination of trauma and impaired circulation. The venous return of blood from the legs depends on efficient working of the calf muscles to act as a pump coupled with the action of valves in the deep veins preventing the reflux of blood. Damage to the valves of the deep veins following deep vein thrombosis may be associated with pregnancy. Injury or immobilization may also lead to valvular incompetence.

The most common types of ulcer are venous ulcers caused by leaking valves in the deep veins, resulting in venous hypertension, and edema of the subcutaneous tissue. An extravascular accumulation of fibrinous material leaked from dermal blood vessels results in a fibrous cuff around the capillaries that prevents diffusion of oxygen and other nutrients through the blood vessel wall as well as causing fibrosis, and sclerosis of the dermal capillaries. Venous ulcers are more common in women than men and result from inadequate provision of nutrients and oxygen to the skin. Venous ulcers are most commonly seen on the medial aspect of the lower leg usually above the medial malleolus (inner aspect of the ankle) (Fig. 7). Large ulcers may encircle the leg.

On examination, the lower leg and ankle and foot may be edematous, there may be prominent varicose veins present. Leakage of blood into the skin may cause deposition of hemosiderin. The ulcers may heal spontaneously or may become chronic and indolent. The ulcers may be complicated by infection, bleeding or by eczema. Varicose eczema, surrounding the ulcer, is common, and in many patients, it may be due to allergic contact hypersensitivity to medicaments, such as neomycin clioquinal (Vioform), lanolin, or ethylene diamine, used in treating the leg ulcer or eczema. Rarely, in long-standing ulcers, malignant change with the development of squamous cell carcinoma may occur, and patients with long-standing ulcers may become anemic. It is important to try and improve venous return (drainage) by

1. Elevation of the legs for regular periods
2. Compression bandaging with elasticated stockings or bandages (which should be graduated so that pressure is greatest at the ankle and least at the top of the dressing)

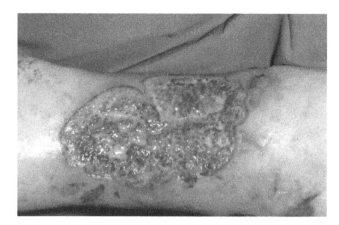

Figure 7   A large venous ulcer on the medial (inner) aspect of the leg.

3.   Regular exercise
4.   Weight reduction

Exudate and slough should be removed and ulcers may be cleaned with normal saline, sodium hypochloride solution, Eusol, or 5% hydrogen peroxide. Topical antiseptics, such as povidone iodine or potassium permanganate, may help reduce the bacterial load. Crust may be loosened by the application of saline- or potassium permanganate-soaked dressings. Local surgical debridement may be necessary for a thick eschar. Hydrogen peroxide cream 1.5%, streptokinase/streptodornase solution (Varidase) may also be useful in helping remove thick slough.

Numerous dressings are available to treat leg ulcers, including nonadherent gauze, paraffin gauze, silver sulfadiazine (Flamazine), and absorbent hydrocolloid dressings. Any ulcerated area of skin will rapidly become colonized with numerous different bacteria, and these should be treated only if they are causing local cellulitis. The main function of leg ulcer dressings should be to provide comfort, protection for the wound, and an optimum local environment to allow reepithelization. Hydrocolloid dressings have the advantage that they may be left in situ for several days before changing. They have the disadvantages of an unpleasant odor and an unpleasant fluid may collect beneath the dressing. Surrounding (varicose) eczema may be treated with topical corticosteroids.

Ischemic leg ulcers are due to reduced circulation secondary to atherosclerosis, vasculitis, or other causes of arterial obstruction. Arterial ulcers tend to be more sharply defined and painful than venous ulcers. In the leg, they are more common on the anterior aspect of the shin, rather than the medial aspect of the ankle. Compression bandaging will impair arterial blood supply further and thus it is important that arterial pulses are examined and if necessary Doppler ultrasound or other investigations are performed so that arterial ulcers are not incorrectly treated with compression. Ulcers may also occur in diabetes, and vasculitis, or may be due to pyoderma gangrenosum, as well as secondary to infection, trauma, or malignant disease.

Decubitus ulcers (pressure sores) are the result of localized ischemia secondary to prolonged pressure in patients who are immobile. They are most common over the sacrum ischial tuberosities, heels, occiput, shoulders, and elbows. The ulcers are often deep and sloughy. Neuropathic ulcers result from decreased cutaneous inner-vation, resulting in diminished sensation, and are most commonly seen on the feet. Leg ulcers may also be due to sickle cell disease, idopathic thrombocytopenic pur-pura, tuberculosis, syphillis, and deep fungal infections.

XI. INFECTION OF THE SKIN

A. Yeast Infections

1. Pityriasis Versicolor

This disease is caused by *Pityrosporum orbiculare*, a gram-positive yeast-like mi-croorganism that is usually a skin commensal. In some individuals, the organism can become pathogenic. *Pityrosporum versicolor* often affects young adults, causing brown scaly macules on the trunk and sometimes on the limbs. Carboxylic acids released by the organisms inhibit melanogenesis and thus the affected areas may appear relatively pale following exposure to sunlight. The infection may be dem-onstrated by microscopy of skin scrapings suitably treated with potassium hydroxide solution.

Most patients will respond to a topical imidazole drug, such as miconazole, clotrimazole, or econazole creams, applied once daily for 6 weeks. Ketoconazole shampoo can also be used to wash the affected areas daily for 4–5 days. Other effective topical treatments include Whitfield's ointment (6% benzoic acid and 3% salicylic acid in emulsifying ointment), selenium sulfide shampoo, and 20% sodium thiosulfate solution.

2. Candida

The yeast *Candida albicans* may cause vulvovaginitis in women, especially during pregnancy, in those taking oral contraceptives, or those who are receiving systemic antibiotics for acne. It may also cause stomatitis in infants, and may exacerbate intertrigo in the body folds of obese individuals and the napkin area during infancy. The nail plate may also be infected, and the organism may cause chronic paronychia in those involved with wet-work such as bar workers or housewives. Topical treat-ments with imidazole creams is often effective, although more serious infections may require systemic therapy.

B. Dermatophyte Infections

Dermatophyte infection (ringworm) is restricted to invasion of the stratum corneum, nails, and hair. The dermatophytes, *Trichophyton, Epidermophyton*, and *Microsporum* species may infect humans. *Microsporum* species are usually acquired from infected cats or dogs (*M. canis*) and are a frequent cause of tinea capitis (ringworm affecting the head) in children. Infections from farm or other animals tend to cause more vigorous inflammation than those from other sources. The infection may be diag-nosed from microscopy of skin, nail, or hair treated with potassium hydroxide. Al-ternatively, the fungus may be cultured.

Tinea corporis (ringworm affecting the skin of the trunk or limbs) often presents as a pruritic, annular, erythematous, scaling plaque, which may resemble a patch of eczema or psoriasis, but is often solitary. Tinea cruris (ringworm affecting the groin) presents as a well-demarcated pruritic erythematous scaling rash affecting the groins. The rash may extend onto the thigh and genitalia. *Trichophyton rubrum* and *Epidermophyton floccosum* are the most common causative fungi. Tinea pedis (ringworm affecting the feet) may affect the skin of the toe web spaces, sole, or may extend onto the sides and dorsal aspect of the feet. *Trichophyton rubrum, T. mentagrophytes*; and *E. floccosum* are the most common causative organisms.

Tinea manuum is a chronic form of ringworm affecting the hands (frequently only one palm will be affected); *T. rubrum* is the most commonly identified organism. Tinea unguium (ringworm affecting the nail plate and nail bed) may be caused by *T. rubrum, T. mentagrophytes*, or *E. floccosum*. Affected nails are often thickened and have a yellowish discoloration. Onycholysis (separation of the nail plate from the nail bed) may also be seen.

Tinea incognito is the term used to describe dermatophyte infections treated inappropriately with topical corticosteroids, which suppress the inflammatory response, but allow the fungus to proliferate.

Topical imidazole creams (e.g., miconazole, econozole, or clotrimazole), which interfere with ergosterol synthesis and thereby impair fungal cell wall permeability, when used twice daily for 2 weeks, are adequate for most limited areas of fungal infection. The more recently introduced topical allylamine terbinafine is also very effective (20). More extensive infections, or involvement of the nail or scalp will require systemic therapy.

C. Bacterial Infections

Various acute bacterial infections may affect the skin. These include impetigo, erysipelas, cellulitis, furuncles, carbuncles, anthrax, diphtheria, and various mycobacterial infections, including tuberculosis and leprosy. Of these only impetigo (in which a small area is infected) or furuncles, which are both caused by *Staphylococcus aureus*, are amenable to topical treatment. In impetigo, which is more common in young children, an inflamed erythematous area with a yellow crust may develop on exposed skin. Local treatment with antibiotic washes, such as Phisomed or Hibiscrub, and topical mupirocin or fucidic acid (Fucidin) ointment may be sufficient. More extensive areas, larger than a few centimetres in diameter, will require treatment with systemic antibiotics.

1. Antibiotics

Furuncles are hair follicles infected with *S. aureus* and present as yellow-headed pustules. They are commonly seen on the back of the neck in men or in patients treated with ointments or tar (particularly if the skin has been occluded). Extensive areas of folliculitis (furuncles) will require systemic floxacillin (flucoxacillin), but solitary or isolated lesions will respond to topical mupirocin or sodium fusidate (Fucidin). Mupirocin (pseudomonic acid) interferes with bacterial protein synthesis, has the advantage of no cross-resistance with other antibiotics and is available only as a topical preparation. It is effective against both staphylococci and streptococci and may be used in the treatment of folliculitis, infected eczema, and as prophylaxis against nasal carriage of staphylococci. Fucidic acid inhibits bacterial protein synthesis and is particularly effective against staphylococcal skin infections. Topical application may lead to a hypersensitivity reaction. Metronidazole inhibits DNA synthesis and is active against anaerobic bacteria and protozoa, and it can be used topically in the treatment of rosacea. It has also been used to reduce the smell of infected, sloughy ulcers.

2. Antiseptics

Antiseptics have bactericidal activity and may be used as cleansing agents on the skin, as an adjunct to antibiotic therapy, or to prevent secondary bacterial infection. Povidone iodine is a powerful bactericide and may also be effective against viruses. It is commonly used preoperatively to minimize the chance of sepsis, and it may also be of help in the treatment of leg ulcers and herpetic lesions. Potassium permanganate 1:8,000 or 1:10,000 aqueous solution is an effective antiseptic that may be used as a soak in the treatment of infected or weeping eczema. Potassium permanganate also acts as an astringent which helps "dry up" exudative or weeping rashes. Chlorhexidine is an effective skin disinfectant and is available in a wide range of preparations.

D. Viral Infections

1. Herpes Simplex

Herpes simplex type I commonly causes herpetic lesions on the face and oropharynx whereas herpes simplex type 2 affects the genitalia. Following a primary infection, the virus may become latent within nerve ganglia and recurrent episodes may occur periodically. In children, primary herpes simplex type 1 infection presents as an acute gingivostomatitis with vesicles (which may subsequently ulcerate) scattered on the lips and buccal mucosa. The infection is accompanied by malaise, headache, and fever. Primary herpes simplex type 2 infections are usually sexually transmitted and cause multiple painful genital or perianal blisters and ulcers. The virus can also be inoculated to other areas, or other individuals during contact sports such as rugby or wrestling. Recurrent infections occur more or less at the same site each time, and are often precipitated by ultraviolet stress, respiratory tract infections, or menstruation. There may be a preceeding discomfort or tingling followed within a few hours by the development of erythema and vesicle formation. The episodes are usually self-limiting. A crust often develops within 48 h and the lesions resolves within 5 or 6 days. Some patients with recurrent herpes simplex develop a widespread rash known as erythema multiforme.

   Topical antiseptics or antibiotics may prevent secondary bacterial infections, and topical 5-idoxuridine lotion or acyclovir cream applied five times daily at the first sign of discomfort may be effective in reducing the length and severity of attacks (21). Following entry into herpes-infected cells, acyclovir is phosphorylated to the active compound, acyclovir triphosphate, by herpesvirus-coded thymidine kinase. Acyclovir triphosphate inhibits herpes-specific DNA polymerase, thereby preventing further viral synthesis (22).

   Herpes zoster (shingles) is caused by a reactivation of the varicellar–zoster virus which causes childhood chickenpox. Following an attack of chickenpox, the virus may remain dormant in a sensory root ganglion and may become reactivated at a later date, often many years later. Shingles is more common in the elderly, those with lymphoma, AIDS, or other causes of immunosuppression. The attacks often start with unilateral paresthesiae or pain and there may be accompanying systemic upset and fever. Vesicular lesions similar to those seen in varicella (chickenpox) develop within the course of the cutaneous nerve. Approximately 25% of patients develop a distressing postherpetic neuralgia that may persist for months, or sometimes years, after the rash has settled. Although topical therapy has little role in treating acute herpes zoster, topical antiseptics, such as Betadine paint may prevent secondary bacterial infection. Recently, a counterirritant cream containing capsaicin at a concentration of 0.75% (Axain) has been introduced for the treatment of postherpetic neuralgia, and this should be applied three or four times daily after the herpetic lesions have healed (23).

2. Viral Warts

Viral warts are caused by infection with one of the many papillomaviruses, and are spread by direct contact from infected individuals or possibly from shed skin on changing room floors. Warts present as horny nodules in which small black thrombosed capillaries may be seen. Wart infections may persist for many months and even years, but most resolve spontaneously. Topical therapies, including salicylic acid (10–50%), lactic acid (4–20%), podophyllin (up to 15%), or glutaralderhyde (10%), may be effective. Topical podophyllin (15%) in compound benzoin tincture may be applied weekly to external genital warts and should be washed off 6 h after application. The preparation is irritant and care should be taken not to apply the paint to nonaffected skin. Severe toxicity has been reported to be caused by treatment of extensive lesions. Podophyllotoxin 0.5% may be applied to genital warts for 3-consecutive days and repeated weekly for up to 5 weeks if necessary.

3. Molluscum Contagiosum

Molluscum contagiosum is a common viral infection, caused by a poxvirus, and is common in childhood. Individuals with atopic eczema are particularly prone to infection. Small pink umbilicated papules occur on the skin of the trunk and limbs. The lesions usually resolve spontaneously within a few months but may be treated by curettage, cryotherapy, or topical salicylic acid preparations as used for viral warts. An alternative is to squeeze the papule with a forceps so as to express the contents and to then apply silver nitrate, phenol, or iodine.

E. Scabies

Scabies is an infestation with the mite *Sarcoptes scabei* var. *hominis*, which is spread by close skin-to-skin contact. The female mite burrows into the human stratum corneum and lays her eggs. Affected individuals become sensitized to the mite and to their waste products. Individuals who are immunosuppressed may develop extensive infestation. Scabies presents as a severe generalized pruritus, affecting the trunk and limbs, which is often worse at night. The patient may have a generalized eczematous rash which may become impetiginized, and small linear streaks of scabies burrows may be seen, in particular on the flexural aspect of the wrists, on the palm, on the sides of the fingers, and on the soles of the feet. In men, papular lesions on the genitalia are common. Norwegian scabies is characterized by infestation with large numbers of mites and is most commonly seen in immunosuppressed patients, such as those with leukemia or HIV infection. Patients develop crusted water lesions on the hands and feet which contain hundreds of mites. This form of scabies is highly contagious owing to the large number of mites present.

Benzyl benzoate (25% lotion), 10% sulfur in yellow soft paraffin, lindane (1% lotion), malathion (0.5% lotion), and permethrin (5% cream), all are effective. However, lindane should be avoided during pregnancy or in nursing mothers. It is important that all members of an affected household and any person having close physical contact should be treated whether symptomatic or not. Treatment should be applied to the whole body and in infants and younger children this should include the scalp, neck, face, and ears. It is important that all areas, particularly the finger and toe web spaces, and the genital areas are adequately treated. Two "coats" should be applied (using a 2-in. paint brush for lotions if possible) to ensure all areas are treated. The treatment should be washed off after 24 h and all clothing changed. The most common causes of treatment failure are failure to treat all members of a household or failure to treat all areas of the body. It is important to warn patients that the pruritus associated with infestation may persist for several days after treatment, and topical steroids applied after successful treatment of the mites can help to settle the pruritus.

F. Insect Bites

Some individuals seem particularly sensitive to bites from flea and other insects. They develop pruritic papules and sometimes blister at the site of injury. Topical corticosteroids may be of benefit, and any potential animal source should be treated.

GLOSSARY OF DERMATOLOGICAL TERMS

Abscess	A localized collection of pus greater than 1 cm in diameter
Angioedema	A transient acute swelling caused by tissue edema
Annular	Ring-like
Arcuate	Forming part of a circle
Atrophy	Thinning of the skin owing to reduction in the thickness of the epidermis, dermis, or subcutis
Bulla	A larger fluid-containing lesion of more than 1 cm in diameter
Circinate	Circular
Comedone	A plug of keratin and sebum obstructing the follicle
Ecchymosis	Larger purpura greater than 1 cm in diameter
Erosion	Complete or partial loss of the epidermis that will heal without scarring
Excoriation	Ulcer or erosion produced by physical trauma
Fissure	A split through the surface of the skin
Hematoma	A deep swelling in the skin secondary to hemorrhage
Lichenification	A thickening of the skin with increased prominence of skin markings
Macule	A flat area of discolored skin
Nodule	A firm raised lesion larger than 1 cm in diameter
Nummular	Coin-like or discoid
Papilloma	A pendulant mass projecting from the skin
Papule	A small, firm, raised lesion of less than 0.5 cm in diameter

Petechiae	A small macule caused by presence of blood in the skin
Plaque	A circumscribed elevated area of skin, usually larger than 1 cm in diameter
Purpura	Larger macules or papules caused by the collection of blood in the skin
Pustule	A small visible collection of pus
Reticulate	Net-like
Scale	Small flake arising from the horny layer (stratum corneum) or crust of adherent dry blood or tissue fluid
Scar	The permanent replacement of normal structures of the skin with fibrous tissues secondary to healing
Stria	A linear streak of atrophic skin
Telangectasia	Dilated small cutaneous blood vessel
Tumor	Large nodule
Ulcer	An area of complete loss of the epidermis and sometimes of the dermis and subcutis; ulcers heal to leave a scar
Vesicle	A fluid-containing lesion of less than 1 cm in diameter
Weal	An elevated off-white area of skin often with surrounding erythema (redness), which is usually transient

BIBLIOGRAPHY

For further information and good pictorial examples of the skin conditions described in this chapter the reader is referred to the following compendia.

Champion RH, Burton JL, Ebling FJG, eds. Textbook of Dermatology. Oxford: Blackwell Scientific, 1992.
Lawrence CM, Cox NH. Physical Signs in Dermatology. London: Wolfe, 1993.
du Vivier A. Atlas of Clinical Dermatology. 2nd ed. London: Gower Medical, 1993.

REFERENCES

1. Brain S, Camp R, Dowd P, Kobza-Black A, and Greaves M. The release of leukotriene B_4 like material in biologically active amounts from the lesional skin of patients with psoriasis. J Invest Dermatol 83:70–73, 1984.
2. Woolard PM. Stereochemical differences between 12-hydroxy-5,8,10,14-eicosatetraenoic acid in latelets and psoriatic lesions. Biochem Biophys Res Commun 136:169–76, 1986.
3. Gearing AJ, Fincham NJ, Bird CR, Wadhwa M, Meager A, Cartwright JE, Camp RD. Cytokines in skin lesions of psoriasis. Cytokine 2:68–75, 1990.
4. Schröder J–M, Christophers E. Identification of C5a *des arg* and an anionic neutrophil activating peptide (ANAP) in psoriatic scales. J Invest Dermatol 87:53–58, 1986.
5. Griffiths WAD, Wilkinson JD. Topical therapy. In: Champion RH, Burton J, Ebling FJG, eds. Textbook of Dermatology. Oxford: Blackwell Scientific, pp 3054–3055, 1992.
6. Griffiths WAD, Wilkinson JD. Topical therapy. In: Champion RH, Burton J, Ebling FJG, eds. Textbook of Dermatology. Oxford: Blackwell Scientific, pp 3055–3056, 1992.
7. Milde P, Hauser U, Simon T, Mall G, Ernst V, Haussler MR, Frosch P, Rautertberg EW. Expression of 1,25-dihydroxy-vitamin D_3 receptors in normal and psoriatic skin. J Invest Dermatol 97:230–239, 1991.

8. Hansen AE. Serum lipid changes and therapeutic effects of various oils in infantile eczema. Proc Soc Exp Biol Med 31:160–161, 1933.
9. Marks R. Topical corticosteroids: treatment of eczematous disorders. In: Munson PL, ed. Principles of Pharmacology: Basic Concepts and Clinical Applications. New York: Chapman & Hall, pp 1223–1226, 1995.
10. Hay R, Roberts SOB, and Mackenzie DWR, Mycology. In: Champion RH, Burton J, Ebling FJG, eds., Textbook of Dermatology. Oxford: Blackwell Scientific, pp 1176–1179, 1992.
11. Burton JL, Schuster S. The relationship between seborrhoea and acne vulgaris. Br J Dermatol 84:600–601, 1971.
12. Cunliffe WJ, Schuster S. Pathogenesis of acne. Lancet 1:685–687, 1969.
13. Ebling FJG, Cunliffe WJ. Disorders of the sebaceous glands. In: Champion RH, Burton J, Ebling FJG, eds., Textbook of Dermatology. Oxford: Blackwell Scientific, pp 1699–1744, 1992.
14. Dobson RL, Belknap BS. Topical erythromycin solution in acne: results of a multi-clinic trial. J Am Acad Dermatol 3:478–482, 1980.
15. Feucht CL, Allen BS, Chalker BK, Smith JG. Topical erythromycin with zinc in acne: a double-blind controlled study. J Am Acad Dermatol 3:483–491, 1980.
16. Gloor M, Kraft H, Franke M. Effectiveness of topically applied antibiotics on anaerobic bacteria in the pilosebaceous duct. Dermatologica 157:96–104, 1984.
17. Stoughton RB. Topical antibiotics for acne vulgaris. Arch Dermatol 115:486–489, 1979.
18. Nazzaro–Porro M, Passi S, Picardo M. Beneficial effect of 15% azelaic acid cream on acne vulgaris. Br J Dermatol 109:371–374, 1983.
19. Spickett SG. Aetiology of rosacea. Br Med J 1:1625–1626, 1962.
20. Ryder NS. Terbinafine: mode of action and properties of squalene epoxidase inhibition. Br J Dermatol 126(suppl 39):2–7, 1992.
21. Fiddian AP, Yeo JM, Stubbings R, Dean D. Successful treatment of herpes labialis with topical acyclovir. Br Med J 286:1699–1701, 1983.
22. Motley RJ. Viral skin disease and its treatment. In: Munson PL, ed. Principles of Pharmacology: Basic Concepts and Clinical Applications. New York: Chapman & Hall, pp 1255–1257, 1995.
23. Westerman RA, Roberts RG, Kotzmann RR, Westerman DA, Delaney C, Widdop RE, Carter BE. Effects of topical capsaicin on normal skin and affected dermatomes in herpes zoster. Clin Exp Neurol 25:71–84, 1988.

3

Basic Mathematical Principles in Skin Permeation

ADAM C. WATKINSON* and KEITH R. BRAIN

An-eX Analytical Services Ltd., Cardiff, Wales

I. INTRODUCTION

sound knowledge of the underlying mathematical principles of membrane transport is essential if we are to expand our understanding of how membrane barriers fulfill their function and how we can alter their properties to our advantage. The subject of the mathematics of diffusion are enough to fill entire books (1), but in this chapter we have attempted to pick out those mathematical solutions and descriptions that are both commonly used and most appropriate in the field of percutaneous absorption. It is the purpose of this work to attempt to present these equations in a manner that will enable readers to apply them to real numbers generated in their laboratories.

At its simplest and most ideal a membrane can be described as a homogeneous slab of an inert material, with a finite and uniform thickness. This is a convenient theoretical picture and, although it is somewhat removed from the reality of such complex biological membranes as the stratum corneum, it is a logical model with which to begin when attempting to construct any sort of mathematical treatise of the process of membrane permeation.

Much of the early mathematics relating to transmembrane diffusion had its origins in the theoretical description of heat transfer and conductance. Indeed, the most basic of the diffusion equations, Fick's first law, has its roots here.

---

*Current affiliation: Strakan Pharmaceuticals Ltd., Galashiels, Scotland.

II. MOLECULAR DIFFUSION

The transport of molecules across any membrane, including the skin, occurs by the process of passive diffusion. Molecular motion within an isolated system is indiscriminate: molecules are said to follow "random walks" and, therefore, are not subject to a net force in any particular direction. Initially, this may seem anomalous when considering the process of diffusion. If molecules within a membrane move randomly why can we always predict that some proportion of them will traverse such a barrier?

A simple way to explain this phenomenon is by considering the diffusion of a dye in water. We begin with the situation shown in Figure 1a in which a planar boundary separates two sides (A and B) of a container, one containing a solution of dye in water and the other containing pure water. Next we remove this boundary without creating any disturbance at the interface of the two liquids. On a molecular level the dye molecules move randomly and during this process some fraction of them will cross over from side A to side B of the container. The same fraction of water molecules will cross over from side B to side A. However, as there are more dye molecules in side A the random nature of their motion dictates that there will be a net movement of dye from side A to side B of the container. The same is true for the water molecules moving from B to A. There will follow a transfer of molecules down the concentration gradients (see Fig. 1b) between the two sides of the container until the dye is thoroughly mixed with the water to produce a homogeneous solution (see Fig. 1c). A more succinct way of expressing this phenomenon is to say that diffusion of both species has occurred in the direction of decreasing concentration (activity) of that species.

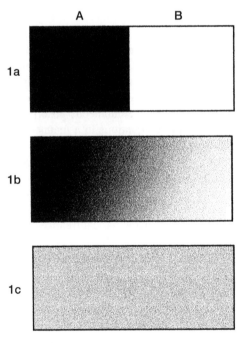

Figure 1   Representation of dye diffusion.

The foregoing series of diffusional events occurs because the system in question is not at equilibrium, and the laws of thermodynamics dictate that it must move toward such a state. This irreversible tendency toward the achievement of a lower-energy state arises as the result of increased entropy within the system and can be expressed in terms of a net decrease in the Gibbs free energy of the system where, under isothermal conditions,

$$\Delta G = \Delta H - T\Delta S$$

where ΔG represents the free energy change, ΔS is the decreasing entropy of the system, T the temperature (in this case constant), and ΔH the change in enthalpy. The process of diffusion is driven primarily by the increase in entropy associated with movement toward a more disordered (mixed) system, but in nonideal cases (which are very common) is accompanied by some change in enthalpy.

In the example discussed thus far we have examined the interdiffusion of two species (i.e., the movement of two mobile phases into each other). The purpose of this chapter is to examine the diffusion of one *mobile* phase (the permeant) into a second *stationary* phase (the membrane). Here, we may view the membrane as a fixed plane of reference and consider only the flux of the permeant into it. To understand this in terms of the bidiffusional process, described earlier, we can consider the void space and other mobile molecules within a membrane to be the second diffusing substance.

A. Fick's First Law of Diffusion

In transport, the flow (or flux, J_i in mol cm^{-2} s^{-1}) is related to the velocity of molecular movement (v in cm s^{-1}) and the concentration (C_i in mol cm^{-3}) of the molecules in motion, Eq. (1).

$$J_i = C_i v \tag{1}$$

A fundamental principle of irreversible thermodynamics is that "the flow, at any point in the system, at any instant, is proportional to the appropriate potential gradient." This was outlined briefly in the consideration of the foregoing dye experiment, and it can be expressed mathematically for a species i as shown in Eq. (2) where $\partial\mu_i/\partial x_i$ is the gradient and L_i is some proportionality constant.

$$J_i = -L_i \left(\frac{\partial\mu_i}{\partial x} \right) \tag{2}$$

Equation 2 is a general form of Fick's first law of diffusion.

If we assume constant temperature and pressure then we can write Eq. (3) and begin to see how the concentration gradient ($\partial C_i/\partial x$) will determine flow.

$$\left(\frac{\partial\mu_i}{\partial x} \right) = \left(\frac{\partial\mu_i}{\partial C_i} \right) \left(\frac{\partial C_i}{\partial x} \right) \tag{3}$$

From classic thermodynamics (the reader is referred to any physical chemistry textbook) we have Eq. (4)

$$\mu = \mu_i^0 + RT \ln \gamma_i C_i \tag{4}$$

and

$$\frac{\partial \mu_i}{\partial C_i} = RT \left[1 + C_i \left(\frac{\partial \ln \gamma_i}{\partial C_i} \right) \right] \bigg/ C_i \tag{5}$$

Substitution of Eq. (5) and Eq. (3) into Eq. (2) gives

$$J_i = -\frac{L_i RT}{C_i} \left[1 + C_i \left(\frac{\partial \ln C_i}{\partial C_i} \right) \right] \left(\frac{\partial C_i}{\partial x} \right) \tag{6}$$

The general expression for the force F_i, acting on a molecule is given by Eq. (7).

$$F_i = -N_A^{-1} \left(\frac{\partial \mu_i}{\partial x} \right) \tag{7}$$

Rearrangement and substitution of this into Eq. (2) gives,

$$J_i = -N_A L_i F_i = C_i v \tag{8}$$

which, on further rearrangement yields,

$$L_i = \frac{C_i}{N_A f_i} \tag{9}$$

where $f_i = v/F_i$ and is defined as the frictional coefficient.

Substitution of Eq. (9) into Eq. (6) produces

$$J_i = -\frac{RT}{N_A f_i} \left[1 + C_i \left(\frac{\partial \ln \gamma_i}{\partial C_i} \right) \right] \left(\frac{\partial C_i}{\partial x} \right) \tag{10}$$

that is,

$$J_i = -D_i \left(\frac{\partial C_i}{\partial x} \right) \tag{11}$$

where

$$D_i = \frac{RT}{N_A f_i} \left[1 + C_i \left(\frac{\partial \ln \gamma_i}{\partial C_i} \right) \right] \tag{12}$$

Equation 11 is the classic form of Fick's first law of diffusion.

As pointed out in the discussion of the dye experiment, Eq. (11) shows that diffusion will cease when the concentration gradient is zero (i.e., when the concentration in the system is uniform). Note, that D_i, the diffusion coefficient, is a function of RT/N_A which is equal to the molecular kinetic energy ($k_B T$) of the system. Also, although there is a dependency of D_i on concentration owing to solute–solute interactions, it is, in practice, small, as the effect of concentration decreases with progressively dilute solutions (as $C_i \to 0$, $D_i \to RT/N_A$).

B. Fick's Second Law of Diffusion

Fick's second law relates the rate of change in concentration with time at a given point in a system to the rate of change in concentration gradient at that point. Under non–steady-state conditions we must consider the principle of conservation of mass to describe the transport taking place. Consider a thin section of cross-sectional area

A and thickness Δx (see Fig. 1) that has a concentration C at position x and time t. The amount of diffusing substance (moving from left to right) that enters the slab per unit time is $J_{in}A$, where J_{in} is the flux and, therefore, the increase in concentration inside the section owing to this influx of material (which has a volume $A\Delta x$) is given by Eq. (13).

$$\frac{\mathrm{d}C}{\mathrm{d}t} = \frac{J_{in}A}{A\Delta x} = \frac{J_{in}}{\Delta x} \tag{13}$$

However, material is also leaving this section with a finite flux that we will call J_{out} and the resulting change in concentration can be expressed as shown in Eq. (14).

$$\frac{\mathrm{d}C}{\mathrm{d}t} = -\frac{J_{out}A}{A\Delta x} = -\frac{J_{out}}{\Delta x} \tag{14}$$

The difference between Eqs. (13) and (14) is equal to the net rate of change of concentration in the section,

$$\frac{\mathrm{d}C}{\mathrm{d}t} = \frac{J_{in} - J_{out}}{\Delta x} \tag{15}$$

If we now use Fick's first law [see Eq. (12)] to describe these fluxes, we can now write Eq. (16).

$$\begin{aligned}
J_{in} - J_{out} &= -D\frac{\mathrm{d}C_{in}}{\mathrm{d}x} + D\frac{\mathrm{d}C_{out}}{\mathrm{d}x} \\
&= -D\frac{\mathrm{d}C_{in}}{\mathrm{d}x} + D\frac{\mathrm{d}}{\mathrm{d}x}\left(C_{in} + \left(\frac{\mathrm{d}C_{in}}{\mathrm{d}x}\right)\Delta x\right) \\
&= D\Delta x\frac{\mathrm{d}^2C}{\mathrm{d}x^2}
\end{aligned} \tag{16}$$

Substituting this into Eq. (15) yields Fick's second law of diffusion [Eq. (17)].

$$\frac{\partial C}{\partial t} = D\frac{\partial^2 C}{\partial x^2} \tag{17}$$

Equation 17 contains partial derivatives because C is a function of both x and t.

C. Solutions to Fick's Laws

Fick's laws are of more applicability if we can specify certain parameters or boundaries within which to apply them. These boundary conditions allow us to be more exacting in defining the problem with which we are dealing and further enable us to solve some interesting diffusion problems. There are many published solutions to these laws of diffusion that cover a wide variety of different diffusional problems. It is impossible to cover such a plethora of solutions in a single chapter, but the solutions presented in the following section are some of those more widely applicable in the laboratory. It is important that the reader is familiar with the need that boundary conditions be obeyed experimentally, or at least, be aware of those conditions used in the derivation of any model they choose to use. This will enable them to use these solutions to their full extent; but at the same time, to appreciate their limitations.

1. Diffusion Through a Homogeneous Membrane with a Constant Activity Difference and a Constant Diffusion Coefficient

Of the solutions determined for Fick's second law, this situation is possibly the closest to that used experimentally for the determination of diffusional phenomena. The mathematical boundary conditions imposed are those of a well-designed diffusion experiment when a permeant is at a high, fixed activity on one side of an inert homogeneous membrane through which it diffuses into a sink on the other side. Before the start of the experiment the membrane is entirely devoid of permeant.

This solution was demonstrated by Daynes (2) and Barrer (3) for which the diffusive flow begins at the high-concentration side (the donor side) of the membrane where $C = C_0$ and $x = h$ at all time t. There is no diffusant material within the membrane before ingress of the permeant being modeled, implying that at $t = 0$ we have $C = 0$ for all x. Diffusion occurs in the direction of decreasing x toward the opposite side of the membrane where $x = 0$ and $C = 0$ (sink receptor phase) for all time t. In this model the diffusion coefficient of the permeant is set as a constant D, and the concentration C, of material at any point x, within the membrane can be calculated as a function of time using Eq. (18).

$$C = \frac{C_0 x}{h} + \frac{2}{\pi} \sum_{n=1}^{\infty} \frac{C_0}{n} \cos(n\pi)\sin\left(\frac{n\pi x}{h}\right) \exp\left(\frac{-Dn^2\pi^2 t}{h^2}\right) \tag{18}$$

In the literature the form of diffusion equations, such as Eq. (18) is often simplified by normalizing the concentration and distance variables relative to their maxima. In Eq. (18), this involves normalizing x (distance) relative to h (membrane thickness) where $\chi = x/h$ and C (concentration in the membrane at any point x) relative to C_0 (concentration in the outer layer of the membrane at $x = h$) where $u = C/C_0$. It is also the norm to introduce the term $\tau = Dt/h^2$. These simplifications yield Eq. (19) in which we can see that, for any value of τ, if $x = 0$ (i.e., at the distal side of the membrane), we obtain $u = 0$.

$$u = \chi + \frac{2}{\pi} \sum_{n=1}^{\infty} \frac{1}{n} \cos(n\pi)\sin(n\pi\chi)\exp(-n^2\pi^2\tau) \tag{19}$$

If we now take a value of $\tau = 0.1$ and generate a plot of u (normalized concentration) against χ (normalized distance) that Eq. (19) produces, we obtain the graph depicted in Figure 2.

Although Eqs. (18) and (19) are often presented in these forms (4,5), they are more easily understood if rewritten using slightly different boundary conditions.

The boundary conditions used in the derivation of Eq. (18) state that diffusion occurs in the direction of decreasing x, resulting in the graphic form of this equation shown in Figure 2. This mathematical quirk makes the application of Eq. (18) to real situations a little difficult to visualize. To make the situation more "palatable," we can simply reverse the form of Eq. (19), making the values of χ increase from zero to unity by expressing the function as in Eq. (20). This situation is shown in Figure 3 and has obvious conceptual advantages over the graph produced by Eq. 19 (i.e., we normally think of diffusion occurring in the direction of positively increasing, rather than negatively decreasing, distance into a membrane.

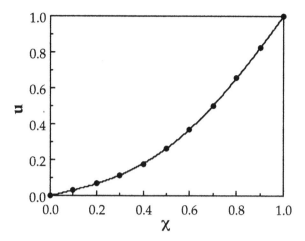

Figure 2 The concentration–depth profile of a permeant within a membrane, as described by Eq. (19) where $\tau = 0.1$.

$$u = 1 - \chi + \frac{2}{\pi} \sum_{n=1}^{\infty} \frac{1}{n} \cos(n\pi)\sin[n\pi(1 - \chi)]\exp(-n^2\pi^2\tau) \qquad (20)$$

Equation (20) is often presented in the literature (6,7) in the form shown in Eq. (21).

$$u = 1 - \chi + \frac{2}{\pi} \sum_{n=1}^{\infty} \frac{(-1)^n}{n} \sin[n\pi(1 - \chi)]\exp(-n^2\pi^2\tau) \qquad (21)$$

The difference between Eqs. (20) and (21) arises because the two equations use different methods of making the sign of the alternate terms in the summation flip from positive to negative [Eq. (22)].

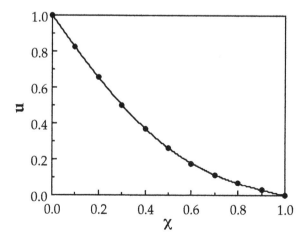

Figure 3 The concentration–depth profile of a permeant within a membrane described by Eq. (20) where $\tau = 0.1$.

$$\sum_{n=1}^{\infty} \frac{(-1)^n}{n} = \sum_{n=1}^{\infty} \frac{1}{n} \cos(n\pi) \tag{22}$$

It is clear from Eq. (21) that as $t \to \infty$ the exponential term will approach zero thus reducing Eq. (21) to the simple form of Eq. (23).

$$u = 1 - \chi \tag{23}$$

This is obviously a linear function and represents the concentration gradient that occurs within a membrane once diffusion has reached steady state. Figure 4 depicts Eq. (21) for increasing values of τ (i.e., it shows the pattern of buildup of a penetrant within a membrane with increasing time; namely, the imperfection in the smoothness of the some of the graphs is simply due to the relatively low number of points calculated. As steady state is reached the distribution pattern becomes linear, as represented by Eq. (23).

Attempts have been made to utilize expressions such as Eq. (21) to analyze data acquired by the skin-stripping technique (7). This method of analysis has been taken a step further by separately modeling concentration gradients across the stratum corneum and viable epidermis (8).

Although useful, equations such as those discussed in the foregoing, are of limited practical use for interpreting permeation data, as they describe the concentration within a membrane at any time point t, and at any position x, within that membrane. A more useful solution that yields the cumulative mass Q, of permeant that passes through a unit area of a membrane in a time t is provided by the following mathematical steps. By differentiating Eq. (21) relative to x, we obtain an expression that describes the instantaneous concentration gradient. The flux dM/dt is then determined at $x = h$ and subsequent integration (between $t = 0$ and $t = t$) of this expression produces Eq. (24), which describes the increase in Q relative to t.

$$Q = C_0 h \left[\frac{Dt}{h^2} - \frac{1}{6} - \frac{2}{\pi^2} \sum_{n=1}^{\infty} \frac{(-1)^n}{n^2} \exp\left(\frac{-Dn^2\pi^2 t}{h^2} \right) \right] \tag{24}$$

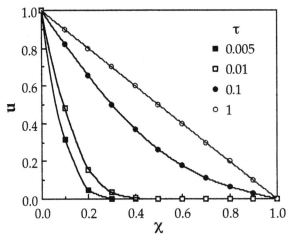

Figure 4 Representation of a concentration–depth profile as predicted by Eq. (21) for different values of τ.

A graphic representation of Eq. (24) (Fig. 5) depicts the time-dependent nature of the total mass transfer Q, through the membrane. As $t \to \infty$, the exponential term in Eq. (24) tends to zero; therefore, Eq. (24) approximates to the line described by Eq. (25) which, on rearrangement, becomes Eq. (26). Figure 5 shows the line described by Eqs. (25) and (26) in relation to the full diffusion curve described by Eq. (24).

$$Q = C_0 h \left[\frac{Dt}{h^2} - \frac{1}{6} \right] \tag{25}$$

$$Q = \frac{DC_0}{h} \left[t - \frac{h^2}{6D} \right] \tag{26}$$

If we put $Q = 0$ into Eq. (26) we can solve for t and this yields the value of the time axis intercept known as the lag time (t_{lag}) as described by Eq. (27), which relates it inversely to the diffusion coefficient and directly to the diffusional pathlength. The use of this extrapolation for the calculation of diffusion coefficients is commonplace, although when skin is concerned, the pathlength is unknown, making calculation of absolute values of D difficult.

$$t_{lag} = \frac{h^2}{6D} \tag{27}$$

If we differentiate Eq. (26) relative to time we obtain Eq. (28), possibly the most well-known form of Fick's first law of diffusion, that describes the flux J, at steady state. Note the link between the steady-state form of this law [Eq. (28)] and the more general form that we derived earlier [Eq. (11)].

$$\frac{dQ}{dt} = J = \frac{DC_0}{h} \tag{28}$$

Figures 6 and 7 demonstrate the effect that changes in the values of the diffusion

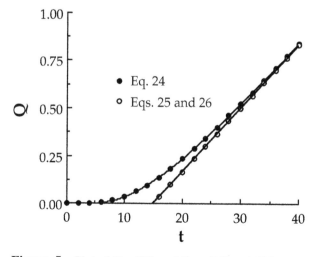

Figure 5 Plot of Eq. (24) and Eqs. (25) and (26) representing diffusion through a membrane with increasing time.

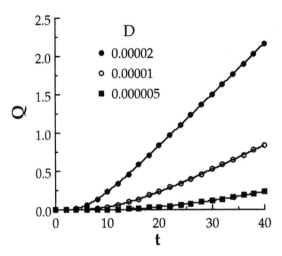

Figure 6 Effect of altering the diffusion coefficient on theoretical permeation profiles generated by Eq. (24) ($h = 0.03$, $C_0 = 100$).

coefficient and the diffusional pathlength have on such plots. It is clear that a reduced diffusion coefficient results in slower mass transfer, as does an increased diffusional pathlength.

It is often impractical to use the forms of Eqs. (25), (26), and (28) as shown because they include a term, C_0 (the concentration of permeant in the outer layer of the membrane), that is extremely difficult to measure. We can replace the value of C_0 with a term that links it to the concentration in the vehicle C_v through the partition coefficient K, as described by Eq. (29) which rearranges to Eq. (30). The partition

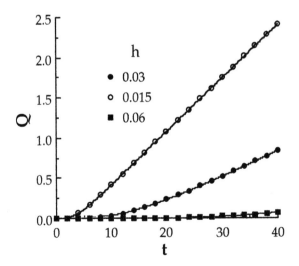

Figure 7 Effect of altering the diffusional pathlength on theoretical permeation profiles generated by Eq. (24) ($D = 0.00001$, $C_0 = 100$).

coefficient is simply a measure of the relative affinity that a diffusant has for the two media involved (i.e., the membrane and the vehicle above it).

$$K = \frac{C_0}{C_v} \tag{29}$$

$$C_0 = KC_v \tag{30}$$

Substitution of Eq. (30) into Eq. (28), for example, produces Eq. (31), which is of more practical use, as it links flux to the concentration of the permeant in the vehicle.

$$\frac{dQ}{dt} = J = \frac{DKC_v}{h} \tag{31}$$

We have already examined one derivation of the steady-state form of Fick's first law but we can approach the solution from a slightly different angle. Consider a plane sheet, of thickness h, for which surfaces, $x = 0$, $x = h$, are maintained at constant concentrations C_0 and C_h, respectively. If the diffusion process has reached steady state, thus producing a constant concentration gradient across it at all points, provided that D is constant, using Fick's second law of diffusion [see Eq. (17)] we obtain the situation described in Eq. (32)

$$\frac{dC}{dt} = D\frac{d^2C}{dx^2} = 0 \tag{32}$$

(i.e., the rate of change of the concentration gradient across the membrane is zero). This situation is known as steady state. If we now integrate Eq. (32) relative to x once, we obtain Eq. (33).

$$D\frac{dC}{dx} = cons\tan t \tag{33}$$

Equation 33 shows us that the concentration across the membrane reduces linearly with distance (here from C_0 to C_h) and, therefore, that the rate of transfer of diffusant (the flux, J) is the same at all positions within the membrane. This rate is given by Fick's first law [see Eq. (11)] and we can now write

$$J = D\frac{dC}{dx} = \frac{D(C_0 - C_h)}{h} \tag{34}$$

If we know the membrane thickness h and the concentrations C_0 and C_h, then we can measure D from the determination of the flux J. Indeed, it is often assumed (by the use of sink conditions in the receptor solution below the membrane) that the value of C_h, the concentration at the inner surface of the membrane is zero. Thus, the problem is simplified further and Eq. (34) reduces to Eq. (35), which is identical with Eq. (28) derived earlier.

$$J = \frac{DC_0}{h} \tag{35}$$

However, as it is difficult to measure the concentration C_0 (the amount of diffusant in the outermost layer of the membrane) we perform the same trick as earlier and use the partition coefficient and the vehicle concentration to give us Eq. (31).

Frequently, particularly in biological membranes, there is a practical difficulty in measuring the diffusional pathlength. Because of this, and that information concerning the individual effects of changes in K and D is often not required, a composite parameter is usually used to replace these values in Eq. (26). The permeability coefficient P, is thus defined as $P = KD/h$, and this simplifies Eq. (31) further to give Eq. (36):

$$J = PC_v \qquad (36)$$

Equation (36) is perhaps the most basic and frequently used expression in the routine assessment of membrane permeability. However, it should always be noted that the principles on which this equation are based stipulate that the donor concentration is constant and that the diffusion has reached steady state. In practice this either means using a saturated donor solution in the presence of excess permeant, or that the change in donor concentration during the course of the experiment is negligible. There are also problems associated with the assumption of steady state and the assessment of when it has been attained.

The use of the steady-state method for the assessment of permeability coefficients and lag times has been questioned (9,10). It is often difficult to assess when this period has been reached and, by using only this steady-state data, we lose out on that data collected at time points before this region. One way to partially solve this is to utilize Eq. (24) and fit it to the entire set of diffusion data using a nonlinear curve-fitting software package. If the values of C_0 (the permeant concentration in the outer layer of the membrane) in Eq. (24) are replaced by the term KC_v (as described earlier) we obtain Eq. (37).

$$Q = KC_v h \left[\frac{Dt}{h^2} - \frac{1}{6} - \frac{2}{\pi^2} \sum_{n=1}^{\infty} \frac{(-1)^n}{n^2} \exp\left(\frac{-Dn^2\pi^2 t}{h^2} \right) \right] \qquad (37)$$

The equation, as shown, is of limited use, as there are three unknown parameters, but if we replace Kh and D/h^2 with P_1 and P_2, respectively, we obtain Eq. (38) with just two variables. By fitting P_1 and P_2 we can obtain a value of the permeability coefficient as $P_1 P_2 = KhD/h^2 = KD/h$, which is equal to the permeability coefficient.

$$Q = P_1 C_v \left[P_2 t - \frac{1}{6} - \frac{2}{\pi^2} \sum_{n=1}^{\infty} \frac{(-1)^n}{n^2} \exp(-P_2 n^2 \pi^2 t) \right] \qquad (38a)$$

Equations (37) and (38a) have been used extensively for the calculation of diffusional parameters (e.g., Ref. 11). A similar use of a non−steady-state approach (12) has been used where the partition coefficients were experimentally measured, and the diffusion coefficient and pathlength estimated using a fitting routine.

It is also possible to use a short-time approximation derived by integrating a Fourier transformation (see Eq. 37) that yields an expression valid at small t (see Eq. 38b) (13,14).

$$\log\left[\frac{Q}{t^{3/2}} \right] = \log\left[\frac{8KC_v}{h^2 \pi^{1/2}} \right] + \frac{3}{2} \log D - \frac{h^2}{9.2Dt} \qquad (38b)$$

By using Eq. (38b) we can construct a plot of $\log(Q/t^{3/2})$ against $1/t$ that has a gradient of $-h^2/9.2D$ (from which D can be calculated) and an intercept on the y-axis that gives an estimate of the remaining parameters. The short-time method is valid up to approximately 2.7 times the lag time. The major drawback of this method is the requirement for excellent analytical sensitivity to collect good data at very short time periods. Note that, as with all the other equations derived in this section, this method is valid only if the diffusion coefficient is constant.

Equation (38b) is identical with that derived by Hadgraft (6), who presented it in a slightly different form [Eq. (39)] where $C_0 = KC_v$ as outlined earlier in Eq. (30).

$$Q = \frac{8C_0 D^{3/2} t^{3/2}}{h^3 \pi^{1/2}} \exp\left(\frac{-h^2}{4Dt}\right) \tag{39}$$

2. Diffusion into a Homogeneous Membrane with a Constant Donor Activity, Constant Diffusion Coefficient, and an Impermeable Distal Side

A further solution to Fick's second law that has essentially the same boundary conditions as those discussed in the derivation of Eq. (18), except the distal side of the membrane is an impermeable surface, has recently been used extensively for the analysis of absorption data collected by attenuated total-reflectance Fourier transform infra red spectroscopy (ATR–FTIR).

The methodology of ATR–FTIR systems has been described by various authors in diffusion studies on polymers (15–19), semisolids (20), glycerogelatin films (21), and stratum corneum (22). Briefly, a membrane is sandwiched between an impermeable zinc selenide ATR crystal, and a reservoir of permeant that provides a constant concentration in the upper surface of the membrane. The membrane is initially devoid of permeant. As diffusion into the membrane occurs, there will be a buildup of permeant at the impermeable membrane–crystal interface until saturation is reached at a concentration C_0 (the solubility of the penetrant in the membrane). An analytical solution describing the buildup of permeant at this interface with time can be obtained, using Fick's second law and the relevant initial and boundary conditions (1), and is given as Eq. (40)

$$\frac{C}{C_0} = 1 - \frac{4}{\pi} \sum_{n=0}^{\infty} \frac{(-1)^n}{2n+1} \exp\left(\frac{-D(2n+1)^2 \pi^2 t}{4h^2}\right) \tag{40}$$

where C is the diffusant concentration at the interface, t is time, D is the diffusion coefficient of the permeant, and h is the membrane thickness. There will be an initial period during which permeant concentration at the interface increases, followed by an exponential rise to a plateau that represents the saturation of the membrane with permeant. Plots of C against time are given in Figures 8 and 9 for different values of D and h, respectively.

For large values of time, the $n = 0$ term in Eq. (40) predominates, and the long-time approximation of Eq. (40) is given by Eq. (41).

$$\ln\left[\frac{\pi}{4}\left(1 - \frac{C}{C_0}\right)\right] = \left(\frac{\pi^2 D}{4h^2}\right) t \tag{41}$$

Equation 41 may be fitted to experimental data in a plot of $\ln[(\pi/4)(1 - C/C_0)]$ against time. The diffusion coefficient D, may then be obtained from the gradient of this plot and the film thickness h.

It is possible, assuming the Beer–Lambert law applies, to replace the concentration terms in Eq. (40) with experimental absorbance values to give Eq. (42) in

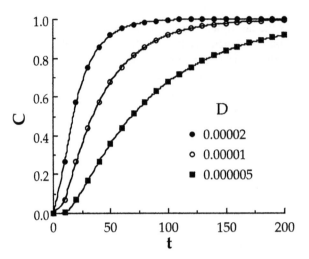

Figure 8 Increase in concentration at the distal face of a membrane supported on an impermeable surface as described by Eq. (40) (effect of altering D, $h = 0.03$).

which A = area of penetrant peak (at time t) of IR absorbance relating to the permeant, and A_0 = area of penetrant peak corresponding to the situation where the membrane is saturated (in the plateau region of the curve).

$$\frac{A}{A_0} = 1 - \frac{4}{\pi} \sum_{n=0}^{\infty} \frac{(-1)^n}{2n + 1} \exp\left(\frac{-D(2n + 1)^2 \pi^2 t}{4h^2}\right) \tag{42}$$

Experimental values of penetrant peak areas against time are fitted using Eq. (42). Values of D/h^2 and A_0 are allowed to vary until a best fit is achieved as measured

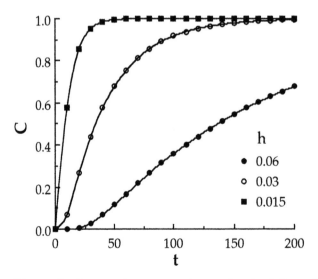

Figure 9 Increase in concentration at the distal face of a membrane supported on an impermeable surface, as described by Eq. (40) (effect of altering h, $D = 0.00001$).

by minimization of χ^2. This technique has great potential for the deconvolution of partitioning and diffusional phenomena in the synergistic action of certain penetration enhancers (22,23). A limitation of this technique is that the diffusant under investigation must have an absorption band in its IR spectrum that can be distinguished from the IR absorption spectrum of the test film. Compounds containing cyano and azo groups are particularly useful in this respect.

3. Diffusion Out of a Precharged Homogeneous Membrane, With a Constant Diffusion Coefficient, With and Without a Reservoir

This scenario is applicable to controlled-release devices that may include transdermal patches, as well as oral and subcutaneous devices. Equations describing release profiles from devices that use a rate-controlling membrane have been derived by Hadgraft (6) for two situations: first, when a rate-controlling membrane is initially full of material, but no reservoir is present and, second, when an infinite reservoir exists.

The solution derived by Hadgraft (6) describing the burst effect release into a receiver fluid from a membrane for the situation without a reservoir is given in Eq. (43) (M_t = amount of drug released at time t, M_∞ = total amount of drug contained in the membrane at $t = 0$, the other parameters are as already described.

$$M_t = M_\infty \left[1 - \frac{8}{\pi^2} \left(\sum_{n=1}^{\infty} \frac{1}{(2n-1)^2} \right) \exp \left(-\frac{(2n-1)^2 \pi^2 Dt}{4h^2} \right) \right] \tag{43}$$

A second, but identical, form of this equation is often quoted (e.g., Ref. 1) and follows as Eq. (44). The only difference between Eqs. (43) and (44) is in the way the summation term is calculated. In Eq. (43) the sum runs from unity upward, whereas in Eq. (44) it runs from zero. this has the effect of producing a slightly different premultiplier and exponential term.

$$M_t = M_\infty \left[1 - \frac{8}{\pi^2} \left(\sum_{n=0}^{\infty} \frac{1}{(2n+1)^2} \right) \exp \left(-\frac{(2n+1)^2 \pi^2 Dt}{4h^2} \right) \right] \tag{44}$$

Note, Eq. (43) is the full solution for the burst effect without a reservoir and may be applied over the complete time range of an experiment.

Hadgraft (6) also simplified this expression to give both short [see Eq. (45)] and long [see Eq. (46)] time approximations (clearly, at long times the amount of diffusant released is equal to that initially in the membrane).

$$M_t = 2M_\infty \left(\frac{Dt}{\pi h^2} \right)^{1/2} \tag{45}$$

$$M_t = M_\infty \tag{46}$$

Equation (47) (6) is the solution describing the buildup of diffusant in the receptor compartment for the case in which the membrane was initially charged with permeant and there was a reservoir present. Here, M_∞ is not a finite quantity of drug, for it represents the content of the unchanging reservoir.

$$M_t = M_\infty \left[\frac{Dt}{h^2} + \frac{1}{3} - \frac{2}{\pi^2} \sum_{n=1}^{\infty} \frac{1}{n^2} \exp \left(\frac{-n^2 \pi^2 Dt}{h^2} \right) \right] \tag{47}$$

Again, short- and long-term approximations can be made. The short-term approximation of Eq. (47) is identical with Eq. (45), whereas to derive the long-term ap-

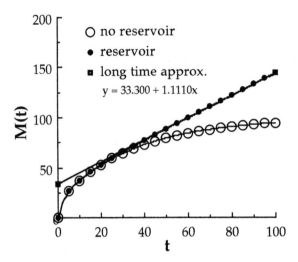

Figure 10 Diffusion from a presaturated slab with and without a reservoir, together with the long-time approximation as in Eq. (48) ($D = 0.00001$, $h = 0.03$, $M_\infty = 100$).

proximation, we simply note that large t will make the negative exponential term in Eq. (47) tend to zero. This solution is given in Eq. (48).

$$M_t = M_\infty \left(\frac{Dt}{h^2} + \frac{1}{3} \right) \tag{48}$$

The depletion of a membrane that is not in contact with a reservoir of permeant [see Eq. (42)] is clearly demonstrated in the graphic representation of Eqs. (43) and (47) (Fig. 10). Note that the long-time approximation of Eq. (48) produces a straight-line with a y-axis intercept equal to one-third of the value of M_∞ used, and has a gradient equal to DM_∞/h^2, as would be predicted from its form.

A point that we have not considered so far is that many of the solutions examined contain summation expressions and that the value of n used within these will affect the result achieved. This is clearly demonstrated in Figure 11, in which n in Eq. (43) has been made equal to 10, 20, and 500. The most dramatic differences are seen at low values of t. This serves to demonstrate that the researcher should assess the effect of differing values of n on the result before applying any of the equations presented herein. The value that is used will alter the computational time required, depending on the simulation software being used. It is suggested that $n = 100$ is a reasonable approximation in most cases.

D. Complex Barriers

1. Diffusion Through Laminates

There are numerous examples for which a membrane is not a single homogeneous system and, indeed, may consist of several barriers. For example, the skin can be viewed at different levels of complexity that, as they are probed, reveal more and more barriers in series with one another. At its simplest the skin might be viewed as a single homogeneous slab, but the structure can also be viewed as three barriers

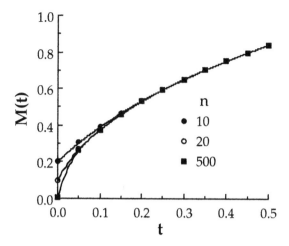

Figure 11 The effect of changing the value of n in the summation term of Eq. (43) ($D = 0.00001$, $h = 0.03$, $M_\infty = 1$).

in series (i.e., the stratum corneum, epidermis, and dermis). Furthermore, the stratum corneum contains bilayer structures that might also be considered to exhibit the properties of barriers in series. The simplest starting point in this argument is to assume that the constituent membranes that make up the laminate are themselves isotropic and that each layer "i" of the laminate contributes a diffusional resistance R_i to the overall resistance of the membrane (24). R is obviously inversely proportional to the diffusivity within the layer and the partition coefficient relative to external phases. The shorter the pathlength, the lower the resistance will be; thus,

$$R_i = \frac{h_i}{D_i K_i} = \frac{1}{P_i} \tag{49}$$

where P_i is equal to the permeability coefficient, as defined earlier. The overall resistance of the laminate system (R_{total}) is equal to the sum of the individual resistances of the n membranes that make up the laminate, giving,

$$R_{total} = \sum_{i=1}^{i=n} R_i = \frac{h_1}{D_1 K_1} + \frac{h_2}{D_2 K_2} + \cdots + \frac{h_n}{D_n K_n} \tag{50}$$

Importantly, the partition coefficients in Eq. (50) relate the concentration of permeant in the ith phase of the laminate to the concentration in the initial phase and not the $(i - 1)^{th}$ phase.

In the skin we may, as alluded to earlier, consider it to be a trilaminate of stratum corneum, epidermis, and dermis. By using Eq. (50) to describe its resistance we would therefore arrive at Eq. (51).

$$R_{skin} = \frac{h_{sc}}{D_{sc} K_{sc}} + \frac{h_e}{D_e K_e} + \frac{h_d}{D_d K_d} \tag{51}$$

If we now take the view that if the stratum corneum is the rate-limiting barrier, then the resistance of the other two layers becomes negligible and we obtain Eq. (52), which is not very surprising because we already know that for a single isotropic membrane we have $P = KD/h$.

$$R_{skin} = \frac{h_{sc}}{D_{sc}K_{sc}} \qquad (52)$$

This approach to diffusion in laminates can be used to look at the situation when we have an isotropic lipophilic membrane sandwiched between two aqueous phases (for example, a donor and receptor phase in a Franz diffusion experiment). Even though it is usual practice to vigorously stir receptor solutions, unstirred aqueous layers will form at stationary surfaces in both donor and receptor compartments (the thickness of these layers being related to the efficiency and rate of stirring). This situation approximates a trilaminate system of aqueous–lipophilic–aqueous barriers. hence, we can write Eq. (53) where R_{aq1}, is the resistance of the stationary aqueous barrier above the membrane, R_{lip} the resistance of the membrane itself and R_{aq2} the resistance of the stationary aqueous layer beneath the membrane.

$$R_{total} = R_{aq1} + R_{lip} + R_{aq2} = \frac{h_{aq1}}{D_{aq1}} + \frac{h_{lip}}{D_{lip}K} + \frac{h_{aq2}}{D_{aq2}} \qquad (53)$$

Only one partition coefficient K, appears in Eq. (53) because the partition coefficient between the bulk aqueous regions and the unstirred diffusion layers is considered to be unity. Thus, K is the partition coefficient between the bulk aqueous phase and the membrane.

This type of system has been rigorously analyzed (4,25,26) and, at steady state, under zero-order boundary conditions, yields Eq. (54).

$$\frac{dQ}{dt} = \left[\frac{KD_m D_{aq}}{h_m D_{aq} + (h_{aq1} + h_{aq2})KD_m} \right] (C_v - C_r) = P_{total}\Delta C \qquad (54)$$

In Eq. (54) we have dQ/dt = steady-state flux; K = membrane/water partition coefficient; subscripts m and aq refer to the membrane and aqueous layers, respectively; h = thickness of layer, as indicated by subscript, C_v = vehicle or donor phase concentration; and C_r = receptor phase concentration. It is clear that in most cases the term $h_m D_{aq} \gg (h_{aq1} + h_{aq2})KD_m$ leading to a simplification of Eq. (54) to the commonly used equation for mass transfer at steady state; that is, we end up with $dQ/dt = KD_m \Delta C/h_m$. However, with a highly lipophilic penetrant, or for extremely thin membranes, this empiricism may not hold. Furthermore, if both of these scenarios occur and we also have large values of D_m that are similar in magnitude to D_{aq} the permeability coefficient will be described by Eq. (55), and the diffusion process will be almost completely controlled by the aqueous boundary layers.

$$P_{total} = \frac{D_{aq}}{(h_{aq1} + h_{aq2})} \qquad (55)$$

In our model we have seen that, as the permeant partition coefficient rises, the diffusion process becomes more and more under the control of the aqueous layers. The structure of skin can be approximated to a laminated structure in which the layers increase in hydrophilicity as the skin is permeated. Thus, in skin our "receptor phase" aqueous boundary layer is replaced by the aqueous epidermal region below the stratum corneum. Hence, Eq. (55) can be applied and we observe a change from lipoidal membrane diffusion control to aqueous diffusion control on increasing the lipophilicity of the permeant. This is a simple demonstration of the situation in which the stratum corneum is no longer the major barrier to percutaneous penetration. This phenomenon has been demonstrated by Roberts et al. (27), who showed that on increasing the lipophilicity of a homologous series of phenolic compounds there was a permeability dependency on partition coefficient (or lipophilicity) only until aqueous diffusion layer control took over.

By using the definition of lag time ($t_{lag} = h^2/6D$) already derived, we can define the lag time for the situation in which we have a very thin membrane (infinitely thin membrane) as shown in Eq. (56).

$$t_{lag} = \frac{(h_{aq1} + h_{aq2})^2}{6D_{aq}} \tag{56}$$

Equation (57) is the result we obtain for a very thick membrane when K is large. If the diffusion layers are of equal thickness we will obtain the expression shown in Eq. (58).

$$t_{lag} = \frac{h_m h_{aq1} h_{aq2} K}{\sum h_{aq} D_{aq}} \tag{57}$$

$$t_{lag} = \frac{h_m h_{aq} K}{2D_{aq}} \tag{58}$$

The use of the laminate principle can be taken further if it is assumed that the diffusional process occurs as a series of point-to-point movements of the diffusing species. Most of the foregoing mathematics view permeability as a rate process that contains contributions from both an equilibrium and a nonequilibrium step, but the assumption of instantaneous partitioning of a solute is not always valid and a treatment based on Eyring's absolute reaction rate theory attempts to account for this (28–30). If the flow of molecules through a membrane is viewed as a series of successive molecular jumps of length λ from one energy minimum to another, the whole diffusion process can be seen in terms of an energy profile (Fig. 12).

Thus, a third factor is introduced into the mathematical approach to diffusion (i.e., the activation energy ΔG_a required to prevent spontaneous partitioning from occurring). If C_i is the concentration (molecules per milliliter) at the ith position in the membrane, then the amount of material in a 1-cm^2 cross-section and length λ_i (the distance between equilibria mazima) is $C_i\lambda_i$, and the velocity of forward diffusion is

$$v_f = k_i C_i \lambda_i \tag{59}$$

where k_i = rate constant for crossing barrier i.

Similarly the rate of backward diffusion over the barrier i will be

$$v_b = k_{i+1} C_{i+1} \lambda_{i+1} \tag{60}$$

If it is assumed that $k_i = k_{i+1} = k$ and $\lambda_i = \lambda_{i+1} = \lambda$ then the net rate of diffusion, or flux is

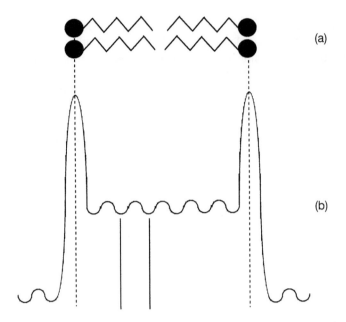

Figure 12 (a) A schematic representation of membrane interfaces and the phases involved in solute transfer; (b) possible potential energy profile for a solute molecule diffusing through a membrane.

$$J = k_i C_i \lambda_i - k'_{i+1} C_{i+1} \lambda_{i+1} \tag{61}$$

$$J = k\lambda(C_i - C_{i+1}) = \frac{-k\lambda(C_{i+1} - C_i)}{\lambda}$$

$$J = \frac{k\lambda^2 dC}{dt} \tag{62}$$

where the diffusion coefficient, $D = k\lambda^2$.

The diffusional process can be viewed as five resistances in series, two at the interfaces, one in the membrane, and two from the bulk solvent phase on either side of the membrane. If it is now assumed that diffusion in the bulk solvent phase is much faster than in the membrane phase or at the interfaces, and that the energy barriers within the membrane are the same height as are those on either side of it, the following equation can be derived for the permeability coefficient P [for an excellent account of this derivation see Ref. 30]:

$$P = \frac{\lambda k_{sm} k_m}{(2k_m + mk_m)} \tag{63}$$

where λ = mean jump distances, m = No. of jumps in the membrane, k_{sm}, k_{ms}, k_m = rate constants of adsorption and desorption at the interfaces, and diffusion in the membrane, respectively.

However, the ratio k_{sm}/k_{ms} can be defined as the partition coefficient K, giving

$$1/P = 2/(k_{sm}\lambda) + m/(k_m\lambda K) \tag{64}$$

As $m\lambda = \Delta x$, the membrane thickness, and $D = k\lambda^2$, then

$$\frac{1}{P} = \frac{2\lambda}{D_{sm}} + \frac{\Delta x}{D_m K} \tag{65}$$

So, if diffusion in the membrane is the rate-limiting step ($k_m \ll k_{ms}$), then

$$P = \frac{D_m K}{\Delta x} \tag{66}$$

(i.e., the same equation as produced by Fick's first law) and, if the slowest step is diffusion through the interface, them $k_m \gg k_{ms}$, so

$$P = \frac{D_{sm}}{2\lambda} \tag{67}$$

that is, the permeability constant will be independent of the partition coefficient and the membrane thickness. If this is true, then the rate of diffusion will be controlled by the nature of the interface and penetrant relative to each other. This can be expressed thermodynamically by considering the expansion of the term k_{sm}:

$$P = \frac{D_{sm}}{2\lambda} = \frac{k_{sm}\lambda}{2} \tag{68}$$

and since $k_{sm} = (kT/h)e^{-\Delta G_{sm}/RT}$, then

$$P = (\lambda kT/2h)e^{-\Delta G_{sm}/RT} \tag{69}$$

where ΔG_{sm} is the free energy of activation necessary for crossing the interface, which will be dependent on the physicochemical relation between penetrant and barrier. Thus, a compound that interacts in a way such that the value of ΔG_{sm} is reduced, will increase the rate of permeation (i.e., enhance penetration). The same is true if the membrane is the rate-limiting medium (i.e., ΔG_m becomes rate-determining) and the creation of free volumes within the acyl chain region by an enhancer may reduce ΔG_m.

It seems likely that it is a combination of these effects that is responsible for the enhancing capacity of many compounds.

2. Diffusion Through Shunt Routes (Barriers in Parallel)

Many membranes can be viewed as containing more than one distinct route through which diffusion can occur. The skin may be thought of as at least possessing the potential to exhibit such parallel routes, in that its structure is pierced by numerous appendages, such as hair follicles and sweat ducts. Indeed, although an issue of some debate, it has been suggested that such routes are important in both passive (31–33) and impassive (34,35) diffusion processes.

The mathematical problem of defining parallel routes of diffusion is less complex than for the laminated structures we examined in the foregoing. If two diffusional pathways through a single membrane are truly parallel, then the total flux through that membrane is simply the sum of the individual fluxes through the parallel routes (36). Thus, for unit area, the total flux (at any time) through n parallel routes is defined by Eq. (70).

$$J_{total} = J_a + J_b + \cdots + J_n \tag{70}$$

Hence, for any situation (whether steady state or otherwise), we can simply sum the expressions describing flux through each route to gain the overall flux. The simplest example would be steady-state flux through two parallel routes (A and B) in which the diffusion coefficients are D_A and D_B, the pathlengths h_A and h_B, and the partition coefficients K_A and K_B. This situation is described by Eq. (71), in which the driving concentration is C_0 in the common donor solution, and P_A and P_B are the permeability coefficients for the two routes. Note that the flux expression is for unit area.

$$J_{total} = J_A + J_B = \frac{K_A D_A C_0}{h_A} + \frac{K_B D_B C_0}{h_B} = C_0 \left[\frac{K_A D_A}{h_A} + \frac{K_B D_B}{h_B} \right] = C_0[P_A + P_B] \tag{71}$$

The situation is more complex if diffusion has not yet reached steady state, but the total flux is still just the sum of the flux through the two routes. Equation (72) demonstrates the increasing degree of complexity that this situation yields for the same situation as that described in Eq. (71), but before a steady state has been reached.

$$Q = \left\{ K_A C_0 h_A \left[\frac{D_A t}{h_A^2} - \frac{1}{6} - \frac{2}{\pi^2} \sum_{n=1}^{\infty} \frac{(-1)^n}{n^2} \exp\left(\frac{-D_A n^2 \pi^2 t}{h_A^2} \right) \right] \right\}$$
$$+ \left\{ K_B C_0 h_B \left[\frac{D_B t}{h_B^2} - \frac{1}{6} - \frac{2}{\pi^2} \sum_{n=1}^{\infty} \frac{(-1)^n}{n^2} \exp\left(\frac{-D_B n^2 \pi^2 t}{h_B^2} \right) \right] \right\} \tag{72}$$

In a manner analogous to that used earlier we can derive the long-time approximation of Eq. (72) as shown in Eq. (73) and use this to define an expression for the lag time [Eq. (74)].

$$Q = \left\{ K_A C_0 h_A \left[\frac{D_A t}{h_A^2} - \frac{1}{6} \right] \right\} + \left\{ K_B C_0 h_B \left[\frac{D_B t}{h_B^2} - \frac{1}{6} \right] \right\} \tag{73}$$

$$t_{lag} = \frac{h_A h_B (K_A h_A + K_B h_B)}{6(K_A D_A h_B + K_B D_B h_A)} \tag{74}$$

If the two diffusional pathlengths for routes A and B are the same then Eq. (74) reduces to Eq. (75).

$$t_{lag} = \frac{h^2(K_A + K_B)}{5(K_A D_A + K_B D_B)} \tag{75}$$

The shape of a plot of amount permeated versus time will obviously be dependent on the relative ease by which a permeant can pass through the parallel routes. If one of the routes is significantly more permeable than the other, there will be an initial period where diffusion appears regular (at which time flux is occurring by a single, more permeable, route only) before an increase in flux occurs as material starts issuing from the second route. This increase in flux will be followed by a period in which steady state is re-established (37).

An interesting example of parallel routes of diffusion are the pores that exist in some membranes. If these are angled or tortuous, then there is an obvious lengthening of the distance that a permeant must travel (i.e., the diffusional pathlength h is increased). This will manifest itself in all the parameters that are dependent on the membrane thickness, which is usually taken as the diffusional pathlength. To deal with this situation a variable, referred to as the tortuosity factor (τ) is introduced as a premultiplier to the membrane thickness (i.e., the new, increased, diffusional pathlength, $H = \tau h$).

3. Barriers with Dispersed Phases

The theory dealing with diffusion through heterogeneous systems has been addressed by Barrer (38) in great detail for numerous types of systems. Many of these complex systems are beyond the scope of this chapter so we will consider only those simple cases in which the dispersion is uniform. A more comprehensive review of complex systems (if particle size and shape are irregular and the dispersion nonuniform) is provided by Barrer (38) and Higuchi and Higuchi (39).

Possibly the simplest case to deal with is that in which the dispersed phase is diffusionally impervious and inert. The effect of such a dispersed phase (the filler) will be to partially obstruct the movement of diffusants through the permeable phase (the matrix). This will result in an increased diffusional pathlength which is addressed mathematically by the introduction of the tortuosity factor, τ. The addition of filler will also reduce the amount of space within the membrane through which diffusion can occur. This will be related to the relative amounts (volume fractions) of matrix (ϕ_1) and filler (ϕ_2) that are present in the system where we have $\phi_1 = 1 - \phi_2$. In a random dispersion the relative cross-sectional areas of the phases are equal to their relative volume fraction. The tortuosity factor is applied to the membrane thickness h, such that the new, increased, diffusional pathlength H equates to the term τh. Hence, for a diffusional system at steady state we can use Eq. (31) and obtain Eq. (76).

$$\frac{dQ}{dt} = J = \frac{KC_0 D}{h} \frac{\phi_1}{\tau} = \frac{KC_0 D(1 - \phi_2)}{H} \tag{76}$$

It is clear that the lag time will reflect the longer pathlength, but it is not affected by the volume of filler present, as demonstrated by Eq. (77).

$$t_{lag} = \frac{h^2 \tau^2}{6D} = \frac{H^2}{6D} \tag{77}$$

Slightly more complex situations arise if the filler has adsorptive properties. This situation has been examined by numerous authors (39–42). Irrespective of how the permeant molecules are adsorbed onto the filler, the steady-state flux for a standard zero-order process is still given by Eq. (76) (because the filler is not involved in any absorption process). Hence, it is only the lag time that is affected and lengthened by such a process. If the amount of permeant adsorbed is linearly related to the concentration, then the lag time is described by Eq. (78) where κ is the adsorptive capacity of the filler.

$$t_{lag} = (1 - \kappa\phi_2) \frac{\tau^2 h^2}{6D} \tag{78}$$

If the situation arises where the concentration is particularly high then the filler can adsorb only a finite amount of permeant and the steady state is still represented by Eq. (76), but the expression for the lag time approximates to Eq. (79). Note that Eq. (77) is an approximate solution only.

$$t_{lag} = \left(\frac{1}{4} - \frac{\kappa \phi_2}{2KC_0} \right) \frac{\tau^2 h^2}{D} \tag{79}$$

A greater complexity arises if the system that we are dealing with contains filler particles that are themselves permeable. An emulsion is a good example of such a system where diffusant can pass through both the filler and its supporting matrix. Higuchi and Higuchi (39) derived an expression [Eq. (80)] for the total permeability of a model system of this nature, in which it was assumed that the dispersed phase was spherical.

$$P_{total} = \frac{2P_1^2(1 - \phi_2) + P_1P_2(1 + 2\phi_2) - GP_1[(P_2 - P_1)/(2P_1 - P_2)]^2(2P_1 + P_2)(1 - \phi_1)}{P_1(2 + \phi_2) + P_2(1 - \phi_2) - G[(P_2 - P_1)/(2P_1 - P_2)]^2(2P_1 + P_2)(1 - \phi_1)} \tag{80}$$

III. VARIABLE BOUNDARY CONDITIONS

In deriving the diffusion equations that we have examined thus far, certain parameters were fixed to simplify the mathematics. These fixed parameters are referred to as *boundary conditions*. In an ideal world it would be feasible to conduct a permeation experiment in which all the boundaries of a system remain constant, except time and mass transported. However, in reality, situations in which these boundaries are not fixed (so-called moving boundary situations) are commonplace. Examples of these moving boundaries include the evaporation of volatile formulation components from a gel or cream as it is applied and during its time of use (which may lead to concentration of the permeant in the vehicle), the swelling or contraction of a delivery device, alterations in the barrier function of the skin itself, owing to the absorption of formulation excipients or some form of occlusion. Changes of this type may include increases or decreases in both the diffusion and partition coefficient of the permeant under investigation.

A. Variations in the Thickness of Applied Medicaments

The mathematics of such situations is complex, but there are partial and full solutions to some of these issues in the literature. Guy and Hadgraft (43) have addressed the issue of the dependence of flux from an applied medicament on its thickness and derived short and long-time approximate solutions [Eqs. (81) and (82), respectively].

$$M_t = \frac{8M_\infty \lambda p D^{3/2} t^{3/2}}{\pi^{1/2} h^3 (K + \lambda p^{1/2})} \exp \left(\frac{-h^2}{4Dt} \right) \tag{81}$$

$$M_t = M_\infty \left(1 - \exp \left(\frac{-\lambda p Dt}{Kh^2} \right) \right) \tag{82}$$

Flynn and colleagues (44) examined this situation and arrived at a solution that describes the situation over the full time period [Eq. (83)]. Using such expressions it is possible to understand how such changes in an applied formation may affect diffusion through the skin.

$$\frac{M_t}{M_\infty} = \frac{\sqrt{D_m}K}{(h_v + h_m K)s^{3/2}}$$

$$\cdot \left\{ \frac{\left(\sinh \sqrt{\frac{s}{D_v}} h_v \cosh \sqrt{\frac{s}{D_m}} h_m + K \sqrt{\frac{D_m}{D_v}} \sinh \sqrt{\frac{s}{D_m}} h_m \cosh \sqrt{\frac{s}{D_v}} h_v \right)}{\left(K \sqrt{\frac{D_m}{D_v}} \cosh \sqrt{\frac{s}{D_v}} h_v \cosh \sqrt{\frac{s}{D_m}} h_m + \sinh \sqrt{\frac{s}{D_v}} h_v \sinh \sqrt{\frac{s}{D_m}} h_m \right)} \right\}$$

$$(83)$$

The case for which the diffusion coefficient is some function of either time, depth, or concentration, may well describe the situation in biological and indeed synthetic membranes more accurately than that for which its value is assumed constant. The movement of a diffusant, and perhaps some constituents of the vehicle in which it is formulated, into a membrane could quite easily change the nature of the membrane itself. One can visualize a situation in skin, for example, where the absorption of penetration enhancers over the period of an experiment may change the value of D. Indeed, the movement of solvent molecules into the stratum corneum may also affect the partitioning behavior of the diffusant. These phenomena will not occur instantaneously, thus making the diffusional and partitioning steps of the penetration process variables that are time-dependent. These are complicated situations that are beyond the scope of this chapter. The interested reader is referred to Crank and Park (45) and Crank (1) for further reading.

The possibility of there being a depth dependency of rates of diffusion through stratum corneum has been addressed (7). If the skin is treated with a penetration enhancer, there will be a concentration gradient of that enhancer across the skin (as a consequence of Fick's second law of diffusion). If the enhancer has the same effect per lipid molecule at all depths; then it follows that the degree of enhancement will reduce as the skin is penetrated farther. If, for the sake of argument, the enhancer action arises from increasing the diffusion coefficient of a penetrant, we will see a consequent reduction in its value with depth, owing to the nature of the enhancer distribution.

B. Effect of Temperature on Diffusion Through Membranes

Temperature does not tend to affect partitioning phenomena, phase volumes, or membrane thicknesses (or diffusional pathlengths) to a great extent. However, in certain membranes containing structures that are subject to phase changes on heating or cooling, there may be a concurrent effect on the diffusion. For example, the flux of water through human stratum corneum increases with temperature and this increase has been linked to an increasingly fluid (or disordered) environment within the stratum corneum lipid bilayers (46). The diffusion coefficient itself may be temperature-dependent, and the nature of this dependence can be empirically expressed in a form analogous to the Arrhenius equations for reaction kinetics.

$$D = D_0 \exp \left[\frac{-E_a}{RT} \right] \tag{84}$$

In Eq. (84) D_0 represents the diffusion coefficient at infinite temperature and can be calculated by back-extrapolating a plot of $\ln D$ versus $(1/T)$ to the y-intercept [Eq. (85)].

$$\ln D = \ln D_0 - \frac{E_a}{RT} \qquad (85)$$

E_a is an activation energy term and is determined by the nature of the barrier. Although Eq. (85) is empirical, it appears to hold in the temperature range over which permeability is generally measured. This relation holds for permeability coefficients where D and D_0 are replaced in Eqs. (86) and (87) with P and P_0, respectively. If a membrane contains constituents that are subject to temperature-dependent phase transitions (as outlined earlier) it is unlikely that these equations will not hold in the temperature region in which the transition occurs.

$$P = P_0 \exp\left[\frac{-E_a}{RT}\right] \qquad (86)$$

$$\ln P = \ln P_0 - \frac{E_a}{RT} \qquad (87)$$

IV. CONCLUDING REMARKS

As stated in the Introduction, the purpose of this chapter was to increase the accessibility of the mathematical approach to the interpretation of diffusion processes and thereby to promote a clearer understanding of better ways to use data in this area of research. We would like to think that the preceding pages have demonstrated the utility with which mathematics and the basic principles of diffusion can be used in examining some of the processes involved in percutaneous penetration.

REFERENCES

1. Crank J. The Mathematics of Diffusion. 2nd ed. London: Oxford University Press, 1975.
2. Daynes HA. The process of diffusion through a rubber membrane. Proc R Soc Lond A 97:286–307, 1920.
3. Barrer RM. Permeation, diffusion and solution of gases. Trans Faraday Soc 35:628–643, 1939.
4. Flynn GL, Carpenter OS, Yalkowsky SH. Total mathematical resolution of diffusion layer control of barrier flux. J Pharm Sci 61:312–314, 1972.
5. Flynn GL. Dermal diffusion and delivery principles. In: Swarbrick J, Boylan JC, eds. Encyclopedia of Pharmaceutical Technology. Vol. 3. New York: Marcel Dekker, pp. 457–503, 1991.
6. Hadgraft J. Calculations of drug release rates from controlled release devices. The slab. Int J Pharm 2:177–194, 1979.
7. Watkinson AC, Bunge AL, Hadgraft J. Computer simulation of penetrant concentration–depth profiles in the stratum corneum. Int J Pharm 87:175–182, 1992.
8. Yamashita F, Koyama Y, Sezaki H, Hashida M. Estimation of a concentration profile of acyclovir in the skin after topical administration. Int J Pharm 89:199–206, 1993.
9. Shah JC. Analysis of permeation data: evaluation of the lag time method. Int J Pharm 90:161–169, 1993.
10. Potts RO, Guy RH. Drug transport across the skin and the attainment of steady-state flux. Proc Int Symp Control Release Bioact Mat 21:162–163, 1994.

11. Lafforgue C, Eynard I, Falson F, Watkinson A, Hadgraft J. Percutaneous absorption of methyl nicotinate. Int J Pharm 121:89–93, 1995.
12. Parry GE, Bunge AL, Silcox GD, Pershing DW, Pershing DW. Percutaneous absorption of benzoic acid across human skin. I. In-vitro experiments and mathematical modelling. Pharm Res 7:230–236, 1990.
13. Rogers WA, Buritz RS, Alpert D. Diffusion coefficient, solubility, and permeability for helium in glass. J Appl Phys 25:868–875, 1954.
14. Short PM, Abbs ET, Rhodes CT. Effect of non-ionic surfactants on the transport of testosterone across a cellulose acetate membrane. J Pharm Sci 59:995–998, 1970.
15. Brandt H. Determination of diffusion specific parameters by means of IR–ATR spectroscopy. Exp Tech Phys 33:423–431, 1985.
16. Brandt H, Hemmelmann K. On the evidence of non-linear sorption at the surface polyethylene/ATR–element for the diffusion of ethylacetate in polystyrene films. Exp Tech Phys 35:349–358, 1987.
17. Hemmelmann K, Brandt H. Investigation of diffusion and sorption properties of polyethylene films for different liquids by means of IR–ATR spectroscopy. Exp Tech Phys 34:439–446, 1986.
18. Van Alsten JG, Lustig SR. Polymer mutual diffusion measurements using infrared ATR spectroscopy. Macromolecules 25:5069–5073, 1992.
19. Potts RO, Doh L, Venkatraman, S, Farinas KC. Characterisation of solute diffusion in a polymer using ATR–FTIR spectroscopy. Macromolecules 27:5220–5222, 1994.
20. Wurster DE, Buraphacheep V, Patel JM. The determination of diffusion coefficients in semisolids by Fourier transform infrared (FT–IR) spectroscopy. Pharm Res 10:616–620, 1993.
21. Tralhão AM, Watkinson AC, Brain KR, Hadgraft J, Armstrong NA. Use of ATR–FTIR spectroscopy to study the diffusion of ethanol through glycerogelatine films. Pharm Res 12:572–575, 1995.
22. Harrison JE, Watkinson AC, Green DM, Hadgraft J, Brain KR. The relative effect of Azone and Transcutol on permeant diffusivity and solubility in human stratum corneum. Pharm Res 13:542–546, 1996.
23. Watkinson AC, Brain KR, Hadgraft J. The deconvolution of diffusion and partition coefficients in permeability studies. Proc Int Symp Controlled Release Bioact Mat 21: 160–161, 1994.
24. Flynn GL, Yalkowsky SH, Roseman TJ. Mass transport phenomena and models: theoretical concepts. J Pharm Sci 63:479–510, 1974.
25. Flynn GL, Yalkowsky SH. Correlation and prediction of mass transport across membranes. I. Influence of alkyl chain length on flux determining properties of barrier and diffusant. J Pharm Sci 61:838–852, 1972.
26. Stehle RG, Higuchi WI. Diffusional model for transport rate studies across membranes. J Pharm Sci 56:1367–1368, 1967.
27. Roberts MS, Anderson RA, Swarbrick J. Permeability of human epidermis to phenolic compounds. J Pharm Pharmacol 29:677–683, 1978.
28. Zwolinski BJ, Eyring H, Reese CE. Diffusion and membrane permeability. J Phys Chem 53:1426–1453, 1949.
29. Danielli JF. The theory of penetration of a thin membrane-derivation of a fundamental equation. In: Davson H, Danielli JF. The Permeability of Natural Membranes. 2nd ed. London: Cambridge University Press, pp. 324–335, 1952.
30. Lakshminarayanaiah N. Transport Phenomena in Membranes. New York: Academic Press, pp. 119–126, 1969.
31. Schaefer H, Watts F, Brod J, Illel B. Follicular penetration. In: Scott RC, Guy RH, Hadgraft J, eds. Prediction of Percutaneous Penetration. Vol. 1. London: IBC Technical Services, pp. 163–173, 1989.

32. Lauer AC, Lieb LM, Ramachandran C, Flynn GL, Weiner ND. Transfollicular drug delivery. Pharm Res 12:179–186, 1995.
33. Hueber F, Wepierre J, Schaefer H. Role of transfollicular routes in percutaneous absorption of hydrocortisone and testosterone: in vivo study in the hairless rat. Int J Pharm 5:99–107, 1992.
34. Jadoul A, Hanchard C, Thysman S, Preat V. Quantification and localisation of fentanyl and trh delivered by iontophoresis in the skin. Int J Pharm 120:221–228, 1995.
35. Craane van Hinsberg WHM, Verhoef JC, Bax LJ, Junginger HE, Bodde HE. Role of appendages in skin resistance and iontophoretic peptide flux: human versus snake skin. Pharm Res 12:1506–1512, 1995.
36. Scheuplein RJ, Analysis of solubility data for the case of parallel diffusion pathways. Biophys J 6:1–17, 1966.
37. Scheuplein RJ, Blank IH, Brauner GJ, MacFarlane DJ. Percutaneous absorption of steroids. J Invest Dermatol 52:63–70, 1969.
38. Barrer RM. Diffusion and permeation in heterogeneous media. In: Crank J, Park GS, eds. Diffusion in Polymers. London: Academic Press, pp. 165–217, 1968.
39. Higuchi WI, Higuchi T. Theoretical analysis of diffusional movement through heterogeneous barriers. J Am Pharm Assoc Sci Ed 49:598–606, 1960.
40. Finger KF, Lemberger AP, Higuchi T, Busse LW, Wurster DE. Investigation and development of protective ointment IV. The influence of active fillers on the permeability of semisolids. J Am Pharm Assoc Sci Ed 49:569–573, 1960.
41. Flynn GL, Roseman TJ. Membrane diffusion II. Influence of physical adsorption on molecular flux through heterogeneous dimethypolysiloxane barriers. J Pharm Sci 60:1788–1796, 1970.
42. Paul DR, Kemp DR. The diffusion time lag in polymer membranes containing adsorptive fillers. J Polymer Sci Polymer Symp 41. Transport Phenomena Through Polymer Films, Kumins CA, ed. pp. 79–93, 1973.
43. Guy RH, Hadgraft J. A theoretical description relating skin penetration to the thickness of the applied medicament. Int J Pharm 6:321–332, 1980.
44. Addicks WJ, Flynn GL, Weiner N, Curl R. A mathematical model to describe drug release from thin topical applications. Int J Pharm 56:243–248, 1989.
45. Crank J, Park GS. Diffusion in Polymers. London: Academic Press, 1968.
46. Potts RO, Francoeur ML. Lipid biophysics of water loss through the skin. Proc Natl Acad Sci USA 87:1–3, 1990.

4

Skin Transport

MICHAEL S. ROBERTS and SHEREE ELIZABETH CROSS

University of Queensland, Princess Alexandra Hospital, Brisbane, Queensland, Australia

MARK A. PELLETT

Whitehall International, Havant, England

I. INTRODUCTION

The skin is a tissue that separates the internal living organism from the external environment. It has a complex structure and performs many physiological functions such as metabolism, synthesis, temperature regulation, and excretion. The outermost layer of this organ, the stratum corneum (SC), is considered to be the main barrier to the percutaneous absorption of exogenous materials. The skin barrier is important in the maintenance of water within the body and in protection of the body from the ingress of compounds, particularly important from an occupational viewpoint for workers in the cosmetic and agrochemical industries (1).

A. Skin as a Delivery Mode

Examples of products targeted to the surface of the skin for protection are shown in Figure 1. In contrast, percutaneous absorption of pharmaceuticals for either systemic or local (appendageal, epidermal, and lower tissue) delivery is a desirable process (see Fig. 1), and can be attained by the combination of appropriate solute properties for skin transport with appropriate dosage form design (e.g., patches, gels, creams, ointments) (Fig. 2). Compounds have been applied to the skin for many centuries (2) and, indeed, drugs in the form of plant or animal extracts have been applied for the relief of a variety of local disorders. In recent years, systemic delivery through the transdermal route has led to the development and successful marketing of various pharmaceuticals in a patch form (e.g., scopolamine, nitroglycerin, clonidine, estra-

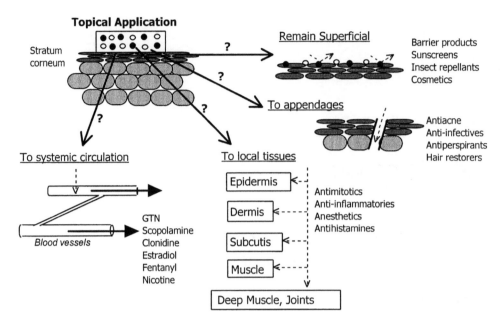

Figure 1   Prevention and management of various conditions using the skin as a delivery site.

diol, testosterone, timolol, fentanyl, and nicotine). The number of compounds being screened for potential transdermal application is increasing.

B.   Advantages of Transdermal Delivery

In contrast to the traditional oral route, first-pass metabolism is minimized, which can often limit the tolerability and efficacy of many orally and parenterally delivered drugs. Furthermore, some drugs degrade in the acidic environment of the stomach, and other drugs, such as NSAIDs, can cause gastrointestinal bleeding or irritation. The mixing of drugs with food in the stomach, and the pulsed, often erratic delivery of drugs to the intestine leads to variability in the plasma concentration–time profiles achieved for many drugs. The transdermal route provides a more-controlled, non-invasive method of delivery, with the added advantage of being able to cease absorption in the event of an overdose or other problems. Furthermore, patient compliance may be improved because of the reduced frequency of administration for short half-life medications or avoidance of the trauma associated with parenteral therapy.

C.   Disadvantages of Transdermal Delivery

As with the other routes of drug delivery, transport across the skin is also associated with several disadvantages, the main drawback being that not all compounds are suitable candidates. A number of physicochemical parameters have been identified that influence the diffusion process, and variations in permeation rates can occur between individuals, different races, and between the old and young (3). Furthermore, diseased skin, as well as the extent of the disease, can also affect permeation rates

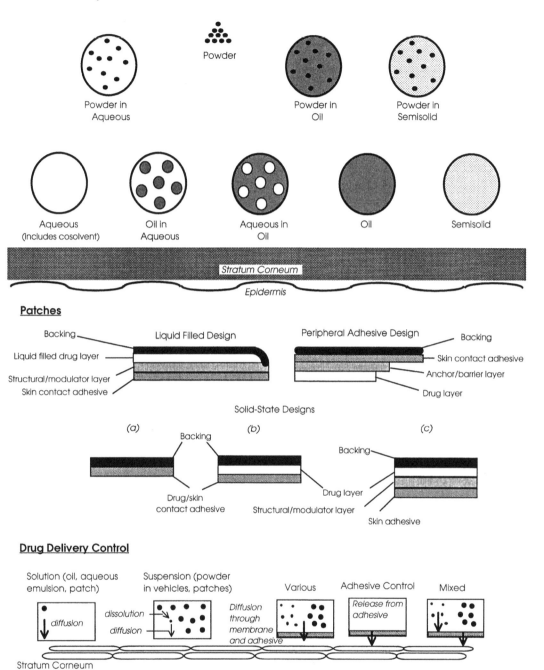

Figure 2 Examples of dosage forms used in topical delivery.

(3). The metabolic enzymes in the skin can also pose a problem, and some drugs are almost completely metabolized before they reach the cutaneous vasculature (4). Another problem that can arise, which is sometimes overlooked, is that some drugs can be broken down before penetration through the SC by the bacteria that live on the skin surface (4).

D.   The Stratum Corneum Barrier

The barrier properties of the SC are now recognized as the major rate-limiting step in the diffusion process of a drug permeating across skin (Table 1). However, as pointed out by Scheuplein and Blank (5), other skin components can contribute to the overall barrier resistance, especially for lipophilic solutes. Scott et al. (6) showed that the permeability to water in vivo and in vitro increased after mild, superficial epidermal alterations: suction blister top removal > adhesive tape stripping > sandpaper abrasion > scalpel blade. After each alteration, the epidermis regenerated in a distinct, biphasic manner. In the rapid first phase, the permeability decreased with the development of a scab. In the second phase, there was a return to normal permeability, with a gradual thickening of the SC. Schaefer et al. (7) showed that in psoriatic skin, the epidermal and dermal concentrations of radiolabeled triamcinolone acetonide were three to ten times higher than in normal skin. A similar increase was reported when the SC was removed by stripping before application.

The structure of the SC, as discussed in an earlier chapter, has been likened to "bricks-and-mortar" (Fig. 3), where the bricks are the component cells, or corneocytes, and the mortar is the intercellular lipids (8). The membrane is interrupted only by appendages such as hair follicles and sweat glands. However, it is still considered to be a predominantly dual-compartment system composed of a matrix of corneocytes tightly packed with keratin, surrounded by a complex array of lipids arranged in bilayers (9–11).

Transport across the SC is largely a passive process, and thus the physicochemical properties of a permeant are an important determinant of its ability to penetrate and diffuse across the membrane. There are generally considered to be three routes by which compounds can diffuse across the SC: intercellular, transcellular, and transappendageal (see Fig. 3). Evidence for and against these routes will be discussed in more detail in the next section (Sec. II) of this chapter. Once it has penetrated through the epidermis, a compound may be carried away by the dermal blood supply or be transported to deeper tissues (see Fig. 3). Therefore, owing to the structure of the skin, the desired physicochemical properties of a permeant are dependent on the route taken to traverse the SC.

E.   Sebum as a Barrier

The surface of the skin is the first point of contact for a topically applied formulation. Under normal circumstances, this is covered by a 0.4- to 10-μm irregular and discontinuous layer of sebum, sweat, bacteria, and dead cells (12–16). The presence of this layer is considered to have a negligible effect on percutaneous absorption, as it allows polar and nonpolar materials to penetrate (12,17–19). Furthermore, no correlation has been found between the hydration state of the SC and the removal of the sebum layer by swabbing with solvents, the total amount of sebhorreic lipids, or their composition (13,14,20). Therefore, the contribution of these endogenous

Table 1 Historical Development of the SC Barrier as a Major Rate-Limiting Step in Drug Permeation Across Skin

Year	Development	Ref.
1853	Determined that the epidermis was more impermeable than the dermis	253
1877	Theorized that intact skin of humans is totally impermeable to all substances	254
1904	Determined that the skin is more permeable to lipid soluble substances than water or electrolytes	255
1909	Discovered that penetration through ichthyotic skin was similar to that of healthy skin not less than, as was expected	256
1919	Determined that mustard gas penetrated into the outer layer of the skin readily, but was unable to rapidly penetrate further	257
1924	A theorized electrical barrier between the stratum corneum and the malpighian layer that reduced the permeability of the skin to ions	258
1930	Outlined the significance of lipid solubility in skin permeability	259
1939	The various layers of the skin can be exposed by stripping with adhesive cellophane tape	260
1945	Suggested that the stratum lucidum was the skin's barrier layer to the penetration of both ions and uncharged molecules	261
1945	Suggested that the entire stratum corneum was responsible for the high DC and AC resistance in the skin	262
1951	Separation of the epidermis from the dermis used for the first time to determine the differences between their permeabilities	33
1951	Determined the horny layer is the barrier to diffusion of water through the skin	33
1953	Determined that the permeability of the skin to water remained unchanged until the lowest lying layers of the stratum corneum was removed, indicating this region must contain the rate-limiting step	263
1953	Conceded that the stratum corneum is uniformly impermeable to water penetration regardless of distance from the surface	264
1954	The stratum corneum was still thought to be a porous membrane through which ions and large molecules could freely permeate	265
1962	Determined the outer layer of the stratum corneum greatly impedes penetration of substances, the concentrations decrease exponentially with distance from the surface	266
1964	Techniques involving drying and staining of skin samples before microscopy alter the appearance and barrier function of the skin	267

Source: Ref. 5.

surface materials to skin transport processes is effectively discounted and will not be discussed further in this chapter.

As stated earlier, the skin is an important barrier to the ingress of undesirable compounds and a potential drug delivery route for therapeutically useful compounds. Therefore, it is important to understand how molecules traverse the skin and how these processes can be influenced to enhance permeation. One aim of this chapter is to define the current understanding of the processes involved in the transport of solutes through the skin from their application site through their eventual diffusion

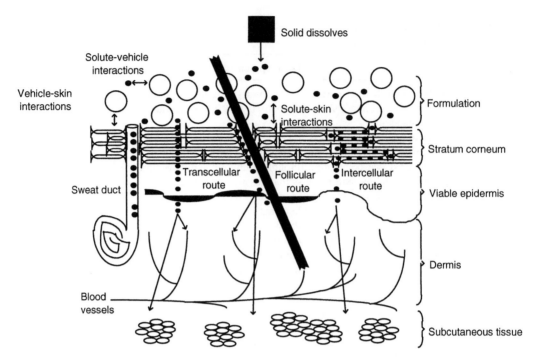

Figure 3 Schematic representation of the processes contributing to the permeability of a solute through the skin to the bloodstream or underlying tissues.

into the systemic blood supply or into deeper tissues. Figure 4 shows the concepts addressed in this chapter. Earlier reviews of the literature listed in Table 2 may be used to provide a more substantial reference list of the historical developments in percutaneous absorption from pharmaceutical preparations.

II. TRANSPORT PATHWAYS THROUGH THE STRATUM CORNEUM

As mentioned previously and shown in Figure 3, there are three pathways postulated for the diffusion of solutes through the SC: transcellular, intercellular (paracellular), and transappendageal. The following sections describe the nature of the transcellular and intercellular pathways as they relate to skin transport and examine the experimental and theoretical evidence for their existence. Transappendageal transport is examined as a separate section toward the end of this chapter. The transport of solutes through the nail plate is also considered later.

A. Transcellular Pathway

It was originally believed that transcellular diffusion mechanisms dominated over the intercellular and transappendageal routes during the passage of solutes through the SC (21). However, transport by the transcellular route would involve the repeated partitioning of the molecule between lipophilic and hydrophilic compartments, including the almost impenetrable corneocyte intracellular matrix of keratin and keratohyaline. Scheuplein further suggested that polar and nonpolar solutes permeate

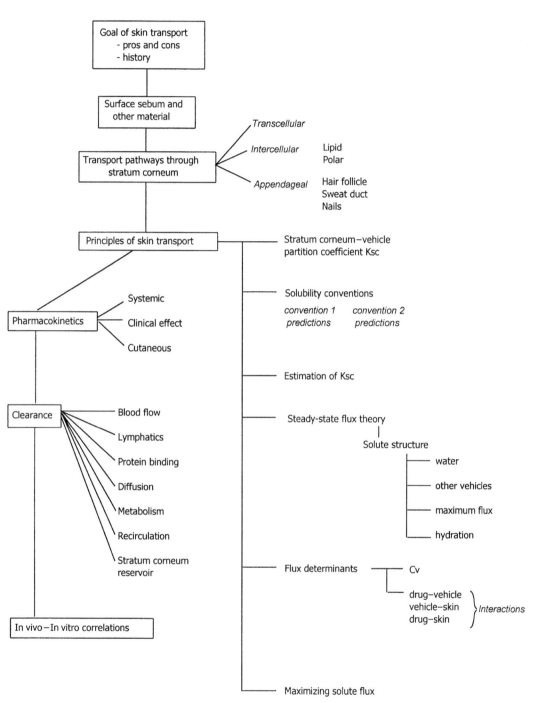

Figure 4 Concepts addressed in this chapter.

Table 2 Early Reviews on Percutaneous Absorption

Year	Author	Year	Author
1950s		*1970s*	
1954	Rothman (265)	1971	Idson (128)
			Katz and Poulsen (281)
			Scheuplein and Blank (5)
1960s		1972	Katz and Poulsen (282)
1961	Wagner (268)	1973	Katz (283)
			Poulsen (284)
		1974	Anderson and Roberts (285)
1962	Barr (269)	1974	Schaefer (286)
1963	Malkinson and Rothman (270)	1975	Idson (287)
1964	Kligman (267)	1977	Dugard (288)
	Malkinson (271)		Higuchi (289)
	Tregear (272)		Malkinson and Gehlmann (290)
	Welles and Lubowe (273)		Webster and Maibach (291)
1965	Stoughton (274)	1978	Scheuplein (22,292)
	Vinson et al. (275)		
1966	Reiss (276)	1979	Flynn (293)
	Tregear (12)		
	Vickers (277)	*1980s*	
1969	Barrett (278)	1980	Scheuplein (294)
	Blank and Scheuplein (279)	1983	Barry (126)
	Winkleman (280)		

the SC by different mechanisms. The polar solutes were thought to diffuse through a high-energy pathway involving immobilized water near the outer surface of keratin filaments. In contrast, the lipid-soluble solutes diffused through a nonpolar (interstitial) lipid pathway (5,22). Our analysis of Scheuplein's data and our own phenol data (23,24), suggested that all solutes were transported through a lipid pathway and ameliorated through the effects of an unstirred water (viable epidermal) layer, as evidenced by a decrease in the energy of activation for permeation. Although the lipid route was thought to be transcellular, evidence for its location was not defined. Scheuplein also recognized that the dermis contributed to the resistance of the more lipophilic solutes. Most experimental evidence now suggests that transport through the SC is by the intercellular route.

B. Intercellular Pathway

The intercellular SC spaces were initially dismissed as a potentially significant diffusion pathway because of the small volume they occupy (5). However, the physical structure of the intercellular lipids was thought to be a significant factor in the barrier properties of the skin (25). Tracer studies (26,27) provided evidence that the intercellular lipid, and not the corneocyte protein, was the main epidermal permeability barrier. Chandrasekaran and Shaw (28) also concluded that the lipid barrier domi= nated. Theoretical evidence, presented by Albery and Hadgraft in 1979 (29), suggested that the tortuous intercellular diffusional pathway around keratinocytes was the preferred route of penetration through the SC, rather than a drug diffusing through the keratinized cells (transcellular route). However, it should be recognized that although theoretical considerations favor this route, there are difficulties in designing appropriate diffusion studies to confirm that this route is the predominant pathway (30).

Most evidence for the existence of the intercellular lipid transport pathway comes from the microscopic organizational structure of the lipid bilayers, the observed histological localization of applied substances within these bilayers following topical application, and the effects of delipidization of the bilayers by appropriate solvents. Histochemical studies have shown that the intracellular spaces of the SC are devoid of lipid (31,32) and that because lipid present in other regions is highly nonpolar, there is no structure suitable to form a lipid diffusional matrix around the intracellular keratin filaments. The intercellular volume fraction is also much larger than originally estimated (33), and experimental evidence using precipitation of percutaneously applied compounds has led to visualization of permeation through intercellular pathways (34).

In 1991, Boddé et al. (35) visualized the diffusion of mercuric chloride through dermatomed human skin samples by using ammonium sulfide vapor to precipitate the compound within the sample and subsequent transmission electron microscopy. Their results indicated that the intercellular route of transport through the SC predominated; however, after longer transport times, the apical corneocytes tended to take up the compound, leading to an apparent bimodal distribution. There was mercury both inside and outside the cells in the apical region of the SC (35), whereas in the medial and proximal region the mercury was located intercellularly. This led to the suggestion of the presence of two types of cells: apical corneocytes that tended to take up mercuric ions relatively easily, and medial and proximal corneocytes, that were less capable of doing so. It has been suggested that, in the corneocytes, the desmosomes may serve to channel material into the cell, especially in the squamous region where the desmosomes are beginning to disintegrate. Hence, the cellular lipid envelopes are leaky, suggesting a reservoir function for the apical zone of the SC.

Elias et al. (36) examined the penetration of [^{3}H]water and [^{14}C]salicylic acid across the same tissue samples and tried to correlate diffusion with the thickness, number of cell layers, and lipid composition of leg and abdominal skin. They found that differences in the thickness and number of cell layers in the SC were insufficient to account for differences observed in percutaneous transport across the leg and abdomen, and that total lipid concentration may be the critical factor governing skin permeability.

The lipid lamellae in the SC play a key role in the barrier function of the skin. The major lipids are ceramides, cholesterol, and free fatty acids (37,38). Figure 5 shows a diagrammatic representation of the lamellae, dimensions of lipid arrangement in an individual lamella, and the types of lipids in an intercellular lipid bilayer. In reality, the lipid composition and arrangement is much more complex. For instance, the lipids present in epidermis could be further classified as phospholipids, monohexosylceramides, ceramides, cholesterol, cholesterol esters, cholesterol sulfate, triglycerides, and fatty acids (39). Furthermore, at least six subclasses of ceramides have been described in pig SC (37,40).

Bouwstra et al. (41), in a summary of the X-ray analysis of the SC and its lipids, concluded that the lamella is the fundamental structure in intercellular domains. An individual lamella is about 13 nm (130 Å) in width and comprises two

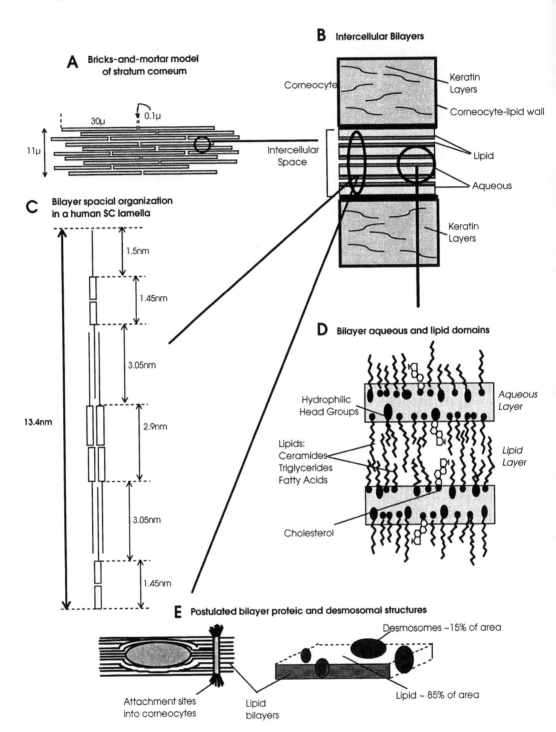

Figure 5 Diagrammatic representation of the structure of the stratum corneum showing (A) the bricks and mortar model of its gross structure; (B) the intercellular bilayers; (C) the spatial organization of lipids within the bilayers; (D) the location of polar and lipid domains; (E) the presence of proteic and desmosomal structures within the lipid bilayers.

or three lipid bilayers (see Fig. 5). Although 13.4-nm (134-Å) lamella is dominant in human SC, an occasional 6.4-nm (64-Å) lamella was observed. In mouse SC, the dominant lamella is 13.1 nm (131 Å), with an occasional 6.0-nm (60-Å) one. The actual organization of (pig) SC lipids is as two lamellar phases with periodicities of approximately 6 and 13 nm, respectively (40). At the molecular level, Bouwstra et al. (40) suggest that the short periodicity phase is composed of only one bilayer, akin to phospholipid membranes. In contrast, the long periodicity phase consists of two broad and one narrow low−electron-dense regions. It is suggested that the two broad regions are formed by partly interdigitating ceramides, with long-chain fatty acids of approximately 24−26 carbon atoms, whereas the narrow low−electron-density region is formed by fully interdigitating ceramides, with a short free fatty acid chain of approximately 16−18 carbon atoms.

C.   Lipid and Polar Pathways Through the Intercellular Lipids

Both diffusional and morphometric data have been presented to support lipid and polar pathways through SC lipids. Southwell and Barry (42) used penetration-enhancing solvents to modify the different diffusional routes through the SC and the partitioning of drugs into these pathways. Steady-state fluxes were measured in vitro for polar methanol, nonpolar octanol, and an intermediate compound, caffeine, selected as model penetrants through human SC conditioned on both sides with water or the two accelerants. They concluded that 2-pyrrolidone enhances permeation through the polar route of the skin by increasing the diffusivity, but reduced nonpolar route transport. Whereas dimethylformamide (DMF) promotes polar route absorption by raising diffusivity and partitioning, but reduces nonpolar absorption by decreasing both parameters. Blank and McAuliffe (43) also suggested the presence of polar and nonpolar pathways in the SC through different routes, on the basis of selective solvent effects on the permeability constants for tritiated water (a polar molecule) and for benzene (a relatively nonpolar molecule). Several investigators recognized the presence of a polar pathway and, through modeling, showed that for lipophilic solutes, such as steroids (44) and β-blockers (45), the contribution of the polar pathway was negligible.

Kim et al. (46) elaborated on an in vitro model for skin permeation in which penetration could occur across the main barrier, the SC by one of two parallel pathways: the lipoidal pathway and the pore pathway, with this barrier existing in series with an epidermal−dermal porous barrier. According to this model very lipophilic molecules are rate-limited by the epidermal−dermal barrier, as described previously by Scheuplein and Blank (5). Extremely polar permeants are rate-limited by the pore pathway of the SC with its limiting permeability coefficient, whereas permeants with intermediate polarity are transported by the lipoidal pathway and exhibit a lipophilicity−dependent permeability coefficient. Such a model has also been proposed by Cooper and Kasting (47) and is discussed by Roberts and Walters (3).

Matsuzaki et al. (48) found that the permeability of very polar solutes through model SC membrane systems was almost constant and similar to that of potassium ions, whereas, for the more lipophilic solutes, permeability increased with solute lipophilicity. This data suggests, therefore, that solutes may be transported through both a polar and nonpolar pathway through the intercellular region. Peck et al. (49,50) examined the in vitro passive transport of urea, mannitol, sucrose, and raffinose across intact and ethanol-treated human epidermal membranes. From the relative permeabilities of these four solutes and hindered diffusion theory, effective pore radii estimates for intact and ethanol-treated human epidermal membrane were between 1.5 and 2.5 nm (15 and 25 Å) and 1.5 and 2.0 nm (15 and 20 Å), respectively. Further studies on the temperature dependence of human epidermal membrane permeability with urea, mannitol, tetraethylammonium ion, and corticosterone strongly support the existence of a porous permeation pathway. Interestingly, the radii estimated is similar to that determined by iontophoretic studies (51,52) for the pore size range attributed for small solutes to transport through the polar intercellular lipid pathway. Sznitowska et al. (53) examined the percutaneous penetration of baclofen, a model zwitterion, in vitro using human cadaver skin with various solvent pretreatments. They concluded that the polar pathway might be intercellular and comprises the aqueous regions surrounded by polar lipids. Finally, Menon and Elias (54) applied hydrophilic and hydrophobic tracers to murine skin in vivo under basal conditions or after permeation enhancement with occlusion, vehicle enhancers, a lipid synthesis inhibitor, sonophoresis, and iontophoresis. Using ruthenium and microwave postfixation methods, tracers were found localized in discrete lacuna domains in the extracellular lamellar membrane system, regardless of their polarity or the enhancement method. Although extracellular lacunar domains were interpreted as being a potential pore pathway for penetration of polar and nonpolar molecules across the SC, the continuity of such a pathway is unclear. Figure 6 is our interpretation of the possible polar and lipid pathways for intercellular transport.

III. PRINCIPLES OF SKIN TRANSPORT

The process of percutaneous absorption involves several individual transport processes, some of which occur in series and others in parallel (see Fig. 3). The two key determinants for a solute crossing a membrane are solubility and diffusivity. The relative solubility of a solute in two phases determines its partition coefficient and, therefore, the likelihood of the solute being taken up into the SC from a vehicle. Also, solubility will determine whether a solute is likely to be desorbed from the SC into deeper layers. The diffusivity is a measure of the speed at which a solute crosses a given barrier and is affected by binding, viscosity of the environment, and the tortuosity of the path.

In the first step of the transport process, molecules must be in solution in the vehicle to partition from the vehicle into the lipids in the outermost part of the SC; they must then diffuse through it; partition back out of the SC and into the viable epidermis. Next, molecules diffuse through the viable epidermis and papillary dermis. At the capillary plexus a high percentage of molecules are transferred into the circulating blood and a lower percentage diffuses into deeper tissues (see Fig. 3). To predict the penetration of a given solute it is necessary (a) to define the skin barrier in terms of a mechanistic model, and (b) to relate transport to a physical property of the solute, such as its organic solvent–water partition coefficient. Scheuplein and Blank (5) suggested that an appropriate skin model is a multilayer barrier consisting of the SC (10 μm) (S_1), the viable epidermis (100 μm) (S_2), and the upper papillary layer of the dermis (100–200 μm) (Fig. 7). We now develop a steady-state model for skin transport consistent with this model and based on the theoretical considerations presented in Chapter 3.

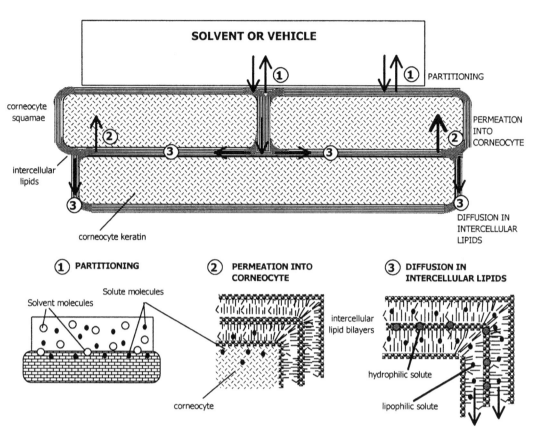

Figure 6 Partitioning and diffusion processes involved in solute penetration through the stratum corneum. (Adapted from Refs. 95 and 123.)

A. Stratum Corneum–Vehicle Partition Coefficients

We first consider the partitioning between the SC and vehicle. The previous chapter recognized that the chemical potential gradient across a membrane is a major determinant of flux (J), the amount of solute passing through a unit area of membrane in unit time. The chemical potential of a solute in a phase is also a major determinant for its partitioning into another adjacent phase. In an ideal solution, the *chemical potential* of a solute μ_i is defined by the standard chemical potential state μ_i^0 for that solute, and its *activity* a_i (defined as the product of its activity coefficient γ_i and concentration C_i, expressed as a mole fraction; i.e., $a_i = \gamma_i C_i$), the gas constant R, and absolute temperature T:

$$\mu_i = \mu_i^0 + RT \ln a_i = \mu_i^0 + RT \ln \gamma_i C_i \tag{1}$$

The partitioning of a solute between the SC and vehicle is defined by the chemical potential difference between the solute in the SC μ_{sc} and that in the vehicle μ_v for which chemical potential is defined by Eq. (1) for each phase. At equilibrium, the chemical potential of the solute in the two phases is equal (i.e., $\mu_{sc} = \mu_v$) such that (55)

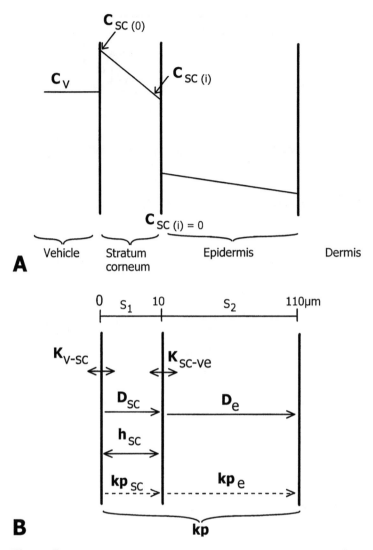

Figure 7 Diagrammatic representation of steady-state concentrations of a solute in the skin after topical application. (From Ref. 5.)

$$\mu_{sc}^0 + RT \ln \gamma_{sc} C_{sc} = \mu_v^0 + RT \ln \gamma_v C_v \tag{2}$$

Rearranging and defining the ratios of $\gamma_i C_i$ as a SC–vehicle partition coefficient K_{sc-v}^a, based on activities, yields

$$K_{sc-v}^a = \frac{a_{sc}}{a_v} = \frac{\gamma_{sc} C_{sc}}{\gamma_v C_v} = \exp\left(\frac{\mu_{sc}^0 - \mu_v^0}{RT}\right) \tag{3}$$

However, we could also define an SC–vehicle partition coefficient K_{sc-v} based on concentrations.

$$K_{sc-v} = \frac{C_{sc}}{C_v} = \frac{\gamma_v a_{sc}}{\gamma_{sc} a_v} = \frac{\gamma_v}{\gamma_{sc}} \exp\left(\frac{\mu^0_{sc} - \mu^0_v}{RT}\right) \tag{4}$$

K_{sc-v} could also be defined in terms of the solubility of the solute in the SC (S_{sc}) and vehicle (S_v):

$$K_{sc-v} = \frac{S_{sc}}{S_v} \tag{5}$$

Solubility, in this chapter, unless otherwise specified, is defined by a phase being saturated with solute such that the chemical potential of the dissolved solute $\mu_{i,\text{saturated}} = \mu_{\text{pure}}$. As a number of studies have applied various approaches used to determine solubility in the prediction of percutaneous penetration flux, we now consider the prediction of solubility using these approaches in further detail.

1. Conventions Used for Prediction of Solubility, Partition Coefficient, and Flux

It should be emphasized that the prediction method is dependent on the definition of the standard state μ^0_i. Two conventions are widely used in defining the standard state of a solute (56). In our experience it is often convenient to use convention 1 and established regular solution theory, to derive solubilities of solutes in different vehicles. This recognizes that, as an approximation, solutes with the same fractional solubility for different solvents have the same activity. Convention 2 is easiest to apply over a range of concentrations.

In the first, or Raoult's law, convention 1, the standard state is the pure substance as a liquid μ^0_{pure}. Hence, $\mu^0_{sc} = \mu^0_v = \mu^0_{\text{pure}}$ and expressing C_i in mole fractions, $\gamma_{sc} \to 1$, $\gamma_v \to 1$ as $X_i \to 1$. Further, applying Eq. (1) and noting that $\mu_{i,\text{saturated}} = \mu_{\text{pure}}$ in saturated systems, it is evident that the (thermodynamic) activity of the solute in the SC (a_{sc}) is equal to that in the vehicle (a_v) and that of a pure liquid solute (a_{pure}); that is, $a_{sc} = a_v = a_{\text{pure}}$. Accordingly, noting from Eq. (1) that the mole fraction solubility (X_i) is related to activity (a_i) and the activity coefficient (γ_i) by:

$$X_i = \frac{a_i}{\gamma_i} \tag{6}$$

then

$$K_{sc-v} = \frac{X_{sc}}{X_v} = \frac{\gamma_v}{\gamma_{sc}} \tag{7}$$

By definition, $\gamma_i = 1$ for an ideal binary liquid mixture. An ideal mixture requires that (a) both phases are mutually soluble in all proportions, and (b) the partial vapor pressure of a given component (p_i) is directly related to its vapor pressure as a pure liquid (p^0_i) by its mole fraction in the mixture (X_i), as defined by Raoult's law. The extent which γ_i deviates from unity can be considered as a measure of deviation from Raoult's law, which is defined in terms of the partial vapor pressure of a solute in solution (p_i), that of the pure component (p^0_i) and X_i (i.e., $p = p^0_i \gamma_i X_i$). Barry et al. (57) shows an example of such a deviation for the vapor pressure of benzyl alcohol plotted against its mole fraction.

The second convention is to use the infinitely dilute solution of the solute in a given phase as the standard state where $\gamma^*_{sc} \to 1$ and $\gamma^*_v \to 1$ as X_i and $C_i \to 0$.

Note that, following the convention of Davis et al. (58), the activity coefficient from convention 2 is given an asterisk to distinguish it from the convention 1 or Raoult's law activity coefficient. The activity coefficient from the second convention γ_i^* has also been referred to as Henry's law activity coefficient, on the basis that this law defines a constant activity coefficient of 1 with varying low concentrations and a γ_i^* of less than 1 is, therefore, a measure of deviation from this law. Hence, under ideal conditions in convention 2, $\gamma_{sc}^* \rightarrow 1$ and $\gamma_v^* \rightarrow 1$, and $K_{sc-v}^a = K_{sc-v}$. It is to be emphasized that, in contrast to convention 1, the activities of solute based on this convention are not equal in saturated solutions (i.e., $a_{sc} \neq a_v \neq a_{pure}$) but are related to the standard chemical potentials, as shown in Eq. (3) in which the standard state is as $\gamma_i^* \rightarrow 1$, $C_i \rightarrow 0$. The activity is also asterisked to distinguish it from activity as defined by convention 1.

2. Prediction of Solubility and Partition Coefficients with the First (Raoult's) Convention

We now examine the estimation of solubility of a solute in a phase based on the first (Raoult's) convention. Solubility and partitioning can be described in terms of the energy required to convert from the solid solute to a molecular form, the energy of dissolution in a vehicle, and the energy of dissolution in the SC (Fig. 8A). As many solutes used in topical delivery are solids, it is necessary to express solubilities and partition coefficients in terms of the activity of the pure solid (s_{solid}), also referred to as its ideal solubility (X_i^0). This ideal solubility varies with the nature of the solute crystal and is related to the energy associated with the formation of the pure liquid form by melting of the crystals at a melting point (T_m) (see Fig. 8A). X_i^0 is a function of the molar heat of fusion (ΔH_f), melting point (T_m), gas constant (R), room temperature (T), and (ΔC_p), the difference in heat capacity of the crystalline and molten states (59):

$$\ln X_i^0 = \frac{-\Delta H_f}{RT} \left[\frac{1}{T} - \frac{1}{T_m} \right] + \frac{\Delta C_p}{R} \left[\frac{T_m - T}{T} - \ln \frac{T_m}{T} \right] \tag{8}$$

Yalkowsky and Valvani (60) have pointed out that the last term can usually be ignored without any significant loss of accuracy so that

$$\ln X_i^0 \cong \frac{-\Delta H_f}{RT} \left[\frac{1}{T} - \frac{1}{T_m} \right] \tag{9}$$

They further note that, as the free energy of fusion is 0 at the melting point, Eq. (9) could be expressed in terms of the entropy of fusion ($\Delta S_f = \Delta H_f / T_m$). Given entropies of fusion are relatively constant across solutes, Grant and Higuchi (61) have suggested Eq. (8) be written in terms of the conventional centigrade temperature (T) and melting point (MP) as follows:

$$\log X_i^0 \cong -0.0099[MP - 25] \tag{10}$$

In a *nonideal* solution, $\gamma_i < 1$ as a result of solute–solvent interactions. The dissolution of a solute in a solvent is characterized by the following processes: (a) dispersion forces associated the transfer of solute molecules from its solution, the formation of a cavity in the solvent to accommodate the solute molecules, and the reorientation of solvent molecules around the solute molecules in the cavity; (b)

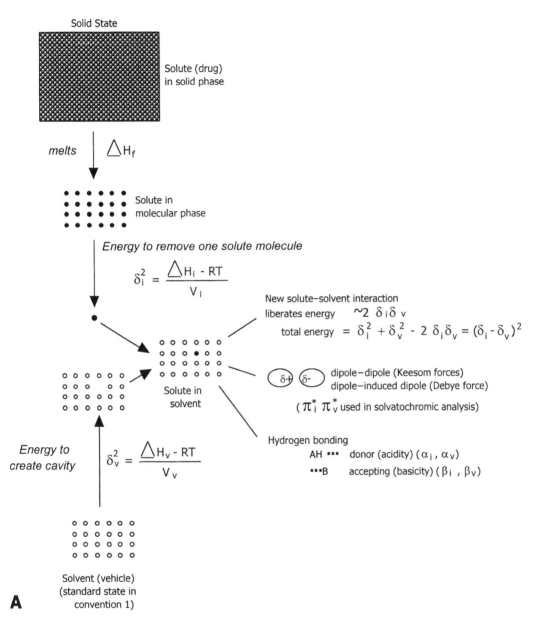

Figure 8A Processes involved in the dissolution of crystalline solute and the related energies.

dipole–dipole and induced dipole–dipole interactions between solute and solvent molecules; and (c) H-bonds between solute and solvent (see Fig. 8A).

In the commonly used Hildebrand solubility parameter approach, it is assumed that nonideality is solely due to dispersion forces. Defining the solubility parameter (δ_i) as the energy required to move a molecule from its solution, $\delta_i = (\Delta H_v - RT/V_i)$; where ΔH_v is the heat of vaporization and V_i is the molar volume of the solute.

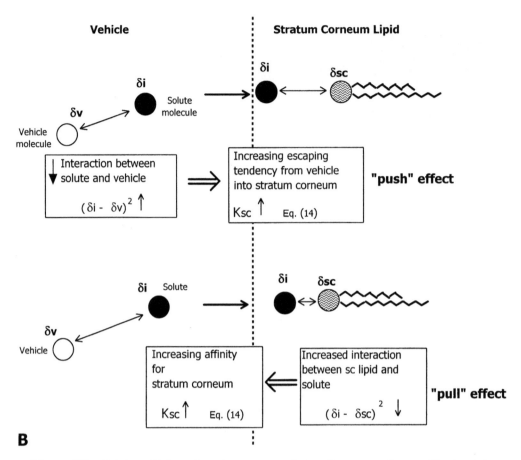

Figure 8B Solute–vehicle and stratum corneum interactions promoting partitioning into the stratum corneum.

Summing the energies required to remove a molecule of solute, to make a cavity in a solvent and reorientate solvent molecules around solute yields the convention 1 or Raoult's law activity coefficient (γ_i) for a solute in the solvent.

$$\ln \gamma_i = [\delta_i - \delta_l]^2 \frac{V_i \phi_l^2}{RT} \tag{11}$$

where V_i is the molar volume of the solute, δ_i and δ_l are the solubility parameters for the solute and solvent, ϕ_l is the volume fraction of the solvent ($\phi_l \rightarrow 1$ for dilute solutions). In a mixed solvent, $\delta_l = \phi_A \delta_A + \phi_B \delta_B + \ldots$. Hence, combining Eqs. (6), (9), and (11), the mole fraction solubility of a solute X_i is given by

$$\ln X_i = \ln X_i^0 - \ln \gamma_i = -\frac{\Delta H_f}{R}\left[\frac{1}{T} - \frac{1}{T_m}\right] - [\delta_i - \delta_l]^2 \frac{V_i \phi_l^2}{RT} \tag{12}$$

Hence, the solute solubility X_i is enhanced either by the solute's lower-melting point [first term in right hand side of Eq. (11) becomes zero when the solute is a liquid;

i.e., $T_m \leq T$] or by choosing a solvent with a solubility parameter close to that of the solute so that $(\delta_i - \delta_l)^2$ is minimized.

We now use the foregoing concepts to derive the SC–vehicle partition coefficient (K_{sc-v}). Substituting Eq. (11) into Eq. (7) yields for the SC and for a vehicle (62):

$$\ln K_{sc-v} = \ln \gamma_v - \ln \gamma_{sc} = [\delta_i - \delta_v]^2 \frac{V_i \phi_v^2}{RT} - [\delta_i - \delta_{sc}]^2 \frac{V_i \phi_{sc}^2}{RT} \tag{13}$$

If low concentrations of solute are present in both the vehicle and the SC, $\phi_v \to 1$ and $\phi_{sc} \to 1$ so that Eq. (12) reduces to

$$\ln K_{sc-v} = ([\delta_i - \delta_v]^2 - [\delta_i - \delta_{sc}]^2) \frac{V_i}{RT} \tag{14}$$

Table 3 shows calculated SC–vehicle partition coefficients derived for theophylline by Sloan et al. (62) based on a SC solubility parameter of 10 $cal^{1/2}$ $cm^{3.2}$. Hence, isopropyl myristate would appear to be the best vehicle for optimizing the partitioning of theophylline into the SC. Sloan et al. (62), have also pointed out that flux of solutes through the membrane is enhanced when the solubility parameter of the vehicle or cosolvent mixture is close to that of the SC.

Kadir et al. (63–65) studied the human skin permeability coefficients of theophylline and adenosine from alkanoic acid solutions and found certain acids promoted penetration through vehicle effects ("push" effect) whereas propionic acid enhances the penetration of theophylline and adenosine by promoting their solubility in the skin–propionic acid medium ("pull" effect). The push effect could be estimated by the solubility parameter approach. A push effect is equivalent to an increase in K_{sc-v} by an increase in the $(\delta_i - \delta_v)^2$ term in Eq. (14), whereas a pull effect is an increase in K_{sc} owing to a decrease in the $(\delta_i - \delta_{sc})^2$ term in Eq. (14) (see Fig. 8B).

Attempts have been made to modify the Hildebrand solubility parameter by including, in addition to the dispersion component (d), the polar component (p) and hydrogen-bonding interactions, to give the so-called three-dimensional solubility parameter $\delta_i^{3D} = (\delta_i^d)^2 + (\delta_i^p)^2 + (\delta_i^H)^2$ (66,67). Groning and Braun (68) used this con-

Table 3 Vehicle Solubility Parameters, Molar Volumes, and Estimates of Theophylline Stratum Corneum–Vehicle Partition Coefficients[a]

Vehicle	Molar volume (cm³/mol)	Solubility parameter (cal/cm³)$^{1/2}$	log K_{sc-v}	Experimental mole fraction solubility × 10³
Isopropyl myristate	317.0	8.5	1.12	0.109
Octanol	157.5	10.3	−0.18	1.49
Dimethylformamide	77.0	12.1	−0.98	14.8
Propylene glycol	73.5	14.8	−1.22	3.30
Ethylene glycol	56.0	16.1	−0.92	2.30
Formamide	39.7	17.9	−0.06	0.346

[a]Based on it having a solubility parameter of 14.0 (cal/cm³)$^{1/2}$ and a molar volume of 110 cm³/mol, the stratum corneum having a solubility parameter of 10 (cal/cm³)$^{1/2}$
Source: Ref. 62.

cept to describe the permeation of solutes through the skin. Ruelle et al. (69) have suggested that both the usual solubility parameter and the three-dimensional (hydrogen bond) solubility parameter are often inappropriate to account for hydrogen bonding, as there is an exothermic reaction with the formation of a solute–solvent hydrogen bond. They suggest that stability constants defining interactions between solute and solvent and between themselves may be more appropriate. Ando et al. (70) assumed that dispersion forces applied to a nonpolar lipid pathway, whereas ion–dipole interactions with keratin applied to a polar pathway. The solvatochromic approach introduced to percutaneous absorption by Roberts et al. (71) and Abraham et al. (72) allows dispersion, dipolar, and hydrogen bonding to be included as separate terms. The solubility of a liquid solute can be expressed in terms of the dispersion, dipole, and hydrogen interactions, using a linear free energy approach:

$$-RT \ln X_i + \text{const} = \Delta G_i = \Delta G_i^d + \Delta G_i^p + \Delta G_i^H = A\delta_i^2 \frac{V_i}{100}$$

$$+ B\pi_i^* \pi_I^* + C\alpha_I\beta_i + D\alpha_i\beta_I \qquad (15)$$

where constants A, B, C, and D are determined by regression; V_i is the molar volume of the solute; $\delta_i^2 V_i/100$ is the energy associated with creating a cavity for a solute molecule of molar volume V_i^*, π_I^*, and π_i^* are the dipole solvatochromic parameters for the solvent and solute, respectively; and α_i, α_i, β_I, and β_i are the solvatochromic parameters for hydrogen-bonding–donating ability of solvent and solute and hydrogen-bonding–accepting ability of the solvent and solute, respectively. Hence, for a solid solute, applying Eqs. (5), (10), and (15):

$$\ln X_i = \text{const} + A\delta_i^2 \frac{V_2}{100} + B\pi_i^* \pi_I^* + C\beta_2 + D\alpha_2 + 0.0099(mp - 25) \qquad (16)$$

Yalkowsky et al. (73) applied Eq. (16) in the estimation of the solubility of 185 solutes in water and obtained ($r^2 = 0.977$):

$$\ln X_w = 0.86 - 0.062V_i + 4.9\beta_i - 0.0099(mp - 25) \qquad (17)$$

According to Eq. (5), $\ln K_{sc-v}$ is simply defined by $\ln S_{sc} - \ln S_v$ and, hence, is of the same form as given in Eq. (16). Abraham et al. (74) showed that for 613 solutes the solvatochromic regression ($r = 0.9974$) for $\log K_{oct}$ was

$$\log K_{oct} = 0.088 + 0.562R - 11.054\pi^H + 0.034\alpha - 3.460\beta + 3.814V_x \qquad (18)$$

where R is an excess molar refraction, π^H the solute dipolarity/polarizability, α and β the effective solute hydrogen bond acidity and basicity, and V_x the characteristic volume of McGowan. It is apparent that $\log K_{oct}$ is dominated by solute hydrogen basicity favoring distribution into water, and solute size favoring distribution into octanol. Yalkowsky et al. (73) argue that the octanol–water partition coefficient method (discussed in the next section) is superior for the estimation of water solubility, as it is two orders of magnitude larger and achieves the same fit with fewer variables.

3. Prediction from Octanol–Water Partition Coefficients

The use of the logarithm of octanol–water partition coefficients ($\log K_{oct}$) for the prediction of biological activity through structure–activity relations originates from the work of Hansch (75). Log K_{oct} has also been used in the evaluation of SC–water partition coefficients (76). Roberts et al. (77) reported the following regression for 45 solutes ($r^2 = 0.839$):

$$\log K_{sc-v} = 0.57 \log K_{oct} - 0.1 \tag{19}$$

In theory, an estimate of the solubility of solutes in SC S_{sc} should be possible by using Eq. (19) in an extended and rearranged form of Eq. (7), providing the solubility of solutes in water X_w can be defined:

$$\log X_{sc} = \log K_{sc-w} + \log X_w = 0.57 \log K_{oct} - 0.1 + \log X_w \tag{20}$$

Yalkowsky and Valvani (60) have suggested that the solubility in water X_w can be estimated using Eqs. (6) and (7) when the activity coefficient for electrolytes in water γ_w can be estimated:

$$\ln X_w = \ln X_i^0 - \ln \gamma_w = -\frac{\Delta H_f}{R}\left[\frac{1}{T} - \frac{1}{T_m}\right] - \ln \gamma_w \tag{21}$$

Rearranging Eq. (7):

$$\log \gamma_w = \log K_{oct} - \log \gamma_{oct} \tag{22}$$

Yalkowsky and Valvani (60) followed an approach similar to the treatment of regular solutions by Hildebrand and Scott (59) to deduce that, as the adhesive interactions between octanol and a solute approximately equals the sum of the cohesive interactions in octanol and in the solute, $\gamma_{oct} \sim 1$ for most solutes. Accordingly,

$$\log \gamma_w \approx \log K_{oct} \tag{23}$$

On substitution into Eq. (21)

$$\log X_w = \log X_i^0 - \log \gamma_w = -\frac{\Delta H_f}{2.303R}\left[\frac{1}{T} - \frac{1}{T_m}\right] - \log K_{oct} \tag{24}$$

Further, by combining Eq. (10) and (24) and assuming rigid solutes, Yalkowsky and Valvani (60) obtained the following regression for 155 solutes ($r^2 = 0.979$):

$$\log S_w = -1.05 \log K_{oct} - 0.0012MP + 0.87 \tag{25}$$

Hence, an expression for the solubility of solutes in SC (X_{sc}) is

$$\log X_{sc} = \log K_{sc-w} + \log X_w = -0.48 \log K_{oct} - 0.0012MP + 0.77 \tag{26}$$

Roberts et al. (24) also observed a decrease in the estimated maximum flux of phenols with octanol water partition coefficient and suggested that it may reflect that the more polar phenols had lower molar volumes. Importantly, Eq. (26) suggests that highest solubilities in the SC will be seen for the lowest-melting point solutes.

4. Use of Group Contributions to Estimate SC–Water Partition Coefficients and Make Deductions about the Naute of the Barrier

Group contributions are now widely used in the estimation of octanol–water partition coefficients, the approach attributed to Hansch and co-workers (75). Recognizing that the free energy for transfer of a solute from water to SC (ΔG_{sc-w}) is related to its SC–water partition coefficient (K_{sc-w}), Scheuplein and Blank (5) assumed that

(ΔG_{sc-w}) was expressed additively by the individual groups in a solute and applied it to their series of alcohols. Written in terms of transfer from water to SC (5):

$$\Delta G_{sc-w} = \Delta G_{sc-w}^{OH} + n\Delta G_{sc-w}^{-CH_2} = -RT \ln K_{sc-w} \qquad (27)$$

They reported a $\Delta G_{sc-w}^{-CH_2}$ of -460 cal mol^{-1} at 25°C, and noted that it was, in absolute terms, lower than for the transfer of —CH$_2$ into olive oil (-740 cal mol^{-1}). Roberts et al. (78) reported group contributions for water to SC and water to octanol as follows (in cal mol^{-1} at 25°C) —OH (670, 710), —CH$_2$ (-410, -710), —Cl (810, 1210), —Br (960, 1550), —NO$_2$ (510, 740), and —COOCH$_3$ (240, 680). Although the hydroxyl group is similar for the transfer from water to octanol, the nonpolar contributions are almost half. Roberts et al. (78), in further recognizing the essential temperature independence of K_{sc} (temperature range 12.6–34.5°C), suggested that the partition was entropy-driven. In further analyses, they compared the relative enthalpies and entropies of solute transfer from water into the SC and into various crystalline states of lecithin. It was concluded that the hydrophobic phase of the SC, with which the solutes are most associated, might be considered to be in a liquid crystalline state or more polar. Anderson's group (79) reported that, at 37°C, the partition coefficients into untreated and delipidized SC were similar. They reported a similar free-energy group contribution for the transfer of the —OH and —COOCH$_3$ group from water to the protein domain of the SC (580, 160 cal mol^{-1} at 25°C). New free-energy group transfers were (in cal mol^{-1} at 25°C): —CONH$_2$ (660); —CON(CH$_3$)$_2$ (160); and —COOH (30). They also reported free-energy group transfers for —CH$_2$ and —COOCH$_3$ from water into octanol similar to Roberts et al. (78), but a somewhat higher value for the —OH group (2350 cal mol^{-1}). Davis et al. (58) report a range of literature values for —OH substituents on an aromatic ring that are all less than 870 cal mol^{-1} (mean 610 calmol^{-1}) for transfer from water into octanol and are comparable with those reported for transfer from water to SC. The SC–water partition coefficients are equilibrium values, which will also occur throughout the SC during the permeation process. The differences in the group contributions for a SC protein domain and transport barrier is consistent with the partitioning involving binding to protein sites, the exact site of which remains ill-defined and may include keratin within the cells and desmosomes in the intercellular space. Interestingly, the solubility parameter for a keratin fragment [11.8 (cal/cm^3)$^{0.5}$] (68) is similar to that for butanol [11.18 (cal/cm^3)$^{0.5}$], advocated by Roberts et al. (24,78) as being of a polarity similar to the apparent partition coefficient domain of the SC. It should be emphasized that group contributions for partition coefficients differ from those for permeation, which are considered later.

A model of partitioning into the protein domains in the intercellular region and into keratinocytes during the permeation process is illustrated in Figure 9. It is apparent that the major effect of partitioning is to act as a buffer in the transport process and, as we will see later, this buffering will decrease the diffusivity and increase the lag time of solutes traversing the SC. As a consequence, the apparent permeability coefficient for a solute through the SC can be shown to depend on both the affinity of solutes for lipids in the diffusion pathway and on binding to other sites [see Eqs. (36) and (37)]. In summary, SC–water partition coefficients are useful in helping define the fraction of unbound solute and its ability to diffuse down the SC intercellular lipid pathway [see Fig. 9; discussed later in Eq. (32)].

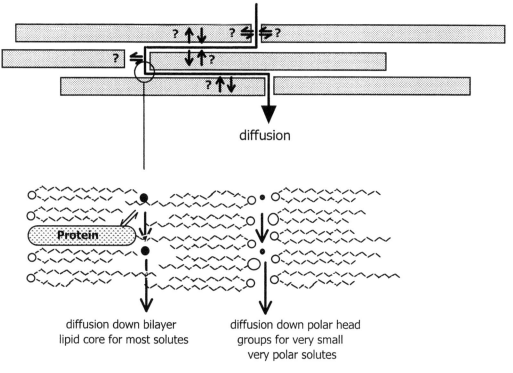

diffusion

diffusion down bilayer
lipid core for most solutes

diffusion down polar head
groups for very small
very polar solutes

Figure 9 Partition and diffusive processes in solute transport through the stratum corneum, assessing an intercellular lipid pathway.

B. Steady-State Flux

As derived in Chapter 3 and presented previously (3), the steady-state solute flux through the SC per unit area of application J_{sc}^{ss} is determined by the concentrations of solute immediately below the outside $C_{sc(o)}$ and inside $C_{sc(i)}$ the SC:

$$J_{sc}^{ss} = \frac{D_{sc}}{h_{sc}} (C_{sc(o)} - C_{sc(i)}) \qquad (28)$$

where D_{sc} is the effective diffusion coefficient in the SC, with a diffusion path length h_{sc}. It should be noted that D_{sc} is defined both by diffusion of free solute and the instantaneous partitioning into immobile sites in the diffusion path. Chandrasekaran et al. (80,81) proposed a dual sorption model using this concept to describe the uptake of drugs by skin. A simple form of their expression can be derived ignoring the differences in the apparent volumes for partitioning and diffusion as well as the rates of distribution [Later work will show that these assumptions are questionable with significance for certain situations (82)]. Under linear binding and "instantaneous" equilibrium conditions, the unbound concentration of solute in SC lipids ($C_{sc,u}$) is related to that partitioned and bound to other SC components ($C_{sc,b}$) by a partition coefficient ($K_{b/u}^{sc}$); that is, $K_{b/u}^{sc} = C_{sc,b}/C_{sc,u}$. If only unbound solute diffuses, the change in unbound concentration in the SC $\partial C_{sc,u}/\partial t$ is given by:

$$\frac{\partial C_{sc,u}}{\partial t} = D_{sc,u} \frac{\partial^2 C_{sc,u}}{\partial x^2} - \frac{\partial C_{sc,b}}{\partial t} = D_{sc,u} \frac{\partial^2 C_{sc,u}}{\partial x^2} - K_{b/u}^{sc} \frac{\partial C_{sc,u}}{\partial t} \tag{29}$$

Rearranging

$$\frac{\partial C_{sc,total}}{\partial t} = (1 + K_{b/u}^{sc}) \frac{\partial C_{sc,u}}{\partial t} = D_{sc,u} \frac{\partial^2 C_{sc,u}}{\partial x^2} \tag{30}$$

since $C_{sc,total} = C_{sc,u} + C_{sc,b}$.

Further recognizing that total concentration C_{sc} in the SC is related to $f_{u,sc}$ the fraction unbound in the SC by $C_{u,sc} = f_{u,sc}C_{sc}$ and that $f_{u,sc} = 1/(1 + K_{b/u}^{sc})$, Eq. (30) can be expressed as

$$\frac{\partial C_{sc,u}}{\partial t} = f_{u,sc}D_{sc,u} \frac{\partial^2 C_{sc,u}}{\partial x^2} = D_{sc} \frac{\partial^2 C_{sc,u}}{\partial x^2} \tag{31}$$

since $D_{sc} = f_{u,sc}D_{sc,u}$. Hence, the measured D_{sc} will reflect a reduction in the free-diffusion coefficient as a result of binding to immobile components in the diffusion path. If we assume that the unbound concentration is in the internal environment of the intercellular lipid bilayer, then

$$f_{u,sc} = \frac{C_{u,sc}}{C_{sc}} = \frac{C_{u,sc}}{C_w} \frac{C_w}{C_{sc}} = \frac{K_{lipid-w}}{K_{sc-w}} = K_{lipid-sc} \tag{32}$$

We now include the SC lipid pathway–vehicle partition coefficient $K_{lipid-v}^{sc}$, following the earlier assumption that the solute in this bilayer is unbound and that the lipid bilayer is the diffusion path:

$$K_{lipid-v}^{sc} = \frac{C_{lipid}^{sc}}{C_v} \tag{33}$$

Substituting Eq. (33) into Eq. (28), recognizing that the transport pathway is the lipid bilayer, therefore, accounting for partitioning differences, yields

$$J_{sc}^{ss} = \frac{K_{lipid-v}^{sc}D_{sc}}{h_{sc}} \left(C_v - \frac{K_{lipid-ve}^{sc}}{K_{lipid-v}^{sc}} C_{ve}^{ss} \right) = k_p^{sc} \left(C_v - \frac{K_{lipid-ve}^{sc}}{K_{lipid-v}^{sc}} C_{ve}^{ss} \right)$$

$$= k_p^{sc} \left(C_v - \frac{1}{K_{ve-v}} C_{ve}^{ss} \right) \tag{34}$$

where C_{ve}^{ss} is the concentration of the solute in the viable epidermis, and $K_{lipid-ve}^{sc}$ is the SC lipid–viable epidermis partition coefficient, and K_{ve-v} is the viable epidermis–vehicle partition coefficient. Similar expressions could be derived for flux through the viable epidermis with the dermis as an adjacent phase as shown in Figure 7. An equation similar in form to Eq. (34) has also been used to describe the flux through SC into the receptor phase of an in vitro penetration study (83). As stated earlier, transport through the SC may occur through various pathways. In earlier work (3,83), we have recognized that k_p^{sc} is a composite parameter and may be more properly expressed as, for instance,

$$k_p^{sc} = k_{p,lipid}^{sc} + k_{p,polar}^{sc} + k_{p,appendages}^{sc} \tag{35}$$

where $k_{p,lipid}^{sc}$, $k_{p,polar}^{sc}$, and $k_{p,appendageal}^{sc}$ are the component SC permeability coefficients for lipid, polar, and appendageal pathways. Kasting et al. (84) have discussed the relative magnitudes of $k^{sc}_{p,lipid}$ and $k^{sc}_{p,polar}$. They suggest that, whereas $k^{sc}_{p,lipid}$ may vary from 0.3×10^{-5} to $13,000 \times 10^{-5}$ cm/h, $k^{sc}_{p,polar}$ varies from 0.1×10^{-5} to 1×10^{-5} cm/h. Thus, $k^{sc}_{p,polar}$ becomes important only for very low permeability solutes. The appendageal component is added to Eq. (35) for completeness, given the later discussion on potential transport through this pathway. For most solutes, penetration appears to occur through the lipid pathway, with a permeability coefficient $k^{sc}_{p,lipid}$. If we assume that diffusion occurs through the lipid pathway and, as stated earlier, that the distribution volumes are the same, then the partition coefficient of relevance will appear to be that from the vehicle into this pathway ($K_{lipid-v}$) and not K_{sc-v} [Eq. (36)]

$$k^{sc}_{p,lipid} = K^{sc}_{lipid-v} \frac{D_{sc}}{h_{sc}} = \frac{K^{sc}_{lipid-v} f_{u,sc} D_{sc,u}}{h_{sc}} = \frac{K^{sc}_{lipid-v} K^{sc}_{lipid-w} D_{sc,u}}{K_{sc-w} h_{sc}} \tag{36}$$

so that when the vehicle is water, then D_{sc} is an apparent value when the partitioning is assumed to be determined by K_{sc-v}; that is,

$$k^{sc}_p = K_{sc-w} \left(\frac{K^{sc}_{lipid-w}}{K_{sc-w}} \right)^2 \frac{D_{sc,u}}{h_{sc}} = K_{sc-w} \frac{D^{app}_{sc}}{h_{sc}} \tag{37}$$

where the apparent diffusion coefficient (D^a_{sc}) is given by

$$D^{app}_{sc} = \left(\frac{K^{sc}_{lipid-w}}{K_{sc-w}} \right)^2 D_{sc,u} \tag{38}$$

It should be reemphasized that this analysis is based on the binding sites being present in the diffusion pathway. In reality, the distribution volumes for lipids and other binding sites differ, and more complex expressions are appropriate (82). Nevertheless, these expressions show that both the SC–vehicle partitioning and diffusion down an exclusively lipid pathway affect the observed permeability coefficients. The diffusion coefficient $D_{sc,u} = k_B T / 6 \pi \eta r$ where k_B is the Boltzmann's constant, T is the temperature, η is the viscosity of the pathway, and r is the radius of the diffusing solute.

Several other heterogeneous skin permeability models have been described. Albery and Hadgraft (29) assumed that impermeable corneocytes were embedded in a permeable, homogeneous lipid phase. In Tojo's model (85), both the lipid and corneocytes phases are permeable, but with a partition coefficient between them. Heisig et al. (86) have suggested that the heterogeneity of the SC precludes an analytical solution. He used a "brick-and-mortar" model of the SC (ten layers with corneocytes 30-μm wide and 1-μm thick and a lipid channel of 0.1 μm, as shown in Fig. 5) and concluded that the long lag times and very small human SC permeabilities can be predicted only for a highly staggered corneocyte geometry and only when the corneocytes are 1000 times less permeable than the lipid phase. Plewig and Marples (87) observed that SC was 15–20 layers of flat cells that are thin squames with a thickness of approximately 0.5 μm and a width of 30–40 μm.

A key issue is then "what is the polarity of the lipid bilayer environment" (see Figs. 5 and 6) in which the solute is diffusing. Anderson and Raykar (88) observed similar group contributions for polar, hydrogen-bonding substituents from permeability and octanol–water data and suggested, as a consequence, that "the barrier microenvironment resembles that of a hydrogen-bonding solvent." It is important to recognize that this barrier phase is more lipophilic than defined by K_{sc-v}, which defines an average polarity of the lipid and "proteinaceous" phases involved in partitioning.

The diagrammatic representation of the steady-state concentration–distance profile in each skin region consistent with this model is shown in Figure 7A. It is evident that there is a favorable distribution of the solute into the SC from the vehicle, as defined by K_{sc-v} being more than 1. Furthermore, the steady-state concentration of solute in each phase declines with distance in accordance with the concentration difference between the boundaries of the phase. It is also evident that the concentration of solute at the inner face of the SC ($C_{sc(i)}$) is not 0, arising from a significant viable epidermal resistance or poor perfusion. When there is no resistance and the viable epidermal phase approaches a perfect sink [$C_{sc(i)} = 0$], as represented in Figure 10A, Eq. (34) reduces to

$$J_{sc}^{ss} = \frac{K_{sc-v}D_{sc}^a C_v}{h_{sc}} = k_p^{sc}C_v \qquad (39)$$

1. Steady-State Function According to Convention 1

The model described here is also based on the assumption that the viable epidermis approaches a perfect sink (see Fig. 10A). Noting $a_v = \gamma_v C_v$, and applying convention 1 for standard state, Eq. (7) could be substituted into Eq. (39) to give the widely quoted expression derived by Higuchi (18):

$$J_{sc}^{ss} = \frac{K_{sc-v}D_{sc}^a C_v}{h_{ss}} = \frac{D_{sc}^a \gamma_v C_v}{h_{sc}\gamma_{sc}} = \frac{D_{sc}^a a_v}{h_{sc}\gamma_{sc}} \qquad (40)$$

Equation (40) has commonly been quoted as the basis for an identical flux of a solute from different saturated solutions through membranes. This basis is readily seen for saturated systems by using convention 1 and equilibrium conditions, in that if the vehicle is in equilibrium with pure solute then the activity of the solute in the vehicle is the same as that for the pure solute. It is apparent, however, that an interaction between the vehicle and the membrane could affect either the solubility in the membrane (defined by γ_{sc} in convention 1) or D_{sc}. Higuchi (18) pointed out that solutes with low-activity coefficients had low-escaping tendencies from the vehicle, and thus, low rates of penetration of solutes through the skin. He reported the following limiting activity coefficients of the nerve gas sarin in various solvents: water 14, diethylene glycol 2.4, isoamyl alcohol 1.07, and benzyl alcohol 0.446. Higuchi also recognized that phenol is less toxic when dissolved in vehicles in which it has a high affinity (also expressed as: low activity or high solubility).

2. Steady-State Flux According to Convention 2

Roberts and Anderson (89) suggested that the expression for J_{sc}^{ss} in terms of activities is given by

$$J_{sc}^{ss} = \frac{D_{sc}^a K_{sc-v}^a a_v^*}{h_{sc}} \qquad (41)$$

The maximum flux is given by Eq. (42) if ($\gamma_{sc}^* \cong \gamma_v^* \cong 1$) and saturated solutions are used (i.e., $a_v^* = \gamma_v^* S_v$ and $K_{sc-v} = S_{sc}/S_v$ (Eq. (3) and Eq. (5))):

$$J_{sc}^{ss} = \frac{D_{sc}^a S_{sc}}{h_{sc}} = \frac{D_{sc}^a K_{sc-v} S_v}{h_{sc}} \qquad (42)$$

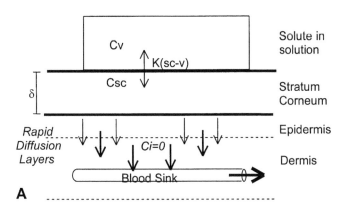

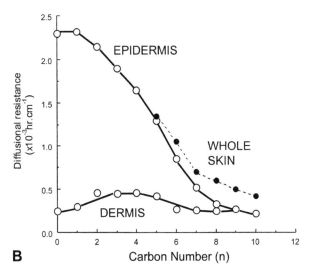

Figure 10 (A) Sink conditions for flux of solute through the stratum corneum; (B) comparison of diffusional resistance of dermis and epidermis for aqueous solutions of alcohols. (From Ref. 5.)

Where K_{sc-v} is the SC–vehicle partition coefficient, S_{sc} is the solubility of the solute in the SC, and S_v is the solubility of the solute in the vehicle.

3. Steady-State Flux Through SC Under Nonsink Conditions

The steady-state flux of solute through the SC (J_{sc}^{ss}) is defined by Eq. (39) when $C_{sc(i)} = 0$. When there is a significant concentration at the viable epidermis interface (Figures 7 and 10A), a lower flux (given by Eq. (34)) is observed. Equation (34) can also be rearranged to obtain an equation similar in form to that given in Eq. (39) (90):

$$J_{skin}^{ss} = k_p'' C_v \qquad (43)$$

where k_p'' is the effective permeability constant for the system, as defined not only by the usual SC permeability coefficient k_p^{sc}, but also, by the viable epidermis permeability coefficient k_p^{ve}, the permeability coefficient associated with the transfer from the vehicle to the SC (not shown here), and an effective removal permeability coefficient k_p^r, reflecting blood perfusion in vivo or receptor sampling in vitro and deeper tissue penetration:

$$\frac{1}{k_p''} = \frac{1}{k_p^{sc}} + \frac{1}{k_p^{ve}} + \frac{1}{k_p^r} \tag{44}$$

Given that resistance to transport through a phase is defined by the reciprocal of the permeability coefficient through that phase, Eq. (44) is simply stating that the total resistance of the skin is given by the sum of its individual component resistances, a similar expression is shown in Eq. (45) [Eqs. (49)–(53) in Chapter 3 express this concept in mathematical terms]. Scheuplein (91) showed that the SC was the major resistance to the skin permeation of water using such an expression for the overall resistance of the skin R_{skin} [see Eq. (51) in Chap. 3].

$$R_{skin} = R_{sc} + R_{ve} + R_D = 9.1 \times 10^6 + 6.3 \times 10^3$$
$$+ 6.3 \times 10^3 = 9.1 \times 10^6 \text{ s cm}^{-1} \tag{45}$$

where R_{sc} is the resistance of the SC, R_{ve} the resistance of viable epidermis, and R_D the resistance of dermis. As shown in Figure 10B, the dermal resistance contributes significantly to the overall skin resistance for the longer-chain alcohols.

C. Solute Structure–Transport Relations

1. Aqueous Solutions

Most solute structure–transport studies have used permeability coefficients of solutes determined using excised human epidermis or animal skin. Various solute structure–epidermal permeability relations have been reported over nearly three decades (Table 4). The logarithm of octanol–water partition coefficient log K_{oct} is often used to define solute structure–transport relations as it is a relevant physicochemical property, which can be readily determined experimentally. log K_{oct} can also be estimated by a fragment addition approach. A reduced form of Eq. (44) expresses the relative importance of the polar and lipid pathways in SC penetration, together with the resistance of the aqueous diffusion layer:

$$k_p'' = \left[\frac{1}{k_{p,lipid}^{sc} + k_{p,polar}^{sc}} + \frac{1}{k_p^{aq}} \right]^{-1} \tag{46}$$

In early work, such as that for the phenolic compounds (24), regressions between log $k_{p,sc}$ and log K_{oct} were made, recognizing the aqueous boundary layer defined in Eq. (46), but not adequately recognizing the potential high dependence on solute size (Fig. 11A). This work is predicated on a linear free energy relation between log $K_{sc-water}$ and log $K_{solvent-water}$, as illustrated by the linear relation between log K_{sc-w} and log K_{oct} given in Eq. (19) (see Fig. 11B). If D_{sc}^a/h_{sc} is relatively constant for the series of solutes chosen, then according to Eq. (39), log k_p^{sc} should be directly related to log K_{oct} when there is no evidence of either a polar pathway or another barrier in series.

As discussed earlier, there is evidence for a polar pathway with a defined radius for transport for very polar solutes. Transport through this pathway is expected to

Table 4 Historical Development of Skin Transport–Structure Relations

Approach	Relationship	Ref.
1970s: Simplistic	k_p related to alcohol carbon chain length and $K_{amyl\ caproate}$	Scheuplein (5,21,157)
	k_p phenylboronic acids, alcohols, and steroids related to $K_{octanol}$, MW, and molar refraction	Lien and Tong (295)
	Flux related to $K_{mineral\ oil-water}$	Michaels et al. (8)
	k_p and K_{sc-v} phenols, aromatic/aliphatic alcohols, and steroids related to $K_{octanol}$, MW, and H bond number	Roberts group (23,24,76,78)
1980s	Maximum flux related to dipole moment, assuming polar and nonpolar pathways	Ando et al. (70)
	Maximum flux related to MW and melting point	Kasting et al. (92)
	Dermal clearance rate related to $K_{octanol}$	Siddiqui et al. (44)
	k_p related to partition coefficients	Various including (88,296)
1990s: Selected multivariate SAR of large k_p datasets $K_{octanol}$ and MW	$\log k_p = -2.74 + 0.71 \log K_{oct} - 0.0061 MW$, $r^2 = 0.67$, $n = 93$	Potts and Guy (94)
Fragmental analysis	$\log k_p = -2.76 + 0.24\,(C^*) - 0.47$ (aromatic rings) $+ 0.46$ (halide) $- 1.27$ (amine) $- 0.64$ (nonaromatic) $- 1.24$ (steroid) $- 0.47$ (OH) $- 0.325$ (O) $- 0.36$ (amide), $r^2 = 0.68$, $n = 90$	Pugh and Hadgraft (297)
MW and H bonds	$\log k_p = -2.170 + 0.07(\log P_{oct})^2 + 0.835 \log K_{oct} - 0.265 H_n + 1.844 \log MW$, $r^2 = 0.956$, $n = ????$	Lien and Gao (298)
Solvatochromic or similar approaches	$\log k_p = -1.29 - 1.72\Sigma\alpha_2 - 3.93\Sigma\beta_2 + 0.026V_x$, $r^2 = 0.94$, $n = 37$	Potts and Guy (299)
	$\log k_p = -0.51 - 0.59\pi_2 - 0.63\Sigma\alpha_2 - 3.42\Sigma\beta_2 + 1.8V_x$, $r^2 = 0.96$, $n = 46$	Roberts et al. (71)
	$\log k_p = -5.24 + 0.44R2 - 0.41\pi_2 - 1.63\Sigma\alpha_2 - 3.28\beta_2 + 2.01V_x$, $r^2 = 0.96$, $n = 47$	Abraham (300,301)

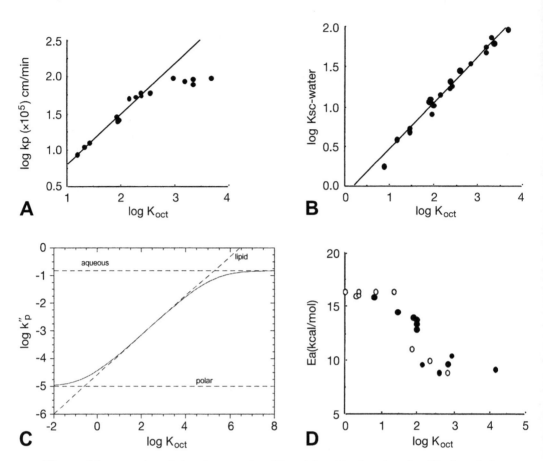

Figure 11 (A) The relation between log k_p^{sc} and log K_{oct} for phenols; (B) the relation between log $K_{sc-water}$ and log K_{oct} for phenols; (C) theoretical prediction of overall log k_p'' versus log K_{oct} for a solute with *MW* of 300 and the contribution of various diffusion pathways to this transport; (D) the relation between activation energies (E_a) for phenols (●) and alcohols (○). (From Refs. 78[A]; 24[B].)

be characterized by a $k_{p,polar}^{sc}$ [upper limit 0.15 cm/h (84)] independent of log K_{oct}. Kasting et al. (84) also refer to an aqueous boundary layer with $k_p^{aq} = 1 \times 10^{-5}$ cm/h. Figure 11C shows a plot of the predicted log permeability coefficient versus log octanol–water, based on these considerations. It is apparent that a sigmoidal curve results, reflecting transport by the polar pathway at low log K_{oct}, a linear portion reflecting log $k_{p,lipid}^{sc}$ versus log K_{oct}, and a limiting aqueous boundary layer permeability at high log K_{oct}. This curvature is consistent with the phenols' results shown in Figure 11A. Roberts et al. (23) interpreted the sigmoidal decrease in the activation energy for permeation for both alcohols and phenols with increasing log K_{oct} (see Fig. 11D) as additional evidence of the aqueous boundary layer effect they used to explain the curvature of the plot of log k_p^{sc} versus log K_{oct} at high log K_{oct} values (see Fig. 11A). Kasting et al. (92) suggested that log D_{sc} may be related to solute molecular weight by Eq. (47):

$$D_{lipid} = D_o(MW)^{-b} \tag{47}$$

in which b should not be 1/3 to 1/2 as assumed for liquid diffusion by Scheuplein and Blank (5), but >3 consistent with diffusion in polymer membranes and lipid bilayers. Anderson and Raykar (88) reported that for the combined sets of methyl-substituted phenols and 21 esters of hydrocortisone, the following relation was found: $\log k_p = 0.83 \log K_{oct} - 4.4 \log MW + 6.4$.

Kasting et al. (92) found that their data sets could be equally well described by a free volume model, in which MV is the molecular volume of the solute and β is a constant:

$$D_{lipid} = D_o \exp(-\beta MV) \tag{48}$$

The importance of molecular size as determinant of SC permeability coefficients has also been recognized by Flynn (93). An extended plot of this data is shown in Figure 12. It is apparent that the study temperature is also a key determinant of k_p. The line divides the data into two. Some anomalies are readily apparent. First, most of the high-permeability solutes above the line were studied at less than 30°C and those below it at greater than 30°C. However, those above the line are low MW and have low hydrogen-bonding capabilities (i.e., one or zero groups), whereas those below the line have a $MW > 300$ and three or more hydrogen-bonding groups. The solutes with two hydrogen-bonding groups are equally distributed.

Potts and Guy (94) suggested that the sigmoidal relations could be linearized when the proper dependence of diffusivity on solute molecular size, as described by the free volume model (see Eq. (48)], is derived. The combination of octanol–water

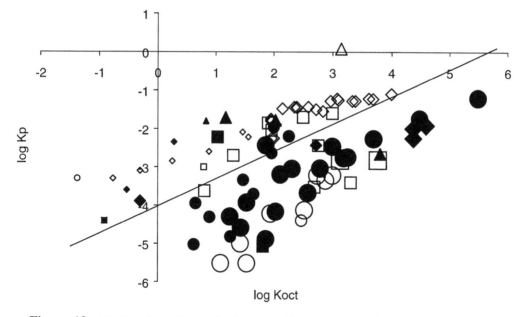

Figure 12 Confounding effects of solute size (datapoints approximately proportional to molecular weight), number of solute hydrogen-bonding groups (where △▲ = 0, ◇◆ = 1, □■ = 2, and ○● = 3) and study temperature (filled datapoints [▲◆■●] $t \geq 30°C$ and open datapoints [△◇□○] $t < 30°C$). The arbitrary line divides the dataset in two. (From Ref. 93.)

partition coefficients and solute size as determinants in skin transport is the basis for perhaps the most widely quoted relation in this area—that of Potts and Guy (94). They related log k_p^{sc} (cm/h) to log K_{oct} and molecular weight MW (a likely determinant of D_{sc}) for 97 solutes and obtained the relation:

$$\log k_p^{sc} = 0.71 \log K_{oct} - 0.0061MW - 2.74 \qquad r^2 = 0.69 \qquad (49)$$

If the slope of log K_{oct} reflects the partitioning from water into SC lipid and that for MW reflects the size effect in diffusion, the slope of 0.71 suggests that the polarity in the intercellular diffusion pathway is less than octanol, but greater than that observed with partitioning [see Eq. (19)]. An alternative analysis is to express k_p^{sc} in terms of groups of fragments from which solutes can be constructed. This analysis recognizes that some different fragments, such as noncarbonyl carbon atoms $C^\#$ may promote absorption by enhancing partitioning, whereas others, such as amine, hydroxyl, amide, and "O" groups slow diffusion by hydrogen bonding effects. Other groups such as aromatic rings, nonaromatic rings, and steroids probably slow diffusion by steric effects. Pugh et al. (95) reported the following relation for Flynn's dataset ($n = 97$):

$$\log k_p^{sc} = -2.76 + 0.241 \ (C^\#) - 0.470 \ (\text{aromatic rings}) + 0.460 \ (\text{halide})$$
$$- 1.27 \ (\text{amine}) - 0.644 \ (\text{nonaromatic rings}) - 1.24 \ (\text{steroid})$$
$$- 0.477 \ (\text{OH}) - 0.325 \ (\text{"O"}) - 0.356 \ (\text{amide}) \qquad r^2 = 0.68 \qquad (50)$$

By using an approach similar to that described earlier for SC–water partition coefficients [see Eq. (27)], group contributions for the free energy for functional group transfer during permeation can be estimated (88). Relatively high free energies for permeation are required for polar groups (—CONH$_2$, 3.05 kcal/mol; —COOCH$_3$, 1.25 kcal/mol; —COOH, 1.95 kcal/mol; —OH, 2.45 kcal/mol). The results of some other sophisticated model-based structure–transport analyses are given in Table 4. Barratt (96) used the Flynn dataset (93) and classified them into steroids (A), other active molecules (B), and (C) the remainder. He reported a high correlation for log k_p^{sc} with log K_{oct}, MP, and MV for steroids and small molecules, but not for the others. Barratt also found that 90% of the variability was explained by the relation ($n = 60$, group C plus "nonhydrocortisone group A"; $r^2 = 0.904$):

$$\log k_p^{sc} = -0.00933MV + 0.82 \log K_{oct} - 0.00387MP - 2.355 \qquad (51)$$

The 1.5 order of magnitude higher value for the 12 hydrocortisone permeability coefficients excluded from the analysis probably reflects that their permeability coefficients were measured at 37°C, whereas phenols, alcohols, and some other solutes were measured at 25°C. This emphasizes the need to ensure that comparable experimental conditions have been employed in aggregating data from different sources. Table 5 shows a listing of the data used in the analyses of Barratt and in work by our group.

Solute structure–transport relations have also been studied by representing the transport process in terms of models, such as parallel pathways of transport, and multiple phases in series (71). Liron and Cohen (97) used regular solution theory (discussed in Fig. 8) to show that the porcine skin permeability coefficient of pure unbranched alkanoic acids (C2–C7) reached a maximum in the solubility parameter range of 9.7–10 cal$^{1/2}$ cm$^{3/2}$. Groning and Braun (68), using O-acylglucosylceramide as a model for the intercellular lipoid matrix, showed that the steady-state flux of three groups of solutes could be related to their three-dimensional solubility parameter differences with the solubility parameter of that intracellular lipid ($[\delta_{solute} - \delta_{O\text{-acylglucosylceramide}}]^2$) (Fig. 13).

An alternative process is to recognize that transport is a series of steps, the first being partitioning into the SC and the second diffusion through the SC. Therefore, it is desirable to separate out the partition and diffusion components of k_p^{sc} to better understand these determinants of epidermal transport. The aqueous SC–water partition coefficient K_{sc-v} was related to the octanol–water partition coefficient K_{oct} for 45 solutes by Eq. (19). As discussed previously, with Eq. (49), there may be a need to assume a higher slope consistent with the more lipophilic transport pathway than the partitioning environment. By using Eq. (19) in an attempt to remove the influence of partitioning, Pugh et al. (98) showed for polyfunctional solutes ($N = 53$) that

$$\log \frac{D_{sc}^a}{h} = -1.50 - 0.91\alpha - 1.58\beta - 0.003MW \qquad r^2 = 0.94 \qquad (52)$$

where α and β are the H-bond donor ability (acidity) and the H-bond acceptor ability (basicity) of the solute. Hence, diffusivity of a solute in the SC is both a function of the hydrogen bonding of a solute and its size. The SC barrier was shown to be a predominantly H-bond donor rather than acceptor with $\alpha_{sc}:\beta_{sc} = 0.6:0.4$. Also, $\log D_{sc}$ is related to the number of hydrogen-bonding groups on a molecule in a nonlinear manner and is suggestive of an adsorption isotherm, being maximal for small, non–hydrogen-bonding molecules and reaching a low minimum with about four hydrogen-bonding groups (98). Comparing Eqs. (49) and (50), it is interesting to speculate whether the melting point term found empirically in Eq. (49) is, in fact, a proxy measure of hydrogen bonding, because hydrogen bond donor ability, and other intermolecular forces and molecular symmetry, are predictors of melting point (99).

2. Other Vehicles

The number of studies on structure–transport relations of solutes from other vehicles is more limited. One of the earliest studies reported is that of Blank (100) in which it was shown that whereas the permeability coefficient of alcohols through human skin from saline increased with the number of carbon atoms, the permeability coefficients from nonaqueous vehicles decreased (Fig. 14A). Similar results have been shown for phenolic solutes (see Figs. 11A and 14B). Roberts (101) attempted to predict the observed relations using data for the epidermal permeability from aqueous solutions and the estimated permeability coefficients. Arachis oil–solvent partition coefficients were measured for a number of phenolic compounds using water–ethanol combinations as solvents (see Fig. 14C). Noting that K_{sc-v} is related to K_{oct} for this series of solutes by Eq. (19) and the partition coefficient between arachis oil and water $K_{oil-water}$ for these solutes are defined by

$$\log K_{oil-water} = 0.98 \log K_{oct} - 0.81 \qquad (53)$$

the apparent SC–arachis oil partition coefficient can be predicted by a suitable substitution of Eq. (53) into Eq. (19):

Table 5 Flynns' Dataset for Skin Permeability Transport Relations

Chemical	log PC (cm/h)	log P (log K_{oct})	Mpt (°C)	Temp (°C)	Receptor[a,b]	Barratt group	log K_{sc}	mv (Å^3)	log Kb_{hex}	α	β	π^*	RCf	Ref.
Aldosterone	-5.52	1.08	164	26	PEN/STR	A	0.83	313.80	*	0.40	1.9	3.47	4.00	302
Amobarbital	-2.64	1.96	158	32	Buff 7.4	B		204.40						303
Atropine	-5.07	1.81	192	30	Ring 6.5	B		266.10	1.81	*	*	*	*	8
Barbital	-3.95	0.65	192	32	Buff 7.4	B		155.60						303
Benzyl alcohol	-2.22	1.10	25	a	a	C	0.61	89.24	-0.62	0.33	0.5	0.87	1.36	21,5
4-Bromophenol	-1.44	2.59	68	25	DDH$_2$O	C		91.95	-0.20	0.67	0.2	1.17	1.28	24,76
2,3-Butanediol	-4.40	-0.92	25	30	a	C		89.64						21,5
Butanoic acid	-3.00	0.79	25	25	a	C	0.18	81.56	-0.96	0.60	0.45	0.62	1.64	21,5
n-Butanol	-2.60	0.88	25	25	0.9% NaCl	C	0.40	81.95	-0.70	0.37	0.48	0.42	1.38	21,5
Butan-2-one	-2.35	0.28	25	30	a	C		75.20	-0.25	0.00	0.51	0.70	0.93	21,5
Butobarbital	-3.71	1.65	127	32	Buff 7.4	B		188.10						303
4-Chlorocresol	-1.26	3.10	48	25	DDH$_2$O	C		103.90	0.36	0.65	0.22	1.02	1.29	24,76
2-Chlorophenol	-1.48	2.15	25	25	DDH$_2$O	C		87.90						24,76
4-Chlorophenol	-1.44	2.39	45	25	DDH$_2$O	C		88.17	-0.12	0.67	0.2	1.08	1.28	24,76
Chloroxylenol	-1.28	3.39	116	25	DDH$_2$O	C		119.30	1.08	0.64	0.21	0.96	1.26	24,76
Chlorpheniramine	-2.66	3.81	25	30	Ring 6.5	B		246.20	3.39	*	*	*	*	8
Codeine	-4.31	0.89	145	37	CPB 7.4	B		254.30	0.89	*	*	*	*	304
Cortexolone	-4.13	2.52	208	26	PEN/STR	A	1.36	317.00	-1.00	0.35	1.57	3.45	3.36	302
Cortexone	-3.35	2.88	138	26	PEN/STR	A	1.57	309.80	0.48	0.15	1.13	3.39	2.26	302
Corticosterone	-4.22	1.94	183	26	PEN/STR	A	1.23	316.50	-1.62	0.40	1.63	3.43	3.51	302
Cortisone	-5.00	1.42	228	26	PEN/STR	A	0.93	320.80	-0.55	0.35	1.84	3.50	3.38	302
o-Cresol	-1.80	1.95	34	25	DDH$_2$O	C	1.03	88.42	0.25	0.52	0.3	0.86	1.26	24,76
m-Cresol	-1.82	1.96	25	25	DDH$_2$O	C	1.03	88.97	-0.35	0.57	0.34	0.88	1.40	24,76
p-Cresol	-1.75	1.95	34	25	DDH$_2$O	C	1.03	88.87	-0.19	0.57	0.31	0.87	1.34	24,76
n-Decanol	-1.10	4.00	25	a	a	C		178.30	*	0.37	0.48	0.42	1.38	21,5
2,4-Dichlorophenol	-1.22	3.08	60	25	DDH$_2$O	C		102.40						24,76
Diethylcarbamazine	-3.89	-0.31	49	30	Ring 6.5	B		195.50						8
Digitoxin	-4.89	1.86*	240	30	Ring 6.5	B		682.60						8

Compound														Ref
Ephedrine	-2.22	1.03	39	30	Ring 6.5	B		156.00						8
b-Estradiol	-3.52	2.69	179	26	PEN/STR	A	1.66	255.30		0.88	0.95	3.30		302
Estriol	-4.40	2.47	282	26	PEN/STR	A	1.36	262.40		1.40	1.22	3.36		302
Estrone	-2.44	2.76	254	26	PEN/STR	A	1.66	249.50		0.56	0.91	3.10		302
Ethanol	-3.10	-0.31	25	25	0.9% NaCl	C	-0.31	50.72	-2.10	0.37	0.48	0.42	1.38	21,5
2-Ethoxyethanol	-3.60	-0.54	25	30	a	C		90.21	*	0.30	0.83	0.50	1.92	21,5
Ethyl benzene	0.08	3.15	25	25	DDH$_2$O	C		97.78	3.00	0.00	0.15	0.51	0.28	305
Ethyl ether	-1.80	0.83	25	30	a	C		83.31	0.60	0.00	0.45	0.25	0.82	21,5
4-Ethyl phenol	-1.46	2.40	45	25	DDH$_2$O	C		104.60	0.23	0.55	0.36	0.90	1.41	24,76
Etorphine	-2.44	1.86	215	37	Tris 7.4	B		368.20	1.86	*	*	*	*	306
Fentanyl	-2.25	4.37	84	37	CPB 7.4	B		314.70	4.37	*	*	*	*	304.307
Fentanyl (2)	-2.00	4.37	84	30	Ring 6.5	B		314.70	4.37	*	*	*	*	3
Fluocinonide	-2.77	3.19	311	37	Succ 4	B		412.00						$
Heptanoic acid	-1.70	2.50	25	25	a	C	1.78	129.30	0.45	0.60	0.45	0.60	1.64	21,5
n-Heptanol	-1.50	2.72	25	25	0.9% NaCl	C	1.48	130.10	1.01	0.37	0.48	0.42	1.38	21,5
Hexanoic acid	-1.85	1.90	25	25	a	C	1.08	113.80	0.24	0.60	0.45	0.60	1.64	21,5
n-Hexanol	-1.89	2.03	25	25	0.9% NaCl	C	1.00	112.80	0.45	0.37	0.48	0.42	1.38	21,5
Hydrocortisone (2) (HC)	-3.93	1.53	214	32	Buff 7.4	A	0.85	326.00	-2.04	0.70	1.87	3.49	4.36	303
HC Dimethylsuccinamate	-4.17	2.03	223	37	Succ 4	A		437.50						79,309
HC Hemipimelate	-2.75	3.26	112	37	Succ 4	A		449.20						79,309
HC Hemisuccinate	-3.20	2.11	171	37	Succ 4	A		401.80						79,309
HC Hexanoate	-1.75	4.48	152	37	Succ 4	A		424.00						79,309
HC 6-OH-hexanoate	-3.04	2.79	144	37	Succ 4	A	1.58	432.10		—*	—*	—*	*	79,309
HC Octanoate	-1.21	5.49	115	37	Succ 4	A		455.90						79,309
HC Pimelamate	-3.05	2.31	185	37	Succ 4	A		452.70						79,309
HC Propionate	-2.47	3.00	196	37	Succ 4	A		375.80						79,309
HC Succinamate	-4.59	1.43	227	37	Succ 4	A		405.30						79,309
Hydromorphone	-4.82	1.25	267	37	CPB 7.4	B		250.70	1.25	*	*	*	*	304
Hydroxypregnenolone	-3.22	3.00	150	26	PEN/STR	A		317.60						302
17-Hydroxyprogesterone	-3.22	2.74	220	26	PEN/STR	A	1.60	311.60	0.40	0.25	1.31	3.35	2.73	302
Isoquinoline	-1.78	2.03	28	32	NaOH 7.4	C		96.32						303
Me-4-hydroxy benzoate	-2.04	1.96	128	25	PBS 7.4	C		111.90	-0.52	0.69	0.45	1.37	1.76	24,76

Table 5 Continued

Chemical	log PC (cm/h)	log P (log K_{oct})	Mpt (°C)	Temp (°C)	Receptor[a,b]	Barratt group	log K_{sc}	mv (Å³)	log Kb_{hex}	α	β	π*	RCf	Ref.
Meperidine	-2.43	2.72	25	37	CPB 7.4	B		224.60	2.72	*	*	*	*	304
Methanol	-3.30	-0.77	25	25	0.9% NaCl	C	-0.22	33.21	-2.80	0.43	0.47	0.44	1.44	21,5
Methyl HC Succinate	-3.68	2.58	143	37	Succ 4	A		418.00						79,309
Methyl HC Pimelate	-2.27	3.70	142	37	Succ 4	A		465.30						79,309
Morphine	-5.03	0.62	200	37	CPB 7.4	B		237.70						79,309
2-Naphthol	-1.55	2.84	123	25	DDH₂O	C	1.52	106.20	0.30	0.61	0.4	1.08	1.56	24,76
Naproxen	-3.40	3.18*	155	a	PBS 7.4	B		188.50	3.18	*	*	*	*	309
Nicotine	-1.71	1.17	25	32	NaOH 9.2	B		150.80						303
Nitroglycerin	-1.96	2.00	25	30	Ring 6.5	C		135.20						8
3-Nitrophenol	-2.25	2.00	98	25	DDH₂O	C		91.32	1.23	0.79	0.23	1.57	1.50	24,76
4-Nitrophenol	-2.25	1.96	115	25	DDH₂O	C		91.53	-2.15	0.82	0.26	1.72	1.59	24,76
NDELA	-5.22	a	25	a	a	C		114.60						21,5
n-Nonanol	-1.22	3.62	25	25	0.9% NaCl	C		159.90	*	0.37	0.48	0.42	1.38	21,5
Octanoic acid	-1.60	3.00	25	25	a	C	2.15	146.40	0.66	0.60	0.45	0.60	1.64	21,5
n-Octanol	-1.28	2.97	25	25	0.9% NaCl	C	3.00	144.50	*	0.37	0.48	0.42	1.38	21,5
Ouabain	-6.11	a	190	25	Ring 6.5	B	0.48	490.30						8
Pentanoic acid	-2.70	1.30	25	25	a	C		98.75	-0.92	0.60	0.45	0.60	1.64	21,5
n-Pentanol	-2.22	1.56	25	25	0.9% NaCl	C	0.70	96.99	-0.40	0.37	0.48	0.42	1.38	21,5
Phenobarbital	-3.34	1.47	178	32	Buff 7.4	B		180.50						303

Phenol	-2.09	1.46	42	25	DDH$_2$O	C	0.73	73.12	-0.82	0.60	0.3	0.89	1.36	24,76
Pregnenolone	-2.82	3.13	192	26	PEN/STR	A	1.70	310.10	3.77	0.32	1.18	3.29	2.59	302
Progesterone	-2.82	3.77	130	26	PEN/STR	A	2.02	304.30	3.77	0.00	1.14	3.29	2.08	302
n-Propanol	-2.85	0.25	25	25	0.9% NaCl	C	0.30	65.51	-1.52	0.37	0.48	0.42	1.38	21,5
Resorcinol	-3.62	0.80	113	25	DDH$_2$O	C	0.25	79.80		1.10	0.58	1.00		24,76
Salicyclic acid	-2.20	2.26	160	32	Buff 7.4	C		94.75						303
Scopolamine	-4.30	1.24	25	30	Ring 6.5	B		263.90	1.24	*	*	*	*	8
Styrene	-0.19	2.95	25	a	a	C		87.66	0.44	0.00	0.16	0.65	0.29	310
Sucrose	-5.28	-2.25	187	30	Succ 4	C		267.40						$
Sufentanyl	-1.92	4.59	97	37	CPB 7.4	B		346.40						304,307
Testosterone	-3.40	3.31	155	26	PEN/STR	A	1.36	270.20	3.31	0.32	1.19	2.59	2.60	302
Thymol	-1.28	3.34	51	25	DDH$_2$O	C	3.30	135.80	1.62	10.0	0.52	0.44	1.51	24,76
Toluene	0.00	2.75	25	a	a	C		82.84	2.89	0.00	0.14	0.52	0.26	310
2,4,6-Trichlorophenol	-1.23	3.69	66	25	DDH$_2$O	C		117.30						24,76
Water	-3.30	-1.38	25	25	Water	C		16.75						21,5
3,4-Xylenol	-1.44	2.35	68	25	DDH$_2$O	C	1.28	104.60	0.28	0.56	0.39	0.86	1.47	24,76

[a] Data not stated or unavailable

[b] Succ, succinimate phosphate buffer; PEN/STR, aqueous solution of penicillin and streptomycin; Ring, Ringer's buffer; CPB, citrate–phosphate buffer; Buff, buffer (specific constituents not stated); DDH$_2$O, distilled water.

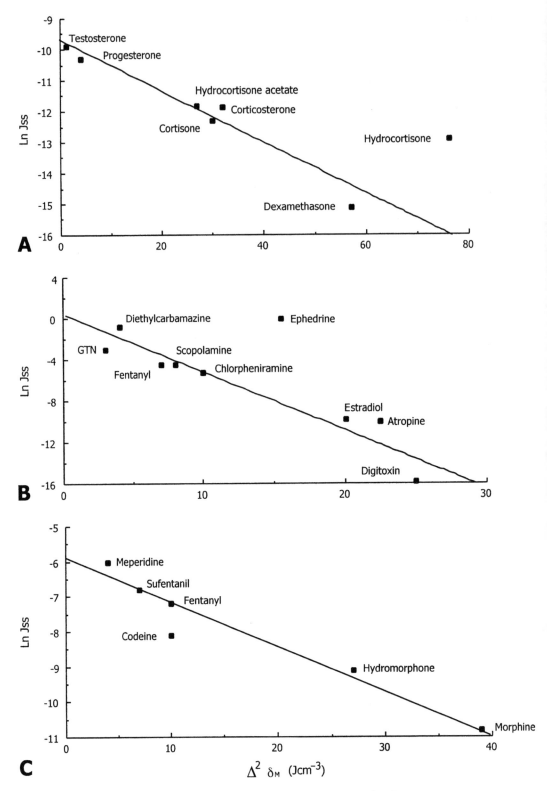

Figure 13 Dependence of transepidermal flux (J_{ss}, mol cm^{-2} h^{-1}) on the difference in solute (δ_{solute}) and o-acylglucosylceramide ($\delta_{o\text{-acylglucosylceramide}}$) solubility parameter [$\delta_{solute} - \delta_{o\text{-acylglucosylceramide}}$] for (A) steroids; (B) a mixture of various drugs; and (C) analgesics. (From Ref. 68.)

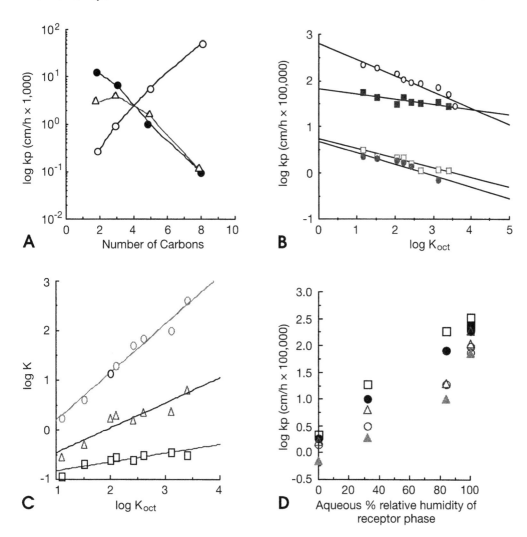

Figure 14 (A) Effect of alkyl chain length on the permeability of alcohols through human skin from olive oil (△), isopropyl palmitate (●), and saline (○); (B) epidermal permeability coefficients (k_p) and octanol–water partition coefficients (K_{oct}) for several phenols from an arachis oil vehicle and an aqueous receptor (○), ethanol 47% in water vehicle and receptor (■), ethanol 95% in water vehicle and receptor (□), and arachis oil in vehicle and receptor (●) at 25°C; (C) aqueous alcohol partition coefficients (K) versus octanol–water partition coefficients (K_{oct}) for a group of phenols. Solvents used were water (○), ethanol 47% in water (△), and ethanol 95% in water (□); (D) relation between permeability coefficients for the group of phenols and the relative humidity generated by the receptor phase used in permeability studies. (From Refs. 100[A]; 101[B–D].)

$$\log K_{sc-arachisoil} = \log \left(\frac{C_{sc}}{C_w} \frac{C_w}{C_{arachisoil}} \right) = \log K_{sc-water} - \log K_{sc-oil}$$

$$= -0.41 \log K_{oct} + 0.71 \tag{54}$$

Thus, a negative slope is predicted. The slope of the observed relation for an arachis oil vehicle to aqueous receptor −0.34 was of a magnitude comparable with that predicted of −0.41. A lower observed slope would be consistent with a higher lipophilicity in the SC diffusion pathway as discussed earlier [see Eq. (34)]. A comparison of predictions for other vehicles with observed values shown in Table 6 shows that observed and predicted slopes are similar in magnitude.

It should be recognized that linear relations between logarithms of the partition coefficients of solutes in different solvents with the logarithm of the octanol–water partition coefficient have been shown for many polar and semipolar solvents (75). Leo et al. (75) point out that for nonpolar solvents, the relation with the logarithm of the octanol–water partition coefficient is poor. A satisfactory correlation is obtained when a hydrogen-bonding constant is added to this relation.

Hagedorn–Leweke and Lippold (102) quantified the transdermal permeabilities and maximum fluxes of various sunscreens and antimicrobial compounds applied as saturated solutions in a propylene glycol–water mixture applied to human skin in vivo. A linear relation was found between the logarithms of permeability coefficients of the penetrants and their corresponding octanol–vehicle partition coefficients. The slope of 0.38 reported in their relation may be explained as being much less than unity as a consequence of the SC being more "polar" than octanol, as deduced from the aqueous partition studies [see Eq. (19)]. An additional reduction in the slope is also expected through the cosolvency effect of propylene glycol. A slope of 0.32 was obtained in a later study on the uptake of homologous esters of nicotinic acid by the skin of healthy volunteers (103).

3. Hydration

It should be emphasized that this approach can be used to predict only slopes and not permeability coefficients, the absolute magnitude of which are also influenced by vehicle effects on membrane properties, such as hydration. Indeed, as shown in Figure 14D, the permeability coefficients in the different vehicle systems used vary by two orders of magnitude and appear to be related to water content, as defined by the relative humidity for the receptor solutions used. Scheuplein and Blank (5) had previously reported that the diffusion of alcohols differs by 100-fold between hydrated and dehydrated SC. The role of skin hydration in percutaneous absorption has been reviewed (104). The mechanism of water enhancement of skin permeability is considered in a later section on vehicle–skin interactions. In their overview of the effects of hydration on solute penetration, Roberts and Walker (104) noted that the reported results were equivocal, with some studies reporting increases of up to tenfold for some substances, and others showing a very small effect. They commented that the major effect of hydration may be on solubility in skin lipids, citing the greater enhancement of pure glycol salicylate relative to methyl or ethyl salicylate by hydration using the human in vivo data of Wurster and Kramer (105). A similar enhancement was shown for methyl ethyl ketone (106). Figures 15A and B summarize these in vivo results. Occlusion has been reported to have a greater effect on

Table 6 Predicted and Observed Slopes of log Permeability Coefficients and Partition Coefficients Versus log Octanol–Water Partition Coefficients for a Series of Phenolic Compounds

Vehicle	log K_{oil-v} vs. log K_{oct}	log k_p^{sc-v} vs. log K_{oct}	
		Observed	Predicted
Water[a,b]	0.57[c]	0.6	0.57
Arachis oil[a]	0.98	−0.34	−0.41
Ethanol 47%[b]	0.51	−0.11	0.06
Ethanol 95%[b]	0.18	−0.21	0.39
Arachis oil[b]	0.98	−0.25	−0.41

[a]Aqueous receptor.
[b]Receptor same composition as donor vehicle.
[c]Stratum corneum–water.

the percutaneous absorption of the more lipophilic radiolabeled tracer steroids in vivo than for nonoccluded conditions.

In their recent summary of the effects of occlusion on percutaneous absorption, Bucks and Maibach (107) noted that occlusion increases the normal water content of the SC from between 5 and 15% to 50%, the temperature from 32°C to as much as 37°C, and the skin pH from 5.6 to 6.7. Hence, a number of mechanisms may be associated with humidity-induced penetration changes. Interpretation of studies in vitro is further complicated by the use of hairless mouse and human epidermal membranes to examine hydration effects. Bond and Barry (108) have shown that, after treatment with saline, the permeability coefficient of 5-fluorouracil through hairless mouse skin sharply increased in permeability after approximately 50 h of hydration, suggesting that the SC had started to disrupt, whereas the flux through human abdominal skin remained unchanged.

4. Maximal Flux

By definition, if neither the vehicle nor the solute affects the membrane, the same maximum flux will be observed for a solute from a range of vehicles irrespective of the range of concentrations [Eq. (55)]. Hence, as $K_{sc} = S_{sc}/S_v$, an increase in S_v for a given vehicle is counterbalanced by a reduction in K_{sc}, giving a constant flux. Figure 6A shows an example of a constant flux for a solute through an inert membrane from a number of saturated solutions in different vehicles, which show no apparent interaction with the membrane. Hence, the maximal flux is defined by the solubility of the solute in the SC (S_{sc}), which could also be expressed as the product of the partition coefficient (K_{sc}) and the solubility in the vehicle (S_v).

$$J_{sc}^{max} = \frac{D_{sc}^a S_{sc}}{h_{sc}} = \frac{D_{sc}^a K_{sc} S_v}{h_{sc}} = k_p^{sc} S_v \tag{55}$$

Consistent with earlier derivations, D_{sc}^a is the apparent diffusivity, S_{sc} is the solubility of the solute in the SC, k_p^{sc} is the epidermal permeability coefficient, and S_v is the solubility in the vehicle. Implicit in Eq. (55) and the assumption that K_{sc-v} is inde-

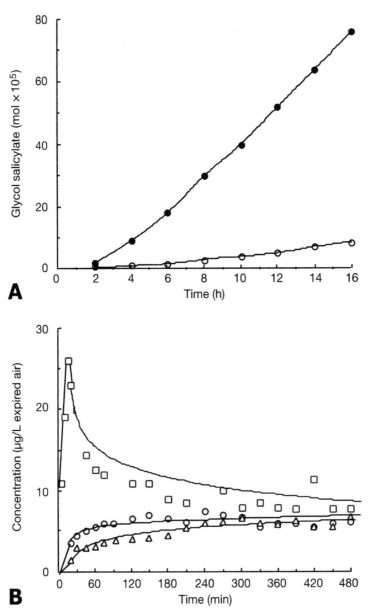

Figure 15 (A) Cumulative urinary salicylate excretion data showing the influence of hydration on the percutaneous absorption rate of glycol salicylate, hydrated system (●) and dehydrated system (○); (B) expired air concentration data for the elimination of methyl ethyl ketone showing the influence of hydration on percutaneous absorption rate: hydrated system (□); normal system (○); and dehydrated system (△); (C) percutaneous absorption of four steroids in humans as a function of penetrant octanol–water partition coefficient and occlusion. (From Refs. 105[A]; 68[B]; 107[C].)

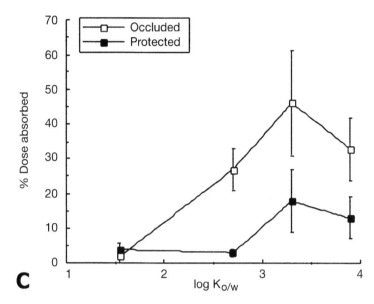

C

Figure 15 Continued

pendent of concentration for any vehicle, is that the flux (J_{sc}) for a given solute concentration (C_v) in a given vehicle [see Eq. (43)] can be predicted from J_{sc}^{max} and S_v; that is, $J_{sc} = J_{sc}^{max} C_v / S_v$. It is emphasized that such a prediction assumes that neither the solute nor the vehicle has affected D_{sc}^a or S_c. By using the data in Table 3 from Sloan et al. (62), Zatz (109) estimated a theoretical $K_{sc}S_v$ (i.e., $=S_{sc}$) and thence $J_{sc}^{max} / (K_{sc}S_v)$. In theory, this latter ratio should be constant [see Eq. (55)]. In practice, the ratio for the various vehicles used to apply theophylline to hairless mouse skin were isopropyl myristate 50, octanol 490, dimethyl formamide 4.7, propylene glycol 3.2, ethylene glycol 2.2, and formamide 2.7. Evidently, the vehicles had caused varying changes in skin permeability. Twist and Zatz (110) and Barry et al. (111) have previously shown that the respective constant steady-state flux for methylparaben through polydimethylsiloxane membrane and the bioavailability for the topical steroid desonide, as measured by vasoconstrictor response, were independent of formation when applied as saturated formulations.

Equation (55) suggests that the maximal flux of a solute through SC J_{sc}^{max} can be enhanced by three mechanisms: *(a) increasing the diffusivity of a solute in the SC; (b) affecting the partitioning between the SC lipids and other SC components, or (c) by increasing its solubility in SC lipid components.* Tojo et al. (112) showed that the permeation steady-state rate of progesterone and its hydroxy substituents across the intact skin and stripped skin of the hairless mouse was approximately proportional to the solubility of drugs in the SC or in the viable skin, respectively.

Note that Eq. (55) is a reduced form of a more general expression, analogous to Eq. (34), which recognizes the potential effects of the viable epidermal resistance:

$$J_{sc}^{max} = k_p^{sc} S_v (1 - f_{S_{ve}}) \tag{56}$$

where $f_{S_{ve}}$ is the solute concentration in the viable epidermis C_{ve}^{ss}, expressed as a fraction of its solubility in the viable epidermis S_{ve} (i.e., $f_{S_{ve}} = C_{ve}^{ss} / S_{ve}$).

One means of increasing its solubility beyond normal saturation is to produce a supersaturated state by processes such as cooling, evaporation of vehicle, solvent additions, and change of pH. Pellet et al. (113) showed that supersaturation of piroxicam in a propylene glycol–water cosolvent vehicle resulted in higher (supersaturated) concentrations of solute in the SC with a resultant higher flux through the SC. Schwarb et al. (114) have recently reported similar results with fluocinonide. Hadgraft (115) has recently reviewed issues associated with using supersaturated systems in transdermal delivery.

Maximal flux is both solute- and vehicle-dependent and can be predicted by a number of approaches. One approach is to apply the solubility parameter approach and estimate the solubility of a solute in SC lipids applying Eq. (12). Assuming a similar mole heat of fusion for solutes and an "ideal" solution, the logarithm of the mole fraction solubility is linearly related to the reciprocal of the melting point [see Eq. (9)]. Several studies have applied this relation to percutaneous absorption, including Guy and Hadgraft (116) (Fig. 16) and Kai et al. (45). An implicit assumption of a constant diffusivity is being made here in relating the logarithm of the maximum flux to the reciprocal of the melting point.

Because maximal flux is both solute- and vehicle-dependent, extending Eqs. (5) and (19) under ideal conditions (i.e., $\gamma_{sc}^* = \gamma_v^* = 1$) in an aqueous system, $\log S_{sc} \simeq 0.59 \log S_{oct}$. Substituting into Eq. (55) yields

$$\log J_{sc}^{\max} = \log \frac{D_{sc}^a}{h_{sc}} + 0.59 \log S_{oct} \tag{57}$$

where S_{sc} is the solubility of the solute in the octanol. Thus, it is expected that there will be a parabolic relation in anticipation of solutes with a polarity similar to 0.59 $\log S_{oct}$, which will be most soluble in the SC (3). Figure 17A shows that Yano and co-workers' (117) in vivo data for nonsteroidal anti-inflammatory solutes has a maximum $\log P_{oct}$ at about 2.5. Consistent with Eq. (55), Tojo et al. (112) also observed that the solubility of progesterone and its hydroxyl derivatives in the SC increased with the lipophilicity of the penetrant, and they reported that the diffusivity of these solutes across the SC and viable skin appeared to be independent of their polarity.

Kasting et al. (92) related $\log J_{sc}^{max}$ to $\log S_{oct}$ and molecular volume for 35 compounds. By using ANOVA, they showed that $\log S_{oct}$ accounted for 53% of the observed variance, and molecular volume accounted for another 21% (i.e., $r^2 = 0.74$ for a multivariate regression). In 15 healthy volunteers, Le and Lippold (103) studied and estimated the uptake of homologous esters of nicotinic acid by the skin. Permeabilities and maximum fluxes J_{sc}^{max} were determined from the concentration decrease of the aqueous solutions after fixed time periods. Although no clear dependence was observed between the maximum flux J_{max} and the octanol solubility S_{oct}

Figure 16 (A) Steady-state flux of hydrocortisone through polydimethylsiloxane membrane from saturated solutions in various vehicles that themselves are not sorbed to any significant extent by the membrane (mean ± SD); (B) an inverse relation between drug flux at steady state through excised human skin and penetrant melting point (MP) (dg, digitoxin; ou, ouabain; es, estradiol; at, atropine; ch, chlorpheniramine; fn, fentanyl; sc, scopolamine; ng, nitroglycerin; dc, diethylcarbamazine; ep, ephedrine). (From Refs. 68[A]; 116[B].)

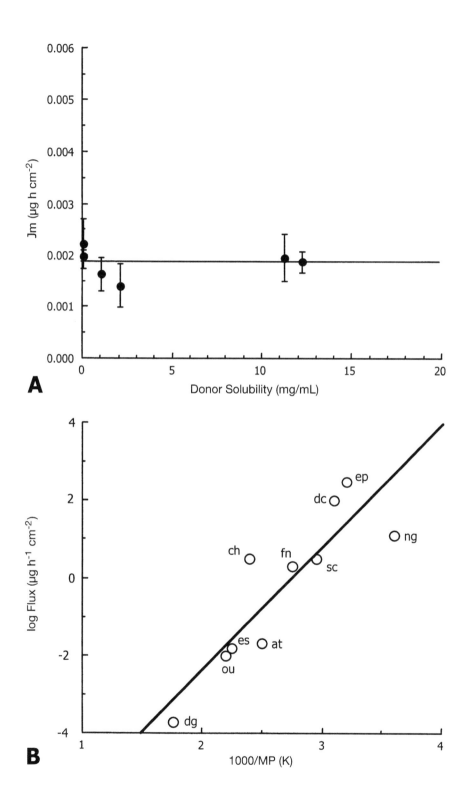

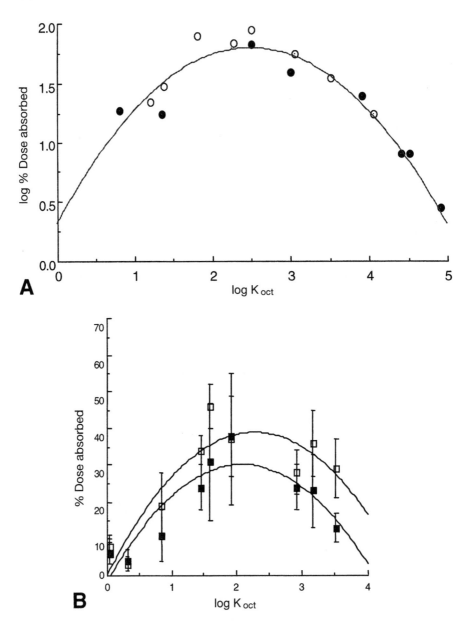

Figure 17 (A) In vivo percutaneous penetration of a series of salicylates (○) and of other nonsteroidal anti-inflammatory drugs (●) plotted as a function of log K_{oct}; (B) in vivo percutaneous absorption of phenols in humans under occluded (□) and protected (■) conditions as a function of penetrant log K_{oct}. (From Refs. 117[A]; 68;324[B].)

of the esters, a linear relation was found between log J_{sc}^{max} + (1 − 0.32)log K_{oct} versus log S_{oct}. The maximum fluxes of transdermally absorbed sunscreens and other solutes were also estimated from the disappearance of the solutes from saturated solutions of a propylene glycol−water mixture which was applied to human skin in vivo (102). The maximum fluxes were then related to the octanol−water partition coefficients and octanol solubilities for the compounds.

Hinz et al. (118) have suggested that the significant parabolic (log J_{sc}^{max} = −0.18 + 1.35 log K_{oct} − 0.30 [log K_{oct}]^2) and bilinear (log J_{max} = −0.17 + 1.08 log P − 1.95 [log($\beta \cdot 10$ log K_{oct} + 1)]) dependencies obtained may reflect a change in the rate-limiting transport step (for compounds of high log P) from diffusion across the SC to partitioning at the SC−viable epidermis interface. Bucks and Maibach (107) also observed a parabolic relation for phenols in vivo using deposited solids (see Fig. 17B). Hostynek and Magee (119) concluded in their analysis of human in vivo skin absorption data for 28 solutes that the maximum flux was increased by occlusion.

Roberts and Sloan (120) modified the Potts and Guy equation [see Eq. (49)] to apply to more lipophilic and more polar vehicles than skin. Maximum fluxes for 39 prodrugs from saturated solutions in isopropyl myristate were best described by the relation:

$$\log J_{sc}^{max} = -0.193 + 0.525 \log S_{ipm}$$
$$+ (1 - 0.525)\log S_w - 0.00364 MW \qquad r^2 = 0.945 \qquad (58)$$

where S_{ipm} is the solubility in isopropyl myristate, S_w is the solubility in aqueous pH 4.0 buffer and MW is molecular weight. The significant difference to the earlier expressions derived to date is the inclusion of the water solubility term, with a slope suggesting that it is almost as important as lipid solubility in predicting flux. A similar model was used to analyze the maximum flux data of Kasting et al. (92) from propylene glycol (PG), but using propylene glycol solubility S_{PG} instead of S_w:

$$\log J_{sc}^{max} = -1.673 + 0.599 \log S_{ipm}$$
$$+ (1 - 0.599)\log S_{PG} - 0.00595 MW \qquad r^2 = 0.852 \qquad (59)$$

This work, using solubilities in polar and nonpolar solvents as predictors of flux, suggests that the bipolar nature of the SC lipids needs to be recognized in modeling maximum fluxes from different vehicles.

To date, our analysis has been limited to the prediction of maximum fluxes for solutes on the basis that, consistent with the theoretical considerations defined by Raoult's or convention 1, the activity of a saturated system from any vehicle should be identical and equal to that for a pure liquid compound. In reality, solutes and vehicles interact with the skin affecting both solubility and diffusivity. Deviations in maximum fluxes between vehicles or in the ratios of a flux through epidermis and an inert membrane between vehicles (121) is evidence of either a drug−skin or a vehicle−skin interaction as discussed in the following section.

IV. FACTORS AFFECTING SKIN FLUX

The key determinants of epidermal flux are solute concentration in the SC (C_{sc}), the effective diffusivity in the SC, and the potential buildup of solute concentrations in the viable epidermis [see Eq. (28) and Fig. 7]. A maximum flux is attained, therefore, at the solubility of the solute in the SC [see Eq. (55)], recognizing that solubility may include the thermodynamically unstable potential supersaturation. The concentrations of solute in the SC may be related to those in the vehicle by a partition coefficient. Our analysis also shows that the apparent diffusivity is a function of both the diffusivity of unbound solute down the intercellular lipid pathway as well as the fraction of solute unbound in this pathway. Finally, solutes may be transported in the vapor phase as has been shown for the alcohols (5) and for a homologous series of acetate esters (122). In the latter, the vapor pressure of the pure solutes decreased as the alkyl chain was increased. The observed SC permeation rate decreased with the decrease in vapor pressure. We now consider factors affecting each determinant.

A. Solute Concentration in Vehicle

Equation (39) suggests that J_{sc}^{ss} should be linearly related to the concentration of solute in the vehicle C_v, up to the solute saturation solubility in the vehicle. Thereafter, at higher solute concentrations, a suspension exists and the solute flux is the maximal flux, which has been discussed earlier. Hence, Barry et al. (57) showed that benzyl alcohol vapor flux was linearly related to benzyl alcohol activity, suggesting that percutaneous absorption is controlled by thermodynamic activity when the vehicle has no effect on the SC barrier. It may be important to recognize that, if a solute activity is defined as fractional solubility (as implied by convention 1), then the flux from different vehicles will be the same for all fractional concentrations. Flux is not necessarily linear with fractional concentrations, as illustrated by the deviations from Raoult's law for benzyl alcohol vapor concentration versus mole fraction (57).

If, on the other hand, solute activities are those measured (as implied by convention 2), for a given concentration, the highest flux will be seen from the vehicle in which the solute is least soluble with identical fluxes being apparent when both vehicles are saturated. These deductions are based on the assumption that the effects of the two vehicles on the skin are the same and there is no nonlinearity in flux versus C_v profiles.

A nonlinearity in flux–convention profiles may also arise if the concentration of the solute used is sufficiently high to affect the integrity of the SC barrier, or if nonsink conditions preclude the attainment of equilibrium during the course of the experiment (90). Figure 18 shows three examples of nonlinearity. In Figure 18A a positive deviation from linearity can be shown to arise as a consequence of solute effects on SC permeability by comparison with a flux through an inert membrane. The proportionality of flux to vehicle concentration through the inert membrane is evidence that the effects do not arise from alterations in the activity coefficients of

Figure 18 (A) Penetration flux of phenol through rat skin (○) and polyethylene film (●) at 37°C for various concentrations of phenol in water; (B) Fluxes of octylsalicylate at various concentrations: (left-hand axis: polyethylene membrane (○) and nylon membrane filter (△); (right-hand axis) human epidermis (▲) and dialysis membrane (●); (C) penetration flux of pentanol from an olive oil vehicle through human skin (○) and polyethylene film (●). (From Refs. 101[A,C]; 325[B].)

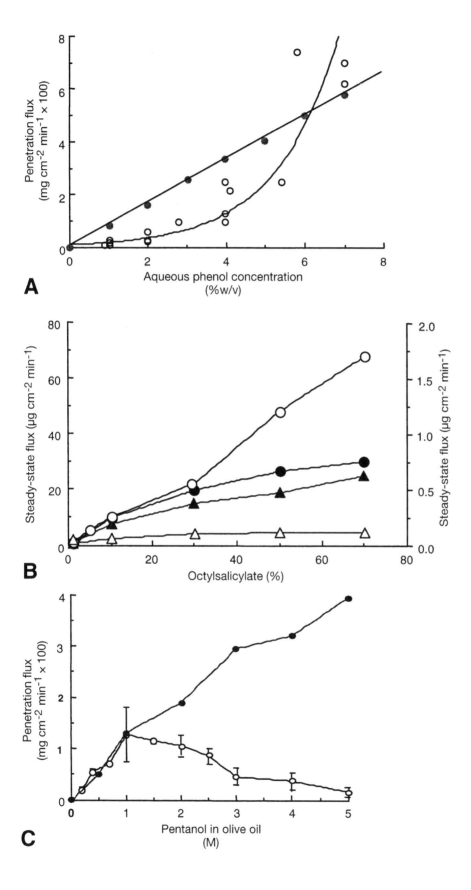

A

B

C

the solutes in the vehicle. The permeability coefficient is relatively constant; but increases abruptly at about 2% phenol as a result of several changes (123). Figure 18B shows a negative deviation from a linear flux versus concentration relation through membranes other than nylon. The negative deviation here arises from octyl salicylate self-association in the vehicle at high concentrations and is accounted for in Eq. (40) by a reduction in K_{sc-v} as a result of a decreased γ_v [see Eq. (7)]. The positive deviation from linearity for the nylon membrane suggests that octyl salicylate has increased flux by plasticization or other effects on the membrane. Figure 18C shows a negative deviation in flux from a linear flux concentration relation as a result of an effect such as dehydration and a reduction in D_{sc}, arising from the high concentration of solute in the vehicle.

Twist and Zatz (124) reported a parabolic relation between flux and solute concentration for methylparaben and propylparaben through polydimethylsiloxane membrane from 1-propanol. They proposed that the propanol vehicle is sorbed by the membrane and creates an environment ("clusters") in which the paraben can dissolve. The resultant paraben membrane concentration and flux is higher than if the propanol was not present in the membrane. At high paraben concentrations, the propanol activity in the vehicle is reduced: less partitions into the membrane. The paraben solubility and flux therefore decreases. Another nonlinearity that may arise is the nonlinear binding of components to SC. Bronaugh and Congdon (125) showed that hair dye binding to human epidermis could be described by a Scatchard plot, and that permeability values followed the rank order of dye permeability and paralleled the partition coefficients only when the binding sites were saturated. Wurster (122) has reported the adsorption isotherm for sarin's uptake on p-dioxane-conditioned callous tissue.

B. Drug–Vehicle Interactions

Figure 19A shows that the penetration flux of phenol decreases through both rat skin and polyethylene owing to a higher affinity of dimethyl sulfoxide (DMSO) than for water, even though DMSO is a very strong penetration enhancer. A similar profile is observed for glycerol, an agent that has less effect on the epidermis (see Fig. 19B) (123). The effect of the DMSO and glycerol relative to water is simply a reduction in K_{sc-v} owing to a greater solubility [see Eq. (42)] or low-activity coefficient owing to the high affinity [see Eq. (40)] of phenol for these vehicles than for water. Indeed when the logarithm of the penetration flux is plotted against the percentage glycerol, a linear relation is observed (see Fig. 19C) consistent with the relation (126).

$$\log J_s = \log J_{s\,\text{water}} + (\log J_{s\,(\text{glycerol})} - \log J_{s\,(\text{water})})(1 - f_g) \tag{60}$$

Where (J_s) is the penetration flux for a given binary composition, $J_{s\,(\text{glycerol})}$ and $J_{s\,(\text{water})}$ are the penetration fluxes of phenol from glycerol and water vehicles, respectively, and f_g is the fraction of glycerol in the glycerol–water vehicle.

Vehicles may also affect drug release by a diffusion limitation in the vehicle with a range of expressions being presented (see Chap. 3). Other effects such as vehicle evaporation, dissolution kinetics, solvent flux through stratum corneum, and changes in vehicle composition with time are dealt with elsewhere in this book and the literature (83,126,127). In the present context, the shape of the cumulative amount versus time profile is often indicative of whether flux is membrane-limited as dis-

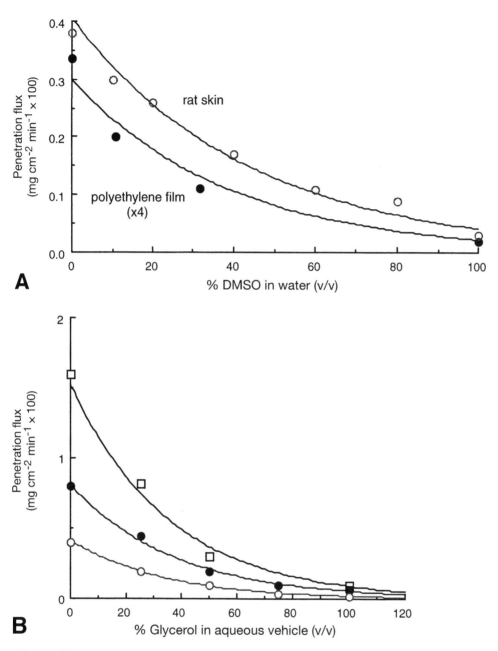

Figure 19 (A) The relation between flux of phenol through polyethylene film and excised skin, and the percentage DMSO in the vehicle; (B,C) penetration flux of phenol through excised skin from a 5% w/v (□) and a 2% w/v (○) aqueous glycerol solution and through a polyethylene membrane from a 2% w/v (●) aqueous glycerol solution. (From Refs. 330[A]; 101[B,C].)

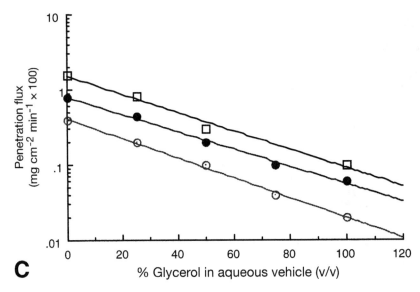

Figure 19 Continued

cussed to date or limited by diffusion in the vehicle. In general, when uptake is limited by diffusion in the vehicle and the percentage of solute released from the vehicle is less than 30%, an approximate form of the diffusion equation from a slab into a perfect sink can be used. For a solute dissolved in the vehicle, the amount released per unit area (M_t) over time t can be related to the diffusivity of the solute in the vehicle D_v and the initial concentration of solute in the vehicle C_0 by:

$$M_t \cong 2C_0 \left(\frac{D_v t}{\pi} \right)^{1/2} \tag{61}$$

The derivation of the expression for a suspension of drug with an apparent concentration C_{total} (total amount of dissolved plus undissolved solid per unit volume) and a solubility in the vehicle S_v, described originally by Higuchi, has been reported by many authors, including Barry (126). When $C_{total} \gg S_v$, M_t is given by

$$M_t \cong (2C_{total}S_vD_vt)^{1/2} \tag{62}$$

Hence, a square root relation for the amount of solute released with time is expected, whether or not the vehicle is saturated with solute. The range of vehicles used in topical applications is discussed in greater detail in the following chapters.

C.   Water Enhancement of Permeability

The state of hydration of the normal skin has been ranked next to the nature of the penetrating molecule as the most important factor in the rate of percutaneous passage of any substance (128). However, the reported effectiveness of hydration on skin penetration appears equivocal (104). It is well recognized that the SC swells continuously on immersion in water, absorbing as much as ten times the dry weight. The water is bound within the intracellular keratin. Permeability increases rapidly initially and then slows down to a steady-state diffusion. Alonso et al. (129,130) used spin-label electron-spin resonance technique to monitor the effect of hydration on the molecular dynamics of lipids at C-5, C-12, and C-16 positions of the alkyl chain. They found that an increased hydration of neonatal SC led to an increase in membrane fluidity near the membrane–water interface region and less so in the deeper hydrophobic core. Solid-state nuclear magnetic resonance (NMR) has also been used to show a greater mobility of the skin components in the presence of water (131).

D. Other Vehicles–Skin Interactions

Vehicles or their components can interact with SC lipids to enhance skin permeability. Two books have been published relatively recently on this subject (132,133). In addition, Williams and Barry (134) and Davies et al. (see Chap. 6) have reviewed this area. Williams and Barry (134) identified the potential modes of actions of accelerants by the lipid–protein-partitioning (LPP) theory, which summarizes the mechanisms of action as the following:

1. Disruption of the intercellular bilayer lipid structure
2. Interaction with the intracellular proteins of the SC
3. Improvement of partitioning of a drug, coenhancer, or cosolvent into the SC

Menon et al. (124) extended the LPP model to recognize

4. Disruption of the corneocyte envelope by caustic agents such as 7% phenol and hydrocarbons
5. Effects on proteic junctions, such as desmosomes, involved in squamae cohesion

In this chapter we have introduced a potential sixth, but as yet unexplored mechanism, namely,

6. Alteration of the partitioning between SC components and the lipid in the diffusion pathway

A diagrammatic illustration of the mechanisms of vehicles and their components on the skin is shown in Figure 20 and Table 7. Polar channels have been suggested to be formed by the actions of a number of lipophilic enhancers (e.g., terpenes) as an additional mechanism of action to their effect of causing disruption of intercellular lipid bilayers (135). Cornwell and Barry (135) determined the conductivity of human skin in vitro before and after treatment with various enhancers. Significant increases in the conductivity following treatment suggested that new polar channels were being opened up in the SC, which is considered the major barrier to ion transport through human skin (136). Cornwell and Barry (135) were also able to show a correlation between the observed increases in conductivity following enhancer treatment and the flux of the polar, nonelectrolyte 5-fluorouracil (5-FU), suggesting the polar channels created allowed the passage of both ions and 5-FU. It was concluded from these studies that because differential scanning calorimetry (DSC) had demonstrated that terpene enhancers disrupt the intercellular lipids in the SC (137), the most likely site of pathway formation is through the lipid bilayers. The creation of polar channels through the SC has also been suggested by Francoeur et al. (138) following the application of oleic acid when, contrary to pH-partitioning theory, piroxicam flux

Solvent/Vehicle	Effect	
Lipoidal (cholesterol, phospholipid) Azone	Insertion of solvent molecules into lipid tail regions causing "fluidization" by loosening of bilayer structure.	
Polar solvents	Insertion of polar molecules into polar headgroup regions causing "fluidization" by loosening of packing structure.	
Organic solvents	Solubilization and extraction of lipids from bilayers causing loosening and disorganization of structure.	
Water DMSO	Formation of water "pools" around polar head groups resulting in a facilitated polar pathway or channel through the bilayers.	
Liposomes	Facilitated transport by the movement of liposome structures within the lipid tail region of the bilayers.	

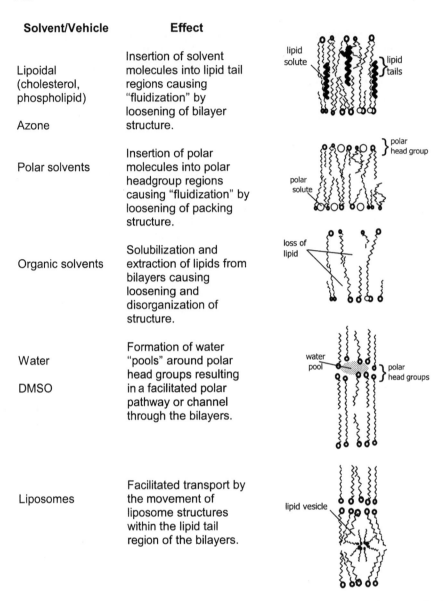

Figure 20 Diagrammatic representation of the possible mechanisms of action of skin penetration enhancers. (Adapted from Ref. 123.)

increased with the proportion of ionized drug. Fourier transform infrared (FTIR) studies by Ongpipattanakul et al. (139) later suggested that polar channels were formed by the lateral phase separation between oleic acid and indigenous lipids. Enhancement of the skin permeability of ionized salicylic acid by DMSO (140), another enhancer shown by DSC to cause intercellular lipid disruption (141), has also been suggested to occur by the creation of polar channels in the intercellular lipids (137). The effect of surfactants on skin permeation has been reviewed (109). As discussed earlier, one mechanism for nonionic surfactants is an increase in a_v

Table 7 The Effect of Vehicles on Structures of the Stratum Corneum

Solvent/Vehicle	Effect	

Water
Occulsive agents
  increasing
  hydration
	Increase in water localization
	and swelling of the
	intracellular keratin region
	of corneocytes, creating
	polar pathways
		Corneocyte keratin
		
		Water pool
		
		corneocytes in stratum corneum
		swelling

Caustic solvents Acids Phenols	Breaking of desmosome junctions and separation of corneocytes loosening stratum corneum, together with the disruption of intracellular keratin organization	corneocytes in stratum corneum swelling

[see Eq. (40)] and a "push" mechanism (see Fig. 8B) (65). Ionic surfactants affect D_{sc} with peak effects for a given series of surfactants often occurring at C_{12} or C_{14}.

Several methods have been used to assess the effects of enhancers on membrane permeability. A low flux or permeability constant through the skin from a given vehicle does not necessarily imply that enhancement has occurred. The ratio of fluxes through skin and an inert membrane from a given vehicle is independent of the activity of solute in the vehicle (121) and defines the ratio of the permeability coefficients through the skin and the membrane. Table 8 shows the results obtained for phenol from different vehicles. Also shown are the partition coefficients between light liquid paraffin and the vehicle. It is apparent that the moderately polar hydrogen-bonding vehicles, and known penetration enhancers, of dimethyl formamide and dimethyl sulfoxide have a high affinity for phenol, as is evident by their low, light-liquid paraffin–vehicle partition coefficients and low permeability coefficients through the inert membrane. Their permeability ratios between skin and the inert membrane show, however, that they have markedly enhanced skin permeability.

Table 8 Effects of Vehicles on the Penetration of Phenol Through Polyethylene and Excised Skin (37°)

Vehicle	Permeability constant ($N = 5$) (cm min^{-1} × 10^3)		Partition coefficient: light liquid paraffin–vehicle	Permeability ratio skin: polyethylene[a]
	Polyethylene	Skin		
Light liquid paraffin	0.5	1.0	1.0	2
Water	0.044	0.19	0.12	4
Arachis oil	0.014	0.029	0.034[b]	2
Glycerol	0.005	0.010	0.015	2
Ethanol	0.004	0.016	0.030	4
Dimethyl formamide	0.002	0.022	0.008	11
Dimethyl sulfoxide	0.001	0.018	0.003	18

[a]Estimated s.d. = ±20%.
[b]Light liquid paraffin–water partition coefficient/arachis oil–water partition coefficient.

Thus, it is apparent in this instance that *drug–vehicle interactions* strong *outweigh vehicle–skin interactions*

An alternative approach to the assessment of vehicle–skin interactions is to compare the *fluxes of solutes from saturated solutions*, as these should be identical unless the vehicle has affected the skin (D_{sc}^a or S_{sc}) (see Sec. III.A). However, care should be taken with this approach as some solutes, such as phenol, denature the skin in certain vehicles at high concentrations. This is an example of a *drug–skin interaction outweighing a drug–vehicle interaction*. Figure 21 shows the relation between the maximum flux for benzophenone through epidermal and high-density polyethylene membranes and the solubility parameters of the vehicles used. It is evident that maximal fluxes occur for the epidermal membrane with ethanol ($\delta_v = 14.9$ cal cm^{-3}) as a vehicle and for the high-density polyethylene membrane with isopropyl myristate ($\delta_v = 8$ cal cm^{-3}) or C$_{12}$–C$_{15}$ benzoate alcohols ($\delta_v = 7.6$ cal cm^{-3}) as vehicles. The major effects of the solvents appear to be diffusivity changes, and it is apparent that some solvents enhance skin permeability and others polyethylene permeability.

A third method is the assessment of the *skin penetration flux for a solute or penetration enhancer through the skin before and after application* of the solute or enhancer. Hence, the ratio of phenol permeability coefficients at high concentrations from vehicles can be compared with that from a low concentration before and after treatment to give a damage ratio and an irreversible damage ratio, respectively (121). Scheuplein and Ross (142) had previously shown that pretreatment of the skin with ethanol increased the permeability of pure butanol. Williams and Barry (137) compared the drug permeability before and after terpene treatment to assess the effects of terpenes on skin permeability.

A fourth method is to use a technique (or techniques), which allows an *independent assessment of solubility and diffusion effects*. Harrison et al. (143) compared the effects of the enhancers Azone (see Chap. 6) and the solvent Transcutol (diethyleneglycol monoethyl ether) on changes in the diffusivity and solubility of a model

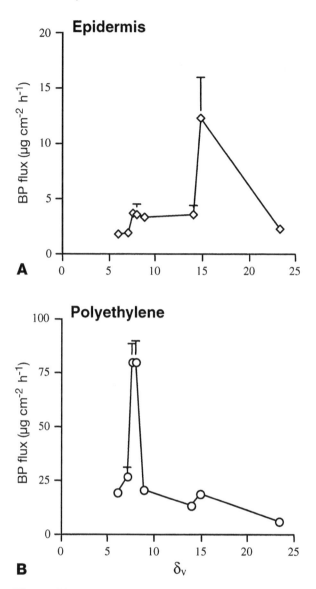

Figure 21 Relation between benzophenone membrane flux and vehicle solubility parameter (δ_v) for (A) epidermis and (B) polyethylene. (From Ref. 330.)

permeant (4-cyanophenol) in human SC using attenuated total reflectance Fourier transform infrared (ATR-FTIR) spectroscopy. They suggested that Azone acts by reducing the diffusional resistance of the SC, whereas Transcutol increases the solubility of the penetrant in the SC barrier (Fig. 22). Zhao and Singh (144) investigated the mechanism(s) of percutaneous absorption enhancement of propranolol hydrochloride across porcine epidermis by terpenes (e.g., menthone and limonene) in combination with ethanol using both Franz diffusion cells and by determining the partitioning of propranolol hydrochloride into powdered SC from control and enhancer

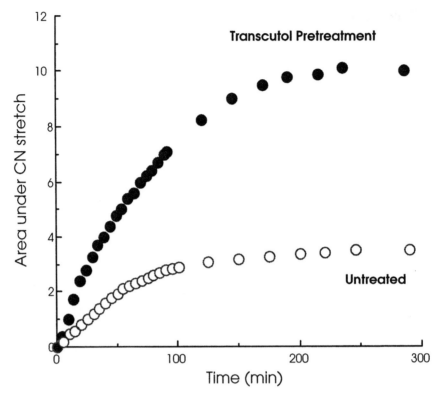

Figure 22 The effect of transcutol pretreatment on the in vitro diffusion of cyanophenol (CN) through the stratum corneum. (From Ref. 331.)

solutions. In a more recent review, Bach and Lippold (145) highlighted the quantification of enhancing effects on drug penetration that is possible by an indirect determination through the measurement of pharmacodynamic response. They suggested that the enhancing effects may be determined from the activity–response lines obtained with and without enhancer, respectively.

A range of methods have been used to define the mechanisms by which vehicles affect skin permeability. These include fluxes, partition coefficients, various spectroscopic techniques (146), and differential-scanning calorimetry. Lee and Tojo (147) used differential-scanning calorimetry to show that the skin-enhancing effect of vitamin C is through its effects on skin hydration and a "solubilizing action on the protein domain of the SC." Enhancement may also be by lipid extraction (123). Goates and Knutson (148), in examining the influence of alcohol chain length on mannitol permeation in human skin, used FTIR spectroscopy to show that SC lipid conformation and mobility was unaffected, but that there was evidence of a lipid extraction altered SC protein conformation.

E. Drug–Skin Interactions

To date, drug–skin interactions have been examined mainly in terms of the solubility of a drug in the SC and the diffusivity of the drug in SC lipids. Drug–skin interactions are also of interest from two other viewpoints: substantivity and corrosivity.

Substantivity is a measure of the binding of solutes to binding sites in the SC in terms of showing a resistance to being washed off or removed. This requirement is particularly desirable for certain cosmetics, such as sunscreens. Hagedorn–Leweke and Lippold (102) reported that the affinity of ten nonionic compounds, including sunscreens, antioxidants, antimicrobial compounds, and a repellent to animal keratin and human callus, was linear with concentration and that the keratin affinity was directly related to their octanol–vehicle partition coefficients. They suggested that genuine substantivity, associated with specific adsorption, therefore, does not seem to occur for these solutes. As discussed earlier, saturable nonlinear binding has been reported for hair dyes, which have an intended affinity for keratin. Often there is no apparent relation between skin permeation and SC–water partition coefficients, unless the binding sites are saturated (125,149). Dressler (150) has recently reviewed the percutaneous absorption of hair dyes. Triclosan (2,4,4′-trichloro-2′-hydroxydiphenyl ether), a nonionic, broad-spectrum, antimicrobial agent present in many personal care products (deodorant soaps, underarm deodorants, shower gels, and health care personnel handwashes) shows a moderate degree of substantivity to the skin, leading to a remnant antimicrobial effect in many products (151). Early studies on skin binding have been summarized (126).

Substantivity is also important in skin toxicology. Islam et al. (152) mapped the SC substantivity of chloroform in terms of exposure time and depth of penetration into the SC. Eight minutes was required for the steady-state gradient to be established, and substantivity was affected by evaporation. Attempts have also been made to determine the adsorption of surfactants by the human horny layers in vivo (153).

Often the adsorption process onto keratin may take some time. For instance, omoconazole nitrate, a topical antifungal agent, required 10 or more days to reach equilibrium in the skin (154). Nickel and cobalt are also highly adsorbed to human SC (155). Tape-stripping may be an appropriate method to study substantivity. This method has been used to show that the amounts of lindane that were recovered in tape-strippings taken at 6 h (representative of SC content) were substantially greater than in the remainder of the skin, for both an acetone solution and a formulation (156). Desquamation rates may also be important and, for the scalp, ranges from 8 days under normal circumstances to within 3–4 days in pityriasis and dandruff conditions not associated with erythema (150).

Several solutes can affect skin permeability, and many of these effects are most evident for pure solutes. The mechanisms by which many of these solutes affect the skin are similar to those outlined in the previous section. Of particular interest, however, has the the *corrosivity* of solutes, as these have obvious occupational health implications. In our early work, we hypothesized that there was a threshold molar concentration at which phenols altered skin permeability through a caustic effect (24). As a consequence, certain phenols (e.g., phenol and the cresols) had a sufficient skin solubility to be damaging, whereas the more nonpolar phenols did not, owing to solubility limitations (Fig. 23A). Scheuplein and Blank (157) also showed that the greatest extent of irreversible damage with a series of pure alcohols occurred with methanol (see Fig. 23B). The use of pure liquid solutions may involve a solubilizing component. Barry et al. (57,158) found that, at comparable thermodynamic activities, liquid fluxes were often tenfold higher than vapor fluxes for model penetrants (benzyl alcohol, benzaldehyde, aniline, anisole, and 2-phenylethanol) applied in model vehicles (butanol, butyl acetate, isophorone, isopropyl myristate, propylene carbonate,

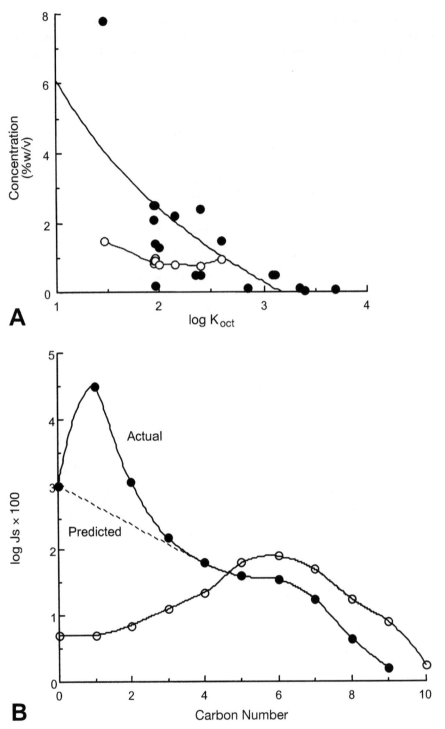

Figure 23 (A) Aqueous solubilities (●) and aqueous threshold concentrations for damage (○) for various phenols and their octanol–water partition coefficients (K_{oct}); (B) permeation rates (J_s) of the alcohols as pure solvents (●) and as solutions (○) through the epidermis. The curves cross each other near a value of log J_s = 1.6, units for J_s = μmol cm^{-2} h^{-1}. (From Refs. 101[A]; 157[B].)

toluene, *n*-heptane, and water). They suggested that the differences were reflected by the partition coefficients and the amount of penetrant entering the SC membrane. They also suggested that whereas liquid fluxes were membrane-controlled, an interfacial effect may account for a low vapor permeation.

Various authors have shown that skin irritation can be related to the solute's pK_a. Berner et al. (159) showed that, for a homologous series of benzoic acid derivatives, which permeate through human skin at comparable rates, skin irritation and pK_a were correlated for pK_as less than or equal to 4. For basic permeants, skin irritation in vivo increases with increasing pK_a (160).

Quantitative structure–activity relations (QSARs) have been derived relating skin corrosivity data of organic acids, bases, and phenols to their log (octanol–water partition coefficient), molecular volume, melting point, and pK_a (161,162). It is apparent that these relations reflect permeability limitations, such as those defined by octanol–water partitioning, molecular volume, and melting point (discussed earlier in Sec. III.C.) flux, together with intrinsic acidity of the solutes as defined by pK_a.

F. Drug–Vehicle–Skin Interactions

Barry (126) has considered several models concerned with percutaneous absorption from binary solutions.

1. The ideal case when neither SC solubility nor diffusivity was affected by either the vehicle or the solute
2. When a vehicle or drug leads to an increase in solute solubility in the SC (the "pull" effect, see Fig. 8B)
3. When an added cosolvent reduced the partitioning of a solute into the skin
4. When alterations occur in the diffusivity of the SC

First, we consider the *ideal case when neither SC solubility nor diffusivity was affected by either the vehicle or the solute.* Maximum flux in this instance is, as described earlier, at the maximum solubility of the solute in a given vehicle, provided release from a given vehicle is not diffusion-limited. A greater effect may be achieved by adding other solutes with the same action to the solvent system. A combination of three corticosteroids exhibiting independent solubility, partitioning, and diffusion behavior resulted in a higher total steroid concentration in solution than was possible for any steroid alone, with evidence of greater in vivo human vasoconstriction than observed for the individual steroids (163).

In the second case (126) a vehicle or drug leads to an increase in solute solubility in the SC, the "pull" effect described earlier in Figure 8B. For instance, Kadir et al. (65) reported that addition of paraffin oil to a propionic acid solution increased the flux of either theophylline or adenosine through enhancing the flux of propionic acid into the skin, and promoted the partitioning of the purine solutes in the modified skin barrier ("pull" effect). Similar effects can be achieved for adenosine using binary vehicles of hexanoic acid and propionic acid or isopropyl myristate and propionic acid (64) and for theophylline (63). Harrison et al. (143) have shown that transcutol enhances cyanophenol's solubility in SC lipids. From binding studies, it was suggested that the enhancement in the permeability coefficient of tamoxifen by 5% w/v menthol and thymol in a 50% ethanol solution was, at least partly due, to improvement in the partitioning of the drug to the SC (164). Indeed, menthol also enhanced the skin permeation of testosterone eightfold by forming eutectic mixtures with testosterone, cholesteryl oleate, and ceramides (165), thereby, increasing the solubility of testosterone in the SC lipids. A further 2.8-fold increase in the flux of testosterone resulted from a corresponding increase in the solubility of testosterone in an aqueous ethanol vehicle. The enhancing effects of 1-methyl-, 1-hexyl-, and 1-lauryl-2-pyrrolidone on the transdermal penetration of 5-fluorouracil, triamcinolone acetonide, indomethacin, and flurbiprofen were also suggested to be by increasing the solubility of penetrants in the SC (166).

It appears that supersaturation of a solute in a vehicle is accompanied by supersaturation in the SC (113). The percutaneous absorption of nifedipine was greatly enhanced from binary solvent systems of acetone and propylene glycol or isopropyl myristate, relative to either saturated nifedipine solution in propylene glycol or isopropyl myristate alone (167). Given the effectiveness of a polymer additive, it is possible that this system leads to enhanced absorption by facilitating supersaturated solutions, a concept discussed in greater depth in Chapter 6.

It is possible that an added agent to a vehicle will partition into the skin and produce a polarity of the skin that will result in a *reduced partitioning of a solute into the skin and a reduced flux*. We are not aware of any specific examples of this effect.

Finally, there are various agents, that can either promote or retard skin penetration. Agents promoting diffusivity were discussed in the earlier section on vehicle–skin interactions. There are also several penetration retarders, which have been identified, including substances such as the Azone analogue N-0915, for which the mode of action is suggested to be by increasing the order of SC lipids (168). Binary cosolvents consisting of isopropyl myristate and short-chain alkanols, such as ethanol (EtOH), isopropanol (*i*PrOH), and tertiary butanol (*t*B*t*OH), in particular a 2:8 combination, produced a marked synergistic enhancement of BZ flux from the mesylate salt, whereas a retarding effect was noticed for permeation of the benztropine base (169). Kim et al. (170) reported that *S,S*-dimethyl-*N*-(benzenesulfonyl) iminosulfurane; *S,S*-dimethyl-*N*-(2-methoxycarbonylbenzene-sulfonyl) iminosulfurane; and *S,S*-dimethyl-*N*-(4-chlorobenzenesulfonyl) iminosulfurane significantly decreased the permeation of hydrocortisone through hairless mouse skin and may be acting as retardants.

G.   Non–Steady-State Solutions

The mathematics for the non–steady-state condition is more complex owing to the need to solve the second-order diffusion equation with various boundary conditions, reflecting the system used (e.g., finite dose, infinite dose, viable epidermal, or clearance limitations, and so on). Many of the solutions for various conditions are presented in our recent work (83) (see also Chap. 3). Even more complex solutions arise when the diffusional processes in each phase are considered simultaneously. In practical, conceptual terms, the most important consequence of the diffusion process is to impose a lag time on the appearance of a solute at one edge of the membrane after application at the other edge. Figure 24A shows typical profiles for the uptake into, accumulation in, and transport of a solute through a membrane after application of a constant concentration. The Laplace and analytical solutions are presented in our earlier work (83) and are not reproduced here as they require appropriate non-

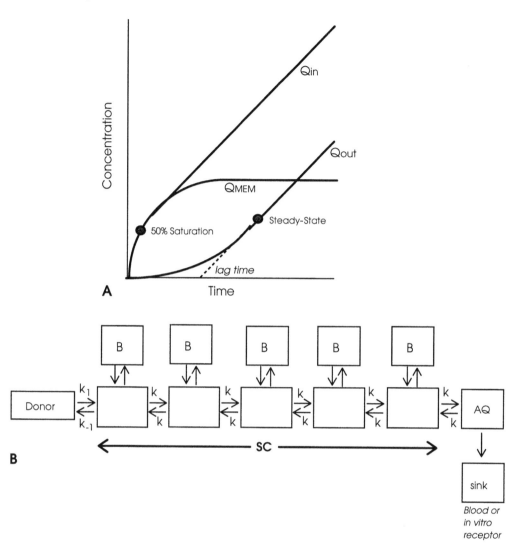

Figure 24 (A) Sorption and permeation curves for a simple membrane showing the total quantities of solute entering the membrane (Q_{in}), exiting (Q_{out}), and accumulating within it (Q_{mem}); (B) a compartmental representation of the SC as suggested by Zatz. (From Refs. 22[A]; 171[B].)

linear regression techniques to fit relevant data. In contrast, the steady-state expressions are straightforward and can be solved by linear regression, noting the concerns raised in the previous chapter. The expressions for the steady-state portions of the profiles for the amount entering the membrane $Q_{in}^{ss}(t)$, amount leaving the membrane $Q_{out}^{ss}(t)$, and the amount remaining in the membrane $Q_{mem}^{ss}(t)$ at different times from a constant donor concentration are

$$Q_{in}^{ss}(t) = k_p^{ss} C_v(t + 2 \text{ lag}) \tag{63}$$

$$Q_{out}^{ss}(t) = k_p^{sc} C_v(t - \text{lag}) \tag{64}$$

$$Q_{mem}^{ss}(t) = \frac{K_{sc-v}C_v h_{sc}}{2} \tag{65}$$

where lag $= h_{sc}^2/6D_{sc}$ [see Eq. (27); Chap. 3]. More complex expressions for lag times, taking into account the resistance of deeper layers and clearance limitations, have been given (90). An alternative approach is to use a compartmental model to represent skin penetration kinetics. Many of these models have been recently reviewed (83). Zatz (171) presented a compartmental representation of the SC, five compartments being the diffusion path, and another five being binding sites in the diffusion path. They suggested that binding affected lag time, but not steady-state flux.

In the next section, we interrelate the principles of skin transport to pharmacokinetic considerations. The important variable for this purpose is the absorption rate or flux of solutes, defined as $J = dQ/dt$.

V. SKIN PHARMACOKINETICS

Pharmacokinetics is the time course of drugs in the body or in individual tissues after input into the body. Relevant to transport of drugs through the skin, solutes are normally applied to the skin for local or for systemic effects. The desirable requirements for the two effects are different. Systemic effects are usually best achieved by the skin providing minimal resistance, binding, and local metabolism of solutes. In contrast, local effects are best achieved by relatively high cutaneous concentrations, with desirably minimal spillover to the systemic circulation so that the body load of the drug is low or barely detectable. Therefore, we will consider both systemic and cutaneous pharmacokinetics in this analysis. Given that a number of mathematical models used to describe various aspects of percutaneous penetration in terms of the underlying physical processes and representation of those processes by diffusion and compartmental models have recently been reviewed (83), our emphasis will be placed on approximate forms useful for interpreting the effects of vehicles and solute structure on systemic and cutaneous pharmacokinetics after topical application. Our earlier work reported model solutions and showed cumulative amount and flux time profiles for a range of boundary conditions and situations, as well as considering topics such as physiological pharmacokinetic models, pharmacodynamics, deconvolution, and methods of pharmacokinetic analysis. To minimize confusion associated with the differing notation used in various published papers, our own earlier work, and the previous chapter, we have adopted a convention of representing J as flux per unit area, and the permeability coefficient as k_p^i. We first consider quasi–steady-state solutions for systemic and cutaneous pharmacokinetics with a constant flux from a vehicle and with depletion of solute in the vehicle. We then consider flux in terms of its determinants so that the role of vehicle and solute structure on pharmacokinetics can be related to the physicochemical properties of the solute and the vehicle. Finally, we consider both biological and physicochemical factors reported to affect the determinants of flux and the resultant pharmacokinetics.

A. Pharmacokinetic Principles

The concentration $C(t)$ of a solute at time t at any site is defined by the input flux to the site ($J(t)$) and the transfer function ($tr(t)$) for this site:

$$C(t) = J(t)*tr(t) \tag{66}$$

where * is the symbol for convolution and $tr(t)$ is the convention–time profile at the site following the clearance of a unit impulse input to that site. In principle, Eq. (66) can be applied to define plasma concentrations (C_p), concentrations at sites of pharmacological action, and solute concentrations in different parts of the skin. Hence, concentrations in the viable epidermis (C_{ve}) are given by the flux of the solute to the viable epidermis (J_{ve}) and the transfer function for the viable epidermis ($tr(t)_{ve}$) into the dermis, deeper tissues, and skin blood circulation:

$$C_{ve}(t) = J_{ve}(t)*tr(t)_{ve} \tag{67}$$

Care needs to be exercised in attempting to extrapolate in vitro data into the in vivo situation using Eqs. (66) and (67). These equations assume that the processes determining concentration are independent of each other. The steady-state solute flux through the SC per unit area of application (J_{sc}^{ss}), for instance, is determined by the concentrations of solute immediately below the outside ($C_{sc(o)}$) and inside ($C_{sc(i)}$) of the SC [see Eq. (28)]. As shown in Figure 7, a significant viable epidermal resistance or poor perfusion will increase $C_{sc(I)}$ and lead to a reduction in J_{sc}^{ss}, the flux through the SC. In these circumstances, appropriate models of skin flux defined by such processes should be used. It may be most appropriate to recognize such processes, for instance, by using a non–steady-state solution (90) corresponding to Eq. (43), an expression for J_{skin}^{ss}, that recognizes its component permeability constants.

Systemic plasma concentrations of a solute after topical application ($C_p^{topical}$) are then appropriately defined by the convolution of the skin flux–time profile for the solute through the skin to the systemic circulation ($J_{skin}^{circ}(t)$) and the plasma concentration–time profile after intravenous administration of a unit dose ($C_p^{iv}(t)$) (i.e., the transfer function for the whole body):

$$C_p^{topical}(t) = J_{skin}^{circ}(t)*C_p^{iv}(t) \tag{68}$$

B. Systemic Pharmacokinetics

When the solute concentration in the vehicle (C_v) can be assumed to be constant (i.e., no depletion in the vehicle at the skin surface with time), the flux for the skin system per unit area ($J_{skin}(t)$) becomes constant (or at steady-state) after a lag time lag (see Fig. 24a). $J_{skin}(t)$ can be expressed in terms of an effective permeability constant for the system (k_p''), defined by Eq. (43) when $t >$ lag.

Roberts and Walters (3) have pointed out that if there is significant skin metabolism or irreversible adsorption, the in vivo flux will be reduced, as defined by the cutaneous availability F defined by the ratio of topical and systemic areas under the plasma concentration–time curves adjusted for dose differences. Hence, Eq. (43) can also be rewritten as

$$J_{skin} = Fk_p''C_v \qquad t > \text{lag} \tag{69}$$

Anissimov and Roberts (90) have commented on the limitations associated with Eq. (69). In most studies, an "instantaneous" transfer of a solute from the dermal blood supply to the systemic circulation is assumed so that $I(t)$ can be assumed to be a value of unity. A range of models have been used to describe plasma concentrations after intravenous administration and include a single exponential (also referred to as one-compartmental), biexponential (also referred to as two-compartmental), and physiological models (Fig. 25A). The plasma concentration–time profile after intravenous administration (C_p^{iv}) for the simplest single-exponential model is given by:

$$C_p^{iv} = \frac{dose_{iv}}{V_{body}} \exp(-k_{el}t) \tag{70}$$

In usual pharmacokinetic terms, a constant flux is the equivalent of a constant rate of administration, such as an infusion rate. Applying Eqs. (69) and (70), $C_p^{topical}$ for the topical absorption of a solute with a constant flux (J_{skin}) of a product applied over an area (A) for a period of time (T), after which it is removed, can be described by Eq. (71) (3,83):

$$C_p = \begin{cases} 0 & t < \text{lag} \\ \dfrac{J_{skin}A}{Cl_{body}} (1 - \exp[-k_{el}(t - \text{lag})]) & t < \text{lag} + T \\ \dfrac{J_{skin}A}{Cl_{body}} (1 - \exp[-k_{el}T])\exp[-k_{el}(t - T - \text{lag})] & t \geq \text{lag} + T \end{cases} \tag{71}$$

Figure 26 shows the extent to which this quasi–steady-state absorption model plasma concentration–time profile after topical application corresponds to an absorption model, defined by the diffusion equation, only when the lag time is small. The greatest deviation between the quasi–steady-state model and the diffusion-predicted model occurs at the termination of the topical application. Hence, although Eq. (71) is most appropriate for iontophoretic delivery with its very small lag times (172), it should be used cautiously in describing passive percutaneous absorption kinetics.

Under steady-state conditions Eq. (71) reduces to Eq. (72) (3,84):

$$C_p^{ss} = \frac{J_{skin}A}{Cl_{body}} \tag{72}$$

Hadgraft and Wolff (173) used Eq. (72) to show that it was possible to predict nitroglycerin plasma levels from in vitro patch-release data. Equation (72) can be rearranged to estimate the flux rate that is desirable for a transdermal delivery system to achieve a desired steady-state plasma concentration (174):

$$F \times \text{dosing rate} = J_{skin}A = C_p^{ss}Cl_{body}$$
$$= \text{target blood concentration} \times \text{clearance} \tag{73}$$

Hence, for clonidine with a clearance of 3.1 mL/min kg^{-1} and a target blood concentration of 0.5 ng/mL, a dosing rate of 0.156 mg/day is needed for a 70-kg person (174). Table 9 gives some examples for other drugs of interest.

Usually, plasma concentration–time profiles of solutes applied to skin are also modified by the significant depletion of solute in the topical product with time. In the simplest case, referred to by Riegelman (175), topical absorption through the skin is assumed to be first-order, with a rate constant k_a. Plasma concentrations are then described by the first part of Eq. (71) up until the product is removed at time T, if at all. The plasma concentrations after product removal are then described by the second part of Eq. (71).

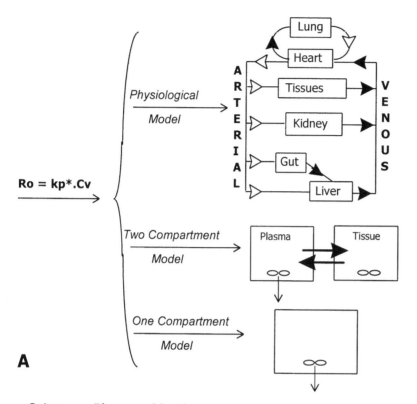

Cutaneous Pharmacokinetics

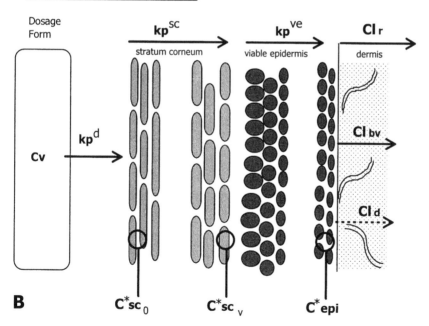

Figure 25 (A) Pharmacokinetic compartment and physiological models of the body; (B) diagrammatic representation of the pharmacokinetic processes involved in cutaneous permeation.

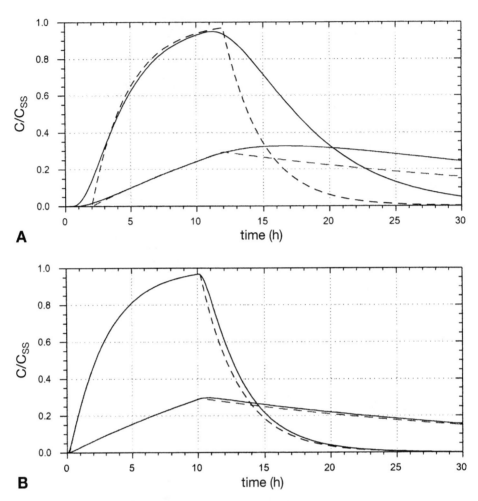

Figure 26 Plasma concentration–time profile after topical patch application and termination at 10 h. Solid line represents modeling to concentration with the diffusion model, and the dashed line shows approximation by the compartmental approach [see Eq. (71)]: (A) a lag time of 1 h; (B) a lag of 0.2 h. The upper curves of both (A) and (B) involve a half-life of 2 h and the lower curves a half-life of 20 h.

$$
C_p = \begin{cases}
0 & t < \text{lag} \\
\dfrac{k_a F dose}{V_{body}(k_{el} - k_a)} (\exp[-k_a(t - \text{lag})] - \exp[-k_{el}(t - \text{lag})]) & \text{lag} < t < \text{lag} + T \\
\dfrac{k_a F dose}{V_{body}(k_{el} - k_a)} (\exp[-k_a T] - \exp[-k_{el} T])\exp[-k_{el}(t - T - \text{lag})] & t \geq \text{lag} + T
\end{cases}
$$

$$(74)$$

A simplified form of Eq. (74) arises over very long times when absorption is much slower than elimination (i.e., $k_a \ll k_{el}$), as occurs when slow-releasing patches are applied to the skin. Hence, during the application phase, the plasma concentration is given by:

Table 9 Effective Plasma Concentrations, Epidermal Permeability Coefficients, Clearances, and Physicochemical Data Used to Predict Required Solute Transdermal Flux ($J_{s.a}$) [from Eq. (16)] for Passive Topical Delivery Systems

Solute	Plasma level ($\mu g/L$)	Cl_{body} (L/h)	Estimated $J_{s.a}$ required ($\mu g/h$)	$t_{1/2}$ (h)	MW	MP (°C)	log K_{oct}
Clonidine	0.2–2.0	13	2.6–26	6–20	230	140	1.77
Estradiol	0.04–0.06	615–790	24.6–47.4	0.05	272	176	2.69
Fentanyl	1	27–75	27–75	3–12	337	83	4.37
Isosorbide dinitrate	22	1.22	26.9	105	236	68	1.31
Nicotine	10–30	77.7	77.7–2231	2	162	~80	1.17
Nitroglycerin	1.2–11	13.5	16.2–148.5	0.04	227	13.5	1.62
Scopolamine	0.04	67.2	2.69	2.9	303	59	1.23
Testosterone	10–100	20.8	208–2080		288	153	3.31
Timolol	15	30.7	460.5	4.1	316	72	2.46
Triprolidine	5–15	43.7	218.5–655.5	2–6	278	60	3.92

$$C_p = \begin{cases} 0 & t < \text{lag} \\ \dfrac{k_a F dose}{V_{body}(k_{el} - k_a)} (\exp[-k_a(t - \text{lag})]) & t \geq \text{lag} \end{cases} \qquad (75)$$

Hence, the logarithm of the plasma concentration should be linearly related to the time after application over very long times. The urinary excretion rate is an alternative pharmacokinetic representation amount of drug in the body with time. Beckett et al. (176) showed that time to peak urinary excretion rate was longer and the actual peak height lower for transdermal applications than for the oral route. A comparison of the loglinear profiles for the oral and transdermal products shows that the terminal phase for the topical product has a half-life of 8.4 h whereas that for the oral dose is 3.3 h (175) (Fig. 27A). This behavior is consistent with the terminal pharmacokinetic phase being controlled by the rate of topical absorption (i.e., $k_a \ll k_{el}$). Such a relation may also be expected with a 7-day patch of timolol, which has an elimination half-life of 4.1 h (3). Consistent with this prediction, McCrea et al. (177) reported an apparent linear decline in plasma timolol concentrations during the application of a patch designed to release 50% of the timolol over 7 days.

The interpretation of k_a in terms of the underlying physicochemical properties of the system is not straightforward. In principle, if the vehicle is assumed to be homogeneous, k_a may be equivalent to the permeability constant k_p'' divided by the thickness of the vehicle film on the skin. However, this approximation applies only when the lag time is very small relative to $1/k_a$ (i.e., k_a lag $\ll 1$). In other cases, a quasi–steady-state assumption is inappropriate, with deviations most likely to be seen on cessation of dosing, as shown for constant input (see Fig. 27). Modeling of the plasma concentration–time profiles using model representation of the processes is more appropriate in these circumstances. Solution of expressions for finite vehicle applications show that, in general, an exponential absorption process exists under pseudo–steady-state conditions (i.e., $t > $ lag) when a diffusion model is used to describe transport across the SC (82), the exponent (equivalent to k_a) being a complex function of vehicle thickness, SC–vehicle partition coefficient, SC diffusion time, and other variables. An alternative approach is to use a compartmental representation of the skin (116,178–181), a multiple compartmental representation of the body, on a combination thereof. The pharmacokinetics of hyoscine (scopolamine) in the body after topical application has been described in terms of urinary excretion rate and a two-compartmental representation of the body (81). It is apparent in Figure 27B that the urinary excretion–time profile has four phases: (a) an absorption phase to peak, (b) the absorption nose associated with two-compartment kinetics, (c) a loglinear decline while the topical application continues, consistent with Eq. (75) and (d) a loglinear decline due to hyoscine elimination kinetics from the body as defined by Eq. (71) for $t > $ application time + various lags. Plasma levels after multiple dosing have also been described with such a model (116). Berner (182) has described the pharmacokinetics of drug delivery from transdermal controlled-release devices consisting of a membrane plus a reservoir or a monolithic slab.

When the pharmacokinetics associated with skin absorption are uncertain, it may be more appropriate to use some form of deconvolution. One of the earliest forms used in pharmacokinetics is the Wagner–Nelson model based on the assumption that the disposition of a solute after intravenous administration can be described by a single exponential. Birmingham et al. (183) used such a model to suggest that

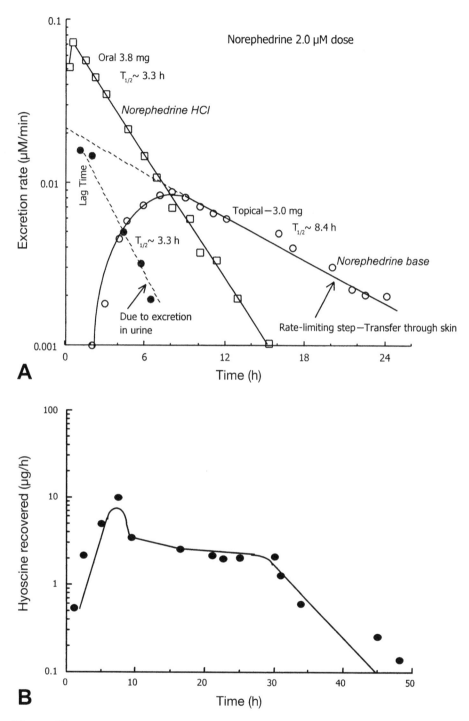

Figure 27 (A) Excretion rate following oral (□) and topical (○) application of norphedrine HCl. Subtraction from the extrapolation line represents a process involved in drug absorption and excretion (●); (B) excretion rate profile of in vivo transdermal hyoscine in humans showing a comparison of theory (solid line) and in vivo data (points) after a single application. (From Refs. 176[A]; 332[A].)

salicylic acid has an apparent first-order absorption rate when applied topically to rabbits in vivo.

The clinical effect (pharmacodynamic response) associated with topical delivery is commonly examined in terms of the subsequent relations between pharmacodynamic response and plasma concentrations of solutes. The E_{max} pharmacodynamic model, defined by Eq. (76) (184) has, for instance, been used to relate postexercise heart rate and plasma timolol concentrations C_p achieved after weekly application of a timolol patch (177).

$$E = E_o - \frac{E_{max} - C_p}{IC_{50} + C_p} \tag{76}$$

where E is the suppression of the postexercise heart rate, E_o is the baseline exercise heart rate before patch application, E_{max} is the maximal suppression for timolol, and IC_{50} is the C_p corresponding to 50% of E_{max}. Dermatological corticosteroid products are usually assessed for their clinical potency by skin blanching, as defined by a vasoconstrictor assay, and is related to the dose absorbed, as defined by dose duration. Singh et al. (185) have recently attempted to validate the vasoconstrictor assay dose–response relation using an area under the effect curve for each dose duration and the E_{max} model [see Eq. (76)]. They found that model fits to all individual subject dose–response data were unacceptable for all dermatological corticosteroid products tested, but that population dose–responses were adequately described by the E_{max} model.

C. Cutaneous Pharmacokinetics

In cutaneous pharmacokinetics, the goal is to target a given region of the skin as shown in Figure 25B. The first approach has been to estimate skin target concentrations based on a knowledge of the flux to the site and clearance from the site. Siddiqui et al. (186) first applied this approach using methotrexate. Later studies applied the approach with steroids (44) and phenols (101). An extension of this approach is the use of skin target-site free-drug concentration (C^*) estimated from in vitro flux data to predict topical in vivo efficacy. Such a model has been applied into examining critical factors that influence topical bioavailability and bioequivalence (187) and in interpreting the activity of solutes such as acyclovir (ACV) in the treatment of cutaneous herpes simplex virus-type 1 (HSV-1) infections (188). In work to date, the target site has presumed to be the basal cell layer of the epidermis. Patel et al. (189) have applied the C^* concept in predicting the topical antiviral efficacies of ACV formulations for the treatment of cutaneous HSV-1 infections, using a hairless mouse model. They found that, over a wide range of efficacies, the predictions based on C^* (estimated from the experimental in vitro fluxes) were in good agreement with in vivo antiviral efficacies measured at the end of a 4-day–treatment protocol. The physical model involved validating a "three-tiered" model for finite-dose drug uptake and transport in skin with experimentally determined input parameters (partition coefficient K, and steady-state permeability coefficients P, for the SC, viable epidermis, and dermis) (189).

Values of the steady-state unbound (or "free") concentration C_{ss}^* at different sites in the skin (see Fig. 25B) are related to the total concentration at that site C_{ss} by the fraction of solute unbound at that site f_u^*, and is defined by the input flux to that site and removal clearance from that site. An expression for the steady-state concentration at a given site C_{ss}, such as the receptor solution in an in vitro situation, has been defined (83). Hence, substituting $C_{ss}^* = f_u^* C_{ss}$ into that expression, C_{ss}^* is given by Eq. (77).

$$C_{ss}^* = f_u^* C_{ss} = \frac{f_u^* k_p' C_v}{(Cl^*/A) + k_p'(K^*/K_m)} \qquad (77)$$

where k_p' is the apparent permeability coefficient to the site; C is the concentration of solute in the vehicle; Cl^* is the clearance from the site divided by the area of application A; k' is the SC–site partition coefficient ($K^* = C_{sc}/C_{ss}$); and K_m is the SC–vehicle partition coefficient ($K_m = C_{sc}/C_v$). Siddiqui et al. (44) and Roberts (101) used in vitro skin permeability coefficients for steroids and phenols together with in vivo dermal clearances to estimate C_{ss}^*. The different estimated concentrations C_{ss}^* for phenols and steroids shown in Figure 28A, reflect the different magnitudes in the clearance per unit area (see Fig. 28B), and the apparent permeability coefficient corrected for partition coefficient effects.

D. Clearance

As shown in Figures 4 and 25B, solutes may either be carried away by the local blood supply on entering the dermis or transported into deeper tissues by perfusion or diffusion. Lymphatic transport is significant for the larger molecular weight solutes (190). Microdialysis studies have shown significant penetration of solutes into deeper tissues of human subjects after topical application (191,192). Muller et al. (193) have found that when a diclofenac foam (5%) was administered epicutaneously at the thigh (80 mg/200 cm² twice daily for a period of 7 days) of healthy volunteers, significantly higher skeletal muscle concentrations of diclofenac (219.68 ± 66.36 ng/mL) were found compared with that found in plasma (18.75 ± 4.97 ng/mL).

We have examined the disposition of a series of solutes assuming a compartmental representation for each tissue (194) (Fig. 29A). In general, the localized targeting of solutes to deeper tissues is seen most readily (a) at early times after application when there are negligible tissue concentrations as a consequence of re-circulation (see Fig. 29B); (b) when there is a high body clearance for the solute; and (c) when vasoactive agents are used appropriately (194). The steady-state concentration in each tissue C_{ss}^i is described by

$$C_{ss}^i = \frac{k_p^{i'} C_v A}{Cl^i + k_p^{i'} A(K^i/K_m)} \qquad (78)$$

The free concentration in each tissue is described by Eq. 77. According to Eq. 78, C_{ss}^i can be increased by one of two methods. The most usual is to reduce the clearance from the tissue and higher tissues by the local blood supply using vasoconstriction or removal of solute (Fig. 30A). The second approach is to facilitate transport to deeper tissues by using the local blood supply (195) (see Fig. 30A). We have shown higher tissue levels at early times with methyl salicylate, consistent with the increase in blood flow induced by methyl salicylate in human cutaneous vessels (196). Figure 30B illustrates the effect of vasoconstriction on the loss of lidocaine from a dermal cell relative to anesthetized blood flow and no blood flow (197). The corresponding tissue levels of lidocaine at various depths of application after 24 h

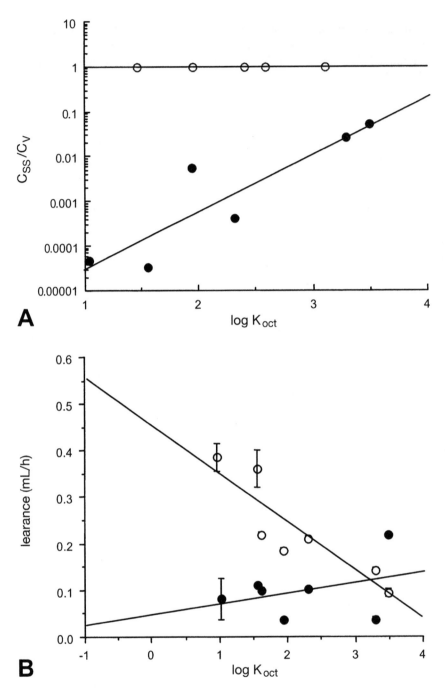

Figure 28   (A) Epidermal concentration (C_{ss})/vehicle concentration (C_v) ratio for phenols (○) and steroids (●) and their octanol–water partition coefficients (K_{oct}); (B) clearance of steroids from dermal diffusion cells against their octanol–water partition coefficients (K_{oct}) for sacrificed (●) and anesthetized (○) rats. (From Refs. 101[A]; 44[B].)

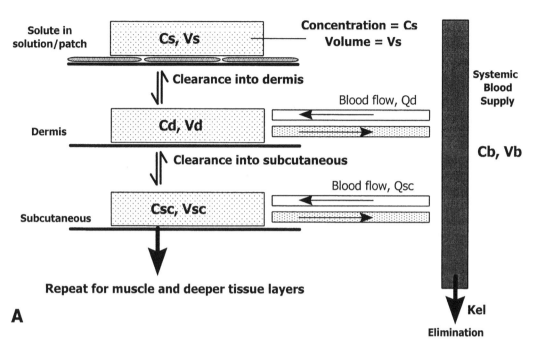

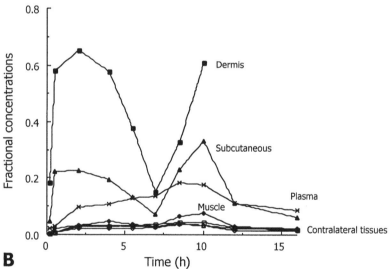

Figure 29 (A) Tissue deposition of dermally applied solute assuming compartmental representation; (B) salicylic acid concentration in different tissues and in the plasma after the application of an aqueous salicylic acid solution to the dermis. (From Ref. 194.)

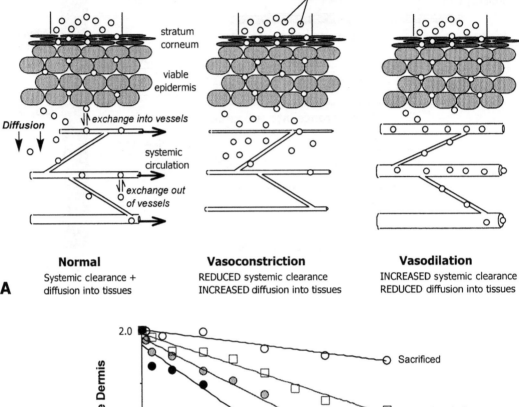

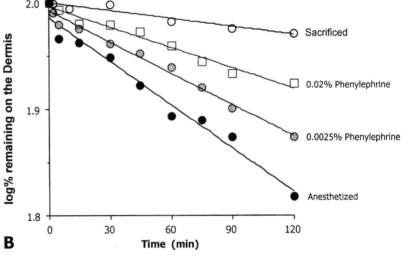

Figure 30 (A) Diagrammatic representation of the effects of dermal blood flow on the clearance of topically applied solutes; (B) effect of phenylephrine on the dermal clearance of lidocaine applied in aqueous solution to rats; (C) effect of phenylephrine on the fraction of dermally applied lidocaine penetrating into deeper tissues in rats; (D) distribution of hydrocortisone in human skin in vitro (○) and in vivo (●). (From Refs. 197[B,C]; 127[D].)

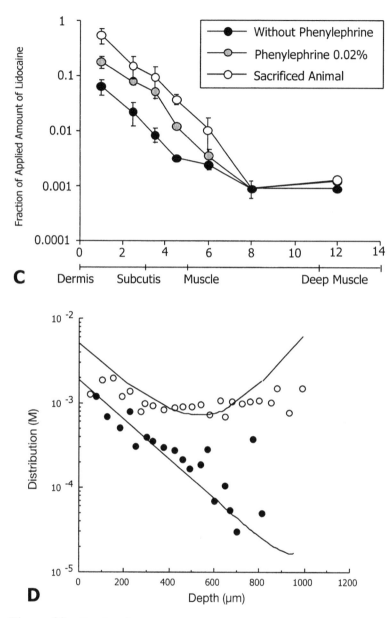

Figure 30 Continued

under such conditions are shown in Figure 30C. In Figure 30D, the distribution of hydrocortisone after application to human skin in vitro and in vivo is shown. An apparent logarithmic profile is observed in vivo consistent with a constant removal by the blood supply through the depth of the dermis. In contrast, the in vitro profile plateaus, as would be expected if there was a clearance limitation (see Fig. 30D).

Protein binding of a solute may also affect tissue levels as shown in Figure 31 (190). Plasma protein binding will facilitate removal into the blood supply (see Fig.

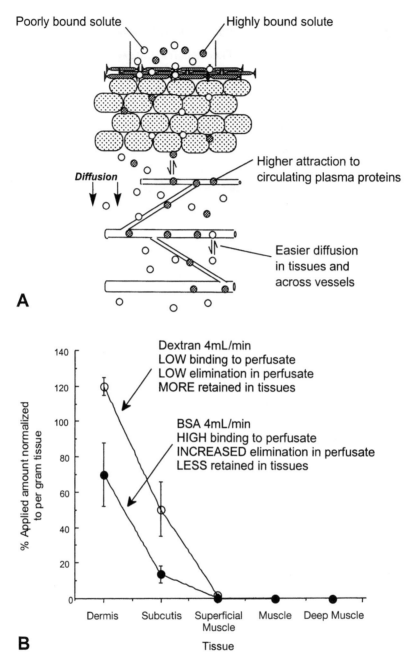

Figure 31 (A) Diagrammatic representation of the effect of protein binding on the diffusion and clearance of topically applied solutes; (B) effect of protein binding in perfusate on the tissue distribution of diclofenac applied dermally to the perfused hindlimb of rats. (From Ref. 198.)

31B), whereas tissue binding will impede binding into lower tissues in a manner similar to the way binding can slow SC diffusion. We have recently described a physiological disposition model for solute disposition below a topical site (198). The model predicts that the half-life for elimination of a solute in an underlying tissue is related to the volume of plasma in a tissue (V_p) with an extravascular water volume (V_E), the perfusate flow rate (Q_p), perfusate protein binding (fu_p), and tissue protein binding (fu_T). An approximate solution applying in most cases is Eq. (79).

$$t_{0.5el} = \frac{0.693 V_D}{Q_P} \simeq \frac{0.693\, fu_P\, V_{TE}}{fu_T Q_P} \tag{79}$$

Hence, the retention of solutes in tissues after topical application is dependent on the relative magnitude of binding in blood (fu_p) and tissue binding (fu_T) as well as on the tissue blood flow Q_p. Interestingly, diazepam and diclofenac are highly bound to underlying topical tissues (198) and, according to Eq. (79) are more likely to be retained in these tissues than the more nonbound solutes.

E. Skin Metabolism

Metabolism of solutes in the SC and epidermis are also clearance mechanisms, which can affect skin permeabilities and resultant pharmacological effects. The metabolism of solutes by skin enzymes has recently been reviewed (4). In relation to skin transport, difficulties may arise from the as yet undefined anatomical distribution of metabolizing enzymes, both in the various layers of the skin and appendages and the variable activity that may arise from processing skin for permeation studies. For instance, the activity of some enzymes is reduced by the process of heating used to separate the dermis from the epidermis (199). Some of the key enzymes involved in skin metabolism include aryl hydrocarbon hydroxylase, deethylases, hydrolases, monoxygenases, esterases, peptidases, and aminopeptidases. This skin enzyme activity can vary among species and may be induced. A major outcome of these enzymatic activities is the *skin first-pass effect* whereby a significant proportion of the solute is metabolized between application to the skin and diffusing to its site of action in regions of the skin or into the systemic circulation. Nakashima et al. (200) used intravenous and transdermal ointment administration of nitroglycerin to estimate that the fraction of nitroglycerin avoiding this first-pass was 0.68–0.76 and was comparable with values reported in rhesus monkeys (0.80–0.84). A higher skin first-pass effect has been reported for methyl salicylate, for which the first-pass availability in both humans (190) and rats (201) is very low. It has been suggested that the new retinoid, tazarotene, is superior to those used orally because of its limited percutaneous penetration as well as its rapid esterase metabolism in the skin to a more water-soluble active metabolite tazarotenic acid. The latter has a resultant systemic absorption of between 1 (normal) and 5% (psoriasis) on repeated applications (202,203).

Bronaugh et al. (204) have also reviewed some aspects of cutaneous metabolism during in vitro percutaneous absorption. From the perspective of skin transport, skin metabolism can be adequately modeled only by using a two-phase model. One of the first studies in this area was that of Ando et al. (205). Higuchi's group have since then reported in several papers on the effects of skin metabolism on solute transport (206–210). In a later study, the influence of low levels of ethanol on the simultaneous diffusion and metabolism of β-estradiol with several enzyme distribution models was determined (211). The best model was that for which the enzyme activity resided totally in the epidermis and near the basal layer of the epidermis. Liu et al. (212) reported there would be less metabolism and that a much smaller amount of the transdermal metabolite would be taken up by the blood capillary owing to the shorter dermis pathlength for permeants in vivo than in vitro, when using dermatomed split-thickness skin.

Recently, Sugibayashi et al. (213) reported the effect of enzyme distribution in skin on the simultaneous transport and metabolism of ethyl nicotinate in hairless rat skin after its topical application. Gysler et al. (214) studied the skin penetration and metabolism of topical glucocorticoids in reconstructed epidermis and in excised human skin. The influence of enzyme distribution on skin permeation was also studied (215). Species differences can also be important (216).

Seko et al. (217) used pretreatment with an esterase inhibitor, diisopropyl fluorophosphate, to study the penetration of propyl and butyl paraben across Wistar rat skin in vitro. A two-layer skin diffusion model predicted an increasing metabolic rate and decreased the lag time for penetration of both the parent and metabolite.

F. Stratum Corneum Reservoir

Malkinson and Ferguson (218) originally proposed an SC reservoir based on their studies with radiolabeled corticosteroids. Several other workers (including Vickers, Barry and Woodford, MacKenzie and Atkinson, Wickrema, Shima, and others) have provided evidence for the reservoir effect—these studies have been summarized by Barry (126). In essence, occlusion for some time after topical application of a steroid promotes absorption of the steroid retained in the horny layer, with vasoconstriction occurring as a result. Barry (126) suggests that the phenomenon is probably a consequence of the high solubility and normal low diffusivity of the steroids in the SC. Hence, occlusion and accelerants may enhance the sorption of steroids into the SC. Because of its low intrinsic diffusivity, the steroid remains relatively trapped in the SC, once these are removed. On reocclusion or application of an accelerant, diffusivity is promoted and the steroid is absorbed. This same effect may be possible for certain solutes by altering dermal blood flow. Hence, high, reservoir-like concentrations are observed in various tissues in the presence of a vasoconstrictor (see Fig. 30C): remove the vasoconstrictor and the reservoir effect is lost.

The assessment of SC concentrations in tape-stripped samples is becoming increasingly advocated in topical bioequivalence assessment (219,220). It can be seen from Figure 7 that the amount of solute in the SC at steady-state M_{sc}^{ss} is given by Eq. 80.

$$M_{sc}^{ss} = \frac{K_{sc-v}h_{sc}}{2}\left(C_v - \frac{C_{ve}^{ss}}{K_{ve-v}}\right) \approx \underset{C_{ve}^{ss}\to 0}{\frac{K_{sc-v}C_v h_{sc}}{2}} \tag{80}$$

Hence, assuming $C_{ve}^{ss} \to 0$ and using Eqs. (63), (64), and (65), results in Eq. (81).

$$J_{sc}^{ss} = \frac{M_{sc}^{ss}D_{sc}^a}{2h_{sc}^2} \tag{81}$$

Hence, the steady-state flux is directly proportional to the amount of solute in the SC, providing that D_{sc}^a [see Eq. (38)] is constant. The observed flux for percutaneous absorption of small solutes, varying 40-fold, was very similar to that predicted from amounts present in the SC after 30 min of contact (221). This relation also exists for variable application time, application dose, vehicle, and anatomical site (222).

G. In Vivo–In Vitro Correlations

Franz (223) conducted one of the first validations of in vitro human skin absorption by correlating the percentage of dose absorbed in vitro and in vivo. Despite the range of data sources a good correspondance was evident. An important component of this study was the application of a finite dose of solutes in vitro in mimicking in vivo results. He highlighted the importance of regional variability in skin absorption and desquamation, pointing out that, with an average of one cell layer of SC lost per day, any material that had not diffused beyond that layer in a day in vivo would be lost.

There appears to be an excellent correlation between in vitro human epidermal penetration flux and in vivo flux, as deduced by residual drug and pharmacokinetic methods, when both groups of studies have been conducted by a single group of investigators (as shown in Fig. 32A for 18 drugs by Shaw's group). The extent of correlation between in vitro and in vivo transport when cutaneous metabolism is involved is not quite so impressive. For instance, esterase activity in vivo results in an almost complete first-pass metabolism of methyl salicylate in human skin, much greater than is seen in vitro (224) (see Fig. 32B). Overall, there is an increasing emphasis on the combined use of in vitro cadaver skin, in vivo animal pharmaco-kinetics, in vivo human pharmacokinetics, and in vitro–in vivo pharmacology and microbiology in dermal bioequivalence assessments (225).

H. Species Differences

There is a substantive and often contradictory literature on species differences in percutaneous absorption. Walters and Roberts (226), who summarized the variation in lipid content and SC thickness among the various species, reported the results from a number of studies in which differences in percutaneous penetration among species have been compared. Figure 33A shows the results obtained for the penetration of paraquat and water permeability coefficients through the skin from various species (227).

I. Other Determinants of Percutaneous Absorption

Roberts and Walters (3) have summarized the effects of age, race, temperature, re-peated applications, disease, and body site on percutaneous absorption. Further de-tails are given in Chapter 5. In general, temperature increases absorption and ab-sorption from different sites is in the order scrotum > scalp > forehead > postauricular > abdomen > arm > palm > plantar. Figure 33B shows the variation of absorption with site as reported for salicylate excretion after topical methyl salicylate application (224).

VI. TRANSAPPENDAGEAL TRANSPORT

The role of appendages in skin transport has been controversial and remains so. The earliest evidence to support the existence of the transfollicular route of transdermal

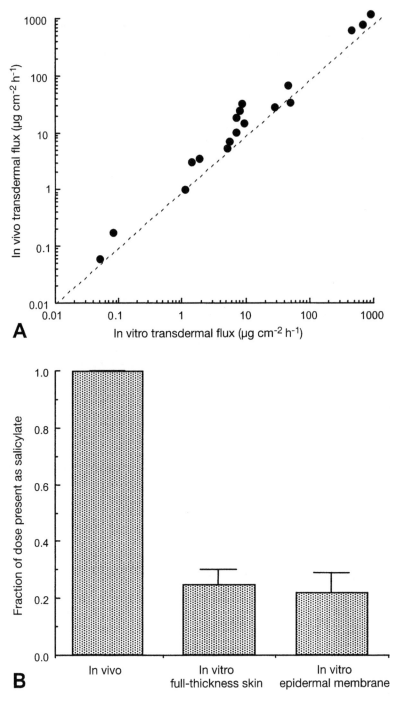

Figure 32 (A) Correlation between in vivo and in vitro transdermal drug flux: each datapoint represents a different drug, dashed line indicates perfect correlation between in vivo and in vitro transdermal drug flux; (B) fraction of total concentration of topically applied methylsalicylate determined in microdialysate or in vitro diffusion cell studies present as salicylate following application of a 20% methylsalicylate formulation. (From Refs. 332[A]; 119[B].)

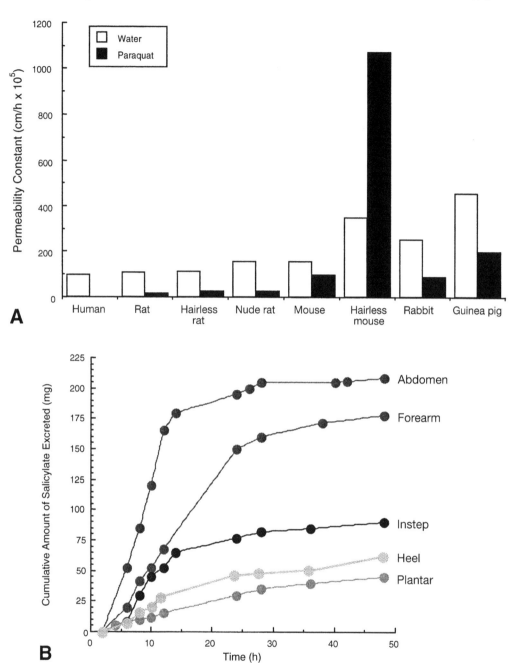

Figure 33 (A) Effect of species on the in vitro absorption of water and paraquat through excised skin; (B) the influence of skin application site on topical bioavailability determined from cumulative urinary recovery of salicylate following application of 5 g Metsal (methyl-salicylate)/50 cm² for 10 h in human volunteers. (From Refs. 227[A]; 224[B].)

transport, back in the 1940s, were largely qualitative histological studies based on stain and dye localization within the appendages (Table 10). Table 11 outlines some of the literature supporting the existence of follicular penetration of topically applied solutes. Scheuplein (91) suggested that appendageal route dominates transport early before the lag time is reached for transcellular transport. However, at longer times, transcellular transport dominates (Fig. 34). Although it remains generally accepted

Table 10 Historical Evidence for the Existence of Transfollicular Penetration Following Topical Application

Evidence	Skin	Ref.
Preferential staining of hair follicles following topical application of iron, bismuth, sulfonamides, and dyes in a number of different vehicles	Guinea pig, human	Mackee et al. (261)
Changes in pharmacological response to epinephrine and histamine applied in propylene glycol observed with changes in follicular density	Human	Shelley and Melton (311)
Follicular deposition of vitamin A observed by quantitative fluorescent microscopy following application in various solvents	Guinea pig	Montagna (232)
^{14}C-Labeled pesticide absorption and urinary excretion increased over follicle-rich areas such as the scalp and forehead, follicular route "possibly" contributing	Human	Maibach et al. (236)
Trichlorocarbanalide compound deposition in follicles and sebaceous glands seen to vary with vehicle	Guinea pig	Black et al. (312)
[^{3}H]hydrocortisone from hydroalcoholic vehicle penetrates normal skin 50-fold faster than follicle-free skin. Retention also 20 to 30-fold higher in normal skin	Rat	Illel and Schaefer (313)
Particle size dependency of follicular penetration, optimum 5 μm	Rat	Schaefer et al. (314)
Greater concentrations of hydrocortisone and testosterone observed in epidermis and dermis of normal skin, particularly at the depth of sebaceous glands, compared with follicle-free skin. In vivo effect less pronounced than in vitro	Rat	Hueber et al. (229)
Flux and absorption of caffeine, niflumic acid, and p-aminobenzoate threefold slower in follicle-free skin	Rat	Illel et al. (315)
Particle size dependency of follicular penetration, optimum 5 μm. Targeting of the antiacne drug adapalene into follicles is achieved using 5-μm microspheres as particulate carriers	Mouse, human	Rolland et al. (230)

Table 11 Summary of Studies Examining Formulation Effects on Transfollicular Transport

Evidence	Skin	Ref.
Deposition of vitamin A into the follicular duct and sebaceous glands was seen within 10 min of application in ethanol or chloroform vehicles, compared with much slower penetration from oleic acid or petrolatum.	Guinea pig	Montagna (232)
Penetration of a trichlorocarbanalide compound into follicles and sebaceous glands seen after application in a nonsoap sodium alkoyl isothionate detergent, compared with a soapy vehicle, which resulted in penetration into the stratum corneum.	Guinea pig	Rutherford and Black (316)
Within 2 h of application of [^{3}H]estradiol in DMSO, ethylene glycol, or sesame oil vehicles, radioactivity could be detected in the epidermis, follicles, and sebaceous glands. At 24 h hair follicles and sebaceous glands still retained high activity—possible depot function.	Rat	Bidmon et al. (317)
Increased deposition of fluorescent beads into follicles following application in lipoidal vehicles.	Rat	Schaefer et al. (314)
Liposomal entrapment of calcein, melanin, and DNA allowed delivery into follicles compared with control aqueous solutions.	Histocultured mouse skin	Li et al. (319,320)
Migration of topically applied steroids through the follicular duct or accumulation in the sebaceous glands varied with the polarity of the steroid and the lipophilicity of the vehicle.	Rat	Hueber et al. (321)
Vehicles favoring the transfollicular penetration of pyridostigmine included ethanol, DMSO, and propylene glycol.	Rat	Bamba and Wepierre (235)
50% ethanol, glyceryl dilaurate-based nonionic liposomes and egg phosphatidylcholine-based liposomes all achieved appreciable deposits of cimetidine into the pilosebaceous unit, although possible ion-pairing of cimetidine to phospholipids reduced its pharmacological activity in these formulations.	Hamster ears	Lieb et al. (322)
Nonionic liposome formulations were superior to phospholipid-based formulations for delivery of interferon-α and cyclosporine to follicles. Both formulations were far superior to aqueous solutions.	Hamster ears	Niemiec et al. (323)

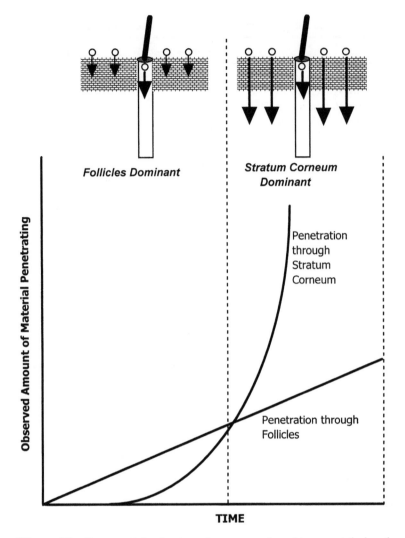

Figure 34 Suggested domination of transappendageal transport during the earlier stages of percutaneous penetration. (From Ref. 91.)

that the intercellular route may dominate during the steady-state penetration of compounds, it has been argued that the skin appendages (hair follicles, pilosebaceous and eccrine glands) may offer an alternative pathway for a diffusing molecule.

A. Hair Follicles and Sweat Ducts

Possible routes of penetration through hair follicles could involve the hair fiber itself, through the outer root sheath of the hair into the viable cells of the follicle, or through the air-filled canal and into the sebaceous gland. In addition, the release of sebum by the sebaceous glands may provide a lipoidal pathway that may influence absorption by this route (228). The route for the sweat duct may involve diffusion through either the lumen or walls to below the epidermis and through the thin ring of keratinized cells. Dense capillary networks closely envelop the bases of both the hair follicles and sweat ducts, providing access to the circulation for most molecules reaching these regions. Hueber et al. (229) used the observation that a higher reservoir and permeability barrier function in appendage-free (scar) SC than in normal SC, as supporting evidence for a significant contribution of the appendageal route to overall skin transport.

There are estimated to be close to 500–1000 pilosebaceous units/square centimeter of skin on areas such as the face and scalp, each with an orifice with a diameter of 50–100 μm and 4×10^{-5} cm^2 surface area. These orifices represent 0.1% of the surface area of the skin in low-density areas and up to 10% in high-density areas, such as those on the face and scalp. The openings lead down to an epithelial surface which does not have a protective SC, and exists only from the ostia of the sebaceous gland upward to the skin surface (Fig. 35). These characteristics have been used to selectively target drugs into the hair follicles and sebaceous glands. Given that the exposed surface area of appendages is much higher than that of the openings used in earlier evaluations and that the current intercellular transport also has a restricted area for transport, the role of the follicle as a pathway for transdermal delivery is being reconsidered. In the hair follicle, for example, the outer

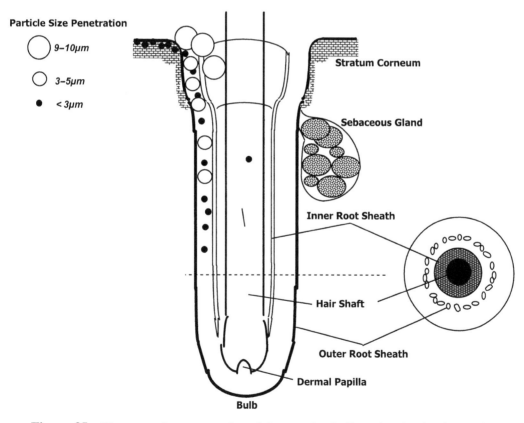

Figure 35 Diagrammatic representation of the postulated effect of molecular size on the penetration of solutes into hair follicles. (From Ref. 231.)

root sheath (see Fig. 35) is thought to be of greatest importance for drug delivery, as this layer is continuous with the epidermis and is indistinguishable from it, which potentially allows for increased surface area for absorption beneath the skin's surface. In addition, there is increasing demand for localized drug delivery to the hair follicle itself, particularly for the treatment of dermatological disorders such as acne, alopecia, areata, and androgenetic hair loss.

One of the most important determinants of targeted follicular transport is the particle size of applied materials. By using fluorescent microspheres, Rolland et al. (230) showed that the degree of penetration into the human hair follicle was inversely related to particle size. The optimum size at which microspheres selectively entered follicles was 3–10 μm; below this size, particles were also seen to be distributed in the superficial layers of the SC (see Fig. 35). The depth of penetration into the follicle was also determined by size, with beads of 5–7 μm reaching the deeper parts of the upper follicle, though rarely penetrating the superficial SC, and those between 9 and 10 μm were observed to concentrate only around the opening of the follicles, but not inside them. No beads larger than 1 μm were observed to penetrate as deep as the hair bulb of the follicles.

Recent studies using fluorescence-labeled oligonucleotides and dextrans applied to fresh human scalp skin confirmed earlier findings that follicular penetration was determined by size and also charge (231). These studies also identified that the primary anatomical structures for the pathway(s) of intrafollicular delivery of these molecules were along the junction of the inner root sheath and outer root sheath (see Fig. 35). Although this pattern of distribution was particularly evident with oligomers formulated with cationic lipids, the molecular features that allow a selected agent to move into and through this region await definition. In the same study, it was also noticed that rhodamine-labeled dextran (3000 MW) applied in a hydroalcoholic formulation (40% ethanol) was present in the center of the hair shaft as well as within the follicle. It was speculated that this region of the hair shaft may be more amorphous relative to keratin content compared with the rest of the hair; therefore, it may be more permeable to certain agents, although whether entry occurred by diffusion across the hair shaft or down the cut end of the hair was unclear.

The concept that vehicle and formulation significantly influence the rate of drug localization within hair follicles, following the application of vitamin A in various vehicles to guinea pigs back, was noticed in 1954 by Montagna (232). Reviews covering formulation effects, in particular the use of liposomes, to optimize transfollicular delivery can be found (233,234). Some of the literature in this area pointing to the favoritism for lipophilic vehicles in follicular targeting is outlined in Table 11.

It can be seen from Table 11 that alcoholic vehicles are among those tending to increase transfollicular penetration. Bamba and Wepierre (235) speculated that, as ethanol is primarily a lipid solvent, as well as increasing the fluidity of lipid areas within the SC and extending the hydrophobic domain between polar head groups, it was also acting on the sebum within the follicles and allowing the more rapid migration of solute in the sebaceous glands, thereby making the transfollicular pathway predominant in the initial stages of absorption. DMSO is thought to act on normal SC by creating a "solvent pathway" through the skin or fluidizing the lipids (140). However, in the studies of Bamba and Wepierre (235) the concentration was too weak to have this enhancing effect through the whole epidermis, and it was suggested that its solvent properties would favor pilosebaceous migration by incorporating the drug in the sebum.

The contribution of transappendageal transport to systemic clearance, rather than local deposition, was considered by Maibach et al. (236) following the topical application of radiolabeled pesticides to human volunteers. A greater urinary recovery was noted after application to follicle-rich areas, such as the forehead and scalp, than after application to less hairy areas such as the forearm. The authors concluded that transfollicular transport could not be ruled out as a contributing factor to the observed differences. However, studies examining the effect of increasing hair folicle density on percutaneous absorption (237) failed to show any correlation with the amount of solute absorbed, suggesting that the follicles' overall contribution to transdermal delivery is negligible.

In the recent review by Lauer (238), it was concluded that the contribution of the pilosebaceous unit to localized and percutaneous absorption may have been underestimated in the past and that a more detailed understanding of formulation factors, such as drug and vehicle physicochemical properties and particle size, may allow optimization of follicular delivery. The potential clinical significance of the ability to selectively deliver drugs to follicles for the treatment of associated dermatological disorders warrants the pursuit of this area of transdermal research.

B. Permeation of the Nail Plate

The nature of the barrier to topical drug delivery of the nail plate as an extension of the skin has been given some consideration, especially relative to the potential for local treatment of fungal infections. The nail plate itself is produced mainly by differentiation of cells in the nail matrix, and it comprises three horizontal layers: a thin dorsal lamina, the thicker intermediate lamina, and a ventral layer from the nail bed (Fig. 36). It contains significant amounts of phospholipid, mainly in the dorsal and intermediate layers, which contribute to its flexibility, and it has essentially the same keratin analysis as hair. The keratins in hair and nail are classified as "hard" trichocyte keratins, unlike the "soft" epithelial keratins found throughout the skin.

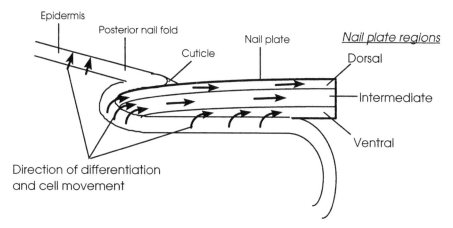

Figure 36 Diagrammatic representation of the structures of the nail plate.

Studies by Walters et al. (239) indicated that there is a marked difference between the permeability characteristics of the nail plate and epidermis. These observed differences have been largely attributed to the relative amounts of lipid and protein regulation within the structures and the possible differences in the physicochemical nature of the respective phases. The SC contains nearly 10% lipid, predominantly intercellular, whereas lipid levels in the nail plate are near 1%, which combined with lower water levels of about only 10% (240) affords the nail plate much different barrier properties to external penetrants than those of the SC. The composition of the nail plate suggests that it would be comparatively less sensitive than the SC to the effects of penetration enhancers that produce their effects by delipidization or fluidization of intercellular lipids. Studies using DMSO have found this to be true (241–245), together with the observation that the nail plate is incapable of absorbing much of the applied DMSO (246). The nail plate is permeable to dilute aqueous solutions of a series of low molecular weight homologous alcohols (239); however, it possesses a unique ability to restrict increasingly the diffusive passage with increases in alkyl chain length. It was suggested that the nail plate possessed a highly "polar" penetration route that is capable of excluding permeants on the basis of their hydrophobicities. Interestingly, the applied concentration of alcohols was a determinant of their penetration velocities, with pure liquid forms of the alcohols giving a fivefold decrease in permeation (247).

The existence of a minor "lipid" pathway through the nail matrix, which could become rate-controlling for hydrophobic solutes, has been suggested based on the significant decrease in the permeation of the hydrophobic entity n-decanol following delipidization of the nail plate by chloroform/methanol (247). Increases in nail plate absorption of the antifungal amorolfine following pretreatment with DMSO have been demonstrated (248); however, a decrease in the absorption of methanol and hexanol applied with DMSO was also noted (247). Overall it can be seen that the permeation characteristics of the nail plate do not correspond to those of the SC and that the effects of skin penetration enhancers, such as DMSO, cannot be extrapolated to the nail plate. The targeting of drugs into this area requires the further understanding of its barrier properties to develop enhancers that specifically interfere with the more keratinous matrix of the nail plate.

Mertin and Lippold (249) have recently examined the role of octanol–water partition coefficients and water solubility on the permeability and maximum flux of a homologous series of nicotinic acid esters through the nail plate and a bovine hoof keratin membrane model. They noticed that the permeability coefficient of the nail plate as well as the hoof membrane did not increase with increasing partition coefficient (range 7 to >51,000) or lipophilicity of the ester, indicating that these barriers behaved similar to hydrophilic gel membranes, rather than lipophilic partition membranes, as seen in the SC. Further penetration studies with acetaminophen (paracetamol) and phenacetin showed that maximum flux was first a function of drug solubility in water or in the swollen keratin. Mertin and Lippold were also able to predict the maximum flux of ten antimycotics through the nail plate on the basis of their penetration rates through the hoof membrane and their water solubilities (250).

One of the major problems with the application of traditional formulations, such as creams or solutions of polar vehicles, to the nail plate is the ease with which they are wiped or washed off the nails. This results in the need for at least once daily application, but patient compliance over the several months that are usually required for treatment of fungal infections, such as onychomycosis, decreases with the length of therapy. Consequently, lipophilic vehicles, especially nail lacquers, have been developed because of their better adhesion. Pittrof et al. (251) used a porcine hoof model membrane to optimize a lacquer formulation of amorolfine for drug release, stability, and ease of application and showed that the drug was able to penetrate the horn barrier with approximately 1.8% of the applied dose available under the horn after a 7-day penetration period. More recently, Mertin and Lippold (252) used the hoof model and the compound chloramphenicol to show that the course of penetration through the nail plate from a lacquer vehicle followed first-order kinetics after a lag time of 400 h. Penetration was seen to be initially membrane-controlled and later matrix-controlled as the drug concentration in the lacquer decreased.

VII. CONCLUDING REMARKS

As seen in this chapter, the skin is an important barrier to the ingress of both therapeutic and potentially toxic compounds. We have attempted to present a review of the current understanding of the factors that affect the ability of a solute to traverse the barrier and how these processes can be influenced to enhance penetration. The increasing availability of data from both in vitro and in vivo studies will, we hope, make the elucidation of the finer details of these transport processes and the predictability of skin permeation kinetics, through mathematical modeling and interpretation of structure–activity relations and local physiology, much easier in the years to come. Until such time, however, we hope that the concepts and studies summarized in this chapter will help bring to many an appreciation of the complex nature of skin transport processes and the amount of work that has been involved in bringing us to the level of understanding we have today.

ACKNOWLEDGMENTS

The authors wish to acknowledge the financial support of the National Health and Medical Research Council of Australia and the continuing encouragement of the Lions Medical Research and PAH Research Foundations. In addition, one of us (MSR) also thanks Jonathan Hadgraft, John Pugh, and Karen Milne for our often lively discussions on which convention is best applied in using chemical potential and activity as determinants of skin transport. The authors would like to thank Brett MacFarlane for help in preparation of the manuscript.

REFERENCES

1. Halkier–Sorensen L. Occupational skin disease. In: Roberts MS, Walters KA, eds. Dermal Absorption and Toxicity Assessment. New York: Marcel Dekker, pp 127–154, 1998.
2. Cross SE. Topical therapeutic agents used in wound care. In: Roberts MS, Walters KA, eds. Dermal Absorption and Toxicity Assessment. New York: Marcel Dekker, pp 415–442, 1998.
3. Roberts MS, Walters KA. The relationship between function and barrier function of the skin. In: Roberts MS, Walters KA, eds. Dermal Absorption and Toxicity Assessment. New York: Marcel Dekker, pp 1–42, 1998.

4. Hotchkiss SAM. Dermal metabolism. In: Roberts MS, Walters KA, eds. Dermal Absorption and Toxicity Assessment. New York: Marcel Dekker, pp 43–101, 1998.
5. Scheuplein RJ, Blank IH. Permeability of the skin. Physiol Rev 51:702–747, 1971.
6. Scott RC, Dugard PH, Doss AW. Permeability of abnormal rat skin. J Invest Dermatol 86:201–207, 1986.
7. Schaefer H, Zesch A, Stuttgen G. Penetration, permeation, and absorption of triamcinolone acetonide in normal and psoriatic skin. Arch Dermatol Res 258:241–249, 1977.
8. Michaels AS, Chandrasekaran SK, Shaw JE. Drug permeation through human skin. Theory and in vitro experimental measurement. Am Inst Chem Eng J 21:985–996, 1975.
9. Elias PM, Menon GK. Structural and lipid biochemical correlates of the epidermal permeability barrier. Adv Lipid Res 24:1–26, 1991.
10. Elias PM, Bonar L, Grayson S, Baden HP. X-ray diffraction analysis of stratum corneum membrane couplets. J Invest Dermatol 80:213–214, 1983.
11. Williams ML, Elias PM. The extracellular matrix of stratum corneum: role of lipids in normal and physiological function. 3:95–122, 1987.
12. Tregear RT. The permeability of mammalian skin to ions. J Invest Dermatol 46:16–23, 1966.
13. Kligman AM. The use of sebum. Br J Dermatol 75:307–319, 1963.
14. Kligman A, Christophers E. Preparation of isolated sheets of human stratum corneum. Arch Dermatol 88:70–73, 1963.
15. Greisemer RD. J Soc Cosmet Chem 11:79, 1960.
16. Miescher G, Schuonberg A. Bull Schweiz Akad Med Wiss 1:101, 1944.
17. Tregear RT. THe permeability of skin to albumin, dextrans and polyvinyl pyrrolidone. J Invest Dermatol 46:24–27, 1966.
18. Higuchi TJ. Physical chemical analysis of percutaneous absorption process from creams and ointments. J Soc Cosmet Chem 11:85–87, 1960.
19. Rothman S. J Soc Cosmet Chem 6:193, 1955.
20. Gloor M, Willebrandt U, Thomer G, et al. Water content of the horny layer and skin surface lipids. Arch Dermatol Res 268:221–223, 1980.
21. Scheuplein RJ. Mechanism of percutaneous adsorption. I. Routes of penetration and the influence of solubility. J Invest Dermatol 45:334–346, 1965.
22. Scheuplein RJ. Permeability of the skin, a review of major concepts. Curr Prob Dermatol 7:172–186, 1978.
23. Roberts MS, Anderson RA, Swarbrick J, Moore DE. The percutaneous absorption of phenolic compounds: the mechanism of diffusion across the stratum corneum. J Pharm Pharmacol 30:486–490, 1978.
24. Roberts MS, Anderson RA, Swarbrick J. Permeability of human epidermis to phenolic compounds. J Pharm Pharmacol 29:677–683, 1977.
25. Sweeney TM, Downing DT. The role of lipids in the epidermal barrier to water diffusion. J Invest Dermatol 55:135–140, 1970.
26. Squier CA. The permeability of keratinised and nonkeratinised oral apithelium to horseradish peroxidase. J Ultrastruct Res 43:160–177, 1973.
27. Elias PM, Friend DS. The permeability barrier in mammalian epidermis. J Cell Biol 65:180–191, 1975.
28. Chandrasekaran SK, Shaw JE. Factors affecting the percutaneous absorption of drugs. Curr Probl Dermatol 7:142–155, 1975.
29. Albery WJ, Hadgraft J. Percutaneous absorption: theoretical description. J Pharm Pharmacol 31:129–39, 1979.
30. Bunge AL, Guy RH, Hadgraft J. The determination of a diffusional pathlength through the stratum corneum. Int J Pharm 188:121–124, 1999.

31. Elias PM, Goerke J, Friend D. Mammalian epidermal barrier layer lipids: composition and influence on structure. J Invest Dermatol 69:535–546, 1977.

32. Elias PM, Brown BE, Fritsch P, Goerke J, Grey GM, White RJ. Localisation and composition of lipids in neonatal mouse stratum granulosum and stratum corneum. J Invest Dermatol 73:339–348, 1979.

33. Berenson GS, Burch GE. Studies on diffusion of water through dead human skin. Am J Trop Med 31:843–853, 1951.

34. Niemanic MK, Elias PM. In situ precipitation in novel cytochemical technique for visualisation of permeability pathways in mammalian stratum corneum. J Histochem Cytochem 28:573–578, 1980.

35. Boddé HE, van den Brink I, Koerten HK, de Haan FHN. Visualisation of in vitro percutaneous penetration of mercuric chloride; transport through intercellular space versus cellular uptake through desmosomes. J Control Release 15:227–236, 1991.

36. Elias PM, Cooper ER, Korc A, Brown BE. Percutaneous transport in relation to stratum corneum structure and lipid composition. J Invest Dermatol 76:297–301, 1981.

37. Bouwstra JA, Dubbelaar FE, Gooris GS, Weerheim AM, Ponec M. The role of ceramide composition in the lipid organisation of the skin barrier. Biochim Biophys Acta 1419:127–136, 1999.

38. Elias PM, Mak V, Thornfield C, Feingold KR. Interference with stratum corneum lipid biogenesis. In: Bronaugh RL, Maibach HI, eds. Percutaneous Absorption: Drugs–Cosmetics–Mechanisms–Methodology. 3rd ed. New York: Marcel Dekker, pp 3–55, 1999.

39. Wertz PW. Epidermal lipids. Semin Dermatol 11:106–113, 1992.

40. Bouwstra JA, Gooris GS, Dubbelaar FE, Weerheim AM, Ijzerman AP, Ponec M. Role of ceramide 1 in the molecular organization of the stratum corneum lipids. J Lipid Res 39:186–96, 1998.

41. Bouwstra JA, Thewalt J, Gooris GS, Kitson N. A model membrane approach to the epidermal permeability barrier: an X-ray diffraction study. Biochemistry, 36:7717–7725, 1997.

42. Southwell D, Barry BW. Penetration enhancers for human skin: mode of action of 2-pyrrolidone and dimethylformamide on partition and diffusion of model compounds water, n-alcohols, and caffeine. J Invest Dermatol 80:507–514, 1983.

43. Blank IH, McAuliffe DJ. Penetration of benzene through human skin. J Invest Dermatol 85:522–526, 1985.

44. Siddiqui O, Roberts MS, Polack AE. Percutaneous absorption of steroids: relative contributions of epidermal penetration and dermal clearance. J Pharmacokinet Biopharm 17:405–424, 1989.

45. Kai T, Isami T, Kobata K, Kurosaki Y, Nakayama T, Kimura T. Keratinised epithelial transport of β-blocking agents. 1. Relationship between physicochemical properties of drugs and the flux across rat skin and hamster cheek pouch. Chem Pharm Bull 40:2498–2504, 1992.

46. Kim YH, Ghanem AH, Higuchi WI. Model studies of epidermal permeability. Semin Dermatol 11:145–56, 1992.

47. Cooper ER, Kasting G. Transport across epithelial membranes. J Controlled Release 6:23–35, 1987.

48. Matsuzaki K, Imaoka T, Asano M, Miyajima K. Development of a model membrane system using stratum corneum lipids for estimation of drug skin permeability. Chem Pharm Bull (Tokyo) 41:575–579, 1993.

49. Peck KD, Ghanem AH, Higuchi WI. Hindered diffusion of polar molecules through and effective pore radii estimates of intact and ethanol treated human epidermal membrane. Pharm Res 11:1306–1314, 1994.

50. Peck KD, Ghanem AH, Higuchi WI. The effect of temperature upon the permeation of polar and ionic solutes through human epidermal membrane. J Pharm Sci 84:975–982, 1995.

51. Yoshida NH, Roberts MS. Solute molecular size and transdermal iontophoresis across excised human skin. J Controlled Release 25:177–195, 1993.
52. Lai PM, Roberts MS. An analysis of solute structure–human epidermal transport relationships in epidermal iontophoresis using the ionic mobility: pore model. J Controlled Release 58:323–333, 1999.
53. Sznitowska M, Janicki S, Williams AC. Intracellular or intercellular localization of the polar pathway of penetration across stratum corneum. J Pharm Sci 87:1109–1114, 1998.
54. Menon GK, Elias PM. Morphologic basis for a pore-pathway in mammalian stratum corneum. Skin Pharmacol 10:235–246, 1997.
55. Florence AT, Attwood D. Physicochemical Principles of Pharmacy. 2nd ed. New York: Macmillan, pp 158–160, 1988.
56. Florence AT, Attwood D. Physicochemical Principles of Pharmacy. 2nd ed. New York: Macmillan, pp 48–56, 1988.
57. Barry BW, Harrison SM, Dugard PH. Correlation of thermodynamic activity and vapour diffusion through human skin for the model compound benzyl alcohol. J Pharm Pharmacol 37:84–90, 1985.
58. Davis SS, Higuchi T, Rytting JH. Determination of thermodynamics of functional groups in solutions of drug molecules. In: Bean HS, Beckett AH, Carless JR, eds. Advances in Pharmaceutical Sciences. Vol 4. London: Academic Press, pp 73–261, 1974.
59. Hildebrand JH, Scott RL. Regular Solutions. Englewood Cliffs, NJ: Prentice Hall, 1962.
60. Yalkowski SH, Vlavani SC. Solubility and partitioning I. Solubility of nonelectrolytes in water. J Pharm Sci 69:912–922, 1980.
61. Grant DJW, Higuchi T. Solubility Behaviour of Organic Compounds. New York: Wiley, pp 24, 1990.
62. Sloan KB, Koch SAM, Siver KG, Flowers FP. Use of solubility parameters of drug and vehicle to predict fluxes through skin. J Invest Dermatol 87:244–252, 1986.
63. Kadir R, Stempler D, Liron Z, Cohen S. Delivery of theophylline into excised human skin from alkanoic acid solutions: a "push–pull" mechanism. J Pharm Sci 76:774–779, 1987.
64. Kadir R, Stempler D, Liron Z, Cohen S. Penetration of adenosine into excised human skin from binary vehicles: the enhancement factor. J Pharm Sci 77:409–413, 1988.
65. Kadir R, Stempler D, Liron Z, Cohen S. Penetration of theophylline and adenosine into excised human skin from binary and ternary vehicles: effect of a nonionic surfactant. J Pharm Sci 78:149–153, 1989.
66. Bagley EB, Nelson TP, Scigliano JM. Three dimensional solubility parameters and their relationship to internal pressure measurements in polar and hydrogen bonding solvents. Paint Res Inst Proc 43:35–42, 1971.
67. Martin A, Newberger J, Adjei A. Extended Hildebrand solubility approach: solubility of theophylline in binary polar solvents. J Pharm Sci 69:487–491, 1980.
68. Groning R, Braun FJ. Three dimensional solubility parameters and their use in characterising the permeation of drugs through the skin. Pharmazie 51:337–341, 1996.
69. Ruelle P, Rey–Merment C, Buchmann M, Nam–Tran H, Kesselring UW, Huyskens PL. A new predictive equation for the solubility of drugs based on the thermodynamics of mobile disorder. Pharm Res 8:840–850, 1991.
70. Ando HY, Schultz TW, Schnaare RL, Sugita ET. Percutaneous absorption: a new physicochemical predictive model for maximum human in vivo penetration rates. J Pharm Sci 73:461–467, 1984.
71. Roberts MS, Pugh WJ, Hadgraft J, Watkinson AC. Epidermal permeability—penetrant structure relationships. 1. An analysis of methods of predicting penetration of monofunctional solutes from aqueous solutions. Int J Pharm 126:219–233, 1995.

72. Abraham MH, Chadha HS, Mitchell RC. Hydrogen bonding part 36. Determination of blood brain distribution using octanol–water partition co-efficient. Drug Des Discov 13:123–131, 1995.

73. Yalkowsky SH, Pinal R, Banjeree S. Water solubility: a critique of the solvatochromic approach. J Pharm Sci 77:74–77, 1988.

74. Abraham MH, Chadna HS, Whiting GS, Mitchell RC. Hydrogen bonding part 32. An analysis of water–octanol and water alkane partitioning and the delta log P parameter of Seiler. J Pharm Sci 83:1085–1100, 1994.

75. Leo A, Hansch C, Elkins D. Partition coefficients and their uses. Chem Rev 71:525–516, 1971.

76. Roberts MS, Anderson RA, Swarbrick J. Permeability of human epidermis to phenolic compounds. J Pharm Pharmacol 29:677–683, 1977.

77. Roberts MS, Pugh WJ, Hadgraft J. Epidermal permeability: penetrant structure relationships. 2. The effect of H-bonding groups in penetrants on their diffusion through the stratum corneum. Int J Pharm 132:23–32, 1996.

78. Roberts MS, Anderson RA, Moore DE, Swarbrick J. The distribution of nonelectrolytes between human stratum corneum and water. Aust J Pharm Sci 6:77–82, 1977.

79. Raykar PV, Fung MC, Anderson BD. The role of protein and lipid domain in the uptake of solutes by human stratum corneum. Pharm Res 5:140–150, 1988.

80. Chandrasekaran SK, Michaels AS, Campbell PS, and Shaw JE. Scopolamine permeation through human skin in vitro. AIChE J 22:828–832, 1976.

81. Chandrasekaran SK, Bayne W, Shaw JE. Pharmacokinetics of drug permeation through human skin. J Pharm Sci 67:1370–1374, 1978.

82. Anissimov Y, Roberts MS. Diffusion modelling of percutaneous kinetics: 2. Finite vehicle volume and solvent deposited solids. J Pharm Sci (submitted for publication).

83. Roberts MS, Anissimov Y, Gonsalvez R. Mathematical models in percutaneous absorption. In: Bronaugh RL, Maibach HI, eds. Percutaneous Absorption: Drugs–Cosmetics–Mechanisms–Methodology. 3rd ed. New York: Marcel Dekker, pp 3–55, 1999.

84. Kasting GB, Smith RL, Anderson BD. Prodrugs for dermal delivery: solubility, molecular size and functional group effects. In: Sloane KB, ed. Prodrugs. New York: Marcel Dekker, pp 117–161, 1992.

85. Tojo K. Random brick model for drug transport across stratum corneum. J Pharm Sci 76:889–891, 1987.

86. Heisig M, Lieckfeldt R, Wittum G, Mazurkevich G, Lee G. Nonsteady-state descriptions of drug permeation through stratum corneum I. The basic brick and mortar model. Pharm Res 13:421–426, 1996.

87. Plewig G, Marples RR. Regional differences in cell sizes of the human stratum corneum. Part 1. J Invest Dermatol 54:13–18, 1970.

88. Anderson BD, Raykar PV. Solute structure–permeability relationships in human stratum corneum. J Invest Dermatol 93:280–286, 1989.

89. Roberts MS, Anderson RA. The percutaneous absorption of phenolic compounds: the effect of vehicles on the penetration of phenol. J Pharm Pharmacol 27:599–605, 1975.

90. Anissimov Y, Roberts MS. Diffusion modelling of percutaneous absorption kinetics: 1 Effects of flow rate, receptor sampling rate, viable epidermal resistance and epidermal metabolism for a constant donor concentration. J Pharm Sci 88:1201–1209, 1999.

91. Scheuplein RJ. Properties of the skin as a membrane. Adv Biol Skin 22:125–152, 1972.

92. Kasting GB, Smith RL, Cooper ER. Effect of lipid solubility and molecular size on percutaneous absorption. Pharmacol Skin 1:138–153, 1987.

93. Flynn GL. Physicochemical determinants of skin absorption. In: Gerrity TR, Henry CJ, eds. Principles of Route to Route Extrapolation for Risk Assessment. Elsevier Science, pp 93–127, 1990.

94. Potts RO, Guy RH. Predicting skin permeability. Pharm Res 9:663–669, 1992.

95. Pugh WJ, Hadgraft J, Roberts MS. Physicochemical determinants of stratum corneum permeation. In: Roberts MS, Walters KA, eds. Dermal Absorption and Toxicity Assessment. New York: Marcel Dekker, pp 245–268, 1998.

96. Barratt MD. Quantitative structure activity relationships for skin permeability. Toxicol In Vitro 9:27–37, 1995.

97. Liron Z, Cohen S. Percutaneous absorption of alkanoic acids II: Application of regular solution theory. J Pharm Sci 73:538–542, 1984.

98. Pugh WJ, Roberts MS, Hadgraft J. Epidermal permeability–penetrant structure relationships. Part 3. Effect of hydrogen bonding interactions and molecular size on diffusion across the stratum corneum. Int J Pharm 138:149–165, 1996.

99. Dearden JC. The QSAR prediction of melting point, a property of environmental relevance. Sci Total Environ 109–110:59–68, 1991.

100. Blank IH. Penetration of low-molecular weight alcohols into the skin. I. The effect of concentration of alcohol and type of vehicle. J Invest Dermatol 43:425–420, 1964.

101. Roberts MS. Structure–permeability considerations in percutaneous absorption. In: Scott RC, Guy RH, Hadgraft J, Bodde HE, eds. Prediction of Percutaneous Penetration—Methods, Measurement and Modelling. Vol 2. IBC Technical Services, pp 210–228, 1991.

102. Hagedorn–Leweke U, Lippold BC. Absorption of sunscreens and other compounds through human skin in vivo: derivation of a method to predict maximum fluxes. Pharm Res 12:1354–1360, 1995.

103. Le VH, Lippold BC. Influence of physicochemical properties of homologous esters of nicotinic acid on skin permeability and maximum flux. Int J Pharm 124:285–292, 1995.

104. Roberts MS, Walker M. Water: the most natural penetration enhancer. In: Walters KA, Hadgraf J, eds. Pharmaceutical Skin Enhancement. New York: Marcel Dekker, pp 1–30, 1993.

105. Wurster DE, Kramer SF. Investigations of some factors influencing percutaneous absorption. J Pharm Sci 50:288–293, 1961.

106. Munies R, Wurster DE. Factors influencing percutaneous absorption of methyl ethyl ketone. J Pharm Sci 54:554–556, 1964.

107. Bucks D, Maibach H. Occlusion does not uniformly enhance penetration in vivo. In: Bronaugh RL, Maibach H, eds. Percutaneous Absorption: Drugs–Cosmetics–Mechanisms–Methodology, 3rd ed. New York: Marcel Dekker, pp 81–105, 1999.

108. Bond JR, Barry BW. Hairless mouse skin is limited as a model for assessing the effects of penetration enhancers in human skin. J Invest Dermatol 90:810–813, 1988.

109. Zatz J. Modification of skin permeation by surface active agents. In: Zatz J, ed. Skin Permeation Fundamentals and Application. Wheaton, IL: Allured Publishing, pp 149–162, 1993.

110. Twist JN, Zatz JL. Influence of solvents on paraben permeation through idealised skin model membranes. J Soc Cosmet Chem 37:429–444, 1986.

111. Barry B, Furand O, Woodford R, Ulshagen K, Hogstad G. Control of the bioavailability of a topical steroid; comparison of desonide creams 0.05% and 0.1% by vasoconstrictor studies and clinical trials. Clin Exp Dermatol 12:406–409, 1987.

112. Tojo K, Chiang CC, Chien YW. Drug permeation across the skin: effect of penetrant hydrophilicity. J Pharm Sci 76:123–126, 1987.

113. Pellet MA, Roberts MS, Hadgraft J. Supersaturated solutions evaluated with an in vitro stratum corneum tape stripping technique. Int J Pharm 151:91–98, 1997.

114. Schwarb FP, Imanidis G, Smith EW, Haigh JM, and Surber C. Effect of concentration and degree of saturation of topical fluocinonide formulations on in vitro membrane transport and in vivo availability on human skin. Pharm Res 16:909–915, 1999.

115. Hadgraft J. Passive enhancement strategies in topical and transdermal drug delivery. Int J Pharm 184:1–6, 1999.

116. Guy RH, Hadgraft J. Physicochemical interpretation of the pharmacokinetics of percutaneous absorption. J Pharmacokinet Biopharm 11:189–203, 1983.

117. Yano T, Nakagawa A, Masayoshi T, Noda K. Skin permeability of nonsteroidal anti-inflammatory drugs in man. Life Sci 39:1043–1050, 1986.

118. Hinz RS, Lorence CR, Hodson CD, Hansch C, Hall LL, Guy RH. Percutaneous penetration of para-substituted phenols in vitro. Fundam Appl Toxicol 17:575–583, 1991.

119. Hostynek JJ, Magee PS. Modelling in vivo human skin absorption. Quant Struct Activ Relation 16:473–479, 1997.

120. Roberts WJ, Sloan KB. Correlation of aqueous and lipid solubilities with flux for prodrugs of 5-fluorouracil, theophylline, and 6-mercaptopurine: a Potts–Guy approach. J Pharm Sci 88:515–522, 1999.

121. Roberts MS, Anderson RA. The percutaneous absorption of phenolic compounds: the effect of vehicles on the penetration of phenol. J Pharm Pharmacol 27:599–605, 1975.

122. Wurster DE. Some physical–chemical factors influencing percutaneous absorption from dermatologicals. Curr Probl Dermatol 7:156–171, 1978.

123. Menon GK, Lee SH, Roberts MS. Ultrastructural effects of some solvents on the stratum corneum and other skin components: evidence for an extended mosaic partitioning model of the skin barrier. In: Roberts MS, Walters KA, eds. Dermal Absorption and Toxicity Assessment. New York: Marcel Dekker, pp 727–751, 1998.

124. Twist JN, Zatz JL. A model for alcohol-enhanced permeation through polydimethylsiloxane membranes. J Pharm Sci 79:28–31, 1990.

125. Bronaugh RL, Congdon ER. Percutaneous absorption of hair dyes: correlations with partition coefficients. J Invest Dermatol 83:124–127, 1984.

126. Barry BW. Dermatological Formulations: Percutaneous Absorption. New York: Marcel Dekker, 1983.

127. Schaefer H, Rieldelmeier TE. Skin Barrier. Basel: Karger, pp 153–212, 1996.

128. Idson B. Hydration and percutaneous absorption. Curr Probl Dermatol 7:132–141, 1978.

129. Alonso A, Meirelles NC, Tabak M. Effect of hydration upon the fluidity of intercellular membranes of stratum corneum: an EPR study. Biochim Biophys Acta 1237:6–15, 1995.

130. Alonso A, Meirelles NC, Yushmanov VE, Tabak M. Water increases the fluidity of intercellular membranes of stratum corneum: correlation with water permeability, elastic, and electrical resistance properties. J Invest Dermatol 106:1058–63, 1996.

131. Wiedmann TS. Influence of hydration on epidermal tissue. J Pharm Sci 77:1037–1041, 1988.

132. Walters KA, Hadgraft J. Pharmaceutical Skin Enhancement. New York: Marcel Dekker, 1993.

133. Smith EW, Maibach HI. Percutaneous Penetration Enhancers. Boca Raton, FL: CRC Press, 1995.

134. Williams AC, Barry BW. Chemical penetration enhancement: possibilities and problems. In: Roberts MS, Walters KA, eds. Dermal Absorption and Toxicity Assessment. New York: Marcel Dekker, pp 297–312, 1989.

135. Cornwell PA, Barry BW. The routes of penetration of ions and 5-fluorouracil across human skin and the mechanisms of terpene skin penetration enhancers. Int J Pharm 94:189–94, 1993.

136. Yamamoto T, Yamamoto Y. Electrical properties of the epidermal stratum corneum. Med Biol Eng 14:151–158, 1976.

137. Williams AC, Barry BW. Terpenes and the lipid–protein-partitioning theory of skin penetration enhancement. Pharm Res 8:17–24, 1991.

138. Francoeur ML, Golden GM, Potts RO. Oleic acid: its effects on stratum corneum in relation to (trans)dermal drug delivery. Pharm Res 7:621–627, 1991.

139. Ongpipattanakul B, Burnette RR, Potts RO, and Francoeur ML. Evidence that oleic acid exists in a separate phase within stratum corneum lipids. Pharm Res 8:350–354, 1991.

140. Cooper ER. In: Solution Behaviour of Surfactants: Theoretical and Applied Aspects. New York: Plenum, pp 1505–1516, 1982.

141. Goodman, Barry B. Action of penetration enhancers on human stratum corneum as assessed by differential scanning calorimetry. In: Bronaugh RL, Maibach HI, eds. Percutaneous Absorption, Mechanisms–Methodology–Drug Delivery. 2nd ed. New York: Marcel Dekker, pp 567–593, 1989.

142. Scheuplein RJ, Ross L. Effect of surfactants and solvents on the permeability of the epidermis. J Soc Cosmet Chem 21:853–873, 1970.

143. Harrison JE, Watkins AC, Green DM, Hadgraft J, Brain K. The relative effect of Azone and Transcutol on permeant diffusivity and solubility in human stratum corneum. Pharm Res 13:542–6, 1996.

144. Zhao K, Singh J. In vitro percutaneous absorption enhancement of propranolol hydrochloride through epidermis by terpenes/ethanol. J Controlled Release 62:359–366, 1999.

145. Bach M, Lippold BC. Percutaneous penetration enhancement and its quantification. Eur J Pharm Biopharm 46:1–13, 1998.

146. Potts RO, Guy RH. Mechanisms of Transdermal Drug Delivery. New York: Marcel Dekker, 1997.

147. Lee AR, Tojo K. Characterization of skin permeation of vitamin C: theoretical analysis of penetration profiles and differential scanning calorimetry study. Chem Pharm Bull (Tokyo) 46:174–177, 1998.

148. Goates CY, Knutson K. Enhanced permeation of polar compounds through human epidermis. I. Permeability and membrane structural changes in the presence of short chain alcohols. Biochim Biophys Acta 1195:169–179, 1994.

149. Wolfram LJ, Maibach HI. Percutaneous absorption of hair dyes. Arch Dermatol Res 277:235–241, 1985.

150. Dressler W. Percutaneous absorption of hair dyes. In: Roberts MS, Walters KA, eds. Dermal Absorption and Toxicity Assessment. New York: Marcel Dekker, pp 489–536, 1998.

151. Bhargava HN, Leonard PA. Triclosan: Applications and safety. Am J Infect. Control 24:209–218, 1996.

152. Islam MS, Zhao L, McDougal JN, Flynn GL. Uptake of chloroform by skin during short exposures to contaminated water. Risk Anal 15:343–352, 1995.

153. Imokawa G, Mishima Y. Cumulative effect of surfactants on cutaneous horny layers: adsorption onto human keratin layers in vivo. Contact Dermatitis 5:357–366, 1979.

154. Hashiguchi T, Ryu A, Itoyama T. Uchida K, Yamaguchi H. Study of the effective dose of topical antifungal agent. Omoconazole nitrate, on the basis of percutaneous pharmacokinetics in guinea pigs and mice. J Pharm Pharmacol 49:757–761, 1997.

155. Fullerton A, Hoelgaard A. Binding of nickel to human epidermis in vitro. Br J Dermatol 119:675–682, 1988.

156. Dick IP, Blain PG, Williams FM. The percutaneous absorption and skin distribution of lindane in man. II. In vitro studies. Hum Exp Toxicol 16:652–657, 1997.

157. Scheuplein RJ, Blank IH. Mechanisms of percutaneous absorption. IV Penetration of nonelectrolytes (alcohols) from aqueous solutions and from pure liquids. J Invest Dermatol 60:286–296, 1973.

158. Barry BW, Harrison SM, Dugard PH. Vapour and liquid diffusion of model penetrants through human skin; correlation with thermodynamic activity. J Pharm Pharmacol 37:226–236, 1985.

159. Berner B, Wilson DR, Guy RH, Mazzenga GC, Clarke FH, Maibach HI. The relationship of pK_a and acute skin irritation in man. Pharm Res 5:660–663, 1998.
160. Nangia A, Andersen PH, Berner B, Maibach HI. High dissociation constants (pK_a) of basic permeants are associated with in vivo skin irritation in man. Contact dermatitis 34:237–242, 1996.
161. Barratt MD. Quantitative structure activity relationships (QSAR's) for skin corrosivity of organic acids, bases and phenols: principal components and neural network analysis of extended datasets. Toxic In Vitro 10:85–94, 1996.
162. Barratt MD. Quantitative structure activity relationships for skin corrosivity of organic acids, bases and phenols. Toxicol Lett 75:169–176, 1995.
163. Poulsen BJ, Chowhan ZT, Pritchard R, Katz M. The use of mixtures of topical steroids as a mechanism for improving total drug bioavailability: a preliminary report. Curr Probl Dermatol 7:107–120, 1978.
164. Gao S, Singh J. In vitro percutaneous absorption enhancement of a lipophilic drug tamoxifen by terpenes. J Controlled Release 51:193–199, 1998.
165. Kaplun–Frischoff Y, Touitou E. Testosterone skin penetration enhancement by menthol through formation of eutectic with drug and interaction with skin lipids. J Pharm Sci 86:1394–1399, 1997.
166. Sasaki H, Kojima M, Mori Y, Nakamura J, and Shibasaki J. Enhancing effect of pyrrolidone derivatives on transdermal penetration of 5-fluorouracil, triamcinolone acetonide, indomethacin and flurbiprofen. J Pharm Sci 80:533–538, 1991.
167. Kondo S, Yamanaka C, Sugimoto I. Enhancement of transdermal delivery by superfluous thermodynamic potential. III. Percutaneous absorption of nifedipine in rats. J Pharmacobiodyn 10:643–749, 1987.
168. Hadgraft J, Peck J, Williams DG, Pugh WJ, Allan G. Mechanism of action of skin penetration enhancers/retarders: Azone and analogues. Int J Pharm 141:17–25, 1996.
169. Gorukanti SR, Li L, Kim KH. Transdermal delivery of antiparkinsonian agent, benztropine. I. Effect of vehicles on skin permeation. Int J Pharm 192:159–172, 1999.
170. Kim N, El-Khalili M, Henary MM, Strekowski L, Michniak BB. Percutaneous penetration enhancement activity of aromatic S,S-dimethyliminosulfurane. Int J Pharm 187:219–229, 1999.
171. Zatz J. Simulation studies of skin permeation. J Soc Cosmet Chem 43:37–48, 1992.
172. Singh P, Roberts MS, Maibach HI. Modelling of plasma levels of drugs following transdermal iontophoresis. J Controlled Release 33:293–298, 1995.
173. Hadgraft J, Wolff M. In vitro/In vivo correlations in transdermal drug delivery. In: Roberts MS, Walters KA, eds. Dermal Absorption and Toxicity Assessment. New York: Marcel Dekker, pp 269–280, 1998.
174. Cleary GW. Transdermal delivery systems: a medical rationale. In: Shah VP, Maibach HI, eds. Topical Drug Bioavailability, Bioequivalence and Penetration. New York: Plenum Press, pp 17–68, 1993.
175. Riegelman S. Pharmacokinetics. Pharmacokinetic factors affecting epidermal penetration and percutaneous absorption. Clin Pharmacol Ther 16:873–883, 1974.
176. Beckett AH, Gorrod JW, Taylor DC. Comparison of oral and percutaneous routes in man for the systemic absorption of "ephedrines." J Pharm Pharmacol 24(suppl):65–70, 1972.
177. McCrea JB, Vlasses PH, Franz TJ, Zeoli L. Transdermal timolol: β blockade and plasma concentrations after application for 48 hours and 7 days. Pharmacotherapy, 10:289–293, 1990.
178. Kubota K. A compartment model for percutaneous drug absorption. J Pharm Sci 80:502–504, 1991.
179. Kubota K, Maibach HI. A compartment model for percutaneous absorption: compatibility of lag time and steady-state flux with diffusion model. J Pharm Sci 81:863–865, 1992.

180. Guy RH, Hadgraft J, Maibach HI. A pharmacokinetic model for percutaneous absorption. J Int Pharm 11:119–129, 1982.
181. Wallace SM, Barnett G. Pharmacokinetic analysis of percutaneous absorption: evidence of parallel penetration pathways for methotrexate. J Pharmacokinet Biopharm 6:315–325, 1978.
182. Berner B. Pharmacokinetics of transdermal drug delivery. J Pharm Sci 74:718–721, 1985.
183. Birmingham BK, Greene DS, Rhodes CT. Percutaneous absorption of salicylic acid in rabbits. Drug Dev Ind Pharm 5:29–40, 1979.
184. Holford NHG, Scheiner LB. Understanding the dose–effect relationship: clinical application of pharmacokinetic–pharmacodynamic models. Clin Pharmacokinet 6:429–453, 1981.
185. Singh GJ, Adams WP, Lesko LJ, Shah VP, Molzon JA, Williams RL, and Pershing LK. Development of in vivo bioequivalence methodology for dermatologic corticosteroids based on pharmacokinetic modelling. Clin Pharmacol Ther 66:346–357, 1999.
186. Siddiqui O, Roberts MS, Polack AE. Topical absorption of methotrexate: role of dermal transport. Int J Pharm 27:193–203, 1985.
187. Borsadia S, Ghanem AH, Seta Y, Higuchi WI, Flynn GL, Behl CR, and Shah VP. Factors to be considered in the evaluation of bioavailability and bioequivalence of topical formulations. Skin Pharmacol 5:129–145, 1992.
188. Mehta SC, Afouna MI, Ghanem AH, Higuchi WI, Kern ER. Relationship of skin target site free drug concentration (C*) to the in vivo efficacy: an extensive evaluation of the predictive value of the C* concept using acyclovir as a model drug. J Pharm Sci 86: 797–801, 1997.
189. Patel PJ, Ghanem AH, Higuchi WI, Srinivasan V, Kern ER. Correlation of in vivo topical efficacies with in vitro predictions using acyclovir formulations in the treatment of cutaneous HSV-1 infections in hairless mice: an evaluation of the predictive value of the C* concept. Antiviral Res 29:279–286, 1996.
190. Cross SE, Roberts MS. Subcutaneous absorption kinetics and local tissue distribution of interferon and other solutes. J Pharm Pharmacol 45:606–609, 1993.
191. Cross SE, Anderson C, Roberts MS. Topical penetration of commercial salicylate esters and salts using human isolated skin and clinical microdialysis studies. Br J Clin Pharmacol 46:29–35, 1998.
192. Tegeder I, Muth–Selbach U, Lotsch J, Rusing G, Oelkers R, Brune K, Meller S, Kelm GR, Sorgel F, Geisslinger G. Application of microdialysis for the determination of muscle and subcutaneous tissue concentrations after oral and topical ibuprofen administration. Clin Pharmacol Ther 65:357–368, 1999.
193. Muller M, Rastelli C, Ferri P, Jansen B, Breiteneder H, Eichler HG. Transdermal penetration of diclofenac after multiple epicutaneous administration. J Rheumatol 25: 1833–1836, 1998.
194. Singh P, Maibach HI, Roberts MS. Site of effects. In: Roberts MS, Walters KA, eds. Dermal Absorption and Toxicity Assessment. New York: Marcel Dekker, pp 353–370, 1998.
195. McNeill SC, Potts RO, Francoeur ML. Local enhanced topical delivery (LETD) of drugs: does it truly exist? Pharm Res 9:1422–1427, 1992.
196. Cross SE, Megwa SA, Benson HAE, Roberts MS. Self promotion of deep tissue penetration and distribution of methylsalicylate after topical application. Pharm Res 16: 427–433, 1999.
197. Singh P, Roberts MS. Effects of vasoconstriction on dermal pharmacokinetics and local tissue distribution of compounds. J Pharm Sci 83:783–791, 1994.
198. Roberts MS, Cross SE. A physiological pharmacokinetic model for solute disposition in tissues below a topical application site. Pharm Res 16:1394–1400, 1999.

199. Wester RC, Christoffel J, Hartway T, Poblete N, Maibach HI, Forsell J. Human cadaver skin viability for in vitro percutaneous absorption: storage and detrimental effects of heat separation and freezing. Pharm Res 15:82–84, 1998.

200. Nakashima E, Noonan PK, Benet LZ. Transdermal bioavailability and first-pass skin metabolism: a preliminary evaluation with nitroglycerine. J Pharmacokinet Biopharm 15:423–437, 1987.

201. Megwa, SA, Benson HA, Roberts MS. Percutaneous absorption of salicylates from some commercially available topical products containing methyl salicylate or salicylic salts in rats. J Pharm Pharmacol 47:891–896, 1995.

202. Marks R. Pharmacokinetics and safety review of tazarotene. J Am Acad Dermatol 39: S134–S138, 1998.

203. Tang-Liu DD, Matsumoto RM, Usansky JI. Clinical pharmacokinetics and drug metabolism of tazarotene: a novel topical treatment for acne and psoriasis. Clin Pharmacokinet 37:273–287, 1999.

204. Bronaugh RL, Kraeling EK, Yourick JJ, Hood HL. Cutaneous metabolism during in vitro percutaneous absorption. In: Bronaugh RL, Maibach HI, eds. Percutaneous Absorption: Drugs–Cosmetics–Mechanisms–Methodology, 3rd ed. New York: Marcel Dekker, pp 57–64, 1999.

205. Ando HY, Ho NF, Higuchi WI. Skin as an active metabolizing barrier I: Theoretical analysis of topical bioavailability. J Pharm Sci 66:1525–1528, 1977.

206. Yu CD, Fox JL, Ho NF, Higuchi WI. Physical model evaluation of topical prodrug delivery—simultaneous transport and bioconversion of vidarabine-5′-valerate I: physical model development. J Pharm Sci 68:1341–1346, 1979.

207. Yu CD, Fox JL, Ho NF, Higuchi WI. Physical model evaluation of topical prodrug delivery—simultaneous transport and bioconversion of vidarabine-5″-valerate II: parameter determinations. J Pharm Sci 68:1347–1357, 1979.

208. Yu CD, Higuchi WI, Ho NF, Fox GL, Flynn GL. Physical model evaluation of topical prodrug delivery—simultaneous transport and bioconversion of vidarabine-5″-valerate III: permeability differences of vidarabine and *n*-pentanol in components of hairless mouse skin. J Pharm Sci 69:770–772, 1980.

209. Yu CD, Fox JL, Higuchi WI, Ho NF. Physical model evaluation of prodrug delivery —simultaneous transport and bioconversion of vidarabine-5″-valerate IV: distribution of esterase and deaminase enzymes in hairless mouse skin. J Pharm Sci 69:772–775, 1980.

210. Yu CD, Gordon NA, Fox JL, Higuchi WI, Fox NF. Physical model evaluation of topical prodrug delivery—simultaneous transport and bioconversion of vidarabine-5″-valerate V: mechanistic analysis of influence of nonhomogeneous enzyme distributions in hairless mouse skin. J Pharm Sci 69:775–780, 1980.

211. Liu P, Higuchi WI, Song WQ, Kurihara–Bergstrom T, Good WR. Quantitative evaluation of ethanol effects on diffusion and metabolism of beta-estradiol in hairless mouse skin. Pharm Res 8:865–872, 1991.

212. Liu P, Higuchi WI, Ghanem AH, Good WR. Transport of beta-estradiol in freshly excised human skin in vitro: diffusion and metabolism in each skin layer. Pharm Res 11:1777–1784, 1994.

213. Sugibayashi K, Hayashi T, Morimoto Y. Simultaneous transport and metabolism of ethyl nicotinate in hairless rat skin after its topical application: the effect of enzyme distribution in the skin. J Controlled Release 62:201–208, 1999.

214. Gysler A, Kleuser B, Sippl W, Lange K, Holtje HD, Korting HC. Skin penetration and metabolism of topical gluticocorticoids in reconstructed epidermis and in excised human skin. Pharm Res 16:1386–1391, 1999.

215. Hatanaka T, Rittirod T, Katayama K, Koizumi T. Influence of enzyme distribution and diffusion on permeation profile of prodrug through viable skin: theoretical aspects for

several steady-state fluxes in two transport directions. Biol Pharm Bull 22:623–626, 1999.

216. Rittirod T, Hatanaka T, Uraki A, Hino K, Katayama K, Koizumi T. Species differences in simultaneous transport and metabolism of ethyl nicotinate in skin. Int J Pharm 178: 161–169, 1999.

217. Seko N, Bando H, Lim CW, Yamashita F, Hashida M. Theoretical analysis of the effect of cutaneous metabolism on skin permeation of parabens based on a two-layer skin diffusion/metabolism model. Biol Bull Pharm 22:281–287, 1999.

218. Malkinson FD, Ferguson EH. Percutaneous absorption of hydrocortisone-4-^{14}C in two human subjects. J Invest Dermatol 25:281–283, 1955.

219. Shah VP, Flynn GL, Yacobi A, Maibach HI, Bon C, Fleischer NM, Franz TJ, Kaplan SA, Kawamoto J, Lesco LJ, Marty JP, Pershing LK, Schaefer H, Sequeira JA, Shrivastava SP, Wilkin J, Williams RL. Bioequivalence of topical dermatological dosage forms—methods of evaluating bioequivalence. AAPS/FDA Workshop on: Bioequivalence of Topical Dermatological Dosage Forms—Methods of Evaluating Bioequivalence. Sept 4–6, 1996. Bethesda, MD. Skin Pharmacol Appl Skin Physiol 11:117–124, 1998.

220. Shah VP, Flynn GL, Yacobi A, Maibach HI, Bon C, Fleischer NM, Franz TJ, Kaplan SA, Kawamoto J, Lesco LJ, Marty JP, Pershing LK, Schaefer H, Sequeira JA, Shrivastava SP, Wilkin J, Williams RL. Bioequivalence of topical dermatological dosage forms—methods of evaluating bioequivalence. Pharm Res 15:167–171, 1998.

221. Rougier A, Rallis M, Krien P, Lotte C. In vivo percutaneous absorption: a key role for stratum corneum partitioning. Arch Dermatol Res 282:498–505, 1990.

222. Rougier A, Dupuis D, Lotte C, Maibach HI. Stripping method for measuring percutaneous absorption in vivo. In: Bronaugh RL, Maibach HI, eds. Percutaneous Absorption: Drugs–Cosmetics–Mechanisms–Methodology. 3rd ed. New York: Marcel Dekker, pp 375–394, 1999.

223. Franz TJ. Percutaneous absorption on the relevance of in vitro data. J Invest Dermatol 64:190–195, 1975.

224. Roberts MS, Favretto WA, Mayer A, Reckmann M, Wongseelasote T. Topical bioavailability of methyl salicylate. Aust NZ J Med 12:303–305, 1982.

225. Jamoulle JC, Schaeffer H. Cutaneous bioavailability, bioequivalence and percutaneous absorption: in vivo methods, problems and pitfalls. In: Shah VP, Maibach HI, eds. Topical Drug Bioavailability, Bioequivalence and Penetration. New York: Plenum Press, pp 129–153, 1973.

226. Walters KA, Roberts MS. Veterinary applications of skin penetration enhancers. In: Walters KA, Hadgraft J, eds. Pharmaceutical Skin Enhancement. New York: Marcel Dekker, pp 345–364, 1993.

227. Scott RC, Walker M, Dugard PH. A comparison of the in vitro permeability properties of human and some laboratory animal skins. Int J Cosmet 8:189–194, 1986.

228. Ebling FJG, Hale PA, Randall VA. Hormones and hair growth. In: Physiology, Biochemistry and Molecular Biology of the Skin. Vol 1. New York: Oxford University Press, pp 660–698, 1991.

229. Hueber F, Wepierre J, Schaefer H. Role of transepidermal and transfollicular routes in percutaneous absorption of hydrocortisone and testosterone: in vivo study in the hairless rat. Skin Pharmacol 5:99–107, 1992.

230. Rolland A, Wagner N, Chatelus A, Shroot B, Schaefer H. Site-specific drug delivery to pilosebaceous structures using polymeric microspheres. Pharm Res 10:1738–1744, 1993.

231. Lieb LM, Liimatta AP, Bryan RN, Brown BD, Krueger GG. Description of the intrafollicular delivery of large molecular weight molecules to follicles of human scalp in vitro. J Pharm Sci 86:1022–1029, 1997.

232. Montagna W. Penetration and local effects of vitamin A on the skin of the guinea pig. Proc Exp Biol Med 86:668–672, 1954.
233. Lauer AC, Lieb LM, Ramachandran C, Flynn GL, Weiner ND. Transfollicular drug delivery. Pharm Res 12:179–186, 1995.
234. Lauer AC, Ramachandran C, Lieb LM, Niemiec S, Weiner ND. Targeted delivery of the pilosebaceous unit via liposomes. Adv Drug Deliv Rev 18:311–324, 1996.
235. Bamba FL, Wepierre J. Role of the appendageal pathway in the percutaneous absorption of pyridostigmine bromide in various vehicles. Eur J Drug Metab Pharmacokinet 18:339–348, 1993.
236. Maibach, HI, Feldman RJ, Milby TH, Serat WF. Regional variation in percutaneous penetration in man. Pesticides. Arch Environ Health 23:208–211, 1971.
237. Rougier A, Lotte C, Maibach HI. In vivo percutaneous penetration of some organic compounds related to anatomic site in humans: predictive assessment by the stripping method. J Pharm Sci 76:451–454, 1987.
238. Lauer AC. Percutaneous drug delivery to the hair follicle. In: Bronaugh RL, Maibach HI, eds. Percutaneous Absorption: Drugs–Cosmetics–Mechanisms–Methodology, 3rd ed. New York: Marcel Dekker, pp 449–427, 1999.
239. Walters KA, Flynn G, Marvel JR. Physicochemical characterisation of the human nail: permeation pattern for water and the homologous alcohols and differences with respect to the stratum corneum. J Pharm Pharmacol 35:28–33, 1983.
240. Baden HP, Goldsmith LA, Flemming B. A comparative study of the physicochemical properties of human keratinized tissues. Biochim Biophys Acta 322:269–278, 1973.
241. Chandrasekaran SK, Campbell PS, Michaels AS. Effect of dimethyl sulphoxide on drug permeation through human skin. AIChE J 23:810–816, 1977.
242. Embery G, Dougard PH. The isolation of dimethyl sulfoxide soluble components from human epidermal preparations: a possible mechanism of action of dimethyl sulfoxide in effecting percutaneous migration phenomena. J Invest Dermatol 57:308–311, 1971.
243. Allenby AC, Creasey NH, Edington JAG, Fletcher JA, Schock C. Mechanism of action of accelerants of skin penetration. Br J Dermatol 81(suppl 4):47–55, 1969.
244. MacGregor WS. The chemical and physical properties of DMSO. Ann NY Acad Sci 141:3–12, 1967.
245. Montes LF, Day JL, Wand CJ, Kennedy L. Ultrastructural changes in the horny layer following local application of dimethyl sulfoxide. J Invest Dermatol 48:184–196, 1967.
246. Kligman A. Topical pharmacology and toxicology of dimethyl sulfoxide. JAMA 193:796–804, 1965.
247. Walters KA, Flynn GL, Marvel JR. Physicochemical characterisation of the human nail: solvent effects on the permeation of homologous alcohols. J Pharm Pharmacol 37:771–775, 1985.
248. Franz TJ. Absorption of amorolfine through human nail. Dermatology 184(suppl 1):18–20, 1992.
249. Mertin D, Lippold BC. In vivo permeability of the human nail and of a keratin membrane from bovine hooves: influence of the partition coefficient octanol/water and the water solubility of drugs on their permeability and maximum flux. J Pharm Pharmacol 49:30–34, 1997.
250. Mertin D, Lippold BC. In vivo permeability of the human nail and of keratin membrane from bovine hooves: prediction of the penetration rate of antimycotics through the nail plate and their efficacy. J Pharm Pharmacol 49:866–872, 1997.
251. Pittrof F, Gerhards J, Erni W, Klecak G. Loceryl nail lacquer—realization of a new galenical approach to onchomycosis therapy. Clin Exp Dermatol 17(suppl 1):26–28, 1992.
252. Mertin D, Lippold BC. In vivo permeability of the human nail and of a keratin membrane from bovine hooves: penetration of chloramphenicol from lipophilic vehicles and a nail lacquer. J Pharm Pharmacol 49:241–245, 1997.

253. Homalle A. Experiences physiologiques sur l'absorption par la tegument externe chez l'homme dans le bain. Union Med 7:462–463, 1853.

254. Fleischer R. Untersuchungen uber das resorptionsvermaogen der menschlichen haut. Habilitationschrift 81, 1877.

255. Schwenkenbecker A. Das absorptionsverniogen der haut. Arch Anat Physiol 121–165, 1904.

256. Oppenheim M. Beitrage zin frage der hautabsorption mit besonderer beruchtigung der enkrankten haut. Arch Dermatol U Syph 93:85–106, 1908.

257. Smith HW, Clawes HA, Marshall EK. Mustard gas. IV. The mechanism of absorption by the skin. J Pharmacol 13:1–30, 1919.

258. Rein H. Experimental electroendosomotic studies on living human skin. Z Biol 81: 125–140, 1924.

259. Collander R, Burland H. Permeability in chara ceratophylla. II. Permeability to non-electrolytes. Acta Botan Fenn 11:1–114, 1930.

260. Wolf J. Die innere strucktur der zellen des stratum desquamans der menchlichen epidermis. Z Mikroskop Anat Forsch 46:170–202, 1939.

261. MacKee GM, Sulzberger MB, Herrmann F, Baer RL. Histological studies on percutaneous penetration with special reference to the effect of vehicles. J Invest Dermatol 6:43–61, 1945.

262. Rosendal T. Concluding studies on the conducting properties of human skin to AC. Acta Physiol Scand 9:39–49, 1945.

263. Blank IH. Factors which influence the water content of the stratum corneum. J Invest Dermatol 18:433–440, 1953.

264. Blank IH. Further observations on factors which influence the water content of the straum corneum. J Invest Dermatol 21:259–269, 1953.

265. Rothman S. Physiology and Biochemistry of the Skin. Chicago: University Press, pp 26–59, 1954.

266. Blank IH, Gould E. Study of mechanisms which impede the penetration of anionic surfactants into skin. J Invest Dermatol 37:311–315, 1962.

267. Kligman A. The biology of the stratum corneum. In: Montagna W, Lobitz WC, eds. The Epidermis. New York: Academic, pp 387–433, 1964.

268. Wagner JG. Biopharmaceutics: absorption aspects. J Pharm Sci 50:359–363, 1961.

269. Barr MJ. Percutaneous absorption. J Pharm Sci 51:395–409, 1962.

270. Malkinson FD, Rothman S. Percutaneous absorption. In: Jadassohn J, ed. Handbuch der Haut- und Geschlecht-skrankherten, Normale und Pathologische Physiologie der Haut. Berlin: Springer, 1:90–156, 1963.

271. Malkinson FD. Permeability of the stratum corneum. In: Montagna W, Lobitz WC, eds. The Epidermis. New York: Academic, pp 435–452, 1964.

272. Tregear RT. The permeability of the skin to molecules of widely differing properties. In: Rook A, Champion RH, eds. Progress In Biological Sciences in Relation to Dermatology 2nd ed. London: Cambridge University Press, pp 275–281, 1964.

273. Welles FV, Lubowe II. Cosmetics and the Skin. New York: Van Nostrand Reinhold, 1964.

274. Stoughton RB. Penetration absorption. Toxicol Appl Pharmacol 7(suppl 2):1–6, 1965.

275. Vinson LJ, Masurat T, Singer EJ. Basic Studies in Percutaneous Absorption: Final Comprehensive Report (no. 10). Edgewood, MD: Army Chemical Center, 1965.

276. Reiss F. Therapeutics: percutaneous absorption, a critical and historic review. Am J Med Sci 252:588–602, 1966.

277. Vickers CFH. In: McKenna RMB, ed. Modern Trends in Dermatology. Vol. 3. London: Butterworths, pp 84, 1966.

278. Barrett CW. J Soc Cosmet Chem 20:487, 1969.

279. Blank IH, Scheuplein RJ. Transport into and within the skin. Br J Dermatol 81(suppl 4):4–10, 1969.
280. Winklemann RK. The relationship of structure of the epidermis to percutaneous absorption. Br J Dermatol 81(suppl 4):11–22, 1969.
281. Katz M, Poulsen JB. Absorption of drugs through the skin. In: Brodic EE, Gillette JR, eds. Concepts in Biochemical Pharmacology. Berlin: Springer-Verlag, 1971.
282. Katz M, Poulsen JB. Corticoid, vehicle and skin interaction in percutaneous absorption. J Soc Cosmet Chem 23:565–590, 1972.
283. Katz M. In: Ariens EJ, ed. Drug Design. Vol 4. New York: Academic Press, pp 93–148, 1973.
284. Poulsen BJ. Design of topical drug products: Biopharmaceutics. In: Ariens EJ, ed. Drug Design, Vol 4. New York: Academic Press, pp 149–190, 1973.
285. Anderson RA, Roberts MS. Absorption of drugs through the skin. Aust J Pharm Sci 3:75–80, 1974.
286. Schaefer H. Skin penetration of drugs. Z Hautkr 49:719, 1974.
287. Idson B. Percutaneous absorption. J Pharm Sci 64:901–924, 1975.
288. Dugard PH. Skin permeability theory in relation to measurements of percutaneous absorption in toxicology. In: Marzulli FN, Maibach HI, eds. Dermatotoxicology and Pharmacology. Washington: Hemisphere, pp 525–550, 1977.
289. Higuchi T. In: Roche B, ed. Design of Biopharmaceutical Properties Through Prodrugs and Analogues. Washington: American Pharmaceutical Association, pp 409, 1977.
290. Malkinson FD, Guhlmann L. In: Drill VA, Lazar P, eds. Cutaneous Toxicology. New York: Academic Press, pp 63, 1977.
291. Wester RC, Maibach HI. Percutaneous absorption in man and animals: a perspective. In: Drill VA, Lazar P, eds. Cutaneous Toxicity. New York: Academic Press, 1977.
292. Scheuplein RJ. Skin permeation. In: Jarrett A, ed. The Physiology and Pathophysiology of the Skin. Vol 5. New York: Academic Press, pp 1659–1752, 1978.
293. Flynn GL. Topical drug absorption and topical pharmaceutical systems. In: Banker GS, Rhodes CT, eds. Modern Pharmaceutics. New York: Marcel Dekker, pp 263, 1979.
294. Scheuplein RJ. Percutaneous absorption: theoretical aspects. In: Mauvais–Jarvis PM, et al., eds. Percutaneous Absorption of Steroids. New York: Academic Press, pp 1–17, 1980.
295. Lien EJ, Tong GL. Physicochemical properties and percutaneous absorption of drugs. J Soc Cosmet Chem 24:371–384, 1973.
296. Ackermann C, Flynn GL, Smith WM. Ether–water partitioning and permeability through nude mouse skin in vitro II. Hydrocortisone n-alkyl esters, alkanols and hydrophilic compounds. Int J Pharm 36:67–71, 1987.
297. Pugh WJ, Hadgraft J. Ab initio prediction of human skin permeability coefficients. Int J Pharm 103:163–178, 1994.
298. Lien EL, Gao H. QSAR analysis of skin permeability of various drugs in man as compared to in vivo and in vitro studies in rodents. Pharm Res 12:583–587, 1995.
299. Potts RO, Guy RH. A predictive algorithm for skin permeability: the effects of molecular size and hydrogen bond activity. Pharm Res 12:1628–1633, 1995.
300. Abraham MH, Chandha HS, Mitchell RC. The factors that influence skin penetration of solutes. J Pharm Pharmacol 47:8–16, 1995.
301. Abraham MH, Martins F, Mitchell RC. Algorithms for skin permeability using hydrogen bond descriptors: the problem of steroids. J Pharm Pharmacol 49:858–865, 1997.
302. Scheuplein RJ, Blank IH, Brauner GJ, MacFarlane DJ. Percutaneous absorption of steroids. J Invest Dermatol 52:63–70, 1969.
303. Hadgraft J, Ridout G. Development of model membranes for percutaneous absorption measurements. I. Isopropyl myristate. Int J Pharm 39:149–156, 1987.

304. Roy SD, Flynn GL. Transdermal delivery of narcotic analgesics: comparative permeabilities of narcotic analgesics through human cadavar skin. Pharm Res 6:825–832, 1989.

305. Dutkiewicz T, Tyras H. A study of the skin absorption of ethylenebenzene in man. Br J Ind Med 24:330–332, 1967.

306. Jolicoeur LM, Nassiri MR, Shipman C, Choi H, Flynn GL. Etorphine is an opiate analgesic physicochemically suited to transdermal delivery. Pharm Res 9:963–965, 1992.

307. Roy SD, Flynn GL. Transdermal delivery of narcotic analgesics: pH, anatomical and subject influences on cutaneous permeability of fentanyl and sufentanil. Pharm Res 7: 842–847, 1990.

308. Anderson BD, Higuchi WI, Raykar PV. Heterogeneity effects on permeability–partition coefficient relationships in human stratum corneum. Pharm Res 5:566–573, 1988.

309. Cowhan ZT, Pritchard R. Effects of surfactants on percutaneous absorption of naproxen I: comparisons of rabbit, rat and human excised skin. J Pharm Sci 67:1272–1274, 1978.

310. Dutkiewicz T, Tyras H. Skin absorption of toluene, styrene and xylene by man. Br J Ind Med 25:243, 1968.

311. Shelley WB, Melton FM. Factors affecting the acceleration of histamine through normal intact skin. J Invest Dermatol 13:61–71, 1949.

312. Black JG, Howes D, Rutherford T. Skin deposition and penetration of trichlorocarbanilide. Toxicology 3:253–264, 1975.

313. Illel B, Schaefer H. Transfollicular percutaneous absorption: skin model for quantitative studies. Acta Dermatol Venereol 68:427–430, 1988.

314. Schaefer H, Watts F, Brod J, Illel B. Follicular penetration. In: Scott RC, Guy RH, Hadgraft J, eds. Prediction of Percutaneous Penetration, Methods, Measurements, Modelling. London: IBC Technical Services, pp 163–173, 1990.

315. Illel B, Schaefer H, Wepierre J, Doucet O. Follicles play an important role in percutaneous absorption. J Pharm Sci 80:424–427, 1991.

316. Rutherford T, Black JG. The use of autoradiography to study the localization of germicide in the skin. Br J Dermatol 81:75–87, 1969.

317. Bidmon HJ, Pitts JD, Solomon HF, Bondi JV, Stumpf WE. Estradiol distribution and penetration in rat skin after topical application, studied by high resolution autoradiography. Histochemistry 95:43–54, 1990.

318. Li L, Margolis LB, Lishko VK, Hoffman RM. Product-delivering liposomes specifically target hair follicles in histocultured intact skin. In Vitro Cell Dev Biol 28A:679–681, 1992.

319. Li L, Lishko V, Hoffman RM. Liposome targeting of high molecular weight DNA to the hair follicles of histocultured skin: a model for gene therapy of the hair growth process. In Vitro Cell Dev Biol Anim 29A:258–260, 1993.

320. Li L, Lishko VK, Hoffman RM. Liposomes can specifically target entrapped melanin to hair follicles in histocultured skin. In Vitro Cell Dev Biol 29A(3 pt 1):192–194, 1993.

321. Hueber F, Schaefer H, Wepierre J. Effect of vehicle on transdermal and transfollicular absorption of four steroids of different lipophilicity. In: Brain KR, James VJ, Walters KA, eds. Prediction of Percutaneous Penetration. Cardiff: STS Publishing, pp 264–271, 1993.

322. Lieb LM, Flynn G, Weiner N. Follicular (pilosebaceous unit) deposition and pharmacological behaviour of cimetidine as a function of formulation. Pharm Res 11:1419–1423, 1994.

323. Niemiec SM, Ramachandran C, Weiner N. Influence of nonionic liposomal composition on topical delivery of peptide drugs into pilosebaceous units: an in vivo study using the hamster ear model. Pharm Res 12:1184–1188, 1995.

324. Wurster DE, Munies R. Investigation of some factors influencing percutaneous absorption. III. Absorption of methyl ethyl ketone. J Pharm Sci 54:1281–1284, 1965.

325. Cross SE, Pugh WJ, Hadgraft J, Roberts MS. Probing the effect of vehicles on topical delivery: understanding the basic relationship between solvent and solute penetration using silicone membranes. Pharm Res (in press).

326. Bucks DAW. Prediction of percutaneous absorption. PhD dissertation. University of California, San Francisco, 1980.

327. Bucks D, Guy R, Maibach H. Effect of occlusion. In: Bronaugh RL, Maibach H, eds. In Vitro Percutaneous Absorption: Principles, Fundamental and Applications. Boston: CRC Press, pp 85–114, 1991.

328. Jiang R, Roberts, MS, Prankerd R, Benson HA. Percutaneous absorption of sunscreen agents from liquid paraffin: self-association of octyl salicylate and effects on skin flux. J Pharm Sci 86:791–796, 1997.

329. Roberts MS. Percutaneous absorption of phenolic compounds. PhD dissertation. University of Sydney, 1976.

330. Jiang R, Benson HA, Cross SE, Roberts MS. In vitro human epidermal and polyethylene membrane penetration and retention of the sunscreen benzophenone-3 from a range of solvents. Pharm Res 15:1863–1868, 1998.

331. Hadgraft J. Recent developments in topical and transdermal delivery. Eur J Metab Pharmacokinet 21:165–173, 1996.

332. Shaw JE, Prevo ME, Amkraut AA. Testing of controlled-release transdermal dosage forms. Product development and clinical trials. Arch Dermatol 123:1548–1556, 1987.

5

Methods for Studying Percutaneous Absorption

KEITH R. BRAIN, KENNETH A. WALTERS, and ADAM C. WATKINSON*

An-eX Analytical Services Ltd., Cardiff, Wales

I. INTRODUCTION

There is an increasing demand for data describing the rate, degree, and route of penetration of compounds across human skin. First, there is a requirement to optimize the delivery of dermatological drugs into various skin strata for maximum therapeutic effect. Second, the transdermal and topical routes have become popular alternatives to more traditional methods of drug delivery. A third stimulus has been the toxicological and risk assessment implications of the everyday use of a wide range of potentially harmful materials in the agrochemical, chemical, cosmetic, household, and pharmaceutical sectors. This has been driven largely by regulatory and safety bodies and a perceived need for improved data on the permeability of the skin to xenobiotics. For example, the U.S. Environmental Protection Agency (EPA) is currently addressing the issue of the dermal absorption testing of 80 compounds designated by the Occupational Safety and Health Administration (OHSA) and the Interagency Testing Committee (ITC) as worthy of particular interest [1,2].

Although drug delivery and toxicological considerations are perhaps the most important factors, it is also clear that increasing the overall database on percutaneous permeation will enhance our understanding of the mechanics of this process. The need for relevant data, produced under reproducible and reliable conditions, has led to an increase in both the development and the standardization of in vitro and in vivo test procedures. There have been numerous recommendations on in vitro and

---

*Current affiliation: Strakan Pharmaceuticals Ltd., Galashiels, Scotland.

Table 1 Skin Permeation Guidelines and Protocols

Date	Organization or author	Type	Ref.
1987	FDA/AAPS	Guideline	3
1989	FDA/AAPS	Guideline	261
1991	Bronaugh and Collier	Protocol	406
1993	ECETOC	Protocol	5
1993	COLIPA	Protocol	6
1996	EPA	Protocol	1
1996	ECVAM	Protocol	4

in vivo methodologies, and many of these have been collated as guidelines by both regulatory bodies and committees of interested parties. Perhaps the most widely known of these guidelines are those produced following Food and Drug Administration and American Association of Pharmaceutical Science (FDA/AAPS) workshop on the performance of in vitro skin penetration studies (3). However, more recent publications also contain useful information. For example, see the European Centre for the Validation of Alternative Methods (ECVAM) workshop report on *Methods for Assessing Percutaneous Absorption* (4); the documents from the European Centre for Ecotoxicology and Toxicology of Chemicals (5); and those from the European Cosmetic Toiletry and Perfumery Association (6) (Table 1).

II. IN VITRO METHODOLOGY

In vitro techniques to assess skin penetration and permeation are used extensively in industry and academia. Although at present, in vitro permeation data are not a requirement of regulatory bodies, there is an increasing trend for such data to be submitted, either alone or together with in vivo data. In some ways, in vitro techniques have advantages over in vivo testing. For example, permeation through the skin is measured directly in vitro, for which sampling is carried out immediately below the skin surface. This contrasts with most in vivo methods, which rely on the measurement of systemic (or at least nonlocal) levels of permeant. Some form of in vitro diffusion cell experiment is often the most appropriate method for assessment of percutaneous penetration in a developmental drug-delivery (transdermal or topical) program or in a dermal toxicology screen.

The following consideration of in vitro determination of dermal absorption and distribution combines published guidelines and the scientific literature with our own practical experience, and aims to provide the reader with an appreciation of the experimental choices and limitations.

A. In Vitro Skin Diffusion Cells

Most common methods for evaluation of in vitro skin penetration use diffusion cells, and there is a plethora of literature on the satisfactory performance of such experiments. The major advantage of in vitro investigations is that the experimental conditions can be controlled precisely, such that the only variables are the skin and the test material. Although a potential disadvantage is that little information on the metabolism, distribution, and effects of blood flow on permeation can be obtained, it has been reported that such procedures were more effective than several other methods for the assessment of differential delivery of hydrocortisone from commercial formulations (7).

It is essential to consider the ultimate use of generated data when developing experimental protocols. Routinely, simple mathematical models, which are based on certain assumptions or boundary conditions, are applied to experimental data. The most commonly used solutions to diffusion equations that are applied to the in vitro situation make the following assumptions:

1. The receptor phase is a perfect sink.
2. Depletion of the donor phase is negligible.
3. The membrane is a homogeneous slab.

None of these assumptions is wholly true in practice, and the potential significance of these imperfections must not be overlooked. Careful experimental design can be used to achieve a close approximation to reality, and the following section discusses how this can be achieved in practice.

1. Diffusion Cell Design

In vitro systems range in complexity from a simple two-compartment "static" diffusion cell (8) to multijacketed "flow-through" cells (9) (Fig. 1). Construction materials must be inert, and glass is most common, although Teflon and stainless steel (10) are also used. Excised skin is always mounted as a barrier between a donor chamber and a receptor chamber, and the amount of compound permeating from the donor to the receptor side is determined as a function of time. Efficient mixing of the receptor phase (and sometimes the donor phase) is essential, and sample removal should be simple. Neither of these processes should interfere with diffusion of the permeant. Comprehensive reviews on diffusion cell design are available (11–13). Continuous agitation of the receptor medium, sampling from the bulk liquid rather than the side arm, and accurate replenishment after sampling, are important practical considerations. It is essential that air bubbles are not introduced below the membrane during sampling.

Static diffusion cells are usually of the upright ("Franz") or side-by-side type, with receptor chamber volumes of about 2–10 mL and surface areas of exposed membranes of near $0.2–2$ cm^2. Cell dimensions should be accurately measured, and precise values should be used in subsequent calculations, with due attention to analyte dilution resulting from sampling and replenishment. The main difference in the application of these two static cell types is that side-by-side cells can be used for the measurement of permeation from one stirred solution, through a membrane, and into another stirred solution. This is of particular advantage when examining flux from saturated solutions in the presence of excess solid if accumulation of solid on the membrane surface must be prevented. This type of cell can also be modified to allow the absorption of permeants in the vapor phase. For example, volatile material may be retained in a small depression in the donor chamber so that the membrane is exposed to only the permeant in the gaseous state. Upright cells are particularly useful for studying absorption from semisolid formulations spread on the membrane

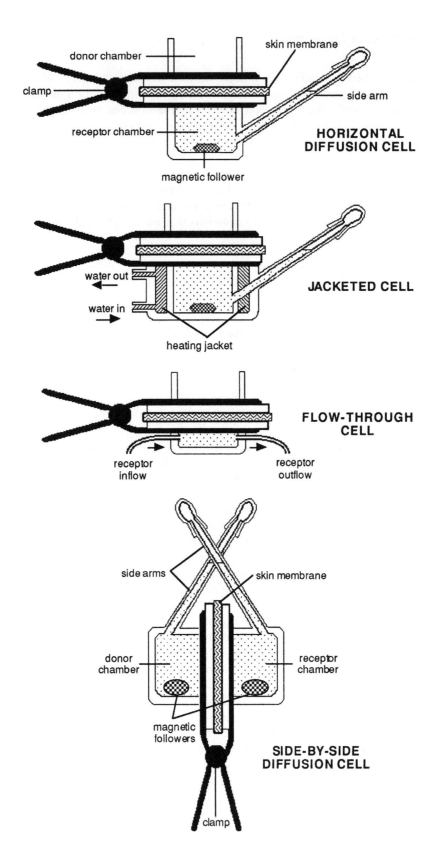

200

surface and are optimal for simulating in vivo performance. The donor compartments can be capped to provide occlusive conditions, or left open, according to the objectives of the particular study.

Flow-through cells can be useful when the permeant has a very low solubility in the receptor medium, and designs are continuously improving (14). Sink conditions are maximized as the fluid is continually replaced using a suitable pump (at a rate of about 1.5 mL/h) (15). However, the dilution produced by the continuous flow can raise problems with analytical sensitivity, particularly if the permeation is low. Flow-through and static systems have produced equivalent results (16,17). Automated flow-through systems can allow unattended sampling, and commercial systems are available. However, these have significant cost implications and are generally limited to small numbers of cells ($\sim$12). For example, Moody (18) has described an automated in vitro dermal absorption (AIVDA) system using only 0.07 cm^2 of skin per cell, held by autosampler vial inserts an autosampler carousel, with samples taken robotically. Although such methods are undoubtedly elegant, adequate validation of automated methods is essential.

Standard upright static diffusion cells, which offer a simple, low-cost, and very versatile system, can be employed on a large scale ($\sim$144 cells) and adapted to meet the particular requirements of a wide range of studies.

To summarize, a well-designed skin diffusion cell should

1. Be inert
2. Be robust and easy to handle
3. Allow the use of membranes of different thicknesses
4. Provide thorough mixing of the receptor chamber contents
5. Ensure intimate contact between membrane and receptor phase
6. Be maintainable at constant temperature
7. Have precisely calibrated volumes and diffusional areas
8. Maintain membrane integrity
9. Provide easy sampling and replenishment of receptor phase
10. Be available at reasonable cost

2. Receptor Chamber and Medium

Receptor chamber dimensions are constrained by the conflicting requirements of guaranteeing that the receptor phase can act as a sink, while ensuring that sample dilution does not preclude analysis. A large receptor volume may ensure sink conditions, but will reduce analytical sensitivity unless large samples can be taken and subsequently concentrated. Concentration of permeant in an aqueous receptor phase may be possible by lyophilization, or by techniques such as solid-phase extraction.

The ideal receptor phase provides an accurate simulation of the conditions pertaining to in vivo permeation of the test compound. As a general rule the con-

---

Figure 1 Basic diffusion cell designs: static horizontal cells may be jacketed (as in the Franz-type) or unjacketed (and temperature-controlled using a water bath or heating block). Flow-through cells usually have a small receptor chamber to maximize mixing. Side-by-side cells are used mainly for solution vehicles.

centration of the permeant in the receptor fluid should not be allowed to exceed approximately 10% of saturation solubility (3). Excessive receptor-phase concentration can lead to a decrease in the rate of absorption, which may result in an underestimate of bioavailability. The most commonly used receptor fluid is pH 7.4 phosphate-buffered saline (PBS), although this is not always the most appropriate material. It has been postulated that if a compound has a water solubility of less than about 10 μg/mL, then a wholly aqueous receptor phase is unsuitable, and the addition of solubilizers becomes necessary (19).

Receptor fluids described in the literature range from water alone to isotonic phosphate buffers containing albumin and preservatives. Albumin increases solubility of the permeant (20), whereas preservatives inhibit microbial growth in the receptor fluid. Microbial growth can produce problems by partitioning of the permeant into, or metabolism of the permeant by, the microbes. One particularly useful fluid is 25% (v/v) aqueous ethanol, which provides a reasonable "sink" for many permeants, while removing the need for other antimicrobial constituents. Other examples of modified receptor phases include 1.5–20% Volpo N20 (see later), rabbit serum, 3% bovine serum albumin, 50% aqueous methanol, 1.5–6% Triton-X100, and 6% Poloxamer 188 (15). It has been suggested that there is a need for formal protocols to determine the suitability of receptor-phase composition (21).

It is important to recognize the possibility that solubilizers may interfere with the barrier function of the skin itself. Bronaugh (19) examined several commonly used receptor phases and made several recommendations. One of the most useful receptor phases was an aqueous solution containing 6% Volpo N20 (Croda Inc.), a nonionic polyethylene glycol(PEG)-20-oleyl ether surfactant. This did not influence the flux of either water or urea across rat skin, when compared with normal saline, suggesting that Volpo N20 did not disrupt the barrier function of rat skin to hydrophilic compounds. Rat skin usually responds to penetration enhancers to a greater degree than human skin and thus one would predict that Volpo N20 should have a negligible effect on the flux of hydrophilic compounds across human skin. Commercial Volpo N20 is a mixture of PEG-oleyl derivatives that averages at PEG-20, and it has been our experience, and that of others (22), that this diversity of components can produce complications in sample analysis.

The problem of very lipophilic permeants has been addressed by the use of nonaqueous (23) and nonliquid receptor media. Sheets of silicone rubber (0.02-in. thick) were used to collect pesticides with low water solubilities, which were then desorbed from the rubber with an appropriate solvent and subsequently analyzed (24). Other solutions have included the use of flowing gaseous receptor phases for volatile permeants (25).

It is also important to appreciate that the pH of an aqueous buffered receptor solution may markedly affect the apparent "flux" of a permeating weakly ionizable compound. The pH of the hydrophilic viable epidermal layers may be "altered" by the receptor solution, and this can theoretically result in modulation of the partitioning tendencies of ionizable species. This was well illustrated experimentally by Kou et al. (26), who determined the permeation of a weak acid and a weak base (nicardipine) through human skin into receptor solutions of varying pH (Fig. 2). Quite clearly the data show a receiver fluid pH dependency on flux, and the authors caution against indiscriminate use of nonphysiological receptor pHs in diffusion experiments.

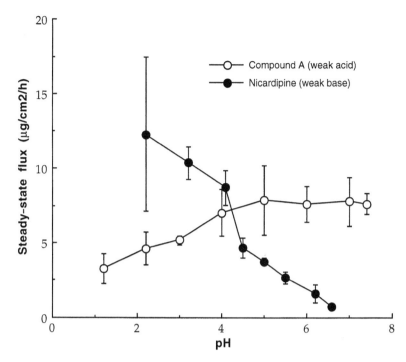

Figure 2 The effect of receptor fluid pH on the flux of a weak acid (pK_a 2.60) and a weak base (pK_a 5.68) across human skin in vitro: skin permeation of the weak acid increased with increasing pH, whereas that for the base decreased. (From Ref. 26.)

3. Selection, Variation, and Preparation of Skin Membranes

A major potential variant in the design of in vitro skin diffusion experiments is the nature of the skin membrane. Animal skins are widely used as substitutes for human skin (27–29), primarily owing to difficulties in obtaining human tissue. The suitability of animal skin as a model for human skin is discussed later in this chapter (see Sec. III.A) and elsewhere (30).

a. Intra- and Intersubject Variability. Few studies have addressed the issues of intra- and intersample variability in human skin permeability, because of the few sufficiently large datasets available for accurate statistical evaluation. There are at least two issues in this area that need to be addressed. The first is the degree and nature of the distribution patterns of skin permeabilities in vitro. This was addressed to some extent by Southwell et al. (31), who investigated both in vitro and in vivo variation in the permeability of human skin between different specimens (interspecimen) and the same specimens (intraspecimen). From the permeation characteristics for a series of compounds, they concluded that in vitro interspecimen variation was $66 \pm 25\%$ and intraspecimen variation $43 \pm 25\%$. The pattern in vivo was similar, although the overall level of variation was somewhat smaller. This analysis has been extended by assessing the statistical distribution of human skin permeabilities. Kastings et al. (32) and Cornwell and Barry (33) concluded that there was evidence that the permeability of human skin in vitro is lognormally, rather than normally distributed. Williams et al. (34) examined the permeation of 5-fluorouracil (644 determinations from 71 specimens) and estradiol (221 determinations from 28 specimens) through human abdominal skin. Here, where site variability was excluded, the data were lognormally distributed. A lognormal distribution implies that the use of normal gaussian statistics is inappropriate, and that use of geometric (rather than arithmetic) means should be considered.

  b. Age and Sex Differences.   The questions of how age and sex affect the permeability of human skin have been rather poorly addressed to date. Some studies have concluded that, in vitro in humans, there was no discernible dependence of skin permeability on age, sex, or storage conditions (34–39). The literature on the effect of age and sex on percutaneous absorption in other species, for which more data is available, is confused. For example, it was reported (40) that the dermal absorption of certain marker compounds was lower in older rats, but that, for different compounds, the reverse was true (41) or indeed, that age did not influence dermal absorption in rats (42). Hairless mouse skin permeation generally increases with age, corresponding to the single-hair cycle, but decreases with a return to the hairless state (43,44).

  The effect of age on percutaneous absorption has been examined in vivo in humans, with variable results. It was postulated (45) that reduced hydration levels and lipid content of older skin may be responsible for a demonstrated reduction in skin permeability if the permeants were hydrophilic (no reduction was seen for model hydrophobic compounds) (Table 2). The reduced absorption of benzoic acid demonstrated in the elderly (46) was in line with this suggestion, but not the reduction in absorption of testosterone (lipophilic) (47), or lack of change in the absorption of methyl nicotinate (more hydrophilic) (48), with age. A number of potential physiological changes that may be responsible for age-related alterations in skin permeability have been suggested. These include a noted increase in the size of individual stratum corneum corneocytes throughout life, increased dehydration of the outer layers of the stratum corneum with age, decreased epidermal turnover, and decreased microvascular clearance (49). The issue of age-related variability, however, is far

Table 2 Age-Related Differences in Percutaneous Absorption

Permeant	log K[a]	Applied dose permeated over 7 days (%)[b]	
		22–40 yr	65 yr
Testosterone	3.32	19.0 ± 4.4	16.6 ± 2.5
Estradiol	2.49	7.1 ± 1.1	5.4 ± 0.4
Hydrocortisone	1.61	1.5 ± 0.6	0.54 ± 0.15
Benzoic acid	1.83	36.2 ± 4.6	19.5 ± 1.6
Acetylsalicylic acid	1.26	31.2 ± 7.3	13.6 ± 1.9
Caffeine	0.01	48.2 ± 4.1	25.2 ± 4.8

[a]Octanol/water partition coefficient.
[b]Compounds (4 $\mu g/cm^2$) were applied in 20 μL acetone to the ventral forearm ($n = 3$–8).
Source: Ref. 45.

from resolved. If the variability of in vivo data is actually slightly lower than that of in vitro data (31) then the small increase in variability in vitro may make any such differences indiscernible.

  c. Racial Differences. Several authors have documented differences between the permeabilities of skin based on its racial origin. White skin is slightly more permeable than black skin (50–55), which correlates with observations that black skin has both more cell layers within the stratum corneum (SC) (56) and a higher lipid content (57). A recent study (52) of white, Hispanic, black, and Asian skin ranked them in order of permeability to methyl nicotinate as black < Asian < white < Hispanic. It has been reported (58,59) that the corneocytes of black, white, and Oriental skin are of a similar size, but that there are differences in spontaneous desquamation. It has also been reported (60,61) that there may be differences between black and white skin in microvascular reactivity, following dermal application of vasoactive agents, but this is possibly due to subjective, rather than objective, measurement (48). No differences in the permeation of water through black or white skin in vitro were observed (35) and, similarly, there was no racial difference in the in vivo percutaneous absorption of diflorasone diacetate (62). More recently, Lotte et al. (63) have determined the penetration and permeation of several compounds into (skin-stripping; see Sec. II.B.1) and through (24-h urinary excretion) Asian, black, and white skin. There were no statistical differences in penetration or permeation of benzoic acid, caffeine, or acetylsalicylic acid among the races (Table 3). These equivocal findings highlight the necessity for further systematic research in this area.

  d. Storage Conditions. It is unclear whether the proposed lognormal distribution in vitro is an experimental artifact caused by excision and isolation from the body, or by storage conditions before use (e.g., freezing). Some authors concluded that freezing had no measurable effect on permeability (36,37). Yazdanian (64) re-

Table 3 Race-Related Differences in Percutaneous Absorption

Permeant	Race	Amount of permeant recovered (nmol/cm²)	
		Urine at 24 h	Stratum corneum at 30 min[a]
Benzoic acid	Caucasian	9.0 ± 1.5	6.8 ± 1.0
	Black	6.4 ± 0.9	6.1 ± 1.0
	Asian	9.7 ± 1.2	8.1 ± 1.5
Caffeine	Caucasian	5.9 ± 0.6	5.5 ± 0.6
	Black	4.5 ± 1.0	5.8 ± 1.0
	Asian	5.2 ± 0.8	6.1 ± 0.9
Acetylsalicylic acid	Caucasian	6.2 ± 1.9	11.9 ± 1.9
	Black	4.7 ± 0.9	9.0 ± 1.7
	Asian	5.4 ± 1.7	10.1 ± 1.7

[a]Amount in stratum corneum determined by tape-stripping (n = 6–9).
Source: Ref. 63.

ported an effect for cattle, although there was no general pattern of differences between frozen and fresh skin. Wester et al. (65) have cautioned against the use of frozen, stored human skin for studies in which cutaneous metabolism may be a contributing factor. It is, however, important to appreciate that the state of hydration of the tissue before freezing may influence subsequent permeation characteristics (66). As a general rule tissues should not be hydrated when placed into frozen storage.

e. Anatomical Site Variations.   Perhaps the clearest data available on variation in skin permeability deals with anatomical site-to-site variation (46,67–75). Site-to-site variation of skin permeability has been examined using the tape-stripping method (71,72) and correlated with corneocyte diameter (46) and, hence, diffusional pathlength. Although, in practice, skin permeation of compounds follows a different pattern in different skin regions, it is generally agreed that some body sites (the head and genital region) are uniformly more permeable than others (extremities). For example, transepidermal water loss and skin permeation of benzoic acid, caffeine, and acetylsalicylic acid decreased in the order forehead > postauricular > abdomen > arm (74). Similarly, the permeation of hydrocortisone decreased in the order scrotum > jaw > forehead > scalp > back > forearm > palm (Fig. 3) (67,6), and abdomen was more permeable to methyl salicylate than either the arms or feet (75).

For in vitro permeation studies, it is advantageous for donor skin to be from anatomical sites relevant to the objectives of the study, although this may be logistically impossible. Because it is difficult to reach any definitive conclusions about the variance and distribution of skin permeabilities, the best advice is that experimenters should be aware of the possibilities and test their data appropriately. An adequate number of replicates, with skin from different donors or sites spread evenly throughout all test groups, should be used.

f. Membrane Preparation.   Different methods can be used to prepare human skin. The membrane is one of the following:

1. Full-thickness skin, incorporating the SC, viable epidermis, and dermis
2. Dermatomed skin, in which the lower dermis has been removed
3. Epidermal membranes, comprising the viable epidermis and the SC (prepared by heat separation)
4. SC alone (prepared from step 3 by enzyme treatment)

The most suitable type of tissue is dependent on the nature of the permeant. The environment of skin in vivo differs somewhat from that in vitro. In vivo the continuously perfused subcutaneous vasculature, which penetrates the dermis to a significant degree, can rapidly remove permeants reaching the epidermal–dermal junction. These vessels, if still present, are not perfused in simple in vitro models. In vitro the relatively aqueous environment of the dermis will inhibit the penetration of lipophilic compounds, whereas in vivo this barrier is circumvented by the capillary bed. Hence, the use of dermatomed, epidermal, or SC membranes is more appropriate for particularly lipophilic permeants.

Other considerations may justify the use of epidermal membranes, even where the dermis does not present an artificial barrier to a permeant. For example, if a study involves an assessment of the skin content of permeant, it is much easier to extract or solubilize epidermal membranes, or SC, than full-thickness skin. Con-

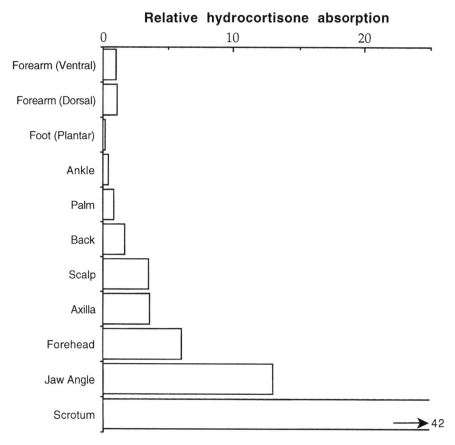

Figure 3 Regional variation in the in vivo absorption of hydrocortisone through human skin: absorption was normalized to that through the ventral forearm. Hydrocortisone was applied in acetone to a marked site. (From Ref. 67.)

versely, if an experiment includes tape-stripping, it may be easier to perform on full-thickness skin, rather than epidermal membranes. However, the latter have been successfully used to correlate in vivo and in vitro tape-stripping (76).

The preparation of epidermal membranes and SC is time-consuming, and the necessary processing increases the possibility of damage to the skin membrane. Careful consideration of the most appropriate type of skin preparation is required, and this should address the physicochemical nature of the penetrating species, the data required, tissue availability, and the time scales involved. With animal skin, full-thickness membranes are usually used, because it is difficult to isolate intact epidermis or SC owing to the presence of numerous hair follicles, which may also compromise dermatomed tissue.

Dermatomed skin can be prepared in vitro as follows: most of the subcutaneous fat is removed, leaving only a small quantity in place; the skin is placed, dermal side down, onto a metal plate; the residual fat adheres to the metal; a thin sheet of plastic is placed over the SC to protect it before a second metal plate is applied; the two plates are clamped together and cooled to −20°C until completely frozen (about

1–3 h); on removal from the freezer, the upper plate is gently warmed to ease removal from the SC; the plastic sheet is removed and a dermatome used to remove strips of skin of the desired thickness (usually ranging from ~200–600 μm). Care must be taken in the preparation of dermatomed skin to ensure that damage to hair follicles is minimized because, if these are severed, erroneously high penetration figures result. The use of "hairless' varieties of animal and relatively "hair-free" types of human skin may reduce, but not eliminate, this problem.

For human skin, the separation of the dermis from the epidermis (SC and viable epidermis) is a relatively simple technique (35,77). First, the subcutaneous fat is removed by blunt dissection. The full-thickness skin membrane is then totally immersed in water at 60°C for about 45 s. Following removal from the water, the skin is pinned, dermal side down, to a dissecting board and the epidermis gently peeled back using a pair of blunt curved forceps. The epidermal membrane can next be floated onto warm water and taken up on a support membrane (membrane or paper filter). It is then ready to be mounted in a diffusion cell. Alternatively, a microwave technique has been proposed for the separation of epidermis from the dermis (78). To isolate the SC from epidermal membranes, the latter are placed in trypsin solution (0.0001%), incubated at 37°C for 12 h, rubbed (with a cotton bud) to remove the epidermal cells, rinsed in distilled water, and air-dried on a surface from which they can be easily removed (79).

4. Permeant and Application Technique

The manner in which a substance is applied to the skin surface can be a major determinant of its subsequent absorption. Several factors must be considered in selecting a suitable application procedure, including the nature of the vehicle, the permeant concentration, the amount of vehicle applied, the mechanism of application, the exposure time, and the method for removing an applied vehicle (if required). Many of these issues may be intrinsic to the purpose of the study. For example, risk assessment involving the study of the skin penetration of an ingredient in a cosmetic or agrochemical formulation should be performed with the material in the formulation as it is marketed or used, and with a regimen that mimics, as closely as possible, the "in use" situation (e.g., 80,81).

a. Application Method. There are two basic approaches to applying substances to the skin. Infinite-dose techniques involve application of sufficient permeant to make any changes in donor concentration during the experimental timeframe, caused by diffusion or evaporation, negligible (i.e., the dose is effectively infinite). This is desirable if the experimental objectives include calculation of diffusional parameters, such as permeability coefficients; or for investigation of mechanisms of penetration enhancement. Finite dose techniques (82), designed to model in-use conditions, involve application of a dose that may show marked depletion during an experiment. Depletion occurs where the proportion of permeant entering the membrane is large, relative to the amount applied. Alternatively, the permeant may be removed from the skin surface during, for example, the simulation of a rinsing or washing procedure. With finite dosing the permeation profile may exhibit the characteristic plateauing effect that accompanies donor depletion (Fig. 4). The finite dose technique may involve application of permeants or enhancers in small volumes of volatile solvents (e.g., acetone or ethanol). This allows assessment of the gross effects

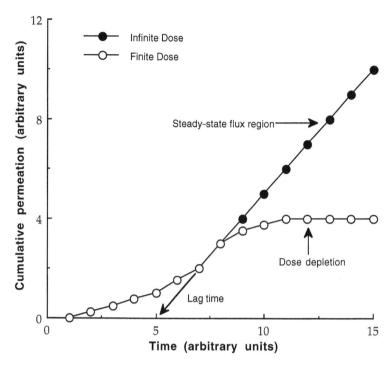

Figure 4   Sample cumulative permeation patterns following finite- and infinite-dosing regimens: with infinite dose, permeation normally reaches a steady-state flux region, from which it is possible to calculate permeability coefficients and diffusional lag times. In finite dosing the permeation profile normally exhibits a plateau effect as a result of donor depletion.

of enhancers, but results are more difficult to interpret mechanistically. Direct comparisons of finite- and infinite-dose applications are relatively rare, but the predicted effects have been investigated (83,84).

   b. Dose Level.   There are conflicting reports on the effect of dose level on the degree of permeation through skin. For example, increased dosage did not produce proportional increases in flux of ibuprofen or flurbiprofen across human skin in vitro when deposited as a thin film from acetone (85). Three concentrations of each drug were applied in only 50 μL, and the authors concluded that as the acetone evaporated from the skin surface the thermodynamic activity of the drugs increased until saturation was reached, at which concentration maximum flux would be expected. When all of the solvent had evaporated, leaving a plug of solid drug with poor dissolution properties, the flux always dropped. The effect of volume on the permeation of minoxidil from ethanol was linear between 10 and 50 μL (86), but it was concluded that it was the total drug loading, rather than the application volume, that was important. Note that this is only true for a volatile vehicle that evaporates rapidly from the skin surface. In practice, the actual skin permeation of volatile compounds, such as dimethylnitrosamine (Fig. 5) and 2-phenoxyethanol, is significantly reduced by evaporation (80,87). Other literature covers the effects of formulation, application time, dose variation, and occlusion on ibuprofen permeation (88),

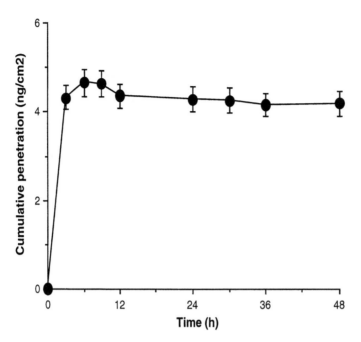

Figure 5   Permeation profile for dimethylnitrosamine (DMN) across human skin in vitro: DMN was applied at finite dose levels in an oil-in-water emulsion vehicle. Note that permeation was significantly reduced by evaporation following a 6-h exposure. (From Ref. 80.)

the effect of application formulation (89), theories of finite dosing (90,91), thickness of the application vehicle (92,93), and contact time effects (94).

There are several published recommendations on both the expression of dose levels and the specific quantities involved. The FDA/AAPS guidelines (3) suggested a universal application weight of approximately 5 mg/cm^2 of formulation. COLIPA (6) proposed 5 μL/cm^2 for liquid formulations, and 2 mg/cm^2 for semisolid formulations (or 5 mg/cm^2 if these are being compared with a liquid). Distribution of accurately measured amounts of such small quantities of semisolid formulations as an even film over the surface of skin membranes presents considerable practical challenges (93). Pretreatment of the skin before mounting in a diffusion cell may be easier than treating premounted membranes. Semisolid materials can be applied and spread with a small preweighed formulation-coated spatula. The precise weight applied is determined by difference, and all test materials should be applied by the same operator. Evaluation of spray-on formulations raises even more complications. Practical evaluation of the effects of enhancers is best made by their incorporation into appropriate formulations, rather than the more common procedure of pretreatment of the skin, which does not model the in-use scenario.

The issue of dosage levels is of particular interest, although there is a dearth of appropriate nonclinical data concerning dose–response curves for topical formulations. Numerous topical formulations have shown clinical equivalence despite containing different concentrations of the same drug (e.g., 95; see also Chap. 8). This can be rationalized if each of the formulations were saturated with drug because, provided that the excipients had no effect on the skin, similar flux would be predicted from each of them (96; see also Chap. 6). Alternatively, if all of the formulations were very poorly bioavailable, then increases in drug content may not have any significant effect on clinical efficacy. The dosage form, level, and exposure time must be carefully considered to ensure that a study will produce appropriate data, particularly in the view of any regulatory bodies involved in subsequent interpretation.

c. Use of Radiolabeled Permeants. The use of radiolabeled permeants is attractive, as this greatly simplifies sample analysis, particularly for mass balance experiments. Guidelines for the choice and use of radiolabels in permeation experiments are provided by COLIPA (6); although not all problems are addressed. It is essential that radiolabeled material is homogeneously mixed with nonlabeled permeant. Where a radiolabel is used to evaluate a formulated product it is prudent to incorporate the radiolabel at the same stage of manufacture as the cold permeant. This may be possible for formulations for which exact constituents and method of manufacture can be reproduced. A less rigorous alternative is to manufacture a small quantity of labeled formulation and blend this with a larger quantity of cold formulation. If the test formulation is a marketed product, for which exact manufacturing process and ingredients are unknown, incorporation of label in these ways is impossible, limiting the options to spiking the market formulation with radiolabel dissolved in the most appropriate solvent. A test for homogeneity of distribution of the label in the final formulation is always essential. It must be appreciated that in a complex, formulation containing discrete phases, homogeneous distribution at a (sub)microscopic level may not be achievable by spiking and that solvent used to incorporate a label may alter the performance of a formulation.

The greatest problem in the use of radiolabels is the lack of specificity. It is essential to ensure that the integrity of the permeant is maintained during the experiment and that counts attributed to permeation of the target molecule are not due to impurities, degradation, exchange, or metabolism. This can be achieved by chromatographic separation of the analyte before counting, although this may remove the major advantage of the radiolabeled approach.

5. The Permeation Experiment

a. Duration. Although some workers have extended the duration of experiments to 120 h (85), it is recommended that they be restricted to 24 (6) or 48 h (4). It has been suggested that 48 h may be too short to establish a steady-state flux from an infinite dose, leading to misinterpretation of the data (97), and the ECVAM (4) report suggests experiments may be extended to 72 h (in the presence of antimicrobial agents) in such instances. From electrical resistance measurements and permeation parameters (Table 4), human epidermal membranes were shown to retain integrity for up to 5 days (98), provided they were supported on suitable non–rate-limiting membranes (e.g., Millipore GSWP filters). In the absence of the filter membrane support, tissue integrity was compromised by the physical stress accompanying sample withdrawal and skin washing. Investigators should, however, be aware of possible barrier degradation over extended time frames.

b. Sample Interval. Sample intervals should be of an appropriate frequency to allow realistic assessment of such parameters as lag time and steady-state, or pseudo–steady-state, flux (if possible). For a compound with unknown permeation

Table 4 Human Epidermal Membrane Integrity During Prolonged Experiments Based on Tissue Electrical Resistivity and Permeation of Mannitol

Time (h)	Unsupported membrane		Supported membrane	
	Resistance (KOhm·cm²)	Permeability (×10⁸ cm/s)	Resistance (KOhm·cm²)	Permeability (×10⁸ cm/s)
25	44	0.48	42	—
50	35	—	40	0.64
75	27	0.94	38	0.78
100	23	—	35	0.70
125	—	—	37	0.76
150	—	—	37	0.80

Source: Ref. 98.

characteristics, samples should ideally be taken at 2-h intervals for the duration of the experiment. Early samples (1–4 h) may be important in identifying compromised cells showing anomalously high early permeability.

c. Number of Replicates. Given the high intra- and intersubject variability in human skin permeability (see Sec. II.A.3.a), a large number of replicates for each dosage regimen is recommended. The most widely quoted recommendation for numbers of replicates in in vitro studies on human skin is 12 (3), and comparisons should be matched. Fewer replicates may be employed, if cost, time, or skin availability is a problem, provided that the limitations of replicate reduction are recognized. The permeability characteristics of laboratory animal skin are, in general, more uniform than those of human skin, and fewer replicates may be successfully employed.

d. Temperature. In vitro skin diffusion experiments are normally conducted with a skin temperature of 32°C (the in vivo value). This is achieved by maintaining the receptor solutions at 37°C, either by immersing cells in a water bath or by using cell jackets perfused at the correct temperature. An apparent nonlinear dependence of skin permeability on temperature was demonstrated for flurbiprofen (99). The skin accumulation of flurbiprofen decreased with a rise of temperature in an in vivo experiment, but not in an equivalent in vitro experiment, suggesting participation of increased blood flow in vivo. Over the range of 22–90°C the in vitro permeability coefficient for water through SC rose in a sigmoidal manner by a factor of about 70 (100). Similar increases in skin permeability have been shown for sodium lauryl sulfate, nickel chloride (101), and ketoprofen (102). This temperature dependence is probably a function of lipid fluidity. Skin lipids undergo major phase transitions between 40° and 70°C (103), which may explain the sigmoidal pattern.

e. Skin Integrity. Skin integrity can be addressed in a qualitative manner by simple visual examination of specimens or, more quantitatively, by measurement of transepidermal water loss (104), or by the flux of marker compounds, such as tritiated water (32,35,84,105) or sucrose (80,81,106). The generally accepted permeability coefficient for water diffusion through human skin is 1.5×10^{-3} cm/h, or less, although an upper limit of 2.5×10^{-3} cm/h has also been used (35). Samples showing particularly high permeability are often rejected as outliers with questionable integrity, but may actually represent the real population spread if their distribution is indeed lognormal (see earlier discussion).

f. Permeant Analysis. The quality of the data derived from any experiment is ultimately dependent on the integrity of the analytical method employed, and all aspects of the analytical procedure should be included in the overall experimental design (107). The ideal analytical procedure provides accurate assessment of both the quantity and nature of the material present at a given time, and the detection limit of the method must be capable of producing data of practical significance. Preliminary prediction from existing data or physicochemical modeling can give "ballpark" estimates of requirements, and potential routes of degradation and metabolism should be taken into consideration. When no permeation is detected, this should be reported as "less than" the detection limit and not "zero." As replication and multiple time point sampling are common, analysis should not be unnecessarily complex, although the integrity of the method must be beyond doubt. High-performance liquid chromatography (HPLC) is particularly useful as relatively large (~ 200 μL) aqueous samples can often be handled without preliminary processing or concentration, although the inherent properties of the permeant can limit detection. Ultraviolet detection is commonly used, provided that the permeant has significant absorption, and interference from extracted skin or formulation components can cause problems, particularly at low wavelengths. Native fluorescence is less subject to interference than absorption, but is also a less common phenomenon. Derivatization may be necessary for detection, but is generally an undesirable complication that increases both cost and variability. Investigations of permeation of commercial materials consisting of a range of related compounds pose particular problems (108). Interference can arise from leaching of material from the skin or components of the test material or components of the diffusion system. The magnitude and composition of skin leachate depends on the type of skin membrane and the time frame. It may interfere directly with detection, or it may bind permeant, thereby making it unavailable to detection. These effects can be adequately investigated only by real-time comparison with appropriate controls, including permeation from placebo vehicles, and by confirming recovery of permeant from spiked samples.

Determination of permeants in tissue samples necessitates some form of extraction or a solubilization process. When radiolabeled permeants are used, tissue samples are routinely taken up in commercially available solubilizers. Such aggressive products are often not applicable in other cases for which more traditional extraction methods are required. Recovery of permeant from SC tape strips is less demanding and can often be accomplished by vortexing and sonicating with relatively small volumes of solvent.

6. Assessment of In Vitro Skin Metabolism

Metabolism of a xenobiotic is a detoxification and elimination process involving formation of molecules that are more hydrophilic and easily excreted. It can be summarized as a two-phase process. The first stage is the exposure or addition of functional groups to form a primary product by, for example, the production of free hydroxyl, carboxyl, or amino groups. The second step usually involves conjugation with a polar molecule (e.g., glucuronic acid) to form hydrophilic compounds that are readily excreted (109). It is widely established that there is potential for biotransformation of molecules within the skin (110,111). The nature of skin enzymes differs quantitatively and qualitatively from those in the liver (112). In general, the activities of many metabolic processes are much lower in skin than in liver (111,113,114) although certain enzymes, such as N-acetytransferases and those involved in reductive processes, have demonstrated fairly high activity (115,116) (Table 5).

The specific activities (percentage of hepatic) of several phase-1 and phase-2 enzyme–substrate systems in skin have been assessed. Phase-1 conversions ranged from 27% (for the cytochrome P450–7–pentoxyresorufin system) to less than 1% (for the cytochrome P450–coumarin system). Phase-2 examples include the glutathione S-transferase–cis-stilbene oxide system (49% of hepatic activity) and the UDP-glucuronyltransferase–3-hydroxybenzo[a]pyrene system (0.6–2% of hepatic activity) (117). Induction of phase-1 enzymes in skin has been studied using topically applied classic enzyme inducers, such as polycyclic aromatic hydrocarbons (PAHs). Many cutaneous enzymes respond substantially (e.g., cytochrome P450 metabolism of retinoic acid is 500–1000 times higher in the presence of a PAH inducer).

Most metabolic investigations on skin have employed epidermal homogenates (118–121) or epidermal cell cultures (122). These are useful for the study of enzyme activity per se, but have little predictive value for the in vivo situation in which permeants may not contact cellular systems (123). A more appropriate in vitro model uses metabolically active intact skin mounted in a diffusion cell under conditions that maintain viability. Bronaugh (123) emphasizes the importance of using fresh skin. Full-thickness skin must be dermatomed ($\sim 200 \ \mu$m) to ensure an adequate supply of nutrients and oxygen (9). The receptor fluid is usually HEPES-buffered Hanks' balanced salt solution or Dulbecco's modified PBS (124).

Table 5 Comparisons Between Specific Activities of Cutaneous Enzymes and Hepatic Enzymes

Enzyme system	Substrate	Cutaneous specific activity (% hepatic)
Cytochrome P-450s	Aldrin	0.4–2.0
	Aminopyrine	1.0
	Diphenyloxazole	2.0–3.0
	Ethylmorphine	0.5
	7-Pentoxyresorufin	20–27
Epoxide Hydrolases	cis-Stilbene oxide	9–11
	$trans$-Stilbene oxide	24–25
	Styrene oxide	6.0
Glutathione transferases	cis-Stilbene oxide	49
	Styrene oxide	14
Glucuronosyltransferases	Bilirubin	3.0
	1-Naphthol	2–50
Sulfotransferases	1-Naphthol	10
Acetyltransferases	p-Aminobenzoic acid	18
	2-Aminofluorene	15

Source: Ref. 111.

These techniques have been used to investigate in vitro permeation and metabolism of several compounds (114), including estradiol and testosterone (124); acetyl ethyl tetramethyltetralin and butylated hydroxytoluene (113); benzoic acid, *p*-aminobenzoic acid, and ethylaminobenzoate (125); benzo[*a*]pyrene and 7-ethoxycoumarin (126); azo dyes (127); butachlor (128); atrazine (129), and retinyl palmitate (130). The use of cultured skin for metabolic work has also been investigated (131,132).

B. Other In Vitro Methods for Studying Percutaneous Absorption

1. Skin-Stripping

Skin-stripping with adhesive tape is commonly used in vivo, and also in vitro. Tape-stripping experiments are performed as follows: (a) a permeant is applied to the skin surface for a fixed time period; (b) permeant remaining on the skin surface is removed (where possible) by wiping or washing; (c) a succession of SC layers are removed by sequential tape strips; (d) the permeant contents of the tape strips are determined. These experiments evaluate how the concentration of a permeant applied to the skin surface changes with depth into the SC. The shape of this concentration–depth profile will be time-dependent and vary with permeant, according to the rapidity, degree, and nature of uptake by the SC.

If steady-state diffusion is achieved, then the distribution of a permeant across a homogeneous membrane should be represented by a linear decline from the outside to the inside (133). Before steady state, the distribution of a permeant across a homogeneous membrane will be represented by a series of concave curves converging to a straight-line as time progresses (133). Hence, in a tape-stripping experiment, assuming that the SC is homogeneous and that the depth profiled with each tape strip is the same, a linear concentration–strip number profile in the steady-state region of any diffusion experiment would be expected. In practice, neither of these assumptions generally appears to be correct, resulting in nonlinear profiles (Fig. 6).

A linear relation between the mass of SC removed and the number of tape strips performed was reported in vivo (134,135) using 0.6-cm–diameter circles of tape (Transpore; 3M; St. Paul, MN). However, Schaefer et al. (136) found an exponential relation between strip number and mass of skin removed in vivo and this has been reproduced (76). An increase in structural integrity of SC with increasing depth has been suggested (137,138), which has been correlated with increasing lipid content (139) and modeled mathematically (140). Such data support a nonlinear relation between strip number and mass of skin removed (141). If the external cells of the horny layer are less well bound, then they will obviously be more easily removed than deeper layers. Alternative techniques have also been applied to assessment of the amount of SC removed by repetitive stripping and both strip protein content analysis and spectrophotometric methods (141) showed nonlinear dependencies on strip number.

If the membrane is not homogeneous, then distribution of a permeant within the skin at steady state will not be a linear function of depth. The majority of both in vivo and in vitro reports show the distribution of compounds within the skin to be related to strip number in a nonlinear fashion. There is usually an approximate exponential decay in the amount of permeant from the outside to the inside of the SC [e.g., erythromycin (142), lanolin (143), fentanyl (144), alniditan (145), and hair-

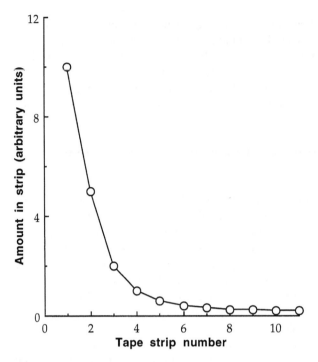

Figure 6 Typical pattern of permeant content in tape strips removed from the stratum corneum following a fixed period of exposure. Note that most of the permeant is found in the first few strips.

dyes (146)] (Fig. 7). However, "steady state" may not have been attained or the nonlinear dependence of concentration on strip number may be a result of nonlinear removal of SC.

There is little published data on the use and validation of in vivo and in vitro skin-stripping techniques. Pershing et al. (134) claimed that although stripping skin in vitro produced the same net loss in weight of SC (approximately 1 mg/cm^2 over ten strips), the pattern of skin removal was different in vitro from that in vivo. They reported a linear relation between tape strip number and weight of SC removed in vivo, but a nonlinear pattern in vitro when proportionately more was removed in the first five tape strips than in vivo. In contrast, Trebilcock et al. (76) found similar nonlinear relations between strip number and mass of SC removed in vivo and in vitro and concluded that there was no significant difference between in vitro and in vivo tape stripping for assessing skin distribution after percutaneous penetration.

The reasons for these disparate results are unclear, although a linear correlation can be closely approximated by an exponential one, depending on the decay constant (147). It is also probable that the particular tape used and precise experimental protocol, will affect SC removal. The tape used can affect the resultant concentration–depth profile (148), as can the time of application of a particular vehicle (149). Monitoring changes in SC concentration–depth profiles as a function of time requires appropriate controls conducted for each time point so that results can be normalized.

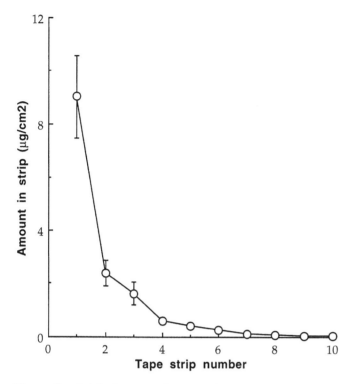

Figure 7 Aciclovir content in tape strips removed from human stratum corneum following 24-h exposure to Zovirax Cream (5% aciclovir): The cream was applied to skin in vitro at a dose of 5 mg/cm² formulation. Data are mean ± SE ($n = 12$).

Much remains to be achieved in skin tape-stripping technology (150), and method validation will need to be rigorous.

2. Attenuated Total Reflectance–Fourier Transform Infrared (ATR–FTIR) Spectroscopy

One of the most important uses of data describing diffusion through skin pertains to the analysis of mechanisms of passive and enhanced transport. Successful solutions to such mechanistic questions require that diffusion profiles are deconvoluted to allow separate evaluation of the contributions arising from the alternative phenomena of diffusion and solubility or partitioning. Attempts (151) to deconvolute these processes in permeation through both synthetic and biological membranes by using diffusion cells, achieved limited success. The problem of deconvoluting diffusion–cell-derived data arises because of the need to demonstrate the attainment of steady state. This is a prerequisite to calculation of flux and other permeation parameters, but the common practice of application of linear regression to subjectively selected data has been criticized (97). Similarly, assessment of lag times (hence, diffusion coefficients) must also be questioned.

The ATR–FTIR technique is fairly straightforward: (a) a membrane is placed in direct contact with the surface of a ZnSe ATR crystal, mounted on a spectrometer; (b) a shallow trough is placed on top of the membrane; (c) the trough is sealed to the membrane (e.g., with petroleum jelly); (d) the solvent–solute system under study is placed in the trough; (e) the spectrometer is linked to a computer equipped with the appropriate software and FTIR spectra taken at the crystal–membrane interface over a time period. The IR beam penetrates the membrane to a depth of only about 2–3 μm; therefore, only the interfacial region is probed. As permeant enters the interfacial region there is an increase in the IR peak areas associated with the penetrating species. The major limitation is that the permeant must have an IR absorption in the transparent region of the membrane (e.g., compounds containing cyano- and azido-groups or deuterated compounds). The data obtained is the concentration–time profile at the crystal–membrane interface, which builds up at a rate related to the diffusion coefficient and gradually plateaus at a level related to the solubility of the permeant in the membrane under study. These two parameters can be estimated by a nonlinear curve fit to the data by using a solution to Ficks' laws of diffusion.

ATR–FTIR spectroscopy was originally used to study diffusion through simple homogeneous synthetic systems, such as silicone membranes (152,153). The technique was applied to examination of diffusion into semisolids (154) and to the investigation of ethanol diffusion in glycerogelatin films (155). In conjunction with bulk transport techniques, it was used to show that values of diffusion coefficients (in synthetic membranes) calculated using the two methods were similar (156). The methodology has been specifically refined (157,158) to permit its use with human SC (159), and it has also been applied to the investigation of morphological differences between the upper and lower layers of the SC (160,161). Data from regular diffusion cells and ATR–FTIR spectroscopy showed a high correlation (160,161) and has been used to predict diffusional pathlengths in SC.

3. Isolated Perfused Tissue Models

Perfused tissue models use excised regions of skin—complete with their associated microvasculature—immediately after sacrifice, and with continuous perfusion [e.g., with Krebs–Ringer buffer, glucose, and albumin aerated with oxygen and carbon dioxide (162). Several variations on the theme involve both different species and areas of skin. Isolated perfused porcine skin flaps (163–165) (Fig. 8), bovine udders (166,167), and rabbit ears (168,169) have been used. The perfused pig-ear model uses an isolated ear perfused with oxygenated blood from the same pig (170). These techniques permit the investigation of the effect of local blood circulation on the accumulation and removal of topically applied materials.

4. Artificial (Cultured) Skin

The technology behind the construction of human skin equivalents (artificial skin) is derived predominantly from research into the treatment of burns. A classification and evaluation of the numerous different types of human skin equivalents concluded that the technique is limited to the reconstitution of the epidermis with a SC (171). Such models have been used to investigate both cutaneous metabolic events (131,132,172) and dermal irritation (173), with varying degrees of success. The use of artificially cultured skin in permeation experiments is still of limited application because the methodology is both expensive and not clearly predictive of in vivo results (174–179). In vitro skin equivalents; per se, are generally approximately ten times more permeable than human skin (180). On the other hand, reconstructed epidermis transplanted onto a nude athymic mouse had a permeability similar to normal human skin after 1 month (181).

5. Autoradiography

Autoradiographic techniques have been used extensively to visualize and quantify penetration through, and distribution within, animal (182–187) and human (405) skin. The general procedure is as follows: (a) a radiolabeled permeant is applied to the skin surface; (b) after a suitable time, a skin sample is excised and microtomed perpendicular or parallel to the surface to produce a section of the skin; (c) the section is placed in contact with a photographic emulsion; (d) exposure to radiation produces a latent image showing the pattern of distribution of the radiolabel within the sample; (e) photographic development reveals the image; (f) evaluation of the intensity of the image allows qualitative and quantitative evaluations of routes and degrees of permeation. The technique is particularly useful in the assessment of attempts to target drugs to specific regions of the skin. For example, compounds that act against acne are targeted at the pilosebaceous unit (186).

6. Laser-Scanning Confocal Microscopy (LSCM)

A major problem with fluorescence visualization using a conventional microscope is that the images are often blurred owing to light originating from out-of-focus sources above and below the region of interest. LSCM allows clear visualization of fluorescent probes within tissue samples (188) because it collects light from only a small in-focus volume so that contributions to the image from fluorescence above and below the in-focus point are reduced. A confocal image is constructed using light from a common plane within a structure to form an optical section. LSCM has been used to examine skin samples that are sufficiently transparent, do not scatter light strongly, and are relatively free from autofluorescence. There are no fixing processes involved in sample preparation; therefore, artifact formation can be largely avoided. Specimens can be rapidly prepared and examined, and permeation of fluorescent probes can be monitored kinetically in the same sample (188). Optical-sectioning methods have been used to examine the permeation routes of compounds across the cornea (189,190), the skin (191–195), the buccal epithelium (196–198), and the nasal mucosa (199). There are, however, some limitations to the technique. The images produced are not absolutely real-time (1-4 s are needed to collect an image, and several images must be averaged to gain a good quality picture), the laser excitation wavelengths available will not excite all fluorophores, and problems of photobleaching and photochemistry can occur. Photobleaching has been utilized as a tool to examine diffusion within controlled-release devices (200) and gels (201). A region of fluorescent molecules within a gel is photobleached using a high-power laser and, as unbleached molecules move into the bleached region, the fluorescence is restored. The rate at which this restoration of fluorescence occurs is related to the diffusion coefficient of the molecule in the matrix. The latter technique may have potential in the study of diffusion in SC.

7. Model Membranes

A range of synthetic artificial membranes have been evaluated as models for percutaneous penetration. These may offer the possibility of providing a less variable system for formulation evaluation and for investigation of permeation phenomena.

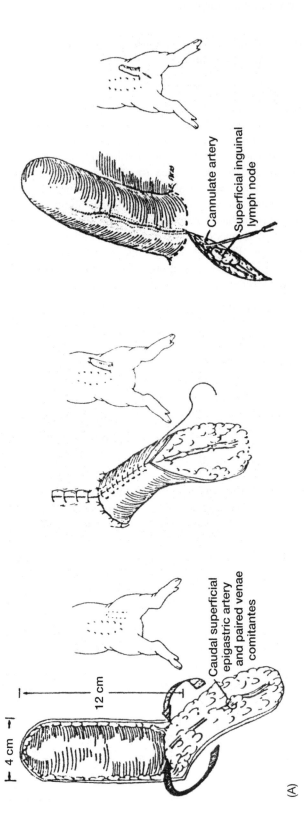

Cannulate artery

Superficial inguinal lymph node

Caudal superficial epigastric artery and paired venae comitantes

4 cm

12 cm

(A)

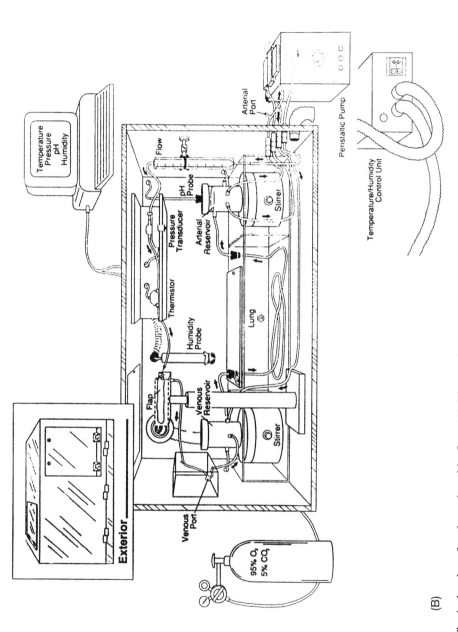

(B)

Figure 8 The isolated perfused porcine skin flap: (A) The surgical procedure involves raising a skin flap on the ventral surface and creating a tube. The superficial epigastric artery is cannulated and the tubed flap is transferred to the perfusion chamber. (B) The perfusion chamber used to maintain viability of the isolated skin flap: The chamber is temperature-, humidity-, and pressure-regulated. The perfusate is a modified Krebs–Ringer bicarbonate buffer (pH 7.4) containing albumin and glucose. (Courtesy of Dr. J.E. Riviere.)

The penetration of lipophilic compounds through solid, supported liquid membranes containing a range of simple lipids (202,203), more complex mixtures (204), and membranes containing Azone (205) has been examined using a rotating diffusion cell (206,207). Liquid membranes are frequently unstable to excipients, severely restricting their application to formulation evaluation. Polymers, such as polydimethylsiloxane (PDMS, Silastic) offer a nonporous, hydrophobic, relatively inert, and reproducible barrier, which has been used to evaluate factors such as donor concentration on permeation. For example, Pellett et al. (208) demonstrated enhanced penetration from supersaturated solutions containing piroxicam. Combinations of different polymers, such as PDMS, poly-(2-hydroxyethylmethacrylate; pHEMA) and cellulose acetate, have been used to try to model the combination of hydrophilic and lipophilic domains present within the SC (209,210). Trilamellar cellulose acetate–silicone (equilibrated with IPM) laminates have been used to compare delivery of piroxicam from gel and cream formulations (211). A silicone–cellulose ester multilaminate was reported to reproduce SC permeation for four formulations containing methyl nicotinate (212). A six-layer composite membrane, containing dodecanol in a collodion matrix, was unusual in that it included a protein phase (210). Although composite systems have been defined that may provide data with a good correlation with actual SC values in specific cases, it appears that, in general, each system is applicable only for comparison of the permeation of a small range of related compounds. Such systems have an application in quality control of existing products, under carefully defined conditions.

Lamellar sheets, which are more representative of the structure in the lipid domains of the SC, can be constructed, by fusing liposome vesicles prepared from a mixture of SC lipids, onto a support membrane (213). A good correlation (r = 0.880) with guinea pig SC was reported for the permeation of a range of compounds, including nonsteroidal anti-inflammatory agents (NSAIDs), cyclobarbitol, and lignocaine (214), and the effects of several penetration enhancers were reproduced (215). Comparison of the permeation of certain NSAIDs through several different model membrane systems has been compared (216). Structured lipid matrices prepared from cholesterol, water, and fatty acids have been proposed as suitable models for investigations of SC barrier function and enhancer interactions (217), and have been used to examine the permeation of 5-fluorouracil and estradiol (218). Lecithin organogels formed by the addition of a critical quantity of water to a solution of pure lecithin in an appropriate organic solvent (e.g., cyclooctane) contain bicontinuous intercalated networks of aqueous and organic phases that may have some structural similarity to the possible pathways within the SC (219). The diffusion of naproxen, diclofenac, and ibuprofen through the lecithin organogels was representative of their relative rate of permeation through full-thickness human skin (Fig. 9). Comparison was also made between the permeation of ibuprofen, from eight commercial formulations, through full-thickness human skin, 700-μm–thick Silastic membranes, and a cyclooctane–lecithin–water organogel (220). Permeation of ibuprofen through human skin was higher from gel systems than from creams, but neither synthetic model was capable of accurately reproducing formulation performance through skin. Permeation modulation, using laurocapracm (Azone), was compared in three model membrane systems; liposome-coated membranes, solid-supported liquid membranes, and a lecithin organogel (221). Although the data were equivocal, there were indi-

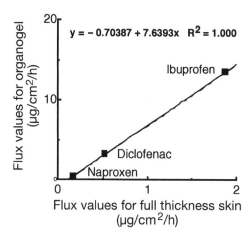

$y = -0.70387 + 7.6393x$ $R^2 = 1.000$

Figure 9   Relation between the flux of various nonsteroidal anti-inflammatory agents across human skin in vitro and across a lecithin–cyclooctane organogel held between cellulose nitrate membranes. (From Ref. 219.)

cations that the organogels provided a more structurally representative model of the skin barrier than the other two membrane systems.

III.   IN VIVO SKIN PERMEATION METHODS

This section discusses the methods used for the investigation of skin penetration and percutaneous absorption in vivo. Many of the methods are based on animal models that are often used in the early stages in product development (i.e., before the availability of extensive toxicological data). Although the relevance of animal models to the human situation has not been fully evaluated and is often questioned (27–30,222,223), there is little doubt that well-planned investigations using laboratory animals can provide useful information for future developmental work. However, the use of animals in testing is becoming less acceptable in certain sectors, and this is reflected in the 6th Amendment to the Cosmetics Directive (93/95/EEC), which includes a future ban on the use of animals in testing cosmetic products and their ingredients. In the development of topical or transdermal pharmaceutical products containing new chemical entities, full-scale clinical studies must be performed. The development of generic dermatological or transdermal products ultimately requires proof of bioequivalence with the innovator product. In either event, it is prudent to have available prototype products optimized for drug delivery using human skin in vitro or laboratory animal–human in vivo determinations.

A.   Animal Models for In Vivo Percutaneous Penetration Studies

It is undeniable that the most appropriate data relating to percutaneous penetration is that generated in vivo in human volunteers (224). However, it is not always ethically or practically possible to conduct pharmacokinetic studies in humans, and animal models, therefore, are often used as a substitute. Numerous species have been used in such percutaneous absorption studies (Table 6).

Table 6 Some Animal Models Used for In Vivo Percutaneous Absorption Studies

Species	Ref.
Mouse	223, 234, 270, 368, 375, 391
Rat	18, 41, 62, 183, 223, 243, 246, 247, 250, 265, 266, 271, 324, 335, 336, 379, 391, 393
Guinea pig	223, 237, 267, 369, 390, 391, 393
Rabbit	223, 237, 243, 247
Dog	234, 397
Pig	185, 223, 234, 243, 393
Monkey	41, 69, 223, 246, 247, 249, 268, 272, 381, 403

1. Species Variation in Skin Structure

Most species in which skin permeability has been investigated are mammals, and their skin is macroscopically separated into three layers, the stratum corneum, the viable epidermis, and the dermis. The most important considerations in terms of barrier function are differences in the sebum, SC, and follicles. Small laboratory animals, such as rats, mice, and rabbits, lack sweat glands, but have more hair follicles than human skin. For example, guinea pig skin contains about 4,000–5,000 follicles/cm^2, rat skin about 8,000 follicles/cm^2, and rabbit skin more than 10,000 follicles/cm^2, whereas human skin contains about 6 follicles/cm^2 (225,226).

There is little data on skin permeation in larger animal species. However, as it has been estimated that there are some 2,000 follicles/cm^2 in cattle and up to 10,000 follicles/cm^2 in some skin regions of Merino sheep, it is likely that the follicular route for skin permeation is significant in these species.

Furthermore, many chemicals can bind strongly to the keratin of hair and wool and, when this occurs, it reduces the amount of applied chemical reaching the SC and, hence, the amount available for absorption. It is important, therefore, that this factor is taken into consideration in experimental design for comparisons between shaved and chemically depilated skin sites within the same animal species.

The chemical nature of the intercellular spaces of the SC has been well established in human, pig, and several laboratory animals (227–229). Those mammalian species for which SC lipids have been studied extensively show that there is not a great variation in general type among different species. The lipids present are predominantly ceramides, cholesterol (plus cholesteryl esters), and free fatty acids. There are some interspecies differences in the quantity of lipid(s) present, and this may have considerable relevance in differences in skin permeation (230).

The SC thickness in most animal species is between 15 and 30 μm but, in general, this tends to increase with animal size. Thus, the thickness of the SC of rats is about 20 μm, whereas that of pig and human are about 30 μm. Although the morphological structure of the SC shows reasonable consistency among species, there are some infrequent deviations. In sheep, for example, the distal layers of the SC separate from the basal layer in a disorganized manner (231) and become embedded in the surface sebum layer.

In the context of evolutionary development, it is perhaps not surprising that larger animals generally share more anatomical and physiological features with humans than smaller laboratory animals (232). This is also true for the integument; therefore, it is predictable that the pig and rhesus monkey (233,234) are preferable to such species as the hairless mouse in the prediction of human skin penetration characteristics. If skin permeability properties of different species were related to the body size and average physiological life time, skin permeability would be expected to be in the order: mouse > rat > guinea pig > rabbit > monkey > dog > goat > sheep > pig > cattle > human. As stated previously, SC thickness increases with animal size, whereas the lipid content decreases with size. The lower amounts of SC lipid are reflected in the observation that the permeability of many un-ionized and lipid-soluble solutes through skin appears to be lower in the larger animal species.

2. Ranking in Skin Permeability

Investigations on the usefulness, and predictability of animal models have been of two types: (a) those in which percutaneous absorption of one or more permeants is measured in several species; and (b) those in which absorption of one or more chemicals is compared between the experimental animal and human. If the data from animal models are to be used for extrapolation to human, it is important that the investigators appreciate the differences in the behavior of the animal model and human tissue. The skin permeability of the animal tissue model must be the same as that of the human, or easily related to it by a constant ratio (4). Furthermore, any response to permeation modulators (physical or chemical), formulation excipients, and occlusion must mimic the response in humans (235,236).

   Norgaard (237) reported that the species ranking of skin permeability rates for cobalt ions was rabbit > guinea pig > human. Tregear (238) ranked the species permeabilities for metal ions as rabbit > rat > guinea pig > human, whereas that given by McCreesh (239) (for two ill-defined organophosphorus solutes) was rabbit, rat > guinea pig > cat, goat, monkey > dog > pig. Scott et al. (240) reported the permeability coefficients of the dicationic herbicide paraquat in a range of species. The observed permeabilities for paraquat relative to water are shown in Table 7. Two features of paraquat permeability are readily apparent: (a) human skin is much

Table 7 The Observed Permeabilities for Paraquat and Water Through the Skin of Various Species as Measured Using In Vitro Techniques

| Species | Type | Permeability coefficient ($cm^2/h \times 10^{-5}$) | | Animal/human ratio for paraquat |
		Water	Paraquat	
Man		92.97	0.73	—
Rat	Wistar Alpk/AP	103.09	26.68	40
	Hairless	103.08	35.51	50
	Nude	151.72	35.34	50
Mouse	Alpk/AP	143.75	97.16	135
	Hairless	350.70	1066.39	1460
Guinea pig	Dunkin–Hartley	442.09	195.63	270
Rabbit	NZ White	252.61	79.92	110

Source: Ref. 240.

less permeable to paraquat than any other species examined, and (b) the hairless mouse is particularly susceptible to paraquat penetration. Given that the skin of most of the laboratory species studied lack sweat glands, but contain more hair follicles than humans, these different permeability coefficients may reflect differences in follicular transport for this permanently ionized compound. Unfortunately, data on the penetration of ions through larger species, such as sheep and cattle, appears to be lacking, which precludes a more complete analysis.

Durrheim et al. (241) and Huq et al. (242) reported that the permeability of alcohols and phenols through hairless mouse and human skin was similar. The aqueous concentration of phenol required to damage the human epidermis, rat, and hairless mouse skin was also similar at about 2%. However, the effects of hydration on the permeability of hairless mouse and human skin differed markedly.

Table 8 summarizes some of the in vivo data reported on the penetration of several un-ionized solutes through the skin of a variety of species (243–245). In general, the magnitude of difference in skin permeability between the species is less than fivefold, with a rank order of rabbit > rat > pig > monkey > human. More recently, Moody et al. showed that absorption of the insecticide lindane was similar in rats and rhesus monkeys (246) and also suggested that animal models for dermal absorption of phenoxy herbicides may be useful in predicting human dermal absorption (247).

Sato et al. (230) have investigated species difference in the percutaneous absorption of nicorandil, using hairless rat, guinea pig, hairless mouse, dog, pig, and human, and have attempted to relate this to the amount of surface lipid in these species. As part of this study the influence of the penetration enhancers Azone and isopropylmyristate on nicorandil permeation was also investigated. There was a clear difference between the amounts of lipid extracted by acetone from the skin of smaller laboratory animals (hairless mouse: 212 $\mu g/cm^2$; hairless rat: 273 $\mu g/cm^2$; and guinea pig: 225 $\mu g/cm^2$) and that from man (60.5 $\mu g/cm^2$). This difference was reflected in the permeability studies. Both permeation and permeation enhancement of nicorandil

Table 8 The Observed Permeabilities for Several Compounds Through the Skin of Various Species, Including Humans, Measured Using In Vivo Techniques

	Applied dose absorbed (%)				
Permeant	Human	Pig	Monkey	Rabbit	Rat
Acetylcysteine	2.4	6.0	—	2.0	3.5
Butter yellow	21.6	41.9	—	100.0	48.2
Caffeine	47.6	32.4	—	69.2	53.1
Cortisone	3.4	4.1	—	30.3	24.7
DDT	10.4	43.4	1.5	46.3	—
Haloprogin	11.0	19.7	—	113.0	95.8
Lindane	9.3	37.6	16.0	51.2	—
Malathion	8.2	15.5	19.3	64.6	—
Parathion	9.7	14.5	30.3	97.5	—
Testosterone	13.2	29.4	—	69.6	47.4

Source: Ref. 30.

was much greater through the skin of the hairless rat, hairless mouse, and guinea pig than through human or pig skin. These data reflect the significance of SC lipids for both the permeation and penetration enhancement process. Similarly, Roberts and Mueller (248) have compared the in vitro flux of glyceryl trinitrate across hairless mouse, pig (Yucatan), and human skin, and concluded that mouse skin was an unacceptable model for the prediction of human skin permeation behavior.

It is apparent, from the overall data available, that the preferred animal models for human skin in vivo are the rhesus monkey (223,249) and pig (234). It has been reported that permeability across hairless rat skin is similar to that of humans (250) (for certain permeants) although the more general suitability of this species has been questioned (251). There are certainly differences between the behavior of penetration enhancers on the skin of hairless rats and humans (252).

To conclude, several general observations can be made on the use of animal skin in percutaneous absorption studies:

1. Animal skin with high follicular density is poorly representative of human skin (225,258).
2. Rat and rabbit do not give reliable estimations of human penetration (243,253–255).
3. Pig and rhesus monkey reasonably approximate absorption of several compounds in humans (223,234,238,243,256,257).
4. Shaving or depilation of hairy skin may alter the barrier function (259,260).

B. Experimental Procedures

This section is based on the recommendations and guidance offered following the FDA–AAPS workshop (Washington, 1989) on in vivo percutaneous penetration (261) and the ECVAM workshop report (4).

1. General Principles and Procedures

Application procedures for in vivo studies are broadly similar, with slight variations depending on the situation. Permeants are applied to a designated area of skin, in their pure form (262), as a solvent-deposited solid (244,263,264), in solution (265–267), in a formulation (266,268–274), as a vapor (275), or applied to a piece of material that is then placed on the skin surface (276). Following application of the test permeant, various techniques may be used to monitor percutaneous absorption. There are five approaches to permeant sampling (261): (a) surface residue; (b) in the skin; (c) in the venous blood draining the application site; (d) in the systemic circulation; and (e) in the excreta.

The most commonly used in vivo method is the assay of urinary and fecal excretion over time (224) following topical dosing of a small amount of radiolabeled permeant. The method is useful for assessing total absorption, is noninvasive, and can be performed in humans. However, the resulting data must be corrected for incomplete elimination by measurement of radiolabel in excretions following intravenous administration (263). It is also important that the collection period is sufficiently long to include all excreted material (82) and that a full mass balance is conducted (277–279). The application site must be protected with a chamber that can be either occlusive or protective.

At the end of the administration period, the chamber is removed and assayed for residual adhering permeant. The skin site is then washed and the washings analyzed. A new chamber is again applied to the application site and the process repeated. The skin may also be tape-stripped at this time to determine drug residence in the SC. These techniques are simple to perform and urinary elimination can be a useful mirror of the plasma concentration if the clearance of the penetrant is rapid.

Blood sampling has been used widely to determine the systemic availability of transdermal products (280–285) (Fig. 10). However, for locally acting topical preparations, blood sampling is of little use because, usually, blood (but not tissue) levels will be very low (286). It is also questionable whether systemic levels are clinically relevant. This issue was addressed by Singh and Roberts (287–289), who showed that (in an anesthetized rat) topical coadministration of a vasoconstrictor (phenylephrine) with other permeants significantly increased the uptake of solutes by the local tissues (Fig. 11). This effect was probably due to reduced local blood flow and therefore reduced uptake of the permeants into the systemic circulation (290). In the absence of vasoconstrictors, direct penetration of solutes into local tissue is more evident after short periods because recirculation of permeated substances returning to the dermal vasculature from the systemic blood supply tends to dominate at later periods (287). An isolated perfused limb model (291,292) has been used to remove the contribution of the systemic circulation and, thereby, permit examination of local vasculature uptake and tissue distribution of permeants (293).

2. Skin Grafting

Recent developments in animal modeling for skin absorption have involved the use of skin-grafting techniques (234,294–299). Human normal or cultured skin is grafted onto congenitally athymic (nude) laboratory rodents (usually mouse or rat) and the area of graft used to determine in vivo absorption. To avoid rejection of the graft the host is immunosuppressed before and after surgery, using specific therapy, such as the cyclic polypeptide cyclosporine (300,301). The surgical technique is straightforward: full-thickness sections of host skin are removed, together with the underlying fat layer, and split-thickness human graft skin or supported cultured skin is attached to this site using sutures or surgical tape. It was estimated, based on surface electrical capacitance measurements, that in grafted cultured skin, a stable skin tissue developed by 8–12 weeks following surgery (296).

Transepidermal water loss and tritiated water permeability through human skin grafted onto mice are similar to those for human skin in situ (181,302). Similarly, the absorption of several compounds of diverse physicochemical properties across grafted skin showed good correlation with human in vivo data (29,234,294,303).

A refinement of this approach involves the surgical preparation of a skin sandwich flap on an independent vascular supply (304–306). This procedure results in the host animal (usually athymic nude rat) supporting grafted human skin. The surgical procedures are fully described (304). There are three stages in the procedure. Briefly, split-thickness skin is grafted to the subcutaneous surface of skin on the abdomen of the rat to create a sandwich. Graft and host skin are then sutured in place in the normal anatomical position. Two weeks later, the sandwich is lifted, together with the associated vasculature, on three sides to create a sandwich flap, and the exposed wound is covered with a syngenic split-thickness rat skin graft. After a further week, the sandwich flap is relocated to the rat dorsal surface, but the

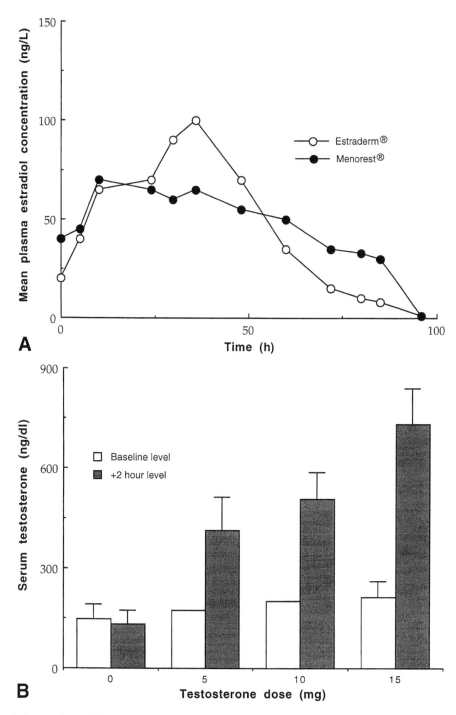

Figure 10 (A) Mean plasma levels of estradiol following application of two different transdermal systems to healthy women. (B) Serum drug levels (mean ± SE) before and after application of various concentrations of testosterone to the scrotum of hypogonadal men. (From: A, Ref. 407; B, Ref. 40.)

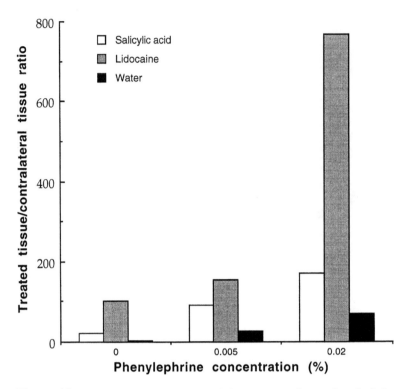

Figure 11 Topical coadministration of the vasoconstrictor, phenylephrine, to anesthetized rats significantly increased the concentration of various permeants in the underlying dermis, compared with that in the contralateral tissue. (Ref. 289; courtesy MS Roberts.)

vasculature remains intact and accessible. The sandwich flap was healed and is ready for use, 2 weeks after the final surgical procedure. Thus, the flap is viable and has functional skin on both sides: rat skin on one side and human skin on the other. Because both skin types have a common vasculature absorption across host or graft, skin can be assessed (Fig. 12). The ability to sample and analyze blood exiting the flap before dilution in the systemic pool minimizes the effect of postabsorption metabolic activity and considerably enhances sensitivity.

The permeation of benzoic acid (304,306) and estradiol (305) has been measured using the skin sandwich flap. Although this technique can provide useful data, particularly in comparing local and systemic effects, there are limitations. The procedure is time-consuming and costly; graft survival rate is low, and there is a limited experimental duration because of the anesthetic requirements. Furthermore, as for standard grafts, animals with surgically produced skin flaps must be treated systemically with immunosuppressive drugs, and the effect of this treatment on skin barrier function is unclear (261).

3.   Residual Analysis

Techniques that assess absorption by residual analysis or difference monitor the disappearance of material from the surface of the skin. Such methods are often referred to as noninvasive because they do not involve surgical procedures or physical sam-

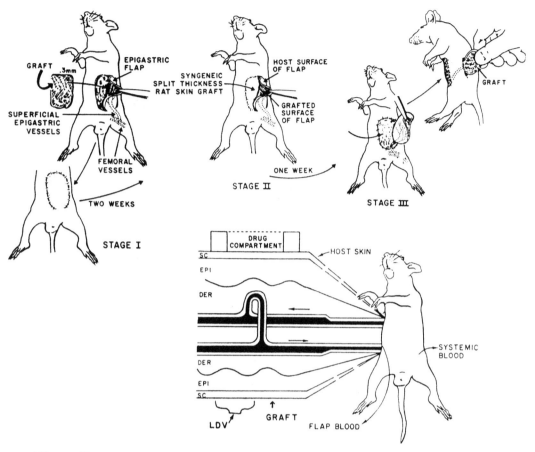

Figure 12 The grafted skin sandwich flap: The surgical procedure is described in the text. The vein draining the flap is cannulated to facilitate collection of blood samples. Percutaneous absorption can be determined through both host and grafted skin, as both are served by a common vasculature. (From Ref. 304.)

pling of body tissues or fluids. Two general approaches have been described: single-point (307) and continuous monitoring of permeant uptake (308). In the single-point method, the test chemical is applied for a fixed time, and the residual formulation is subsequently removed from the skin surface and analyzed. The technique requires only small quantities of permeant, making the use of radiolabel both cost-effective and ethically feasible. The main disadvantage of the single-point method is that it provides only one assay per site per application. If absorption kinetics are required, multiple application sites are necessary. Validity of the method is enhanced if it can be demonstrated that the test chemical is not removed from the skin surface by means other than percutaneous penetration (e.g., through contact with clothing). The amount permeated is calculated as the difference between the amount applied and the amount recovered and, if penetration and permeation are low, there may be a large error associated with the result.

There are various techniques for application of materials. The simplest method is to apply the test material to a defined area of skin as a solvent-deposited solid.

Once the solvent has evaporated, then the area can be covered with a protective material that can be either occlusive or nonocclusive. This approach can also be used for semisolid formulations such as creams or gels. The application of materials as solutions requires a means of restricting the material to the defined area of skin. This can be achieved using wells attached to the skin with acrylic or silicone adhesive. Commercially available devices, such as the Susten Skin Depot or the Bronaugh In Vivo Ring (Fig. 13) may be modified to provide adequate restrictive devices.

Collection of donor–residual concentration data over time allows kinetic analysis of permeation. Following application of the test chemical to the skin surface, disappearance is monitored, either spectroscopically or using radioisotope techniques (308–311). The technique is very simple and can be performed easily using human test subjects. For infrared spectroscopic techniques the test permeants must have distinctive infrared absorbencies in the transparent region of the skin spectrum. Higo et al. (310) used ATR–FTIR to follow the permeation of 4-cyanophenol across human SC. The compound was applied for periods up to 3 h in solution in propylene glycol or propylene glycol containing oleic acid. At the end of the exposure period, an ATR–FTIR spectrum of the treated surface was recorded. The SC was then removed sequentially by tape-stripping, and an ATR–FTIR spectrum recorded following each tape strip. The presence of cyanophenol at various depths in the SC was determined and quantitatively validated by ^{14}C-spiking (Fig. 14). The data demonstrated that ATR–FTIR could be used as a technique to quantitate skin levels of permeants and also as an in vivo method for evaluating skin permeation and penetration enhancement in humans.

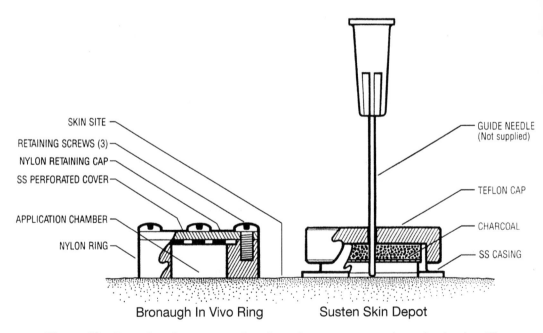

Figure 13 Examples of systems used to determine percutaneous absorption in vivo: The Bronaugh in vivo ring is a protective device that may be used in studies for which evaporation is not a factor. The Susten skin depot is particularly useful for determining the dermal absorption of volatile materials. (Courtesy Crown Glass Company, Somerville, NJ.)

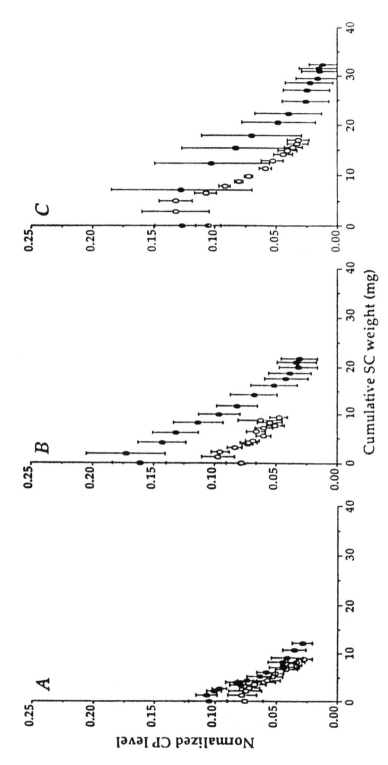

Figure 14 The use of ATR–FTIR to determine the permeation of 4-cyanophenol across human SC: the compound was applied for periods up to 3 h in a solution of propylene glycol (open circles) or propylene glycol containing 5% (v/v) oleic acid (closed circles). The SC was removed sequentially by tape-stripping, and an ATR–FTIR spectrum was recorded following each tape strip. Application periods were (A) 1 h, (B) 2 h, and (C) 3 h (From Ref. 310.)

Spectroscopic techniques have also been used widely in skin lipid biophysics (312) and to assess the water content of skin in vivo and in vitro (313–315). Other spectroscopic methods that have been used in the investigation of the properties of skin include fluorescence (316), impedance (317), and neutron (318) spectroscopy.

Optothermal-imaging techniques (such as optothermal transient emission radiometry; OTTER) have been used to monitor disappearance of material from the skin surface (Fig. 15). The technique is nondestructive and noninvasive and has been used to characterize skin condition (319,320), to follow the disappearance of sunscreens and other materials from the skin surface (321,322), and to monitor emulsion breakdown on the skin (323).

4. Skin-Stripping In Vivo

In Section III.B.1, the basics of tape-stripping and its use in vitro were outlined, and here, the in vivo applications are discussed. The in vivo technique is based on the reservoir principle (324–328). It is hypothesized that if a compound is applied to the skin for a limited time (e.g., 0.5 h) and then removed, the amount of drug in the upper layers of the SC will be predictive of the overall bioavailability of the compound. It follows that determination of the SC content of a permeating material following a short-term application will predict in vivo bioavailability from a corresponding administration protocol. Data obtained in studies of this type have shown reasonable predictability for several compounds (Fig. 16).

The use of in vivo skin-stripping in dermatopharmacokinetic evaluation was discussed at an AAPS–FDA workshop on bioequivalence of topical dermatological dosage forms (Bethesda, USA, September 1996). Although opinion was divided, it was concluded that SC skin-stripping "may provide meaningful information for comparative evaluation of topical dosage forms" (150). Furthermore, it was established that a combination of dermatopharmacokinetic and pharmacodynamic data may provide sufficient proof of bioequivalence "in lieu of clinical trials." However, much remains to be validated in skin-stripping protocols.

An outline protocol for skin-stripping bioequivalence studies has been suggested (150). The basic protocol has two phases: uptake and elimination.

Uptake:

1. Test and reference drug products are applied concurrently at multiple sites.
2. After exposure for a suitable time (determined by a pilot study), excess drug is removed by wiping three times with tissue or cotton swab.
3. The adhesive tape is applied with uniform pressure. The first strip is discarded (skin surface material). Repeat if necessary to remove excess surface material.
4. Collect nine successive tape strips from the same site. If necessary collect more than nine strips.
5. Repeat the procedure for each site at designated time intervals.
6. Extract the drug from the combined tape strips for each time point and site and determine the content of drug using an appropriate validated analytical method.
7. Express the data as amount of drug per cubic centimeter of tape.

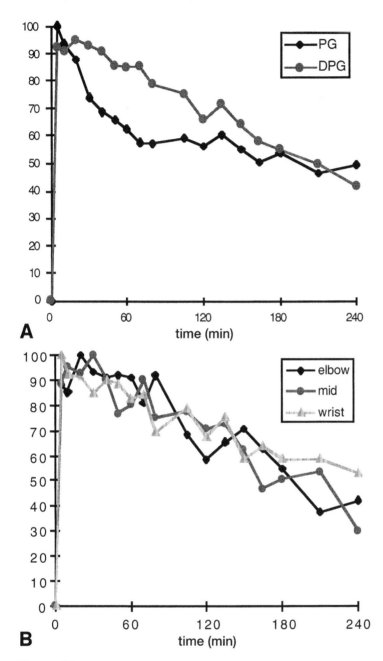

Figure 15 The use of optothermal transient emission radiometry (OTTER) to monitor disappearance of propylene glycol (PG) and dipropylene glycol (DPG) from the skin surface in vivo: disappearance is shown as, (A) the relative concentrations of the two permeants remaining on the skin surface with time (mean values from three sites); (B) disappearance of DPG from three sites on the ventral forearm. (Ref. 322.)

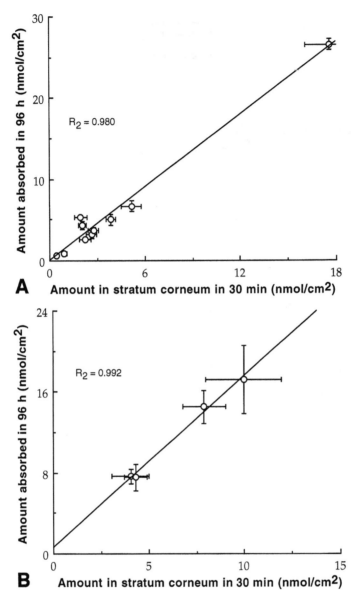

Figure 16 Correlation between the amount of compound absorbed through the skin, as determined from total body analysis or total excretion over 4 days, and the amount of compound recovered from stratum corneum tape strips following 30-min exposure. (A) Data obtained using hairless rat (200 nmol/cm² applied to dorsal skin). (B) Data obtained using human skin (1000 nmol/cm² applied to abdominal skin). (From: A, Ref. 410; B, Ref. 411.)

Elimination:

1. As for the foregoing steps 1–3.
2. After a predetermined time interval (e.g., 1, 3, 5, and 21 h postdrug removal) perform foregoing steps 4–7.

The results may then be expressed as shown in Fig. 17 that plots the amount of drug recovered from the tape strips against time. Uptake and elimination phases are observed and "bioavailability" may be predicted from the AUC.

There are several sources of variability in such studies, all of which must be considered in standard-operating procedures. The major causes of concern are variability in the following:

1. Drug application procedure
2. Type of tape
3. Size of tape
4. Pressure applied by investigator
5. Duration of application of pressure
6. Drug removal procedure
7. Drug extraction procedure
8. Analytical methods
9. Temperature
10. Relative humidity
11. Skin type
12. Skin surface uniformity

Nonetheless, following further validation, the technique will have several advantages. For example, basic pharmacokinetic parameters such as AUC, C_{max}, T_{max}, and half-life (329) may be approximated from the data obtained. In addition, the approach could be applicable to all types of topical preparation.

Pershing et al. (134) validated an in vivo skin-stripping protocol by correlating the SC strip data obtained for betametasone dipropionate with a skin-blanching bioassay experiment. Brief details of the stripping protocol were as follows: 180 mg of formulation were placed in a 1.2-cm–diameter Hilltop chamber (Hilltop Research Inc., Cincinnati, OH) and attached to the volar aspect of the forearm, with an exposed skin area of 1.13 cm^2 (i.e., a dose of 159 mg/cm^2). A maximum of three chambers per forearm were used, each being 2 cm apart, at least 6 cm above the wrist and 6 cm below the elbow. At the end of a 24-h period the chamber was removed, residual formulation on the skin surface wiped off, and the skin wiped with three separate dry cotton applicators. After air drying for 2 min, the skin was stripped ten times using 0.6-cm–diameter disks of Transpore tape. The amount of SC removed by each tape strip was assessed by weighing each tape before and after stripping. The relation between tape strip number and weight of SC removed was linear ($r = 0.996$), with an average of about 30 μg of skin removed per strip. Drug was extracted from the tape strips by vortexing in 200 μL of acetonitrile, followed by centrifugation and analysis of the supernatant by HPLC. Skin blanching was assessed at 1, 24, and 48 h following removal of formulations applied under occlusion for 24 h. The correlation between the amount of betamethasone dipropionate in skin and the skin-blanching score was good ($r = 0.994$) (Fig. 18), although the skin-blanching scores were not entirely objective because they were assessed visually, rather than with a

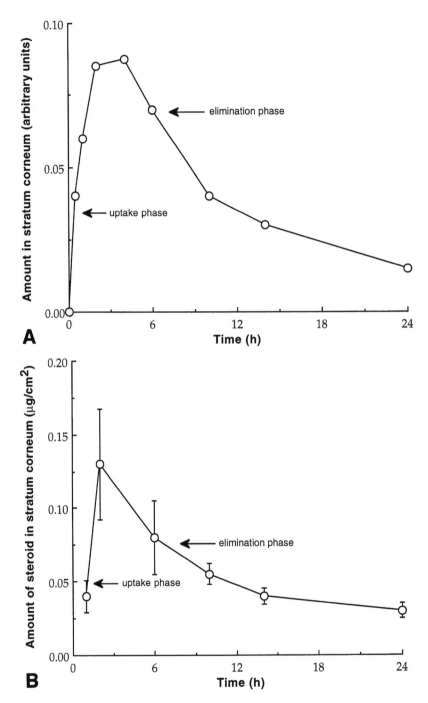

Figure 17 The use of stratum corneum tape-stripping to determine topical bioavailability or bioequivalence: (A) theoretical profile illustrating uptake and elimination phases, "bioavailability" may be predicted from the AUC; (B) experimental data for betamethasone dipropionate. (From Ref. 409.)

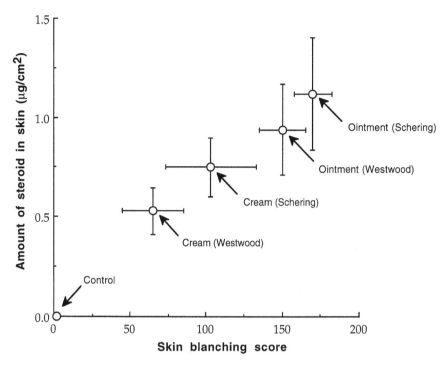

Figure 18 Correlation between the amount of betamethasone dipropionate in skin (determined by tape-stripping) and skin-blanching score (mean ± SE: $n = 5$) for some commercial formulations. (From Ref. 134.)

chromameter. Nonetheless, differences in responses between formulations (cream and ointment) and manufacturers could be discerned with both the pharmacokinetic and the pharmacodynamic techniques.

5. Microdialysis

A further technique that is currently being evaluated for the measurement of in vivo percutaneous absorption is microdialysis (330–338). Cutaneous microdialysis describes the in vivo measurement of substances in the extracellular space of the dermis by using a perfused dialysis membrane system. The technique involves insertion of microdialysis tubing below the skin surface into the dermis and back out again (Fig. 19). The dialysis tube is then perfused with a receiver fluid that removes permeant from the local area. The amount of permeant removed by the receiver medium is then related to the total amount absorbed. Disadvantages of microdialysis include small sample size, requirement of sensitive analytical methods, and the invasive nature of the technique. Detection of lipophilic or highly protein-bound substances are problematic. Nonetheless, cutaneous microdialysis may have a role in the assessment of dermal toxicity and in the development of an understanding of the pathogenesis of dermatological disease.

6. Pharmacodynamic Measurements

The use of pharmacodynamic changes (i.e., bioassays), can provide a method for assessing local absorption. The most widely used technique is the vasoconstriction

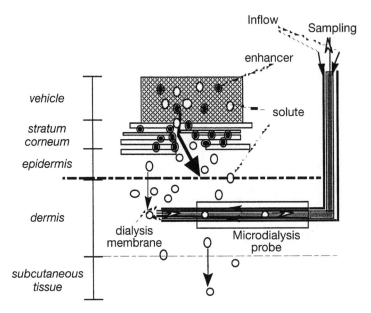

Figure 19 Schematic illustrating the technique of cutaneous microdialysis: the amount of substance migrating across the epidermis and into the extracellular space of the dermis is estimated from the concentration of permeant appearing in the perfusion fluid. (From Ref. 412; courtesy M.S. Roberts.)

assay (or skin-blanching assay) used for measuring the dermal absorption of gluco-corticosteroids (134,339–341). The basis of the assay is the correlation between the vasoconstriction produced by the compounds and their overall clinical effect. Although such methods are inexpensive, internally consistent, and qualitatively reliable (4), objective methods for quantifying the response are essential. Assessment of color change can be quantitatively determined using chromameters (342,343). Pharmaco-dynamic measurements are fully discussed in Section V of Chapter 8.

7. Clinical Evaluation

Clinical efficacy testing will always be necessary to demonstrate the effectiveness of any new transdermal or topical therapy. For example, an extensive review (344) of estrogen replacement therapy that addressed the key pharmacokinetic and metabolic differences between oral and transdermal forms, with particular emphasis on their effects on gonadotropins, hemostasis and coagulation, lipid metabolism, hepatobiliary function, and bone, concluded that the transdermal route was a viable alternative to the oral route of administration. Lin et al. (345) described the development of a transdermal nicotine delivery device and generated clinical evaluation to demonstrate the efficacy of their product. This approach was also used by Shah et al. (346) in the development of a transdermal verapamil system.

The clinical assessment of transdermally delivered substances is often coupled with the generation of pharmacokinetic data. However, it is more difficult to generate pharmacokinetic data during evaluation of the clinical efficacy of dermatological products, such as antifungal preparations, or locally acting topical products, such as nonsteroidal anti-inflammatory agents (e.g., ibuprofen and diclofenac). Furthermore, in the latter example, the demonstration of efficacy is difficult because the indication is often self-resolving, and there is a significant placebo effect. Although there are many studies that describe the bioavailability and efficacy of topical NSAIDs (88,347–362), there still exists some scepticism over their therapeutic value, especially in comparison with simple oral dosing. Recently, several articles have addressed this issue by examining "local enhanced topical drug delivery" and concluded that, for the compounds studied, this does indeed occur (288,289,363,364).

IV. IN VIVO–IN VITRO COMPARISONS

It is important to consider the ability of in vitro skin penetration techniques to predict skin permeation in vivo. Unless it can be demonstrated that in vitro skin penetration data is reasonably similar to absorption across skin in situ, there is little value in obtaining this data. It is also beneficial to compare in vitro skin penetration data with clinical observations (see, e.g., Chap. 6, Sec. II.B.3), although the information available is sparse.

A. Correlation Between In Vitro and In Vivo Measurements

A review of the literature reveals that, in most cases, extrapolation between in vitro and in vivo human skin permeation is reasonably accurate (8,18,82,264,365–386). If, however, a permeating compound undergoes extensive dermal metabolism in vivo, then this may not be reflected by valid in vitro results unless freshly excised skin is used (378). Overall, however, the ranking of prototype formulations or the evaluation of the skin permeation of homologous compounds achieved using in vitro methods will be a reflection of the in vivo scenario.

Many such studies have been performed using small laboratory animals, and their relevance to human skin has always been questionable, especially in terms of risk assessment. However, in terms of system validation, the comparison of in vitro and in vivo skin penetration data obtained using animal models may be useful. A comprehensive study of in vitro–in vivo relations for animal skin was carried out (371) using rat tissue. Some of the data from this set is shown in Table 9 together with other in vivo–in vitro data for different animal models.

Scott et al. (371) determined the percutaneous absorption in vivo and the skin permeation in vitro of eight pesticides (ICI A through ICI G) varying in molecular weight from 187.3 to 416.3, in water solubility from 0.01 to 880 mg/mL, and in log $P_{O/W}$ (log octanol–water partition coefficient) from 2.1 to > 5.0. In this study comparisons were also made between the amount of test chemical remaining on or in the skin after in vivo and in vitro experimentation. In general, despite a difference in rinsing procedure between the in vivo and in vitro experiments, there was good agreement between the amount of test chemical that could be removed from the surface of the skin at the end of the experiments. Similarly, in most cases, the amount of compound recovered from the skin was similar. For all test compounds there was qualitative and quantitative agreement in ranking of absorption between the in vitro and in vivo data (see Table 9). Usually, in vitro absorption values were higher than in vivo values. Nonetheless, for this dataset, the predictability of the in vivo results based on in vitro data, measured by linear regression over all time points, was reasonable ($r^2 = 88.5\%$) and at later time points was good ($r^2 = 95.5\%$).

Table 9 Comparison Between Some In Vitro and In Vivo Penetration Data Using Rat, Mouse, and Monkey Skin

Species	Test compound	In vivo[a]	In vitro[a]	Ratio[b]	Ref.
Rat	Anthracene	18	24	1.3	379
	Benzyl acetate	2	4	2.0	380
	Cypermethrin	1	1.7–2.7	1.7–2.7	370
	ICI A	12	24	0.5	371
	ICI B	86	76	0.9	371
	ICI C	23	22	1.0	371
	ICI D	56	79	1.4	371
	ICI E	4		0.8	371
	ICI F	32	78	2.4	371
	ICI G	2	2	1.0	371
	ICI H	14	9	0.6	371
Mouse	Permethrin	26	18	0.7	368
	DDT	25	27	1.1	368
	2,4-D	44	28	0.7	368
	Fenvalerate	19	20	1.0	368
Monkey	4-Amino-2-nitrophenol	64	48	0.8	381
	2,4-Dinitrochlorobenzene	52	48	0.9	381
	p-Nitroaniline	76	62	0.8	381
	2-Nitro-*p*-phenylene diamine	30	30	1.0	381
	Nitrobenzene	4.2	6.2	1.5	381

[a]Data are expressed as % applied dose absorbed.
[b]In vitro–in vivo.

In a similar study, Surber et al. (369) examined the in vitro–in vivo relation in the penetration and permeation of structurally related phenols and steroids across hairless guinea pig skin. The *para*-substituted phenols varied in log $P_{O/W}$ from 0.3 to 3.5 and in water solubility from 0.3 to 12.6 mg/mL. The steroids varied in log $P_{O/W}$ from 1.5 to 3.9 and in water solubility from 0.003 to 0.39 mg/mL. Penetration into, and permeation through, the skin were determined following application of the test compound in either water or isopropyl myristate. The data are given in Table 10 and shown in Figure 20. As is usual for such studies, a parabolic relation between permeation and test compound lipophilicity was observed for both the in vivo and in vitro experiments. For the phenol analogues, the in vitro measurement overestimated the values obtained in vivo, but this was reversed for the steroids.

Kasting et al. (377) determined the in vitro and in vivo rat skin permeation of three lipophilic capsaicin-like vanilloids, varying in log $P_{O/W}$ from 3.74 to 8.02. The degree of correlation between the data sets was dependent on the nature of the receptor fluid in the in vitro experiment and the time over which comparisons were made (Table 11). At the 24-h timepoint, there was closer agreement for the lipophilic permeants when the receptor fluid was designed to accommodate such permeants. Mostly, however, at the 72-h timepoint oleth 20-containing receptor fluids tended to overestimate the in vivo absorption. The authors rationalized the time-related differences between in vivo and in vitro absorption by proposing that they were the con-

Table 10 Comparison Between Some In Vitro and In Vivo Skin Permeation and Penetration Data Using Guinea Pig Skin

Test compound	Permeation (through skin)		Penetration (into skin)	
	In vitro (μg/cm^2)	In vivo (μg/cm^2)	In vitro (μg/g)	In vivo (μg/g)
A.				
4-Acetamidophenol	23.9	17.3	15.5	7.6
4-Cyanophenol	408.6	317.2	154.3	122.7
4-Iodophenol	242.6	156.7	103.6	90.5
4-Pentyloxyphenol	60.8	89.9	34.1	83.2
Hydrocortisone	1.4	2.2	1.1	1.0
Estradiol	0.8	3.6	0.9	1.1
Testosterone	4.0	5.3	1.3	1.6
Progesterone	0.6	1.1	0.3	0.5
B.				
4-Acetamidophenol	33.1	24.7	7.3	6.9
4-Cyanophenol	2221.8	1373.4	345.6	237.6
4-Iodophenol	537.3	280.0	113.7	84.2
4-Pentyloxyphenol	79.6	56.8	94.3	79.5
Hydrocortisone	2.0	8.6	2.0	15.3
Estradiol	4.0	13.9	9.2	64.7
Testosterone	9.3	41.3	16.7	104.2
Progesterone	2.4	15.6	7.4	64.8

Compounds were applied as saturated solution in either water (A) or isopropyl myristate (B).
Source: Ref. 369.

sequence of an observed gradual decline in the in vivo absorption rates, not observed in vitro, possibly occurring as a result of skin turnover.

The most comprehensive studies on in vitro–in vivo correlations using human skin are those of Franz (8,82). In these experiments conditions of application of the compound to the skin were as similar as possible in the two protocols. The data obtained in an initial study (8) are summarized in Table 12. There was a rank order correlation between the two sets of data (p = 0.01) which was remarkable, given that the in vivo data was obtained using forearm skin (263) and the in vitro data using abdominal skin, in different laboratories. Furthermore, the correlation was considerably better for those compounds that exhibited greater penetration. From the data given in Table 12, it is obvious that for poorly permeating compounds (< 1.0% absorption in vivo), the in vitro system significantly overestimates absorption. This may be rationalized to some extent by a consideration of the differences in the dynamics of the two systems. In vivo there is an average loss, through desquamation, of about one cell layer of SC per day and it can be assumed that, for slow penetrants, a significant proportion of the applied dose of permeant will be lost in this process. Because of a lack of desquamation in vitro this loss does not occur, and the permeating compound will remain in the SC over a longer period and thus continue to be available for diffusion into deeper epidermal layers.

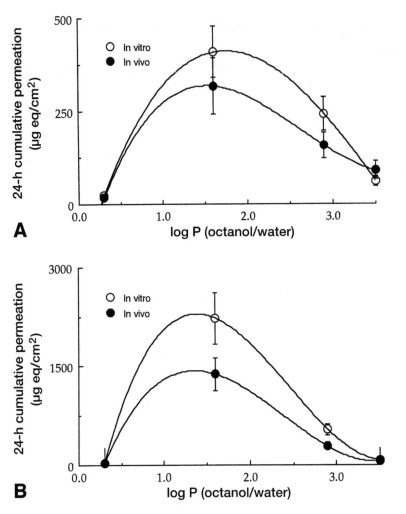

Figure 20   The in vitro–in vivo relation for the permeation of structurally related phenols across hairless guinea pig skin: permeation was determined following application of the test compound in either (A) water or (B) isopropyl myristate. A parabolic relation between permeation and lipophilicity is observed for both the in vivo and in vitro experiments. As can be seen, here, the in vitro measurements overestimate the values obtained in vivo. (From Ref. 369.)

In a follow-up study (82), the compounds that showed the greatest difference between in vivo and in vitro absorption (nicotinic acid, hippuric acid, thiourea, and caffeine) were reevaluated, with a slightly different protocol. In this instance, abdominal skin was used for both in vivo and in vitro application, and in vivo urine was collected until the permeant levels had reached background levels. In addition the area of application was washed 24 h following application, which should minimize the differences that may have been caused by desquamation. The data obtained in the follow-up study are shown in Table 13. The modified protocol in this follow-up experiment produced a much-improved correlation between in vivo and in vitro

Table 11 Influence of In Vitro Receptor Fluid Composition on the Correlation Between In Vitro and In Vivo Skin Permeation of a Series of Vanilloids

	Test permeant					
	Vanillylnonanamide			Olvanil	NE-21610	
Log P	3.74			8.02	7.16	
In vivo[a] (24 h)	4.5			1.4	0.85	
In vivo[a] (72 h)	12.7			3.6	2.01	
In vitro receptor fluid[b]	%[a]	R[c]	%[a]	R	%[a]	R
0.5% BSA (24 h)	2.3	0.51	0.9	0.64	0.14	0.16
0.5% BSA (72 h)	6.8	0.54	3.4	0.94	0.45	0.22
1% Oleth 20 (24 h)	3.2	0.71	1.5	1.07	0.38	0.45
1% Oleth 20 (72 h)	11.6	0.91	6.7	1.86	1.70	0.85
6% Oleth 20 (24 h)	3.7	0.82	1.5	1.07	0.53	0.62
6% Oleth 20 (72 h)	14.4	1.13	7.9	2.19	2.63	1.31

[a] % Applied dose Penetrated.
[b] Dulbecco's phosphate buffered saline (pH 7.4) containing preservative and the additives noted.
[c] In vitro/in vivo ratio.
Source: Ref. 377.

Table 12 Comparison of In Vitro Human Skin Penetration Data with Published In Vivo Values

Compound	Absorption in vivo (%)	Absorption in vitro (%)	In vitro/in vivo ratio
Hippuric acid	0.2	1.2	6.0
Nicotinic acid	0.3	3.3	11.0
Thiourea	0.9	3.4	3.8
Chloramphenicol	2.0	2.9	1.5
Phenol	4.4	10.9	2.5
Urea	6.0	11.1	1.9
Nicotinamide	11.1	28.8	2.6
Acetylsalicylic acid	21.8	40.5	1.9
Salicylic acid	22.8	12.0	0.5
Benzoic acid	42.6	44.9	1.1
Caffeine	47.6	9.0	0.2
Dinitrochlorobenzene	53.1	27.5	0.5

Source: Ref. 263.

Table 13 Comparison of In Vitro and In Vivo Human Skin
Penetration Data Obtained Under Controlled Conditions

Compound	Absorption in vivo (%)	Absorption in vitro (%)	In vitro/in vivo ratio
Hippuric acid	1.0	1.3	1.3
Nicotinic acid	0.3	2.3	7.7
Thiourea	3.7	4.6	1.2
Caffeine	22.1	24.1	1.1

Source: Ref. 82.

results. In most cases, quantitative agreement was excellent. The difference observed
for nicotinic acid is attributable to the observation that only 15% of an intravenous
dose of this compound is excreted in the urine and, when this is factored, the value
for in vivo absorption becomes 2.1%, virtually identical with that obtained in vitro.
It is further evident from the data given in Tables 12 and 13 that a slight alteration
in protocols caused a significant alteration in the in vivo data for each compound,
although the in vitro data (except for caffeine) was comparatively unchanged.

Bronaugh and Franz (365) studied in vivo–in vitro comparisons in the human
skin permeation of benzoic acid, caffeine, and testosterone from a petrolatum vehicle.
Good agreement was obtained (Table 14).

In summary, comparison between in vivo and in vitro data for compounds
penetrating and permeating human skin have shown that a high correlation exists
between the two sets of data. In most cases, for poor penetrants, the in vitro data
has slightly overestimated the values obtained in vivo. However, it is important to
appreciate that, as pointed out by Franz (82) and Kligman (387), when discrepancies
occur between in vivo and in vitro data, it is more likely that the in vivo data have
been measured inaccurately. It can be concluded that, provided the experimentation
is performed using suitably designed protocols and that the in vivo "in-use" conditions are mimicked as closely as possible, valuable and realistic data on skin penetration can be obtained using human skin in vitro.

Table 14 Comparison of In Vitro and In Vivo Human Skin
Penetration Data for Compounds Applied in Petrolatum

Compound	Absorption in vivo (%)	Absorption in vitro (%)	In vitro/in vivo ratio
Benzoic acid	60.6	46.5	0.8
Caffeine	40.6	40.6	1.0
Testosterone	49.5	39.4	0.8

Source: Ref. 365.

B. Correlation Between In Vitro Data and Therapeutic Activity

It is important to realize the limitations of in vitro measurements and to understand that data derived from such experiments should not be overinterpreted. Although, in vitro measurements have undoubtedly generated, and will continue to generate, very useful information for formulation development, and optimization (386), hazard and risk assessment (388), and for bioequivalency testing (see Chap. 8), they are not as useful when attempting to predict onset of therapeutic action. As discussed earlier, neither physiological nor pharmacodynamic responses occur in vitro; thus, for example, a formulation applied to the skin in vivo may be affected by sweat or sebum secretion that cannot be mimicked in vitro. Similarly, a permeating compound that has the property of vasodilation, will increase blood flow to, and increase permeant clearance from, the skin in vivo, but will have no such effect in vitro. In vitro, and indeed in vivo, quantitative measurements of skin penetration, permeation, and absorption lag times are limited by analytical sensitivity. Pharmacodynamic and therapeutic responses are limited by receptor occupancy. Thus, whereas it may take several hours for a permeating compound to reach levels of analytical detection, it may take only a few minutes for sufficient molecules to permeate the skin, attach to receptors, and elicit the pharmacodynamic response.

Nonetheless, there is some evidence that in vitro measurements of skin permeability can qualitatively rank the clinical response. Some examples of this are discussed in Chapter 6 (see Sec. II.B.3). Recently, Foldvari et al. (389) have demonstrated that in vitro measurements of prostaglandin-E_1 permeation across human skin can predict vasodilatory activity. Although the in vitro lag time was 5–10 h (depending on formulation), peak blood flow in vivo, as determined using laser Doppler flowmetry, was attained within 30 min. Nonetheless, formulation ranking of vasodilatory efficacy was predictable by in vitro skin permeation measurement.

V. CONCLUDING REMARKS

There is no doubt that measurement of skin permeation both in vitro and in vivo can play a major role in the optimization of formulations for dermal and transdermal drug delivery. The good correlation between in vitro and in vivo measurements adds support to the usefulness of the former, more cost-effective, option. In addition to the pharmaceutical considerations, regulatory concern about potential dermal absorption leading to possible local or systemic effects, is now a major concern for the cosmetic and chemical industries. For products in these sectors to meet increasingly stringent requirements, conclusive evidence for their efficacy and safety is required. With the recent loss of the option to conduct in vivo animal testing on cosmetics and their ingredients, and the ethical and cost constraints of human in vivo studies, one of the most viable remaining approaches for assessing dermal absorption is that of in vitro human skin permeation evaluation. Although in many respects these techniques may appear straightforward, it is important to appreciate that a generalized protocol may not provide an appropriate framework for a particular study, that it is essential that studies are well-designed to fit their specific purpose, that they are competently executed by experienced scientists, and that the resultant data is interpreted appropriately taking into account all of the relevant factors discussed in the foregoing.

REFERENCES

1. Federal Register. Requests for proposals for enforceable consent agreements; dermal absorption rate testing of eighty OHSA chemicals; solicitation of interested parties; text of protocol. April 3, 61(65); 1996.
2. Walker JD, Whittaker C, McDougal JN. Role of the TSCA interagency testing committee in meeting the U.S. government data needs: designating chemicals for percutaneous absorption rate testing. In: Marzulli FN, Maibach HI, eds. Dermatotoxicology, 5th ed. Washington, DC: Taylor & Francis, pp 371–381, 1996.
3. Skelly JP, Shah VP, Maibach HI, Guy RH, Wester RC, Flynn G, Yacobi A. FDA and AAPS report of the workshop on principles and practices of in vitro percutaneous penetration studies: relevance to bioavailability and bioequivalence. Pharm Res 4:265–267, 1987.
4. Howes D, Guy R, Hadgraft J, Heylings J, Hoeck F, Maibach H, Marty J–P, Merk H, Parra J, Rekkas D, Rondelli I, Schaefer H, Täuber U, Verbiese N. Methods for assessing percutaneous absorption, ECVAM Workshop Report 13. ALTA 24:81–106, 1996.
5. ECETOC Monograph No. 20 Percutaneous absorption. Brussels: European Centre for Ecotoxicology and Toxicology of Chemicals, 1993; 80 pp.
6. COLIPA (The European Cosmetic, Toiletry and Perfumery Association). Cosmetic ingredients: guidelines for percutaneous absorption/penetration, 1993.
7. Lehman PA, Agrawa IN, Franz TJ, Miller KJ, Franz S. Comparison of topical hydrocortisone products: percutaneous absorption vs. tape stripping vs. vasoconstriction vs. membrane rate of release. Pharm Res 13:S310, 1996.
8. Franz TJ. On the relevance of in vitro data. J Invest Dermatol 64:190–195, 1975.
9. Bronaugh RL, Stewart RF. Methods for in vitro percutaneous absorption studies. IV: The flow through diffusion cell. J Pharm Sci 74:64–67, 1985.
10. Walters KA, Flynn GL, Marvel JR. Physicochemical characterization of the human nail: I pressure sealed apparatus for measuring nail plate permeabilities. J Invest Dermatol 76:76–79, 1981.
11. Flynn GL, Smith EW. Membrane diffusion. I. Design and testing of a new multifeatured diffusion cell. J Pharm Sci 60:1713–1717, 1971.
12. Nugent FJ, Wood JA. Methods for the study of percutaneous absorption. Can J Pharm Sci 15:1–7, 1980.
13. Gummer C. The skin penetration cell: a design update. Int J Pharm 40:101–104, 1987.
14. Tanojo H, Roemelé PEH, van Veen GH, Stieltjes H, Junginger HE, Boddé HE. New design of a flow-through permeation cell for studying in vitro permeation studies across biological membranes. J Controlled Release 45:41–47, 1997.
15. Bronaugh RL. Methods for in vitro percutaneous absorption. In: Marzulli FN, Maibach HI, eds. Dermatotoxicology, 5th ed. Washington, DC: Taylor & Francis, pp 317–324, 1996.
16. Hughes MF, Shrivasta SP, Fisher HL, Hall LL. Comparative in vitro percutaneous absorption of p-substituted phenols through rat skin using static and flow-through diffusion systems. Toxicol In Vitro 7:221–227, 1993.
17. Clowes HM, Scott RC, Heylings JR. Skin absorption: flow-through or static diffusion cells. Toxicol In Vitro 8:827–830, 1994.
18. Moody RP, Ritter L. An automated in vitro dermal absorption procedure. II. Comparative in vivo and in vitro dermal absorption of the herbicide fenoxaprop-ethyl (HOE-33171) in rats. Toxicol In Vitro 6:53–59, 1992.
19. Bronaugh RL. In vitro methods for the percutaneous absorption of pesticides. In: Honeycutt R, Zweig G, Ragsdale NN, eds. Dermal Exposure to Pesticide Use. Washington, DC: American Chemical Society, pp 33–41, 1985.
20. Dick IP, Blain PG, Williams FM. Improved in vitro skin absorption for lipophilic compounds following the addition of albumin to the receptor fluid in flow-through

cells. In: Brain KR, James VJ, Walters KA, eds. Prediction of Percutaneous Penetration. Vol. 4b. Cardiff: STS Publishing, pp 267–270, 1996.

21. Sclafani J, Liu P, Hansen E, Cettina MG, Nightingale J. A protocol for the assessment of receiver solution additive-induced skin permeability changes. An example with gamma-cyclodextrin. Int J Pharm 124:213–217, 1995.

22. Jones SP, Greenway MJ, Orr NA. The influence of receptor fluid on in vitro percutaneous penetration. Int J Pharm 53:43–46, 1989.

23. Smith EW, Haigh JM. In vitro diffusion cell design and validation, Acta Pharm Nord 4:171–178, 1992.

24. Ehntholt DJ, Cerundolo DL, Bodek I, Schwope AD, Royer MD, Nielsen AP. A test method for the evaluation of protective glove materials used in agricultural pesticide operations. Am Ind Hyg Assoc J 51:462–468, 1990.

25. Leinster P, Bonsall JL, Evans MJ, Lloyd GA, Miller B, Rackham M. The development of a standard test method for determining permeation of liquid chemicals through protective clothing materials. Ann Occup Hyg 30:381–395, 1986.

26. Kou JH, Roy SD, Du J, Fujiki J. Effect of receiver fluid pH on in vitro skin flux of weakly ionizable drugs. Pharm Res 10:986–990, 1993.

27. Wester RC. Relevance of animal models for percutaneous absorption. Int J Pharm 7:99–110, 1980.

28. Bronaugh RL, Stewart RF, Congdon ER. Methods for in vitro percutaneous absorption studies. II: Animal models for human skin. Toxicol Appl Pharmacol 62:481–488, 1982.

29. Reifenrath W. Evaluation of animal models for predicting skin penetration in man. Fundam Appl Toxicol 4:S224–S230, 1984.

30. Walters KA, Roberts MS. Veterinary applications of skin penetration enhancers. In: Walters KA, Hadgraft J, eds. Pharmaceutical Skin Penetration Enhancement. New York: Marcel Dekker, pp 345–364, 1993.

31. Southwell JD, Barry BW, Woodford R. Variations in permeability of human skin within and between specimens. Int J Pharm 18:299–309, 1984.

32. Kasting GB, Filloon TG, Francis WR, Meredith MP. Improving the sensitivity of in vitro skin penetration experiments. Pharm Res 11:1747–1754, 1994.

33. Cornwell PA, Barry BW. Effect of penetration enhancer treatment on the statistical distribution of human skin permeabilities. Int J Pharm 117:101–112, 1995.

34. Williams AC, Cornwell PA, Barry BW. On the non-gaussian distribution of human skin permeabilities. Int J Pharm 86:69–77, 1992.

35. Bronaugh RL, Stewart RF, Simon M. Methods for in vitro percutaneous absorption studies VII: Use of excised human skin. J Pharm Sci 75:1094–1097, 1986.

36. Kasting GB, Bowman LA. Electrical analysis of fresh excised human skin: a comparison with frozen skin. Pharm Res 7:1141–1146, 1990.

37. Harrison SM, Barry BW, Dugard PH. Effects of freezing on human skin permeability. J Pharm Pharmacol 36:261–262, 1984.

38. Marzulli FN, Maibach HI. Permeability and reactivity of skin as related to age. J Soc Cosmet Chem 35:95–102, 1984.

39. Roskos KV, Bircher AJ, Maibach HI, Guy RH. Pharmacodynamic measurements of methyl nicotinate percutaneous absorption: the effect of aging on microcirculation. Br J Dermatol 122:165–171, 1990.

40. Banks YB, Brewster DW, Birnbaum LS. Age-related changes in dermal absorption of 2,3,7,8-tetrachlorodibenzo-p-dioxin and 2,3,4,7,8-pentachlorodibenzofuran. Fundam Appl Toxicol 15:163–173, 1990.

41. Hall LL, Fisher HL, Sumler MR, Hughes MF, Shah PV. Age-related percutaneous penetration of 2-sec-butyl-4,6-dinitrophenol (dinoseb) in rats. Fundam Appl Toxicol 19:258–267, 1992.

42. Dick IP, Scott RC. The influence of different strains and age on in vitro rat skin permeability to water and mannitol. Pharm Res 9:884–887, 1992.

43. Behl CR, Flynn GL, Linn EE, Smith WM. Percutaneous absorption of coritcosteroids: age, site, and skin-sectioning influences on rates of permeation of hairless mouse skin by hydrocortisone. J Pharm Sci 73:1287–1290, 1984.

44. Behl CR, Flynn GL, Kurihara T, Smith WM, Bellantone NH, Gatmaitan O, Higuchi WI, Ho NFH, Pierson CL. Age and anatomical site influences on alkanol permeation of skin of the male hairless mouse. J Soc Cosmet Chem 35:237–252, 1984.

45. Roskos KV, Maibach HI, Guy RH. The effect of ageing on percutaneous absorption in man. J Pharm Biopharm 17:617–630, 1989.

46. Rougier A. Percutaneous absorption–transepidermal water loss relationship in man in vivo. In: Scott RC, Guy RH, Hadgraft J, Boddé HE, eds. Prediction of Percutaneous Penetration. Vol. 2. London: IBC Technical Services, pp 60–72, 1991.

47. Roskos KV, Guy RH, Maibach HI. Percutaneous absorption in the aged. Dermatol Clin 4:455–465, 1986.

48. Guy RH, Tur E, Bjerke S, Maibach HI. Are there age and racial differences to methyl nicotinate-induced vasodilation in human skin? J Am Acad Dermatol 12:1001–1006, 1983.

49. Roskos KV, Maibach HI. Percutaneous absorption and age: implications for therapy. Drugs Aging 2:432–449, 1992.

50. Wedig JH, Maibach HI. Percutaneous penetration of dipyrithione in man: effect of skin color (race). J Am Acad Dermatol 5:433–438, 1981.

51. Berardesca E, Maibach HI. Racial differences in pharmacodynamic response to nicotinates in vivo in human skin: black and white. Acta Derm Venereol 70:63–66, 1990.

52. Leopold CS, Maibach HI. Effect of lipophilic vehicles on in vivo skin penetration of methyl nicotinate in different races. Int J Pharm 139:161–167, 1996.

53. Weigand DA, Gaylor JR. Irritant reaction in negro and Caucasian skin. South Med J 67:548–551, 1974.

54. Kompaore F, Tsuruta H. In vivo differences between Asian, black and white in the stratum corneum barrier function. Int Arch Occup Environ Health 65:S223–S225, 1993.

55. Kompaore F, Marty J–P, Dupont C. In vivo evaluation of the stratum corneum barrier function in blacks, Caucasians and Asians with two noninvasive methods. Skin Pharmacol 6:200–207, 1993.

56. Weigand DA, Haygood C, Gaylor JR. Cell layers and density of negro and Caucasian SC. J Invest Dermatol 62:563–568, 1974.

57. Rienertson RP, Wheatley VR. Studies on the chemical composition of human epidermal lipids. J Invest Dermatol 32:49–59, 1959.

58. Rougier A, Lotte C, Corcuff P, Maibach HI. Relationship between skin permeability and corneocyte size according to anatomic site, age, and sex in man. J Soc Cosmet Chem 39:15–26, 1988.

59. Corcuff P, Lotte C, Rougier A, Maibach HI. Racial differences in corneocytes. A comparison between black, white and Oriental skin. Acta Derm Venereol 71:146–148, 1991.

60. Berardesca E, Maibach HI. Racial differences in sodium lauryl sulphate induced cutaneous irritation: black and white. Contact Dermatitis 18:65–70, 1988.

61. Berardesca E, Maibach HI. Cutaneous reactive hyperaemia: racial differences induced by corticoid application. Br J Dermatol 120:787–794, 1989.

62. Wickrema Sinha AJ, Shaw SR, Weber DJ. Percutaneous absorption and excretion of tritium-labeled diflorasone diacetate, a new topical corticosteroid in the rat, monkey and man. J Invest Dermatol 71:372–377, 1978.

63. Lotte C, Wester RC, Rougier A, Maibach HI. Racial differences in the in vivo percutaneous absorption of some organic compounds: a comparison between black, Caucasian and Asian subjects. Arch Dermatol Res 284:456–459, 1993.

64. Yazdanian M. The effect of freezing on cattle skin permeability. Int J Pharm 103:93–96, 1994.

65. Wester RC, Christoffel J, Hartway T, Poblete N, Maibach HI, Forsell J. Human cadaver skin viability for in vitro percutaneous absorption: storage and detrimental effects of heat-separation and freezing. Pharm Res 15:82–84, 1998.

66. Fares HM, Zatz JL. Dual-probe method for assessing skin barrier integrity: effect of storage conditions on permeability of micro-Yucatan pig skin. J Soc Cosmet Chem 48: 175–186, 1997.

67. Feldman RJ, Maibach HI. Regional variation in the percutaneous penetration of ^{14}C cortisol in man. J Invest Dermatol 48:181–183, 1967.

68. Maibach HI, Feldman RJ, Mitby TH, Serat WF. Regional variation in percutaneous penetration in man: pesticides. Arch Environ Health 23:208–211, 1971.

69. Wester RC, Noonan PK, Maibach HI. Variations in percutaneous absorption of testosterone in the rhesus monkey due to anatomic site of application and frequency of application. Arch Dermatol Res 267:229–235, 1980.

70. Wester RC, Maibach HI. Cutaneous pharmacokinetics: 10 steps to percutaneous absorption. Drug Metab Rev 14:169–205, 1983.

71. Rougier A, Roguet R. Regional variation in percutaneous absorption in man: measurement by the tape stripping method. Arch Dermatol Res 278:465–469, 1986.

72. Rougier A. In vivo percutaneous penetration of some organic compounds related to anatomic site in humans: stripping method. J Pharm Sci 76:451–454, 1987.

73. Tur E, Maibach HI, Guy RH. Percutaneous penetration of methyl nicotinate at three anatomical sites: evidence for an appendageal contribution to transport? Skin Pharmacol 4:230–234, 1991.

74. Lotte CA, Wilson DR, Maibach HI. In vivo relationship of transepidermal water loss and percutaneous absorption in man: effect of anatomical site. Arch Dermatol Res 279: 351–356, 1987.

75. Roberts MS, Favretto WA, Meyer A, Reckmann M, Wongseelashote T. Topical bioavailability of methyl salicylate. Aust NZ J Med 12:303–305, 1982.

76. Trebilcock KL, Heylings JR, Wilks MF. In vitro tape stripping as a model for in vivo skin stripping. Toxicol In Vitro 8:665–667, 1994.

77. Scheuplein RJ. Mechanism of percutaneous absorption I. Routes of penetration and influence of solubility. J Invest Dermatol 45:334–346, 1965.

78. Kumar S, Behl CR, Patel SB, Malick AW. A simple microwave technique for the separation of epidermis and dermis in skin uptake studies. Pharm Res 6:740–741, 1989.

79. Boddé HE, Ponec M, Ijzerman AP, Hoogstraate AJ, Salomons MAI, Bouwstra JA. In vitro analysis of QSAR in wanted and unwanted effects of azacycloheptanones as transdermal penetration enhancers. In: Walters KA, Hadgraft J, eds. Pharmaceutical Skin Penetration Enhancement. New York: Marcel Dekker, pp 199–214, 1993.

80. Brain KR, Walters KA, James VJ, Dressler WE, Howes D, Kelling CK, Moloney SJ, Gettings SD. Percutaneous penetration of dimethylnitrosamine through human skin in vitro: application from cosmetic vehicles. Food Chem Toxicol 33:315–322, 1995.

81. Walters KA, Brain KR, Howes D, James VJ, Kraus AL, Teetsel NM, Toulon M, Watkinson AC, Gettings SD. Percutaneous penetration of octyl salicylate from representative sunscreen formulations through human skin in vitro. Food Chem Toxicol 35: 1219–1225, 1997.

82. Franz TJ. The finite dose technique as a valid in vitro model for the study of percutaneous absorption in man. Curr Probl Dermatol 7:58–68, 1978.

83. Franz TJ, Lehman PA, Franz SF, North–Root H, Demetrulias JL, Kelling CK, Moloney SJ, Gettings SD. Percutaneous penetration of *N*-nitrosodiethanolamine through human skin (in vitro): comparison of finite and infinite dose applications from cosmetic vehicles. Fundam Appl Toxicol 21:213–221, 1993.

84. Walters KA, Brain KR, Dressler WE, Green DM, Howes D, James VJ, Kelling CK, Watkinson AC, Gettings SD. Percutaneous penetration of *N*-nitroso-*N*-methyldodecylamine through human skin in vitro: application from cosmetic vehicles. Food Chem Toxicol 35:705–712, 1997.

85. Akhter SA, Barry BW. Absorption through human skin of ibuprofen and flurbiprofen; effect of dose variation, deposited drug films, occlusion and the penetration enhancer *N*-methyl-2-pyrrolidone. J Pharm Pharmacol 37:27–37, 1985.

86. Tata S, Flynn GL, Weiner ND. Penetration of minoxidil from ethanol/propylene glycol solutions: effect of application volume and occlusion. J Pharm Sci 84:688–691, 1995.

87. Roper CS, Howes D, Blain PG, Williams FM. Percutaneous penetration of 2-phenoxyethanol through rat and human skin. Food Chem Toxicol 35:1009–1016, 1997.

88. Treffel P, Gabard B. Ibuprofen epidermal levels after topical application in vitro: effect of formulation, application time, dose variation and occlusion. Br J Dermatol 129:286–291, 1993.

89. Hughes MF, Shrivastava SP, Sumler MR, Edwards BC, Goodwin JH, Shah PV, Fisher HL, Hall HL. Dermal absorption of chemicals: effect of application of chemicals as a solid, aqueous paste, suspension, or in volatile vehicle. J Toxicol Environ Health 37:57–71, 1992.

90. Kubota K, Yamada T. Finite dose percutaneous drug absorption: theory and its application to in vitro timolol permeation. J Pharm Sci 79:1015–1019, 1990.

91. Seta AH, Higuchi WI, Borsadia S, Behl CR, Malick AW. Physical model approach to understanding finite dose transport and uptake of hydrocortisone in hairless guinea pig skin. Int J Pharm 81:89–99, 1992.

92. Walker M, Chambers LA, Holingsbee DA, Hadgraft J. Significance of vehicle thickness to skin penetration of halcinonide. Int J Pharm 70:167–172, 1991.

93. Addicks WJ, Flynn GL, Weiner N, Chiang C–M. Drug transport from thin applications of topical dosage forms: development of methodology. Pharm Res 5:377–382, 1988.

94. Ferry JJ, Shepard JH, Szpunar GL. Relationship between contact time of applied dose and percutaneous absorption of minoxidil from a topical solution. J Pharm Sci 79:483–486, 1990.

95. Peltonen L, Solberg VM. A comparison of 0.1% hydrocortisone cream (Milliderm 0.1%; AL) and 1% hydrocortisone cream (hydrocortisone; 1% Orion) in patients with non-infected eczemas. Curr Ther Res Clin Exp 35:78–82, 1984.

96. Davis AF, Hadgraft J. The use of supersaturation in topical drug delivery. In: Scott RC, Guy RH, Hadgraft J, Boddé HE, eds. Prediction of Percutaneous Penetration. Vol. 2. London: IBC Technical Services, pp 279–287, 1991.

97. Potts RO, Guy RH. Drug transport across the skin and the attainment of steady-state fluxes. Proc Int Symp Controlled Release Bioact Mater 21:162–163, 1994.

98. Peck KD, Ghanem AH, Higuchi WI, Srinivasan V. Improved stability of the human epidermal membrane during successive permeability experiments. Int J Pharm 98:141–147, 1993.

99. Sasaki H, Nakamura J, Shibasaki J, Ishino Y, Miyasato K, Ashizawa T. Effect of skin surface temperature on transdermal absorption of flurbiprofen from a cataplasm. Chem Pharm Bull 35:4883–4890, 1987.

100. Potts RO, Francoeur ML. Lipid biophysics of water loss through the skin. Proc Natl Acad Sci USA 87:1–3, 1990.

101. Emilson A, Lindberg M, Forslind B. The temperature effect of in vitro penetration of sodium lauryl sulfate and nickel chloride through human skin. Acta Derm Venereol 73:203–207, 1993.

102. Ohara N, Takayama K, Machida Y, Nagai T. Combined effect of d-limonene and temperature on the skin permeation of ketoprofen. Int J Pharm 105:31–38, 1994.

103. Bouwstra JA, Gooris GS, Salomons–deVries MA, van der Spek JA, Bras W. Structure of human SC as a function of temperature and hydration: a wide-angle X-ray diffraction study. Int J Pharm 84:205–216, 1992.

104. Benech–Kieffer F, Wegrich P, and Schaefer H. Transepidermal water loss as an integrity test for skin barrier function in vitro: assay standardization. In: Brain KR, James VJ, Walters KA, eds. Perspectives in Percutaneous Penetration. Vol. 5a. Cardiff: STS Publishing, p 56, 1997.

105. Hood HL, Wickett RR, Bronaugh RL. In vitro percutaneous absorption of the fragrance ingredient musk xylol. Food Chem Toxicol 34:483–488, 1996.

106. Pendlington RU, Sanders DJ, Cooper KJ, Howes D, Lovell WW. The use of sucrose as a standard penetrant in in vitro percutaneous penetration experiments. In: Brain KR, James VJ, Walters KA, eds. Perspectives in Percutaneous Penetration. Vol. 5a. Cardiff: STS Publishing, p 55, 1997.

107. Brain KR. Adopting an appropriate analytical approach. In: Scott RC, Guy RH, Hadgraft J, Boddé HE, eds. Prediction of Percutaneous Penetration. Vol. 2. London: IBC Technical Services, pp 229–237, 1991.

108. Watkinson AC, Green DM, Brain KR, James VJ, Walters KA, Azri–Meehan S, Dressler W. Nonoxynols: skin penetration of a series of homologues. In: Brain KR, James VJ, Walters KA, eds. Perspectives in Percutaneous Penetration. Vol. 5a. Cardiff: STS Publishing, p 22, 1997.

109. Gibson GG, Skett P. Introduction to Drug Metabolism. 2nd ed. London: Chapman & Hall, 1994.

110. Merk HF, Jugert FK, Frankenburg S. Biotransformations in the skin. In: Marzulli FN, Maibach HI, eds. Dermatotoxicology, 5th ed. Washington: Taylor & Francis, pp 61–73, 1996.

111. Hotchkiss SAM. Dermal metabolism. In: Roberts MS, Walters KA, eds. Dermal Absorption and Toxicity Assessment. New York: Marcel Dekker, pp 43–101, 1998.

112. Sharma R. Xenobiotic metabolizing enzymes in skin. In: Brain KR, James VJ, Walters KA, eds. Prediction of Percutaneous Penetration. Vol. 4b. Cardiff: STS Publishing, pp 14–18, 1996.

113. Bronaugh RL, Stewart RF, Storm JE. Extent of cutaneous metabolism during percutaneous absorption of xenobiotics. Toxicol Appl Pharmacol 99:534–543, 1989.

114. Bronaugh RL, Kraelling MEK. In vitro skin metabolism. In: Brain KR, James VJ, Walters KA, eds. Prediction of Percutaneous Penetration. Vol. 4b. Cardiff: STS Publishing, pp 19–21, 1996.

115. Kawakubo Y, Manabe S, Yamazoe Y, Nishikawa T, and Kato R. Properties of cutaneous acetyltransferase catalysing N- and O-acetylation of carcinogenic arylamines and N-hydroxyarylamine. Biochem Pharmacol 37:265–270, 1988.

116. Fuchs J, Mehlorn RJ, Packer L. Free radical reduction mechanism in mouse epidermis skin homogenates. J Invest Dermatol 93:633–640, 1989.

117. Hotchkiss SAM. Skin as a xenobiotic metabolising organ. Prog Drug Metab 13:217–262, 1992.

118. Andersson P, Edsbäcker S, Ryerfeldt Å, Von Bahr C. In vitro biotransformation of glucocorticoids in liver and skin homogenate fraction from man, rat, and hairless mouse. J Steroid Biochem 16:787–795, 1982.

119. Cheung YW, Li Wan Po A, Irwin WJ. Cutaneous biotransformation as a parameter in the modulation of the activity of topical corticosteroids. Int J Pharm 26:175–189, 1985.

120. Kulkarni AP, Nelson JL, Radulovic LL. Partial purification and some biochemical properties of neonatal rat cutaneous glutathione S-transferases. Comp Biochem Physiol 87B: 1005–1009, 1987.

121. Moss T, Howes D, Blain PG, Williams FM. Characteristics of sulphotransferases in human skin. In: Brain KR, James VJ, Walters KA, eds. Prediction of Percutaneous Penetration. Vol. 4b. Cardiff: STS Publishing, pp 307–310, 1996.

122. Coomes MW, Norling AH, Pohl RJ, Müller D, Fouts JR. Foreign compound metabolism by isolated skin cells from the hairless mouse. J Pharmacol Exp Ther 225:770–777, 1983.

123. Bronaugh RL. Methods for in vitro skin metabolism studies. In: Marzulli FN, Maibach HI, eds. Dermatotoxicology, 5th ed. Washington: Taylor & Francis, pp 383–388 (1996).

124. Collier SW, Sheikh NM, Sakr A, Lichtin JL, Stewart RF, Bronaugh RL. Maintenance of skin viability during in vitro percutaneous absorption/metabolism studies. Toxicol Appl Pharmacol 99:522–533, 1989.

125. Nathan D, Sakr A, Lichtin JL, Bronaugh RL. In vitro skin absorption and metabolism of benzoic acid, *p*-aminobenzoic acid, and benzocaine in the hairless guinea pig. Pharm Res 7:1147–1151, 1990.

126. Storm JE, Collier SW, Stewart RF, Bronaugh RL. Metabolism of xenobiotics during percutaneous penetration: role of absorption rate and cutaneous enzyme activity. Fundam Appl Toxicol 15:132–141, 1993.

127. Collier SW, Storm JE, Bronaugh RL. Reduction of azo dyes during in vitro percutaneous absorption. Toxicol Appl Pharmacol 118:73–79, 1993.

128. Ademola JI, Wester RC, Maibach HI. Absorption and metabolism of 2-chloro-2,6-diethyl-*N*-(butoxymethyl)acetanilide (butachlor) in human skin in vitro. Toxicol Appl Pharmacol 121:78–86, 1993.

129. Ademola JI, Sedik LE, Wester RC, Maibach HI. In vitro percutaneous absorption and metabolism in man of 2-chloro-4-ethylamino-6-isopropylamine-*s*-triazine (atrazine). Arch Toxicol 67:85–91, 1993.

130. Boehnlein J, Sakr S, Lichtin JL, Bronaugh RL. Metabolism of retinyl palmitate to retinol (vitamin A) in skin during percutaneous absorption. Pharm Res 11:1155–1159, 1994.

131. Slivka SR. Testosterone metabolism in an in vitro skin model. Cell Biol Toxicol 8:267–276, 1992.

132. Slivka SR, Landeen LK, Zeigler F, Zimber MP, Bartel RI. Characterisation, barrier function, and drug metabolism of an in vitro skin model. J Invest Dermatol 100:40–46, 1993.

133. Crank J. The Mathematics of Diffusion. 2nd ed. Oxford: Oxford University Press, 1975.

134. Pershing LK, Silver BS, Krueger GG, Shah VP, Skelly JP. Feasibility of measuring the bioavailability of topical betamethasone dipropionate in commercial formulations using drug content in skin and a skin blanching bioassay. Pharm Res 9:45–51, 1992.

135. Pershing LK, Corlett J, Jorgensen C. In vivo pharmacokinetics and pharmacodynamics of topical ketoconazole and miconazole in human SC. Antimicrob Agents Chemother 38:90–95, 1994.

136. Schaefer H, Stuttgen C, Zesch A, Schalla W, Gazith J. Quantitative determination of percutaneous absorption of radiolabelled drugs in vitro and in vivo by human skin. Curr Prob Dermatol 7:80–94, 1978.

137. King CS, Barton PS, Nicholls S, Marks R. The change in properties of the SC as a function of depth. Br J Dermatol 100:165–172, 1979.

138. Guy RH, Mak VHW, Kai T, Bommannan D, Potts RO. Percutaneous penetration enhancers: mode of action. In: Scott RC, Guy RH, Hadgraft J, eds. Prediction of Percutaneous Penetration. Vol. 1. London: IBC Technical Services, pp 213–223, 1990.

139. Bommannan D, Potts RO, Guy RH. Examination of the SC barrier function in vivo by infrared spectroscopy. J Invest Dermatol 95:403–408, 1990.

140. Watkinson AC, Bunge AL, Hadgraft J, Niak A. Computer simulation of penetrant–depth profiles in the SC. Int J Pharm 87:175–182, 1992.

141. Marttin E, Neelissen–Subnel MTA, De Haan FHN, and Boddé HE. A critical comparison of methods to quantify SC removed by tape stripping. Skin Pharmacol 9:69–77, 1996.

142. van Hoogdalem EJ. Assay of erythromycin in tape strips of human SC and some preliminary results in man. Skin Pharmacol 5:124–128, 1992.

143. Clark EW. Short-term penetration of lanolin into human SC. J Soc Cosmet Chem 43:219–227, 1992.

144. Jadoul A, Hanchard C, Thysman S, Préat V. Quantification and localisation of fentanyl and TRH delivered by iontophoresis in the skin. Int J Pharm 120:221–228, 1995.

145. Jadoul A, Mesens J, Caers W, de Beukelaar F, Crabbé R, Préat V. Transdermal permeation of alniditan by iontophoresis: in vitro optimisation and human pharmacokinetic data. Pharm Res 13:1348–1353, 1996.

146. Azri–Meehan S, Grabarz R, Dressler WE. Hair dyes: aspects of risk assessment. In: Brain KR, James VJ, Walters KA, eds. Prediction of Percutaneous Penetration. Vol. 4b. Cardiff: STS Publishing, pp 22–25, 1996.

147. Auton TR. Skin stripping and science: a mechanistic interpretation using mathematical modelling of skin deposition as a predictor of total absorption. In: Scott RC, Guy RH, Hadgraft J, eds. Prediction of Percutaneous Penetration. Vol. 2. London: IBC Technical Services, pp 558–576, 1992.

148. Tsai J–C, Weiner ND, Flynn GL, Ferry J. Properties of adhesive tapes used for SC stripping. Int J Pharm 72:227–231, 1991.

149. Tsai J–C, Cappel MJ, Weiner ND, Flynn GL, Ferry J. Solvent effects on the harvesting of SC from hairless mouse skin through adhesive tape stripping in vitro. Int J Pharm 68:127–133, 1991.

150. Shah VP, Flynn GL, Yacobi A, Maibach HI, Bon C, Fleischer NM, Franz TJ, Kaplan SA, Kawamoto J, Lesko LJ, Marty J–P, Pershing LK, Schaefer H, Sequeira JA, Shrivastava SP, Wilkin J, Williams RL. AAPS/FDA workshop report: bioequivalence of topical dermatological dosage forms—methods of evaluation of bioequivalence. Pharm Res 15:167–171, 1998.

151. Göpferich A, Lee G. Measurement of drug diffusivity in SC membranes and a polyacrylate matrix. Int J Pharm 71:245–253, 1991.

152. Brandt H. Determination of diffusion specific parameters by means of IR-ATR spectroscopy. Exp Tech Phys 33:423–431, 1985.

153. Hemmelmann K, Brandt H. Investigation of diffusion and sorption properties of polyethylene films for different liquids by means of IR–ATR spectroscopy. Exp Tech Phys 34:439–446, 1986.

154. Wurster DE, Buraphacheep V, Patel JM. The determination of diffusion coefficients in semisolids by Fourier transform infrared (FTIR) spectroscopy. Pharm Res 10:616–620, 1993.

155. Tralhão AM, Watkinson AC, Brain KR, Hadgraft J, Armstrong NA. Use of ATR-FTIR spectroscopy to study the diffusion of ethanol through glycerogelatin films. Pharm Res 12:1–4, 1995.

156. Farinas KC, Doh L, Venkatraman S, Potts RO. Characterisation of solute diffusion in a polymer using ATR-FTIR spectroscopy. Macromolecules 27:5220–5222, 1994.

157. Watkinson AC, Hadgraft J, Walters KA, Brain KR. Measurement of diffusional parameters in membranes using ATR-FTIR spectroscopy. Int J Cosmet Sci 16:199–210, 1994.

158. Watkinson AC, Brain K, Hadgraft J. The deconvolution of diffusion and partition coefficients in permeability studies. Proc Int Symp Controlled Release Bioact Mater 21:160–161, 1994.

159. Harrison J, Watkinson AC, Green DM, Brain KR, Hadgraft J. The relative effect of Azone and Transcutol on diffusivity and solubility in human stratum corneum. Pharm Res 13:542–546, 1996.

160. Pellett MA, Watkinson AC, Hadgraft J, Brain KR. Comparison of permeability data from traditional diffusion cells and ATR-FTIR spectroscopy. Part I—synthetic membranes. Int J Pharm 154:205–215, 1997.

161. Pellett MA, Watkinson AC, Hadgraft J, Brain KR. Comparison of permeability data from traditional diffusion cells and ATR-FTIR spectroscopy. Part II—determination of diffusional pathlengths in synthetic membranes and human stratum corneum. Int J Pharm 154:217–227, 1997.

162. Williams PL, Carver MP, Riviere JE. A physiologically relevant pharmacokinetic model of xenobiotic percutaneous absorption utilizing the isolated perfused porcine skin flap. J Pharm Sci 79:305–311, 1990.

163. Williams PL, Riviere JE. Definition of a physiologic pharmacokinetic model of cutaneous drug distribution using the isolated perfused porcine skin flap. J Pharm Sci 78:550–555, 1989.

164. Riviere JE. Isolated perfused porcine skin flap. In: Marzulli FN, Maibach HI, eds. Dermatotoxicology. 5th ed. Washington: Taylor & Francis, pp 337–351, 1996.

165. Riviere JE, Monteiro–Riviere N, Williams PL. Isolated perfused porcine skin flap as an in vitro model for predicting transdermal kinetics. Eur J Pharm Biopharm 41:152–162, 1995.

166. Keitzmann M, Löscher W, Arens D, Maaß D, Lubach D. The isolated perfused bovine udder as an in vitro model of percutaneous absorption: skin viability and percutaneous absorption of dexamethasone, benzoyl peroxide and etofenamate. J Pharmacol Toxicol Methods 30:75–84, 1993.

167. Keitzmann M, Blume B, Unger B. Percutaneous absorption of betamethasone dipropionate from different formulations using the isolated perfused bovine udder. In: Brain KR, James VJ, Walters KA, eds. Prediction of Percutaneous Penetration. Vol. 4b. Cardiff: STS Publishing, pp 262–266, 1996.

168. Henrikus B, Kampffmeyer HG. Metabolism of ethyl 4-amino-benzoate in the isolated single-pass perfused rabbit ear. Skin Pharmacol 6:246–252, 1993.

169. Mura P, Nassini C, Valoti M, Santoni G, Corti P. The single-pass perfused rabbit ear as a model for studying percutaneous absorption of clonazepam. III. Influence of vehicle composition on drug permeation. Eur J Pharm Biopharm 40:90–95, 1994.

170. De Lange J, Van Eck P, Elliott GR, De Kort WLAM, Wolthuis OL. The isolated blood-perfused pig ear: an inexpensive and animal-saving model for skin penetration studies. J Pharmacol Toxicol Methods 27:71–77, 1992.

171. Prunieras M. Human skin equivalents: the state of the art. In: Brain KR, James VJ, Walters KA, eds. Prediction of Percutaneous Penetration. Vol. 3b. Cardiff: STS Publishing, pp 419–427, 1993.

172. Ernesti AM, Swiderek M, Gay R. Absorption and metabolism of topically applied testosterone in an organotypic skin culture. Skin Pharmacol 5:146–153, 1992.

173. Bell E, Parenteau N, Gay R, Nolte C, Kemp P, Bilvo P, Ekstein B, Johnson E. The living skin equivalent: its manufacture, its organotypic properties and its response to irritants. Toxicol In Vitro 5:591–596, 1991.

174. Regnier M, Caron D, Reichert U, and Schaefer H. Reconstructed human epidermis: a model to study in vitro barrier function of the skin. Skin Pharmacol 5:49–56, 1992.

175. Chambin O, Teillaud E, Mikler C. Evaluation of Testskin LSE-100 as a model for human skin: in vitro percutaneous absorption studies. In: Brain KR, James VJ, Walters KA, eds. Prediction of Percutaneous Penetration. Vol. 3b. Cardiff: STS Publishing, pp 458–462, 1993.

176. Kennedy AH, Golden GM, Gay CL, Guy RH, Francoeur ML, Mak VHW. Stratum corneum lipids of human epidermal keratinocyte air–liquid cultures: implications for barrier function. Pharm Res 13:1162–1166, 1996.

177. Lotte C, Hinz RS, Rougier A, Guy RH. A reconstructed skin model (Testskin LSE) for permeation studies. In: Brain KR, James VJ, Walters KA, eds. Perspectives in Percutaneous Penetration. Vol. 5a. Cardiff: STS Publishing, p 61, 1997.

178. Lotte C, Zanini M, Patouillet C, Cabaillot AM, Messager A, Cottin M, Leclaire J. Comparison of different commercial reconstructed skin or epidermis models: permeation studies. In: Brain KR, James VJ, Walters KA, eds. Perspectives in Percutaneous Penetration. Vol. 5a. Cardiff: STS Publishing, p. 62, 1997.

179. Marty P, Faure C, Laroche F, Farenc C. Assessment of human skins obtained by in vitro culture as membrane models for cutaneous permeation tests. In: Brain KR, James VJ, Walters KA, eds. Perspectives in Percutaneous Penetration. Vol. 5a. Cardiff: STS Publishing, p. 64, 1997.

180. Dusser I, Noel–Hudson MS, Wepierre J. Improvement of epidermal differentiation and barrier function in reconstructed human epidermis cultured in insert at air–liquid interface. In: Brain KR, James VJ, Walters KA, eds. Prediction of Percutaneous Penetration. Vol. 4b. Cardiff: STS Publishing, pp. 315–318, 1996.

181. Higounec I, Demarchez M, Regnier M, Schmidt R, Ponec M, Schroot B. Improvement of epidermal differentiation and barrier function in reconstructed human skin after grafting onto athymic nude mice. Arch Dermatol Res 286:1–8, 1994.

182. Bidman HJ, Pitts JD, Solomon HF, Bondil JV, Stumpf WE. Estradiol distribution and penetration in rat skin after topical application, studied by high resolution autoradiography. Histochemistry 95:43–54, 1990.

183. Fabin B, Touitou E. Localization of lipophilic molecules penetrating rat skin in vivo by quantitative autoradiography. Int J Pharm 74:59–65, 1991.

184. Ritchel WA, Panchagnula R, Stemmer K, Ashraf M. Development of an intracutaneous depot for drugs. Binding, drug accumulation and retention studies and mechanism of depot. Skin Pharmacol 4:235–245, 1991.

185. Conti L, Ramis J, Mis R, Forn J, Vilaro S, Reina M, Vilageliu J, Basi N. Percutaneous absorption and skin distribution of ^{14}C-flutrimazole in mini-pigs. Drug Res 42:847–853, 1992.

186. Vingler PF, Bague H, Pruche F, Kermicic M. Direct quantitative digital autoradiography of testosterone metabolites in the pilosebaceous unit—an environmentally advantageous trace radioactive technology. Steroids 58:429–438, 1993.

187. Chu ID, Bronaugh RL, Tryphonas L. Skin reservoir formation and bioavailability of dermally administered chemicals in hairless guinea pigs. Food Chem Toxicol 34:267–276, 1996.

188. Cullander C. Confocal microscopy in the study of skin permeation: utility and limitations. In: Brain KR, James VJ, Walters KA, eds. Prediction of Percutaneous Penetration. Vol. 4b. Cardiff: STS Publishing, pp 5–6, 1996.

189. Rojanasakul Y, Paddock SW, Robinson JR. Confocal laser scanning microscopic examination of transport pathways and barriers of some peptides across the cornea. Int J Pharm 61:163–172, 1990.

190. Imbert D, Cullander C. Assessment of cornea viability by confocal laser scanning microscopy and MTT assay. Cornea 16:666–674, 1997.

191. Cullander C, Guy RH. Visualising the pathway of iontophoretic current flow in real time with laser-scanning confocal microscopy and the vibrating probe electrode. In: Scott RC, Guy RH, Hadgraft J, Boddé HE, eds. Prediction of Percutaneous Penetration. Vol. 2. London: IBC Technical Services, pp. 229–237, 1991.

192. Simoneti O, Hoogstraate AJ, Bialik V, Kempenaar JA, Schrijvers A, Boddé HE, Ponec M. Visualisation of diffusion pathways through the SC of native and of in vitro reconstructed epidermis by confocal laser scanning microscopy. Arch Dermatol Res 287: 465–473, 1995.

193. Prausnitz MR, Gimm JA, Guy RH, Langer R, Weaver JC, Cullander C. Imaging regions of transport across human stratum corneum during high-voltage and low-voltage exposure. J Pharm Sci 85:1363–1370, 1996.

194. Boderke P, Merkle HP, Cullander C, Ponec M, Boddé HE. Localization of aminopeptidase activity in freshly excised human skin: direct visualization by confocal laser scanning microscopy. J Invest Dermatol 108:83–86, 1997.

195. Hakkarainen M, Jaaskelainen I, Suhonen M, Paronen P, Urtii A, Monkonen J. In vitro penetration of liposomes into the skin studied by confocal microscopy. In: Brain KR, James VJ, Walters KA, eds. Prediction in Percutaneous Penetration. Vol. 4b. Cardiff: STS Publishing, pp 42–44, 1996.

196. Hoogstraate AJ, Cullander C, Nagelkerke JF, Senel S, Verhoef JC, Junginger HE, Boddé HE. Diffusion rates and transport pathways of FITC-labelled model compounds through buccal epithelium. Pharm Res 11:83–89, 1994.

197. Hoogstraate AJ, Cullander C, Nagelkerke JF, Spies F, Verhoef JC, Schrijvers AHG, Junginger HE, Boddé HE. A novel in situ model for continuous observation of transient drug concentration gradients across buccal epithelium at the microscopical level. J Controlled Release 39:71–78, 1996.

198. Hoogstraate AJ, Senel S, Cullander C, Verhoef JC, Junginger HE, Boddé HE. Effects of bile salts on transport rates and routes of FITC-labelled compounds across porcine buccal epithelium in vitro. J Controlled Release 40:211–221, 1996.

199. Marttin E, Verhoef JC, Cullander C, Romeijn SG, Nagelkerke JF, Merkus FWHM. Confocal laser scanning microscopic visualization of the transport of dextrans after nasal administration to rats: effects of absorption enhancers. Pharm Res 14:631–637, 1997.

200. Melia CD, Cutts LS, Adler J, Davies MC, Hibberd S, Rajabi–Siahboomi AR, Bowtell R. Visualising and measuring dynamic events inside controlled release matrix systems. Proc Int Symp Controlled Release Bioact Mater 22:32–33, 1995.

201. Cutts LS, Roberts PA, Adler J, Davies MC, and Melia CD. The measurement of diffusion coefficients in gels using the confocal laser scanning microscope. In: Hadgraft J, Kellaway IW, Parr GD, eds. Proceedings of the 3rd UKaps Conference, p 54, 1994.

202. Hadgraft J, Ridout G. Development of model membranes for percutaneous absorption measurements. I. Isopropyl myristate. Int J Pharm 39:149–156, 1987.

203. Hadgraft J, Ridout G. Development of model membranes for percutaneous absorption measurements. I. Dipalmitoyl phosphatidylcholine, linoleic acid and tetradecane. Int J Pharm 42:97–104, 1988.

204. Firestone BA, Guy RH. Prediction of dermal absorption. In: Alternative Methods in Toxicology. New York: Mary Ann Liebert, pp 517–536, 1985.

205. Hadgraft J, Walters KA, Wotton PK. Facilitated transport of sodium salicylate across an artificial lipid membrane by Azone. J Pharm Pharmacol 37:725–727, 1985.

206. Houk J, Guy RH. Membrane models for skin penetration studies. Chem Rev 88:455–471, 1988.

207. Hadgraft J, Guy RH. Synthetic membranes as biological models. In: Ganderton D, Jones T, eds. Advances in Pharmaceutical Sciences, Vol. 6. London: Academic Press, pp 43–64, 1992.

208. Pellett MA, Watkinson AC, Hadgraft J, Brain KR. An ATR-FTIR investigation of the interactions between vehicles and a synthetic membrane. Proc Int Symp Controlled Release Bioact Mater 21:439–440, 1994.

209. Neubert R, Bendas C, Wohlrab W, Glenau B, Furst W. A multilayer membrane system for modelling drug penetration into skin. Int J Pharm 75:89–94, 1991.

210. Neubert R, Wohlrab W, Bendas C. Modelling of drug penetration into human skin using a multilayer membrane system. Skin Pharmacol 8:119–129, 1995.

211. Faustino EPR, Cabral Marques HM, Morais JA, Hadgraft J. Comparison of different models for the in vitro release of topical dosage forms. In: Brain KR, James VJ, Walters KA, eds. Prediction of Percutaneous Penetration. Vol. 4b. Cardiff: STS Publishing, pp 294–298, 1996.

212. Nastruzzi C, Esposito E, Pastesini C, Gambari R, Menegatti E. Comparative study on the release kinetics of methyl nicotinate from topical formulations. Int J Pharm 90:43–50, 1993.

213. Abraham W, Downing DT. Preparation of model membranes for skin permeability studies using stratum corneum lipids. J Invest Dermatol 93:809–813, 1989.

214. Matsuzaki K, Imaoka T, Asano M, Miyajima K. Development of a model membrane system using stratum corneum lipids for estimation of drug skin permeability. Chem Pharm Bull 41:575–579, 1993.

215. Miyajima K, Tanikawa S, Asano M, Matsuzaki K. Effects of absorption enhancers and lipid composition on drug permeability through the model membrane using stratum corneum lipids. Chem Pharm Bull 42:1345–1347, 1994.

216. Little CJ, Brain KR. Development of stratum corneum model membrane systems and comparison of the penetration rates of certain NSAIDs. In: Brain KR, James VJ, Walters KA, eds. Prediction in Percutaneous Penetration. Vol. 4b. Cardiff: STS Publishing, pp 290–293, 1996.

217. Moghimi HR, Williams AC, Barry BW. A lamellar matrix model for stratum corneum intercellular lipids. I. Characterisation and comparison with stratum corneum intercellular structure. Int J Pharm 131:103–115, 1996.

218. Moghimi HR, Williams AC, Barry BW. A lamellar matrix model for stratum corneum intercellular lipids. I. Effect of geometry of the stratum corneum on permeation of model drugs 5-fluorouracil and oestradiol. Int J Pharm 131:117–129, 1996.

219. Little CJ, Brain KR. Stratum corneum models based on lecithin organogels. In: Brain KR, James VJ, Walters KA, eds. Perspectives in Percutaneous Penetration. Vol. 5b. Cardiff: STS Publishing, pp 147–148, 1998.

220. Little CJ, Brain KR. Comparison of some alternative model membrane systems for the estimation of the permeation of ibuprofen from commercial formulations through human skin. In: Brain KR, James VJ, Walters KA, eds. Perspectives in Percutaneous Penetration. Vol. 5b. Cardiff: STS Publishing, pp 149–150, 1998.

221. Little CJ, Brain KR. Investigation of chemical modulation of membrane permeation in different model membrane systems. In: Brain KR, James VJ, Walters KA, eds. Perspectives in Percutaneous Penetration. Vol. 5b. Cardiff: STS Publishing, pp 151–152, 1998.

222. Pershing LK, Kreuger GG. New animal models for bioavailability studies. Skin Pharmacol 1:57–69, 1987.

223. Wester RC, Maibach HI. Animal models for percutaneous absorption. In: Shah VP, Maibach HI, eds. Topical Drug Bioavailability, Bioequivalence, and Penetration. New York: Plenum Press, pp 333–349, 1993.

224. Guy RH, Hadgraft J, Hinz RS, Roskos KV, Bucks DAW. In vivo evaluations of transdermal drug delivery. In: Chien YW, ed. Transdermal Controlled Systemic Medications. New York: Marcel Dekker, pp 179–224, 1987.

225. Scott RC, Corrigan MA, Smith F, Mason H. The influence of skin structure on permeability: an intersite and interspecies comparison with hydrophilic penetrants. J Invest Dermatol 96:921–925, 1991.

226. Scott RC. Personal communication.

227. Gray GM, White RJ. Glycosphingolipids and ceramides in human and pig epidermis. J Invest Dermatol 70:336–341, 1978.

228. Wertz PW, Downing DT. Ceramides of pig epidermis: structure determination. J Lipid Res 24:759–765, 1983.

229. Wertz PW, Downing DT. Glucosylceramides of pig epidermis: structure determination. J Lipid Res 24:1135–1139, 1983.

230. Sato K, Sugibayashi K, Morimoto Y. Species difference in percutaneous absorption of nicorandil. J Pharm Sci 80:104–107, 1991.

231. Lloyd DH, Dick WDB, McEwan–Jenkinson D. The effects of some surface sampling procedures on the stratum corneum of bovine skin. Res Vet Sci 26:250–252, 1979.

232. Davies B, Morris T. Physiological parameters in laboratory animals and humans. Pharm Res 10:1093–1095, 1993.

233. Wester RC, Noonan PK. Relevance of animal models for percutaneous absorption. Int J Pharm 7:99–110, 1980.

234. Reifenrath WG, Chellquist EM, Shipwash EA, Jederberg WW, Kreuger GG. Percutaneous penetration in the hairless dog, weanling pig and grafted athymic nude mouse: evaluation of models for predicting skin penetration in man. Br J Dermatol 111(suppl 27):123–135, 1984.

235. Bond JR, Barry BW. Hairless mouse skin is limited as a model for assessing the effects of penetration enhancers in human skin. J Invest Dermatol 90:810–813, 1988.

236. Choi H–K, Flynn GL, Amidon GL. Transdermal delivery of bioactive peptides: the effect of n-decylmethylsulphoxide, pH, and inhibitors on enkaphalin metabolism and transport. Pharm Res 7:1099–1106, 1990.

237. Norgaard O. Investigations with radiolabeled nickel, cobalt and sodium on the resorption through the skin in rabbits, guinea pigs and man. Acta Derm Venereol 34:440, 1957.

238. Tregear RT. Physical Functions of Skin. New York: Academic Press, 1966.

239. McCreesh AH. Percutaneous toxicity. Toxicol Appl Pharmacol 7(suppl 2):20–26, 1965.

240. Scott RC, Walker M, Dugard PH. A comparison of the in vitro permeability properties of human and some laboratory animal skins. Int J Cosmet Sci 8:189–194, 1986.

241. Durrheim H, Flynn GL, Higuchi WI, Behl CR. Permeation of hairless mouse skin I: experimental methods and comparison with human epidermal permeation by alkanols. J Pharm Sci 69:781–786, 1980.

242. Huq AS, Ho NF, Husari N, Flynn GL, Jetzer WE, Condie L. Permeation of water contaminative phenols through hairless mouse skin. Arch Environ Contam Toxicol 15:557–566, 1986.

243. Barteck MJ, La Budde JA, Maibach HI. Skin permeability in vivo: comparison in rat, rabbit, pig and human. J Invest Dermatol 58:114–123, 1972.

244. Feldman RJ, Maibach HI. Percutaneous penetration of some pesticides and herbicides in man. Toxicol Appl Pharmacol 28:126–132, 1974.

245. Barteck MJ, La Budde JA. Percutaneous absorption in vitro. In: Maibach HI, ed. Animal Models in Dermatology. New York: Churchill–Livingstone, pp 103–112, 1975.

246. Moody RP, Ritter L. Dermal absorption of the insecticide lindane in rats and rhesus monkeys: effect of anatomical site. J Toxicol Environ Health 28:161–169, 1989.

247. Moody RP, Franklin CA, Ritter L, Maibach HI. Dermal absorption of the phenoxy herbicides 2,4-D, 2-4-D amine, 2,4-D isooctyl and 2,4,5-T in rabbits, rats, rhesus monkey and humans: a cross-species comparison. J Toxicol Environ Health 29:237–245, 1990.

248. Roberts ME, Mueller KR. Comparisons of in vitro nitroglycerin (TNG) flux across Yucatan pig, hairless mouse, and human skins. Pharm Res 7:673–676, 1990.

249. Bucks DAW, Hinz RS, Sarason R, Maibach HI, Guy RH. In vivo percutaneous absorption of chemicals—a multiple dose study in rhesus monkeys. Food Chem Toxicol 28:129–132, 1990.

250. Rougier A. The hairless rat: a relevant animal model to predict in vivo percutaneous absorption in humans. J Invest Dermatol 88:577–581, 1987.

251. Morimoto Y, Hatanaka T, Sugibayashi K, Omiya H. Prediction of skin permeability of drugs: comparison of human and hairless rat skin. J Pharm Pharmacol 44:634–639, 1992.

252. Sugibayashi K, Nakagaki D, Hatanaka E, Hatanaka T, Inoue N, Kusumi S, Kobayashi M, Kimura M, Morimoto Y. Differences in enhancing effect of 1-menthol, ethanol and their combination between hairless rat and human skin. Int J Pharm 113:189–197, 1995.

253. Chowhan ZT, Pritchard R. Effect of surfactants on percutaneous absorption of naproxen I: comparisons of rabbit, rat and human excised skin. J Pharm Sci 67:1272–1274, 1978.

254. Niazy EM. Differences in penetration-enhancing effect of Azone through excised rabbit, rat, hairless mouse, guinea-pig and human skins. Int J Pharm 130:225–230, 1996.

255. Hirvonen J, Rytting JH, Paronen P, Urtti A. Dodecyl N,N-dimethylamino acetate and Azone enhance drug penetration across human, snake, and rabbit skin. Pharm Res 8: 933–937, 1991.

256. Marzulli FN, Brown DWC, Maibach HI. Techniques for studying skin penetration. Toxicol Appl Pharmacol 3(suppl):76–83, 1969.

257. Galey WR, Lansdale HK, Nacht S. The in vitro permeability of skin and buccal mucosa to selected drugs and tritiated water. J Invest Dermatol 67:713–717, 1976.

258. Smith FM, Scott RC, Foster JR. An interspecies comparison of skin structure (hair follicle area as determined by a novel technique) and skin permeability. In: Galli CL, Hensby CN, Marinovich M, eds. Skin Pharmacology and Toxicology. New York: Plenum Publishing, pp 299–305, 1990.

259. Scott RC. In vitro absorption through damaged skin. In: Bronaugh RL, Maibach HI, eds. In Vitro Percutaneous Absorption, Principles, Fundamentals and Applications. Boca Raton: CRC Press, pp 129–135, 1991.

260. Scott RC, Dugard PH, Doss AW. Permeability of abnormal rat skin. J Invest Dermatol 86:201–207, 1986.

261. Shah VP, Flynn GL, Guy RH, Maibach HI, Schaefer H, Skelly JP, Wester RC, Yacobi A. In vivo percutaneous penetration/absorption. Int J Pharm 74:1–8, 1991.

262. Opdam JJG, Kimmel JPF, Krüse J, Meuling WJA. The rate of percutaneous absorption in human volunteers; dermal exposure to 1,1,1-tichloroethane. In: Brain KR, James VJ, Walters KA, eds. Prediction of Percutaneous Penetration. Vol. 4b. Cardiff: STS Publishing, pp 199–202, 1996.

263. Feldman RJ, Maibach HI. Absorption of some organic compounds through the skin in man. J Invest Dermatol 54:399–404, 1970.

264. Wester RC, Maibach HI, Melendres J Sedik L, Knaak J, Wang R. In vivo and in vitro percutaneous absorption and skin evaporation of isofenphos in man. Fundam Appl Toxicol 19:521–526, 1992.

265. Vollmer U, Müller BW, Wilffert B, Peters T. An improved model for studies on transdermal drug absorption in vivo in rats. J Pharm Pharmacol 45:242–245, 1993.

266. Franz JM, Gaillard A, Maibach HI, Schweitzer A. Percutaneous absorption of griseofulvin and proquazone in the rat and in isolated human skin. Arch Dermatol Res 271: 275–282, 1981.

267. Bogen KT, Colston BW, Machicao LK. Dermal absorption of dilute aqueous chloroform, trichlorethylene, and tetrachloroethylene in hairless guinea pigs. Fundam Appl Toxicol 18:30–39, 1992.

268. Yu D, Sanders LM, Davidson GWR, Marvin MJ, Ling T. Percutaneous absorption of nicardipine and ketorolac in rhesus monkeys. Pharm Res 5:457–462, 1988.

269. Leonardi GR, Maia Campos PMBG. Influence of glycolic acid as a component of different formulations on skin penetration by vitamin A palmitate. J Cosmet Sci 49: 23–32, 1998.

270. Chatterjee DJ, Li WY, Koda RT. Effect of vehicles and penetration enhancers on the in vitro and in vivo percutaneous absorption of methotrexate and edatrexate through hairless mouse skin. Pharm Res 14:1058–1065, 1997.

271. Megwa SA, Benson HAE, Roberts MS. Percutaneous absorption of salicylates from some commercially available topical products containing methyl salicylate or salicylate salts in rats. J Pharm Pharmacol 47:891–896, 1995.

272. Wrzesinski CL, Feeney WP, Feely WF, Crouch LS. Dermal penetration of 4″-(epimethylamino)-4″-deoxyavermectin-b_{1a} benzoate in the rhesus monkey. Food Chem Toxicol 35:1085–1089, 1997.

273. Davis DAP, Kraus AL, Thompson GA, Olerich M, Odio MR. Percutaneous absorption of salicylic acid after repeated (14-day) in vivo administration to normal, acnegenic or aged human skin. J Pharm Sci 86:896–899, 1997.

274. Bonsall JL, Goose J. The safety evaluation of bendiocarb, a residual insecticide for vector control. In Van Heemstra–Lequin EAH, van Sittert NJ, eds. Biological Monitoring of Workers Manufacturing, Formulating and Applying Pesticides. Amsterdam: Elsevier, p 45, 1986.

275. McDougal JN, Jepson GW, Clewell HJ, Gargas ML, Andersen ME. Dermal absorption of organic chemical vapors in rats and humans. Fundam Appl Toxicol 14:299–308, 1990.

276. Newton M, Norris LA. Potential exposure of humans to 2,4,5-T and TCDD in the Oregon coast region. Fundam Appl Toxicol 1:339–445, 1981.

277. Bucks DAW, McMaster JR, Maibach HI, Guy RH. Bioavailability of topically administered steroids: a "mass balance" technique. J Invest Dermatol 91:29–33, 1988.

278. Bucks DAW, Maibach HI, Guy RH. Mass balance and dose accountability in percutaneous absorption studies: development of a nonocclusive application system. Pharm Res 5:313–315, 1988.

279. Bucks DAW, Guy RH, Maibach HI. Percutaneous penetration and mass balance accountability—technique and implications for dermatology. J Toxicol Cutan Ocul Toxicol 8:439–451, 1989.

280. Noonan PK, Gonzalez MA, Ruggirello D, Tomlinson J, Babcock–Atkinson E, Ray M, Golub A, Cohen A. Relative bioavailability of a new transdermal nitroglycerin delivery system. J Pharm Sci 75:688–691, 1986.

281. Hadgraft J, Hill S, Humpel M, Johnston LR, Lever LR, Marks R, Murphy TM, Rapier C. Investigations on the percutaneous absorption of the antidepressant rolipram in vitro and in vivo. Pharm Res 7:1307–1312, 1990.

282. Le Roux Y, Borg ML, Sibille M, Thebault J, Renoux A, Douin MJ, Djebbar F, Dain MP. Bioavailability study of Menorest, a new estrogen transdermal delivery system, compared with a transdermal reservoir system. Clin Pharmacokinet 10:172–178, 1995.

283. Gorsline J, Gupta SK, Dye D, Rolf CM. Steady-state pharmacokinetics and dose relationship of nicotine delivered from Nicoderm (nicotine transdermal system). J Clin Pharmacol 33:161–168, 1993.

284. Brocks DR, Meikle AW, Boike SC, Mazer NA, Zariffa N, Audet PR, Jorasky DK. Pharmacokinetics of testosterone in hypogonadal men after transdermal delivery: influence of dose. J Clin Pharmacol 36:732–739, 1995.

285. Meikle AW, Arver S, Dobs AS, Sanders SW, Rajaram L, Mazer NA. Pharmacokinetics and metabolism of a permeation-enhanced testosterone transdermal system in hypogonadal men: influence of application site—a clinical research center study. J Clin Endocrinol Metab 81:1832–1840, 1996.

286. Berner G, Engels B, Vogtle–Junkert U. Percutaneous ibuprofen therapy with Trauma-Dolgit gel: bioequivalence studies. Drug Exp Clin Res 15:556–564, 1990.

287. Singh P, Roberts MS. Dermal and underlying tissue pharmacokinetics of salicylic acid after topical application. J Pharmacokinet Biopharm 21:337–373, 1993.

288. Singh P, Roberts MS. Skin permeability and local tissue concentrations of nonsteroidal anti-inflammatory drugs after topical application. J Pharmacol Exp Ther 268:144–151, 1994.

289. Singh P, Roberts MS. Effects of vasoconstriction on dermal pharmacokinetics and local tissue distribution of compounds. J Pharm Sci 83:783–791, 1994.

290. Benson HAE, Schild PN, Cross SE, Roberts MS. Investigation of the effect of vaso-active agents on local skin and underlying tissue blood flow. In: Brain KR, James VJ, Walters KA, eds. Prediction in Percutaneous Penetration. Vol. 4b. Cardiff: STS Publishing, pp 258–261, 1996.

291. Cross SE, Wu Z, Roberts MS. Effect of perfusion flow rate on the tissue uptake of solutes after dermal application using the rat isolated perfused limb preparation. J Pharm Pharmacol 46:844–850, 1994.

292. Wu Z, Rivory LP, Roberts MS. Physiological pharmacokinetics of solutes in perfused rat hindlimb: characterisation of the physiology with changing perfusate flow, protein content and temperature using statistical moments analysis. J Pharm Biopharm 21:653–688, 1993.

293. Roberts MS, Cross SE, Wu Z. Deep tissue penetration after dermal application observed in the isolated perfused limb. In: Brain KR, James VJ, Walters KA, eds. Prediction in Percutaneous Penetration. Vol. 4b. Cardiff: STS Publishing, pp 251–254, 1996.

294. Scott RC, Rhodes C. The permeability of grafted human transplant skin in athymic mice. J Pharm Pharmacol 40:128–129, 1987.

295. Klain GJ. Dermal penetration and systemic distribution of ^{14}C-labeled vitamin E in human skin grafted athymic nude nice. Int J Vitam Nutr Res 59:333–337, 1989.

296. Boyce ST, Supp AP, Harriger MD, Pickens WL, Wickett RR, and Hoath SB. Surface electrical capacitance as a noninvasive index of epidermal barrier in cultured skin substitutes in athymic mice. J Invest Dermatol 107:82–87, 1996.

297. Boyce ST. Cultured skin substitutes grafted to athymic mice. In: Brain KR, James VJ, Walters KA, eds. Perspectives in Percutaneous Penetration. Vol. 5a. Cardiff: STS Publishing, pp 13, 1997.

298. Lane AT, Scott GA, Day KH. Development of human fetal skin transplanted to the nude mouse. J Invest Dermatol 93:787–791, 1989.

299. Riviere JE. Grafted skin and skin flaps. In: Shah VP, Maibach HI, eds. Topical Drug Bioavailability, Bioequivalence, and Penetration. New York: Plenum Press, pp 209–221, 1993.

300. Manning DD, Reed ND, Shaffer CF. Maintenance of skin xenographs of widely divergent phylogenetic origin on congenitally athymic (nude) mouse. J Exp Med 38:488–494, 1973.

301. Biren CA, Barr RJ, McCullough JL, Black KS, Hewitt CW. Prolonged viability of human skin xenographs in rats by cyclosporine. J Invest Dermatol 86:611–614, 1986.

302. Higounenc I, Spies F, Boddé H, Schaefer H, Demarchez M, Shroot B, Ponec M. Lipid composition and barrier function of human skin after grafting onto athymic nude mice. Skin Pharmacol 7:167–175, 1994.

303. Petersen RV, Kislalioglu MS, Liang WQ, Fang SM, Emam M, Dickman S. The athymic nude mouse grafted with human skin as a model for evaluating the safety and effectiveness of radiolabeled cosmetic ingredients. J Soc Cosmet Chem 37:249–265, 1986.

304. Wojciechowski Z, Pershing LK, Huether S, Leonard L, Burton SA, Higuchi WI, Krueger GG. An experimental skin sandwich flap on an independent vascular supply for the study of percutaneous absorption. J Invest Dermatol 88:439–446, 1987.

305. Pershing LK, Lambert LD, Knutson K. Mechanism of ethanol-enhanced estradiol permeation across human skin in vivo. Pharm Res 7:170–175, 1990.

306. Silcox GD, Parry GE, Bunge AL, Pershing LK, Pershing DW. Percutaneous absorption of benzoic acid across human skin II: prediction of an in vivo, skin flap system using in vitro parameters. Pharm Res 7:352–358, 1990.

307. Yano T, Nakagawa A, Tsuji M, Noda K. Skin permeability of various non-steroidal anti-inflammatory drugs in man. Life Sci 39:1043–1050, 1986.

308. Guy RH, Bucks DAW, McMaster JR, Villaflor DA, Roskos KV, Hinz RS, Maibach HI. Kinetics of drug absorption across human skin in vivo. Developments in methodology. In: Maibach HI, Schaefer H, eds. Pharmacology and the Skin. Vol. 1. Skin Pharmacokinetics. Basel: Karger, pp 70–76, 1987.

309. Mak VHW, Potts RO, Guy RH. Percutaneous penetration enhancement in vivo measured by attenuated total reflectance infrared spectroscopy. Pharm Res 7:835–841, 1990.

310. Higo N, Naik A, Bommannan DB, Potts RO, Guy RH. Validity of reflectance infrared spectroscopy. Pharm Res 10:1500–1506, 1993.

311. Sennhenn B, Giese K, Plamann K, Harendt N, Kölmel K. In vivo evaluation of the penetration of topically applied drugs into human skin by spectroscopic methods. Skin Pharmacol 6:152–160, 1993.

312. Naik A, Guy RH. Infrared spectroscopic and differential scanning calorimetric investigations of the stratum corneum barrier function. In: Potts RO, Guy RH, eds. Mechanisms of Transdermal Drug Delivery. New York: Marcel Dekker, pp 87–162, 1997.

313. Potts RO, Guzek DB, Harris RR, McKie JE. A noninvasive, in vivo technique to quantitatively measure water concentration of the stratum corneum using attenuated total-reflectance infrared spectroscopy. Arch Dermatol Res 277:489–495, 1985.

314. Martin KA. Direct measurement of moisture in skin by NIR spectroscopy. J Soc Cosmet Chem 44:249–261, 1993.

315. Gilard V, Martino R, Malet–Martino M, Riviere M, Gournay A, Navarro R. Measurement of total water and bound water contents in human stratum corneum by in vitro proton nuclear magnetic resonance spectroscopy. Int J Cosmet Sci 20:117–125, 1998.

316. Pechtold LARM, Abraham W, Potts RO. Characterization of stratum corneum barrier properties using fluorescence spectroscopy. In: Potts RO, Guy RH, eds. Mechanisms of Transdermal Drug Delivery. New York: Marcel Dekker, pp 199–213, 1997.

317. Burnette RR, DeNuzzio JD. Impedance spectroscopy: applications to human skin. In: Potts RO, Guy RH, eds. Mechanisms of Transdermal Drug Delivery. New York: Marcel Dekker, pp 215–230, 1997.

318. Watkinson AC, Hadgraft J, Street PR, Richards RW. Neutrons, surfaces, and skin. In: Potts RO, Guy RH, eds. Mechanisms of Transdermal Drug Delivery. New York: Marcel Dekker, pp 231–265, 1997.

319. Bindra RMS, Eccleston GM, Imhof RE, Birch DJS. Opto-thermal in vivo monitoring of human skin condition. In: Scott RC, Guy RH, Hadgraft J, Boddé HE, eds. Prediction of Percutaneous Penetration. Vol. 2. Cardiff: STS Publishing, pp 628–635, 1991.

320. Petersen EM. The hydrating effect of a cream and white petrolatum measured by optothermal infrared spectrometry in vivo. Acta Derm Venereol 71:373–376, 1991.

321. Imhof RE, Whitters CJ, Birch DJS. Opto-thermal in vivo monitoring of sunscreens on skin. Phys Med Biol 35:95–102, 1990.

322. Cowen JA, Imhof RE. In vivo surface disappearance measurement of applied chemicals using opto-thermal transient emission radiometry (OTTER). In: Brain KR, James VJ, Walters KA, eds. Perspectives in Percutaneous Penetration. Vol. 5b. Cardiff: STS Publishing, pp 94–97, 1998.

323. Imhof RE, Whitters CJ, Birch DJS. Opto-thermal in vivo monitoring of structural breakdown of an emulsion sunscreen on skin. Clin Mater 5:272–278, 1990.

324. Dupuis D, Rougier A, Roguet R, Lotte C, Kalopissis G. In vivo relationship between horny layer reservoir effect and percutaneous absorption in human and rat. J Invest Dermatol 82:353–358, 1984.

325. Tojo K, Lee AC. A method for predicting steady-state rate of skin penetration in vivo. J Invest Dermatol 92:105–110, 1989.

326. Rougier A. An original predictive method in vivo percutaneous absorption studies. J Soc Cosmet Chem 38:397–417, 1987.

327. Rougier A, Dupuis D, Lotte C. Stripping method for measuring percutaneous absorption in vivo. In: Bronaugh RL, Maibach HI, eds. Percutaneous Absorption: Mechanisms, Methodology, Drug Delivery, 2nd ed. New York: Marcel Dekker, pp 415–434, 1989.

328. Rougier A. Predictive measurement of in vivo percutaneous penetration. In: Scott RC, Guy RH, Hadgraft J, eds. Prediction of Percutaneous Penetration. Vol. 1. London: IBC Technical Services, pp 252–262, 1990.

329. Meibohm B, Derendorf H. Basic concepts of pharmacokinetics/pharmacodynamic (PK/PD) modelling. Int J Clin Pharm Ther 35:401–413, 1997.

330. Anderson C, Andersson T, Molander M. Ethanol absorption across human skin measured by in vivo microdialysis technique. Acta Derm Venereol 71:389–393, 1991.

331. Anderson C, Andersson T, Wårdell K. Changes in skin circulation after insertion of a microdialysis probe visualized by laser Doppler perfusion imaging. J Invest Dermatol 102:807–811, 1994.

332. Anderson C, Andersson T, Boman A. Cutaneous microdialysis for human in vivo dermal absorption studies. In: Roberts MS, Walters KA, eds. Dermal Absorption and Toxicology Assessment. New York: Marcel Dekker, pp 231–244, 1998.

333. Ault JM, Lunte CE, Meltzer NM, and Riley CM. Microdialysis as a dermal sampling technique for percutaneous absorption. In: Brain KR, James VJ, Walters KA, eds. Prediction of Percutaneous Penetration. Vol. 3b. Cardiff: STS Publishing, pp 44–48, 1993.

334. Ault JM, Riley CM, Meltzer NM, Lunte CE. Dermal microdialysis sampling in vivo. Pharm Res 11:1631–1639, 1994.

335. Matsuyama K, Nakashima M, Ichikawa M, Yona T, Saoh S, Groto S. In vivo microdialysis for the transdermal absorption of valproate in rats. Biol Pharm Bull 17:1395–1398, 1994.

336. Matsuyama K, Nakashima M, Nakaboh Y, Ichikawa M, Yona T, Saoh S. Application of in vivo microdialysis to transdermal absorption of methotrexate in rats. Pharm Res 11:686–688, 1994.

337. Muller M, Schmid R, Wagner O, Vonosten B, Shayganfar H, Eichler HG. In vivo characterization of transdermal drug transport by microdialysis. J Controlled Release 37:49–57, 1995.

338. Hegeman I, Forstinger C, Partsch C, Lagler I, Krotz S, Wolf K. Microdialysis in cutaneous pharmacology: kinetic analysis of transdermally delivered nicotine. J Invest Dermatol 104:839–843, 1995.

339. Pepler AF, Morrison JC. The influence of vehicle composition on the vasoconstrictor activity of betamethasone 17-benzoate. Br J Dermatol 85:171–176, 1971.

340. Barry BW, Woodford R. Proprietary hydrocortisone creams. Vasoconstrictor activities and bioavailabilities of six preparations. Br J Dermatol 95:423–425, 1976.

341. Merver E. Comparison of the blanching activities of Dermovate, Betnovate, and Eumovate creams and ointments. Int J Pharm 41:63–66, 1988.

342. Queille–Roussel C, Poncet M, Schaefer H. Quantification of skin colour changes induced by topical corticosteroid preparations using the Minolta chromameter. Br J Dermatol 124:264–270, 1991.

343. Chan SV, Li Wan Po A. Quantitative skin blanching assay of corticosteroid creams using tristimulus colour analysis. J Pharm Pharmacol 44:371–378, 1992.

344. Corson SL. A decade of experience with transdermal estrogen replacement therapy: overview of key pharmacologic and clinical findings. Int J Fertil 38:79–91, 1993.

345. Lin S, Ho H, Chien YW. Development of a new nicotine transdermal delivery system: in vitro kinetics studies and clinical pharmacokinetic evaluations in two ethnic groups. J Controlled Release 26:175–193, 1993.

346. Shah HS, Tojo K, Huang YC, Chien YW. Development and clinical evaluation of a verapamil transdermal delivery system. J Controlled Release 22:125–131, 1992.

347. Bialik W, Walters KA, Brain KR, Hadgraft J. Some factors affecting the in vitro penetration of ibuprofen through human skin. Int J Pharm 92:219–223, 1993.

348. Monteiro–Riviere NA, Inman AO, Riviere JE, McNeill SC, Francoeur ML. Topical penetration of piroxicam is dependent on the distribution of the local cutaneous vasculature. Pharm Res 10:1326–1331, 1993.

349. Treffel P, Gabard B. Feasibility of measuring the bioavailability of topical ibuprofen in commercial formulations using drug content in epidermis and a methyl nicotinate skin inflammation assay. Skin Pharmacol 6:268–275, 1993.

350. Seth PL. Percutaneous absorption of ibuprofen from different formulations. Comparative study with gel, hydrophilic ointment and emulsion cream. Arzneimittelforschung 43:919–921, 1993.

351. Marks R, Dykes P. Plasma and cutaneous drug levels after topical application of piroxicam gel: a study in healthy volunteers. Skin Pharmacol 7:340–344, 1994.

352. Francoeur ML, Monteiro–Riviere NA, Riviere JE. Piroxicam: evidence for local delivery following topical application. Eur J Pharm Biopharm 41:175–183, 1995.

353. Roy SD, Manoukian E. Transdermal delivery of ketorolac tromethamine: permeation enhancement, device design, and pharmacokinetics in healthy humans. J Pharm Sci 84: 1190–1196, 1995.

354. Taburet AM, Singlas E, Glass RC, Thomas F, Leutenegger E. Pharmacokinetic comparison of oral and local action transcutaneous flurbiprofen in healthy volunteers. J Clin Pharm Ther 20:101–107, 1995.

355. Shah AK, Wei G, Lanman RC, Bhargava VO, Weir SJ. Percutaneous absorption of ketoprofen from different anatomical sites in man. Pharm Res 13:168–172, 1996.

356. Bouchier–Hayes TA, Rotman H, Darekar BS. Comparison of the efficacy and tolerability of diclofenac gel (Voltarol Emulgel) and felbinac gel (Traxam) in the treatment of soft tissue injuries. Br J Clin Pract 44:319–320, 1990.

357. Memeo A, Garofoli F, Peretti G. Evaluation of the efficacy and tolerability of a new topical formulation of flurbiprofen in acute soft tissue injuries. Drug Invest 4:441–449, 1992.

358. Hosie G, Bird H. The topical NSAID felbinac versus oral NSAIDs: a critical review. Eur J Rheumatol Inflamm 14:21–28, 1994.

359. Campbell J, Dunn T. Evaluation of topical ibuprofen cream in the treatment of acute ankle sprains. J Accident Emerg Med 11:178–182, 1994.

360. Moore RA, Tramèr MR, Carroll D, Wiffen PJ, McQuay HJ. Quantitative systematic review of topically applied non-steroidal anti-inflammatory drugs. Br Med J 316:333, 1998.

361. Arendtnielsen L, Drewes AM, Svendsen L, Brennum J. Quantitative assessment of joint pain following treatment of rheumatoid arthritis with ibuprofen cream. Scand J Rheumatol 23:334–337, 1994.

362. Dickson DJ. A double-blind evaluation of topical piroxicam gel with oral ibuprofen in osteoarthritis of the knee. Curr Ther Res 49:199–207, 1991.

363. McNiell SC, Potts RO, Francoeur ML. Local enhanced topical delivery (LETD) of drugs: does it truly exist? Pharm Res 9:1422–1427, 1992.

364. Benson HAE, Megwa SA, Roberts MS. Tissue penetration of salicylate following topical application as methyl salicylate and amine salts. In: Brain KR, James VJ, Walters KA, eds. Prediction of Percutaneous Penetration. Vol. 4b. Cardiff: STS Publishing, pp 255–257, 1996.

365. Bronaugh RL, Franz TJ. Vehicle effects on percutaneous absorption: in vivo and in vitro comparisons with human skin. Br J Dermatol 115:1–11, 1986.

366. Scott RC. Percutaneous absorption, in vivo:in vitro comparisons. Pharmacol Skin 1: 103–110, 1987.

367. Al-Khamis K, Davis SS, Hadgraft J. In vitro–in vivo correlations for the percutaneous absorption of salicylates. Int J Pharm 40:111–118, 1987.

368. Grissom RE, Brownie C, Guthrie FE. In vivo and in vitro dermal penetration of lipophilic and hydrophilic pesticides in mice. Bull Environ Contam Toxicol 38:917–924, 1987.

369. Surber C, Wilhelm K–P, Maibach HI. In vitro and in vivo percutaneous absorption of structurally related phenol and steroid analogs. Eur J Pharm Biopharm 39:244–248, 1993.

370. Scott RC, Ramsey JD. Comparison of the in vivo and in vitro percutaneous absorption of a lipophilic molecule (Cypermethrin, a pyrethroid insecticide). J Invest Dermatol 89:142–146, 1987.

371. Scott RC, Batten PL, Clowes HM, Jones BK, Ramsey JD. Further validation of an in vitro method to reduce the need for in vivo studies measuring the absorption of chemicals through the skin. Fundam Appl Toxicol 19:484–492, 1992.

372. Ceschin–Roques CG. Hänel H, Pruja–Bougaret SM, Lagarde I, Vandermander J, Michel G. Ciclopiroxolamine cream 1%: in vitro and in vivo penetration into the stratum corneum. Skin Pharmacol 4:95–99, 1991.

373. Reifenrath WG, Hawkins GS, Kurtz MS. Percutaneous penetration and skin retention of topically applied compounds: an in vitro–in vitro study. J Pharm Sci 80:526–532, 1991.

374. Lien EJ, Gao H. QSAR analysis of skin permeability of various drugs in man as compared to in vivo and in vitro studies in rodents. Pharm Res 12:583–587, 1995.

375. Li B, Birt DF. In vivo and in vitro percutaneous absorption of cancer preventative flavonoid apigenin in different vehicles in mouse skin. Pharm Res 13:1710–1715, 1996.

376. Macpherson SE, Barton CN, Bronaugh RL. Use of in vitro skin penetration data and a physiologically based model to predict in vivo blood levels of benzoic acid. Toxicol Appl Pharmacol 140:436–443, 1996.

377. Kasting GB, Francis WR, Bowman LA, Kinnett GO. Percutaneous absorption of vanilloids: in vivo and in vitro studies. J Pharm Sci 86:142–146, 1997.

378. Bronaugh RL. Methods for in vitro metabolism studies. Toxicol Methods 5:275–281, 1995.

379. Yang JJ, Roy TA, Neil W, Krueger AJ, Mackerer CR. Percutaneous and oral absorption of chlorinated paraffins in the rat. Toxicol Ind Health 3:405–412, 1987.

380. Hotchkiss SAM, Chidgey MAJ, Rose S, Caldwell J. Percutaneous absorption of benzyl acetate through rat skin in vitro. 1. Validation of an in vitro model against in vivo data. Food Chem Toxicol 28:443–447, 1990.

381. Bronaugh RL, Maibach HI. Percutaneous absorption of nitroaromatic compounds: in vivo and in vitro studies in the human and monkey. J Invest Dermatol 84:180–183, 1985.

382. Cline JF, Nichols LD, Soybel JG, Brown LR. Transdermal diffusion: in vivo versus in vitro comparison. Proc Int Symp Controlled Release Bioact Mater 14:182–183, 1987.

383. Sato K, Sugibayashi K, Morimoto Y, Omiya H, Enomoto N. Prediction of the in vitro human skin permeability of nicorandil from animal data. J Pharm Pharmacol 41:379–383, 1989.

384. Altenburger R, Rohr UD, Kissel T. Explanation for the plasma level fluctuations resulting from the transdermal liquid reservoir system delivering β-estradiol. Proc Int Symp Controlled Release Bioact Mater 24:695–696, 1997.

385. Chien T–Y, Gong S–J. Transdermal delivery of levonorgestrel: pilot clinical studies on a matrix-type once-a-week transdermal (monophasic) contraceptive delivery system. Proc Int Symp Controlled Release Bioact Mater 24:715–716, 1997.

386. Rohr UD, Altenburger R, Kissel T. Pharmacokinetics of the transdermal reservoir membrane system delivering β-estradiol: In vitro/in vivo correlation. Pharm Res 15:877–882, 1998.

387. Kligman AM. A biological brief on percutaneous absorption. Drug Dev Ind Pharm 9:521–560, 1983.

388. Gettings SD, Howes D, Walters KA. Experimental design considerations and use of in vitro skin penetration data in cosmetic risk assessment. In: Roberts MS, Walters KA, eds. Dermal Absorption and Toxicity Assessment. New York: Marcel Dekker, pp 459–487, 1998.

389. Foldvari M, Oguejiofor CJN, Wilson TW, Afridi SK, Kudel TA. Transcutaneous delivery of prostaglandin E_1: in vitro and laser doppler flowmetry study. J Pharm Sci 87:721–725, 1998.

390. Bucks DAW, Marty JP, Maibach HI. Percutaneous absorption of malathion in the guinea pig: effect of repeated topical application. Food Chem Toxicol 23:919–922, 1985.

391. Catz P, Friend DR. Transdermal delivery of levonorgestrel. VIII. Effect of enhancers on rat skin, hairless mouse skin, hairless guinea pig skin, and human skin. Int J Pharm 58:93–102, 1990.

392. Chambin O, Vincent CM, Heuber F, Teillaud E, Pourcelot Y, Marty JP. Validation of pig ear skin as a model for in vitro percutaneous absorption studies. In: Brain KR, James VJ, Walters KA, eds. Prediction of Percutaneous Penetration. Vol. 4b. Cardiff: STS Publishing, pp 271–274, 1996.

393. Chow C, Chow AYK, Downie RH, Buttar HS. Percutaneous absorption of hexachlorophene in rats, guinea pigs and pigs. Toxicology 9:147–154, 1978.

394. Dick IP, Scott RC. Pig ear skin as an in vitro model for human skin permeability. J Pharm Pharmacol 44:640–645, 1992.

395. Fujii M, Shiozawa K, Hemi T, Yamanouchi S, Susuki H, Yamashita N, Matsumoto M. Skin permeation of indomethacin from gels formed by fatty acid esters and phospholipids. Int J Pharm 137:117–124, 1996.

396. Howes D, Black JG. Comparative percutaneous absorption of pyrithiones. Toxicology 5:209–220, 1975.

397. Hunziger N, Feldmann RJ, Maibach HI. Animal models of percutaneous penetration: comparison between Mexican hairless dogs and man. Dermatologica 156:79–88, 1978.

398. Imoto H, Zhou Z, Stinchcomb AL, Flynn GL. Transdermal prodrug concepts—permeation of buprenorphine and its alkyl esters through hairless mouse skin and influence of vehicles. Biol Pharm Bull 19:263–267, 1996.

399. Itoh T, Xia J, Magavi R, Nishihata T, Rytting JH. Use of shed snake skin as a model membrane for in vitro percutaneous penetration studies: comparison with human skin. Pharm Res 7:1042–1047, 1990.

400. Moody RP, Nadeau B, Chu I. In vitro dermal absorption of N,N-diethyl-m-toluamide (DEET) in rat, guinea pig and human skin. In Vitro Toxicol J Mol Cell Toxicol 8:263–275, 1995.

401. Rigg PC, Barry BW. Shed snake skin and hairless mouse skin as model membranes for human skin during permeation studies. J Invest Dermatol 94:235–240, 1990.

402. Sloan KB, Beall HD, Weimar WR, Villanueva R. The effect of receptor phase composition on the permeability of hairless mouse skin in diffusion cell experiments. Int J Pharm 73:97–104, 1991.

403. Wester RC, Maibach HI, Bucks DAW, McMaster J, Mobayen M, Sarason R, Moore A. Percutaneous absorption and skin decontamination of PCBs—in vitro studies with human skin and in vivo studies in the rhesus monkey. J Toxicol Environ Health 31:235–246, 1990.

404. Yu HY, Liao HM. Triamcinolone permeation from different liposome formulations through rat skin in vitro. Int J Pharm 126:1–7, 1996.
405. Minter HJ, Shaw A, Howes D, Pendlington RU. The use of microautoradiography to show the distribution of substances topically applied to skin. In: Brain KR, James VJ, Walters KA, eds. Perspectives in Percutaneous Penetration. Vol. 5b. Cardiff: STS Publishing, pp 102–103, 1998.
406. Bronaugh RL, Collier SW. Protocol for in vitro percutaneous absorption studies. In: Bronaugh RL, Maibach HI, eds. In Vitro Percutaneous Absorption: Principles, Fundamentals, and Applications. Boca Raton: CRC Press, pp 237–241, 1991.
407. Le Roux Y, Borg ML, Sibille M, Thebault J, Renoux A, Douin MJ, Djebbar F, Dain MP. Bioavailability study of Menorest, a new estrogen transdermal delivery system, compared with a transdermal reservoir system. Clin Pharmacokinet 10:172–178, 1995.
408. Findlay JC, Place VA, Snyder PJ. Transdermal delivery of testosterone. Clin Endocrinol Metab 64:266–268, 1987.
409. Shah VP, Pershing LK. The stripping technique to assess bioequivalence of topically applied formulations. In: Brain KR, James VJ, Walters KA, eds. Prediction of Percutaneous Penetration. Vol. 3b. Cardiff: STS Publishing, pp 473–476, 1993.
410. Rougier A, Dupuis D, Lotte C, Roguet R, Schaefer H. In vivo correlation between stratum corneum reservoir function and percutaneous absorption. J Invest Dermatol 81: 275–278, 1983.
411. Rougier A, Lotte C, Maibach HI. In vivo percutaneous penetration of some organic compounds related to anatomic site in humans: predictive assessment by the stripping method. J Pharm Sci 76:451–454, 1987.
412. Roberts MS. Recent developments in skin penetration: role of physiology and solute structure. In: Brain KR, James VJ, Walters KA, eds. Perspectives in Percutaneous Penetration. Vol. 5b. Cardiff: STS Publishing, pp 1–4, 1998.

6

Formulation Strategies for Modulating Skin Permeation

ADRIAN F. DAVIS

SmithKline Beecham, Weybridge, Surrey, England

ROBERT J. GYURIK

MacroChem Corporation, Lexington, Massachusetts

JONATHAN HADGRAFT

University of Greenwich, Chatham Maritime, Kent, England

MARK A. PELLETT

Whitehall International, Havant, England

KENNETH A. WALTERS

An-eX Analytical Services Ltd., Cardiff, Wales

I. INTRODUCTION

Because of the barrier properties of the stratum corneum and, also, depending on the physicochemical properties of the drug, transport from simple vehicles will often be insufficient to achieve therapeutic drug concentrations at the site of action. The therapeutic target may be the skin, or the local or distal subcutaneous tissues, depending on the intent for local, regional, or systemic therapy. *Simple vehicles* are defined as those in which the drug is at or close to saturation solubility, and neither the vehicle nor the drug has any interaction with the stratum corneum to reduce its barrier function. From this definition drug–vehicle interactions in subsaturated or saturated solutions will not be reviewed. In these simple systems, depending on the saturated solubility of the drug in each vehicle, large differences in skin penetration can occur between vehicles that are at a fixed drug concentration. Vehicles in which the drug is at or near saturation will show enhanced drug penetration, compared with those in which the drug is subsaturated (1–3).

It is often necessary to increase the amount and rate of dermal or transdermal drug delivery to achieve the required therapeutic drug levels. In these instances skin penetration enhancement strategies may be evaluated. The modeling of skin penetration and permeation as a diffusional process allows consideration of the mechanisms by which enhancement may be achieved. In its simplest form, the steady-state flux of drug per unit area across the skin can be expressed by Fick's first law of diffusion:

$$J = \frac{D\Delta C_{sc}}{L} \tag{1}$$

where D is the diffusion coefficient of the drug in the stratum corneum and ΔC_{sc} is the drug concentration difference across the stratum corneum; L is the apparent thickness of the stratum corneum, accounting for tortuosity. In reality, the concentration in the outermost layers of the stratum corneum is very much greater than that in the innermost, $C_o \gg C_i$ and

$$J = \frac{DC_o}{L} \tag{2}$$

which is equivalent to the more usual form of this equation

$$J = \frac{KDC_f}{L} \tag{3}$$

where C_f is the drug concentration in solution in the vehicle, and K is the partition coefficient of the drug between the stratum corneum and the formulation.

It is apparent, from Eq. (2), that the physicochemical determinants that can be manipulated in an attempt to increase skin penetration are D and C_o (in the following discussion the difference between $C_{o(sat)}$ and $C_{o(supersat)}$ will be outlined). Classic *chemical enhancers*, such as SEPA, laurocapram (Azone), oleic acid, dimethyl sulfoxide (DMSO), propylene glycol, fatty acid derivatives, and others are defined as vehicle components that enter the stratum corneum and increase drug diffusivity within the barrier membrane (D), and increase vehicle drug solubility–partitioning ($C_{o(sat)}$), or both.

Most investigations of penetration enhancement concern the evaluation of materials, such as oleic acid, that penetrate into the stratum corneum, interact with intercellular barrier lipids, and alter D. Structure–activity studies with this class of enhancer indicate that polar head groups linked to long alkyl chains are required. Although chemical enhancers have been used successfully to increase skin penetration, in many instances these materials also possess some potential irritancy (4). A typical example is sodium dodecyl sulfate (SDS).

One mechanism whereby $C_{o(sat)}$ may be increased is enhancer-induced modification of the polarity of the skin, such that drug-saturated solubility in the skin is increased. Here, it is beneficial to use solubility parameters as a measure of polarity, because this value is known for many compounds and has been approximated also for the stratum corneum (5,6). Coenhancer systems, for example fatty acids with propylene glycol (7), are particularly effective, and this is probably due to a complex synergy in which each individual enhancer promotes the delivery of its partner as well as that of the drug.

Equation (2) is a simplification of the exact physicochemical parameters that control flux. To be more exact, the driving force for diffusion is the chemical potential gradient. Higuchi was the first to apply the more rigorous solution [Eq. (4)] to the process of percutaneous penetration (8).

$$J = \frac{Da}{L(\gamma_{sc})} \tag{4}$$

where a is the thermodynamic activity of the drug in the vehicle (and, assuming equilibrium, also in the outer layer of the stratum corneum), and γ_{sc} is the drug activity coefficient in the stratum corneum. As a/γ_{sc} is equivalent to C_o, Eq. (4) corresponds to Eq. (2). From this, Higuchi predicted the potential of supersaturation (a compound having a thermodynamic activity greater than unity, for which 1 represents the compounds activity in saturated solution) to increase percutaneous penetration. Supersaturation is an enhancement mechanism that is specific to the individual drug. Thus, there is no overall reduction in skin barrier properties and less potential for irritancy of excipients. The effect is best described as an increase in "push" of the drug into the stratum corneum. Supersaturation causes an increase in drug solubility in the stratum corneum, $C_{o(supersat)}$ beyond and independent of saturated solubility. From this, enhancer and coenhancer systems affecting $C_{o(sat)}$ or D, and supersaturation, affecting $C_{o(supersat)}$, should work independently (possibly in synergy) and be capable of multiplicative increases in penetration needed to achieve therapeutic levels of some drugs.

This chapter will concentrate on recent developments in the areas of chemical penetration enhancement, supersaturated systems, and vesicles. For the chemical penetration enhancers, only those molecules specifically designed to act as skin penetration enhancers will be discussed. Other chemical enhancers and physical mechanisms for enhancement, including formation of prodrugs, iontophoresis, electroporation, and ultrasound, have been fully discussed elsewhere (9–11), and they will not be included here.

II. CHEMICAL ENHANCERS

A. Azone (1-Dodecylazacycloheptan-2-one)

Azone (1-dodecylazacycloheptan-2-one; laurocapram) and its analogues have probably been one of the most investigated groups of penetration enhancers. Despite the first references to Azone as a penetration enhancer appearing in the literature as long ago as the early 1980s, it is perhaps surprising that there has been no significant commercialization. This is a reflection of the manner in which regulatory authorities apply the same stringent licensing constraints to excipients as they do to new pharmaceutical actives. The extensive research into this group of penetration enhancers has, however, provided a clearer understanding of the mechanisms of skin penetration modulation, and structural activity relations. Many different experiments have been conducted that can be broken down into a number of categories:

1. Interactions with model membranes
2. In vitro experiments
3. In vivo experiments
4. Synergy.

Representative publications will be highlighted to provide a general overview of what is currently known about Azone and its analogues. The information can then be applied to an understanding of other penetration enhancers.

The physicochemical properties of Azone are given in Table 1. Figure 1 provides structural details of some representative analogues that have been synthesized.

1. Interactions with Model Membranes

One of the difficulties in skin penetration studies is attempting to analyze the data and obtain a mechanistic description of the diffusion process. The inherent variability in tissue samples and that the skin is a heterogeneous barrier are the principal reasons for this. To simplify the understanding of the mechanisms of action of Azone, various experiments have been conducted in which simple-structured lipids have been used to simulate the complex bilayer lipid structures that are found in the intercellular regions of the stratum corneum. The structured lipids have been simple monolayers or vesicles composed of materials such as dipalmitoyl phosphatidylcholine (DPPC) or mixed lipids representative of those found in the skin. The simplest experiments that have been conducted involve the incorporation of the Azone analogues into multilamellar vesicles composed of DPPC and then measuring the ability of the modulator to alter the phase-transition temperature of the vesicle. This can be achieved by a light-scattering method (12) or by using differential-scanning calorimetry (13). Azone lowers the phase transition temperature and creates regions within the liposome that are more fluid. It is postulated that the mechanism of action of Azone in skin penetration enhancement is that it inserts itself into the structured lipids of the stratum corneum where it creates a more fluid environment. A diffusing drug, therefore, will encounter a reduced microviscosity, and penetration will be accelerated. The method of insertion has been probed using monolayers in an automated Langmuir trough (14). Azone appears to expand the monolayer and force apart the DPPC molecules more than would be anticipated from ideal mixing. The experiments also allow an estimation of the area per molecule occupied by Azone. Molecular graphics analysis of the end-on area per molecule suggest that Azone

Table 1 Some Physicochemical Properties of Azone

Empirical formula	$C_{18}H_{35}NO$
Molecular weight	281.49
Log octanol/water partition coefficient[a]	7.8
Water solubility[a]	7×10^{-7} g/L
Melting point	$-7°C$
Viscosity	45.2 cp
Refractive index	1.4701
Specific gravity	0.912
Surface tension	32.65 dyne/cm

[a]Estimated values (ACD Inc. software Toronto, Canada).

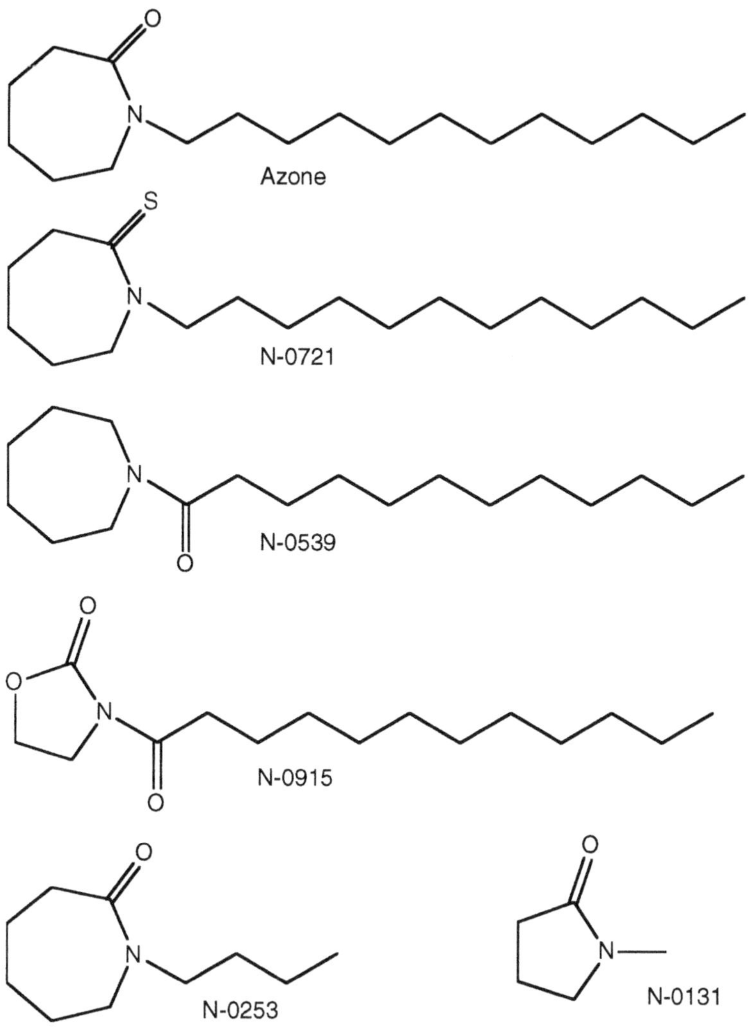

Figure 1 Structure of Azone and some analogues.

exists in the monolayer with its seven-membered ring structure in an orientation that is approximately perpendicular to the alkyl chains that are parallel to the DPPC chains (15). A representation of this is shown in Figure 2.

Further research in which Azone was incorporated into more complex lipid mixtures (more representative of those found in the stratum corneum) indicates that Azone reduces the condensation state of the lipids, which corresponds to increased fluidity (16). X-ray-scattering and differential-scanning calorimetry (DSC) studies on similar complex lipid mixtures show that Azone reduces the phase-transition temperature of the lipids and that it was intercalated into the lipid structures, which induced lateral swelling. These findings could be explained on the basis of the shape of Azone, which has a large interfacial area and head group compared with its alkyl

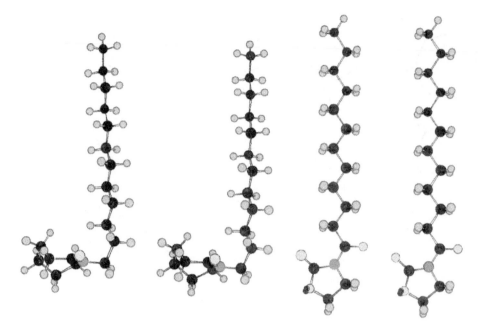

Figure 2   The expected orientation of Azone (left-hand two molecules) and N-0915 (right-hand two molecules).

chain volume (17). X-ray diffraction has also been used to examine the interaction of Azone with partly neutralized fatty acids in water (18). Azone transformed the original reversed hexagonal phase into a reversed micellar phase. This is consistent with similar observations between Azone and soybean lecithin−water or monoolein−water systems (19).

Figure 2 also shows the Azone analogue N-0915. Monolayer experiments on this compound show that its presence in a DPPC monolayer causes condensation of the lipids (14). Additionally incorporation into DPPC liposomes produces a slight increase in the phase-transition temperature (20). This is indicative of a stabilization of the lipid structure; hence, an increase in microviscosity. The results were explained in terms of the ability of N-0915 to hydrogen-bond with adjacent lipid molecules, causing the condensation. Included in the publication (20) was in vitro data on the penetration of metronidazole and the mosquito repellent diethyl-*m*-toluamide through human skin; N-0915 retarded their penetration.

Transfer studies have also been conducted on model solid-supported liquid membranes, such isopropyl myristate (IPM) incorporated into a cellulose acetate membrane. The presence of Azone in the IPM facilitated the flux of sodium salicylate across the membrane. This was attributed to the ability of the nitrogen atom in Azone to protonate at the pH used and to ion-pair with the salicylate anion (21). However, alternative explanations have also been produced (22) in which a model of the extraction coefficients using pK_a, of weak acids was developed to account for the observed variation in partition.

2. In Vitro Experiments

Many in vitro experiments have been conducted on the effects of Azone and its analogues. These can be divided into animal and human studies.

a. Animal Studies. One of the major difficulties in conducting animal experiments is to find a species that can be used to predict effects in human tissue. The indication from the model membrane studies is that it is the ability of Azone to intercalate into the lipids of the stratum corneum that is important. The precise nature of the skin lipids will be important, and it has been documented that these are very species-dependent. The results from animal experiments, therefore, have to be taken with some caution, but could be indicative of what may be expected in human skin. Numerous species have been examined and some comparisons determined. For example the effect of a number of enhancers, including Azone, on the penetration of hydrocortisone through the skin of hairless mouse (HM), hairless rat (HR), and human cadaver (HC) has been measured (23). The enhancement ratios found for Azone were HM: 18.0; HR: 13.1; HC: 5.5, showing that, for hydrocortisone, at least, the effects are far more pronounced in rodent skin. Shed snakeskin has also been a popular choice for enhancer experiments. Experiments on the transport of acetaminophen and ibuprofen through shed snake skin showed that the addition of Azone increased the amount of acetaminophen, but not the amount of ibuprofen (24). The acetaminophen studies showed that the mechanism of enhancement was an increase in partition coefficient K, with no significant increase in the diffusion coefficient D. For the ibuprofen formulation, Azone provided a slight increase in D, which was offset by a decrease in K. These results highlight the importance of partitioning in penetration enhancement. Because Azone can sometimes have a detrimental effect on partitioning of the drug into the skin is often ignored.

A species comparison between snake skin, rabbit skin, and human skin has also been conducted (25). Model drugs, indomethacin, 5-fluorouracil, and propranolol were chosen. The relative permeation improvements in human skin, snakeskin, and rabbit skin were 10–20-fold, 5–50-fold, and 20–120-fold, respectively. Azone significantly increased the permeation of both the lipophilic and hydrophilic compounds. Rabbit skin was considered a poor model for human skin in vitro, whereas snake skin was closer to human skin in terms of transdermal permeability. The relative enhancement of several compounds through snake skin was dependent on the lipophilicity and the molecular size of the permeant, with a greater permeability increase for more hydrophilic and larger-molecular size permeants (26).

Hairless mouse skin is often used in skin permeation studies, and a species comparison has been conducted using dihydroergotamine as the permeant. In this study, Azone increased the flux in the order rabbit > human > rat > guinea pig > hairless mouse (27). Guinea pig skin has also been used in a systematic study of Azone and some analogues (1-geranyl and 1-farnesylazacycloheptan-2-one). Seven penetrants with varying lipophilicities were examined. Large enhancement ratios were found for drugs having octanol–water partition coefficients close to unity (28). Hairless mouse skin has been used in a range of studies examining the enhancing activity of Azone and its analogues (23,29–39).

b. Human Studies. Sophisticated biophysical techniques have been used to probe the mechanism of action of Azone in human skin. For example, deuterium nuclear magnetic resonance (NMR) has shown that a mixture of Azone and propylene glycol causes structural disorder of the intercellular lamellar lipid structures (40). DSC has been used to examine the effect of Azone and its analogues on lowering the phase-transition temperatures of human skin lipids (41). X-ray-scattering experiments have shown that treatment of human stratum corneum with Azone, and its analogues having alkyl chains of more than six carbons in length, results in disordering of the lamellae. These results are similar to those obtained by thermal analysis (42,43). Electron-spin resonance (ESR) has shown that Azone increased both the fluidity and polarity of the environment close to the 5-doxyl stearic acid spin label (44,45). These were similar to the results found for a similar ESR experiment conducted with the enhancer oleic acid, suggesting a related mechanism of action (46).

Fourier transform infrared (FTIR) spectroscopy has indicated that perdeuterated Azone (47) distributes homogeneously within the stratum corneum lipids, in which it induces fluidity (48). This contrasts with the mechanism of action of oleic acid that appears to phase-separate and form liquid pools (49). Clearly, there are differences in the mode of action: the sodium salt of heparin has its flux enhanced by Azone, whereas oleic acid is ineffective as an enhancer for this permeant (50). FTIR has also been used to show two discrete types of enhancer mechanism. The presence of Azone increases the diffusion coefficient of the model permeant 4-cyanophenol, whereas the enhancer Transcutol has no effect on D but increases the solubility of the permeant in the skin (51).

In a structure–activity study, several permeants (alkyl anilines and phenyl alkanols) were examined. The log octanol–water partition coefficients (log K_{oct}) varied from approximately -1 to $+4$. The efficiency of Azone depended on the lipophilicity of the permeant (see foregoing results with snake skin). The effect was dependent on the concentration of Azone used. At 1%, it acted on compounds the log K_{oct} of which was < 1; however, at 5% the Azone threshold was increased to a log K_{oct} of 2.7 (52). Compounds that are hydrophilic (e.g., methotrexate) also have their permeation increased by Azone (53). Because Azone influences skin lipids, the methotrexate must transfer through or across the lipid domains, suggesting that hydrophilic molecules do not permeate through a unique polar route.

Azone and its analogues (see Fig. 1) have been investigated in vitro using human skin and the model permeant metronidazole (20). The results are shown in Fig. 3. Enhancement ratios calculated at 40 min are Azone, 6.7; N-0539, 6.4; N-0253, 3.4; N-0721, 1.4; N-0131, 1.1; and N-0915, 0.2. The sulfur analogue of Azone (N-0721) is significantly less active than Azone, and the short hydrocarbon chain in N-0131 renders it ineffective. The phase transition temperature of multilamellar DPPC liposomes is lowered by the presence of the enhancers in rank order of their enhancing abilities, except N-0915, which increases T_m. This can be explained by alteration of the lateral bonding within the stratum corneum lipid lamella. A model based on molecular graphics of Azone and N-0915 and their hydrogen-bonding capacities to cerebrosides has been proposed. It is possible that models such as this can be used in the future to design more specific and active penetration enhancers.

One of the problems of penetration enhancers is that they tend to be nonspecific in their action. A study on Azone and sodium lauryl sulfate (SLS) showed that the presence of Azone in the skin enhanced the subsequent absorption of the SLS (54,55). This has implications in the use of penetration enhancers and precautions

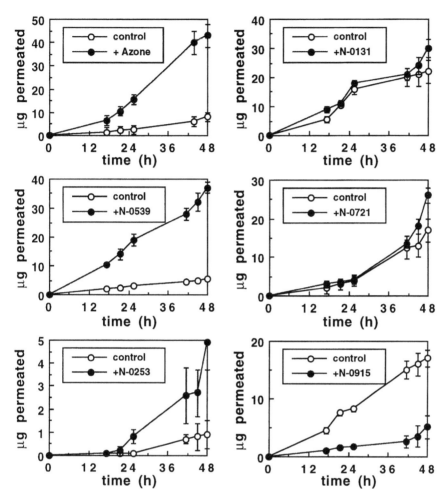

Figure 3 A comparison of the effects of different enhancers on the permeation of metronidazole through human full-thickness skin.

may be necessary to avoid enhanced irritancy caused by subsequent contact of the treated skin with simple household products, such as detergents.

3. In Vivo Experiments

There have been fewer in vivo experiments conducted using Azone particularly from a mechanistic viewpoint. In part this results from the difficulty in interpreting in vivo data. When blood levels are measured, the variability in the data is increased owing to the additional clearance kinetics. In general, the in vivo determinations correlate with what has been found in vitro. A review article (56) considers the clinical experiments in which Azone had been examined, and little additional information has appeared in the literature since then. Some representative publications include the following.

Methotrexate permeation has been examined in vitro and in vivo in hairless mice, and increased blood levels were found using Azone as an enhancer (57). This species has also been used in a study on acyclovir and its ability to reduce the herpes simplex virus-1 (HSV-1) (58), on nitroglycerin absorption (59) and methotrexate (60). In vitro and in vivo experiments have also been conducted using rat skin with the permeant bromhexine. Azone enhanced permeation both in vitro and in vivo (61). A novel dopaminergic D_2 agonist, $S(-)$-2-(N-propyl-N-2-thienylethylamino)-5-hydroxytetralin, was also tested, using a rat model. Azone was effective both in vitro and in vivo (62). Similarly, the β-adenergic−blocking agent pindolol has been examined in rabbits, which comparable results (63). Nitrendipine patches containing Azone have been examined in humans, with encouraging results. There was no apparent irritancy problem in either humans or rabbits (64). Various experiments have been conducted on the topical corticosteroid triamcinolone acetonide (TA). Tritiated steroid was used to quantitate the amount of drug absorbed. In the presence of Azone, higher amounts of TA were absorbed (65).

Several investigations have been conducted into the distribution of Azone in the skin and its subsequent elimination and metabolism (66–70). Owing to the high lipophilicity of Azone (estimated log $K_{oct} = 7.8$) it rapidly penetrates into the lipids of the stratum corneum. Further transit into the viable tissue is hindered by the very poor water solubility of the enhancer.

4. Synergy

The predominant mechanism of action of Azone is to increase the diffusion coefficient of the permeant in the stratum corneum. A simple evaluation of Fick's first law of diffusion [see Eq. (3)] shows that enhancers should be capable of acting on the diffusion coefficient in the skin as well as the solubility of the permeant in the skin lipids. The foregoing experiments (51) showed that Azone influenced D whereas Transcutol altered the solubility of the model permeant cyanophenol. Fick's law demonstrates that, if both enhancement strategies are used, there should be a multiplicative effect (i.e., synergy should be seen).

Many publications indicate the use of Azone in combination with propylene glycol. It is probable that synergistic effects have inadvertently been masked. When relatively systematic studies have been conducted, synergism has been demonstrated. For example metronidazole penetration is enhanced both by Azone and by propylene glycol alone: together synergism is found (71). A similar effect is seen for combinations of Azone and Transcutol for the permeant prostaglandin E_2 (72).

Recent interest in peptide and oligonucleotide delivery into and through the skin has promoted interest in physical methods of penetration enhancement such as iontophoresis, electroporation, and ultrasound (sonophoresis). Theoretically, synergism between physical and chemical enhancement should be possible. In the limited studies reported in the literature, it is difficult to produce guidelines in formulation strategies. Azone appears to act synergistically with iontophoresis in the delivery of metoprolol (73). However, there was no apparent synergistic effect for sotalol (74). In vivo impedance spectroscopy has shown that an Azone−propylene glycol mixture had a profound effect on the postiontophoretic skin impedance, considerably amplifying the effect of current passage (75).

Fewer experiments have been conducted with ultrasound, but synergy has been reported for the combined effect of 150 kHz ultrasound, at 111 mW/cm^2 intensity, and Azone. Aminopyrine was the permeant used in the study with excised hairless rat skin (76). Clearly, more work is indicated to determine the precise mechanism of synergy and how it is influenced by the physicochemical properties of the penetrant.

B.   SEPA (2-*n*-Nonyl-1,3-dioxolane)

SEPA is a registered tradename and an acronym for "soft enhancement of percutaneous absorption." SEPA represents a patented class of compounds containing various substituents (77). The ring system may be dioxane or dioxolane, and the side chains may vary. SEPA 0009 is 2-*n*-nonyl-1,3-dioxolane (Fig. 4)—the congener most widely studied—and is the dioxolane for which the most safety data has been compiled. *"SEPA"* is used in this section to represent SEPA 0009. Because SEPA has only carbon, hydrogen, and oxygen, minimal central nervous system (CNS) or other untoward pharmacological activity is not expected, and none has been observed in safety tests. It is also metabolically benign, being metabolized and readily excreted. These safety properties are the reason for the "soft enhancer" designation.

1.   Safety

SEPA has undergone extensive safety testing in animals, including testing for acute, chronic, and mutagenic effects, with no problems arising. In addition, human safety studies have been conducted as follows:

> Irritation studies: 73 volunteers
> Sensitization studies: 358 volunteers
> Phototoxicity study: 30 volunteers
> Photoallergenicity studies: 156 volunteers
> Phase II and III studies: more than 3500 patients

2.   Efficacy

SEPA's broad-spectrum activity is best demonstrated when compared with other enhancers. Studies were performed in a conventional vertical Franz-type cell using neonatal porcine or human transplantation skin. The formulations studied were standard hydroalcoholic gels, with drug and enhancer present in equivalent amounts. For equivalent formulations that contained the lipophilic drug papaverine, SEPA was approximately twofold more effective than zone over a 24-h period (Fig. 5). Azone, although less effective than SEPA, does show useful transdermal applications for this drug. SEPA also showed good penetration enhancement for estradiol (Fig. 6); although, in this example, Azone showed little enhancement over the control without an enhancer. SEPA was also a more effective enhancer than Azone for ketoprofen (Fig. 7). Therefore, SEPA displayed a broad-spectrum enhancement that was not totally dependent on the molecule being delivered. Both polar and nonpolar drugs are strongly enhanced (78,79).

These example results were generated using in vitro testing methods. It is absolutely crucial to have a transdermal in vitro testing method in place that is

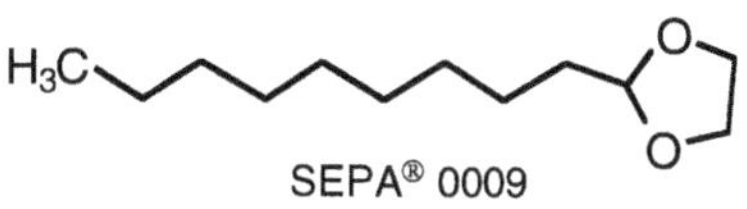

SEPA® 0009

Figure 4   Structure of SEPA.

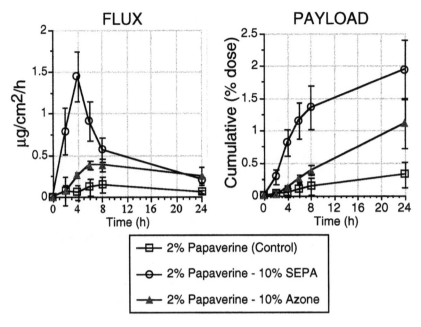

Figure 5   A comparison of the effects of SEPA and Azone on the percutaneous flux of papaverine.

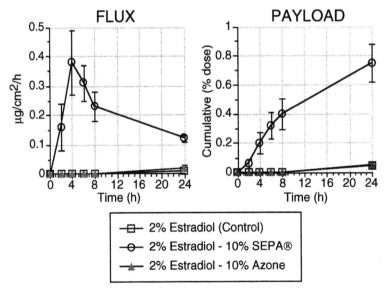

Figure 6   A comparison of the effects of SEPA and Azone on the percutaneous flux of estradiol.

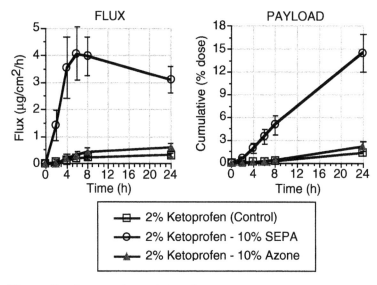

Figure 7 A comparison of the effects of SEPA and Azone on the percutaneous flux of ketoprofen.

predictive of the in vivo condition. The following are two examples of in vitro–in vivo correlation studies that have been carried out using SEPA.

3. In Vitro–In Vivo Correlation

a. SEPA Minoxidil Solutions. Figure 8a shows the results obtained after a series of optimization studies had been performed using neonatal porcine skin. The 10% SEPA formulation shown had been selected earlier through a series of studies to be the best in vitro performer (drug-delivery performance criteria were selected from factors including peak flux, time to peak flux, and total drug delivery) from an array of prototype formulations. As described earlier, the porcine skin model gives results comparable with those of human skin in vitro models. Figure 8a shows how this chosen optimized candidate performs in vitro versus the appropriate unenhanced control. Figure 8b shows the results obtained after minoxidil (Rogaine) solutions were applied twice daily to the scalp of human volunteers. These solutions were of similar composition, except for the amount of enhancer, SEPA, added to the formulation. Serum minoxidil levels were measured at appropriate time points. Figure 8b shows plasma minoxidil values expressed as nanograms per milliliter (ng/mL) of serum. The serum profile clearly shows enhancement above the Rogaine reference treatment; moreover, the optimal SEPA concentration is clearly seen to confirm the choice of the 10% formulation, as forecast by the earlier in vitro optimization studies.

b. SEPA Ibuprofen Gels. Figure 9a shows the results obtained after a series of optimization studies had been performed in human skin with an ibuprofen-containing gel. Similar to the minoxidil approach, the 10% SEPA formulation containing 5% ibuprofen had earlier been selected from a series of studies to be the best in vitro performer from a group of ibuprofen formulations. Figure 9a shows how this optimized candidate performed in vitro compared with several commercially ob-

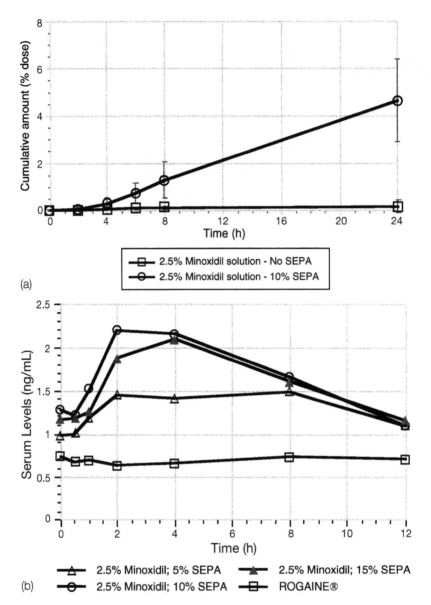

Figure 8 A comparison of the in vitro and in vivo effects of SEPA: (a) in vitro minoxidil permeation enhancement ($n = 6$, porcine skin); (b) in vivo serum levels of minoxidil following topical application ($n = 21$, human volunteers).

tained products; much higher flux and cumulative drug permeation were measured for this one than for any of the other formulations. Figure 9b shows the results obtained after a phase 1-2 clinical study on volunteers with the same composition gel. The delayed onset muscle soreness (DOMS) model for exercise-induced pain in the triceps muscle was employed. Forty-eight hours after heavy exertion of the triceps, 2.5 g of a hydroalcoholic gel, either pooled placebo (placebo with and without

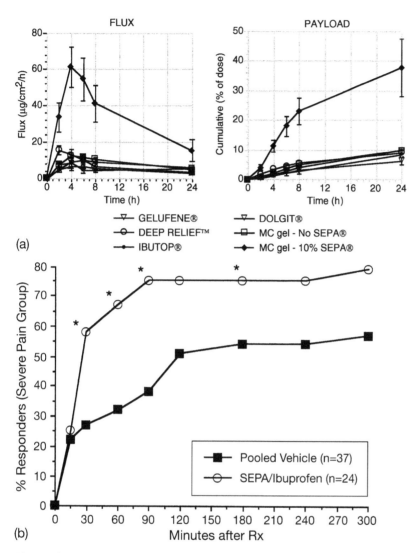

Figure 9 A comparison of the in vitro and in vivo effects of SEPA: (a) in vitro ibuprofen permeation enhancement ($n = 6$, human skin); (b) subject global evaluation of pain relief following exercise (severe pain group).

SEPA) or treatment, was applied as a single dose. Significant clinical alleviation of pain (asterisks; $p < 0.05$) in a higher percentage of the treatment group than that of the placebo (score 0–4; change of 1 unit is considered clinically significant) was seen in the group initially reporting severe pain. No significant difference was seen for the 5% ibuprofen hydroalcoholic gel without SEPA compared with the placebo. Most importantly, statistically significant differences were seen as early as 30 min after application of the gel, and the effect lasted for 5 h. This early onset of action and persistence of activity with SEPA formulations is noteworthy. This muscle soreness model, however, is acknowledged to be a difficult model to interpret because the induced pain is spontaneously resolving during the testing phase.

4. Mode of Action

There are three major areas that are currently under investigation to explain or predict penetration enhancement results. These areas are generally thought to be barrier function modification of the stratum corneum, partitioning effects between the formulation and the stratum corneum, and thermodynamic activity of the drug in the formulation.

There are several changes to the stratum corneum that may influence transdermal penetration. These include alteration of the cellular and intercellular lipid composition (e.g., by increasing fluidization of lipids, such as with SEPA and other lipophilic enhancers); by removing lipids, such as with dimethyl sulfoxide or ethanol; effects on intercellular organization and cohesion; disruption or reordering of the water structure; and alteration of stratum corneum proteins.

Some workers have suggested that there may be two pathways through the stratum corneum barrier: one lipophilic, the other hydrophilic. The lipoidal pore pathway has been proposed on the molecular scale based on observations of the effect of alkyl-pyrrolidones on the hairless mouse skin permeation of steroids (80). Large lipoidal "pores," in a gross morphological sense, have also been described with lipophilic terpene enhancers postulated to form pools, possibly adjacent to the stratum corneum intercellular lipid lamellae (81). Apparent aqueous pores have also been shown after hyperhydration of the stratum corneum (82). Engstrom et al. (83) have suggested a mechanism of action that describes "Azone holes," complex structures that combine both lipophilic and hydrophilic conduits. The phenomenon of disparate pore formation appears logical from a physicochemical perspective, as any reduction in anisotropy to produce a hydrophilic pore must concomitantly produce the corresponding lipophilic pore, akin to a partitioning effect. It may also be the best explanation for the broad-spectrum action of SEPA, which may provide the best combination of the two proposed actions: membrane fluidization anisotropy and disruptive reordering to form conduits that are both aqueous and lipoidal.

The thermodynamic activity of any transdermal delivery system, whether a liquid, a semisolid, or a patch-type system, is of large consequence to the percutaneous absorption of the drug and the enhancer. This will be discussed in Sec. III, particularly relative to supersaturated solutions.

However, there is a more fundamental underlying mechanism that needs to be discussed, an aspect of dermal and transdermal delivery that cannot be overly emphasized and is the facet least explored. The underlying physics, however, are well characterized, based on fundamental physical laws, and have been utilized in the past by other disciplines, such as polymer science, the science of films, film formation, and coating technologies.

As a background for this discussion, recall that Fick's second law of diffusion [see Eq. (5)] describes the way in which concentration changes over time for any region of a solution. It does not, however, apply to non–steady-state conditions, concentrated solutions, or conditions of partial miscibility.

$$\frac{\partial c}{\partial t} = D \ (\partial^2 c/\partial x^2) \tag{5}$$

The Fickian diffusion coefficient is here defined in terms of a concentration gradient. However, the actual driving force for the flux of any component in a more complex system is not the concentration gradient, but the *chemical potential gradient*.

$$D_i = B_i \ (\delta\mu_i/\delta x_i) = 0 \tag{6}$$

Equation (6) describes how the chemical potential (μ_i) of a nonideal solution can be related to the diffusion coefficient D_i at constant pressure and temperature. B_i is a mobility term that is always positive. This relation is well known in the physico-chemical field of film forming, specifically in coacervation technology (84), and it pertains to partially miscible systems, such as those that can be obtained with hydroalcoholic solutions, gels, and emulsions.

Figure 10 shows a hypothetical, but typical example of a partially miscible system containing three components that include a lipophilic phase (which may also contain an enhancer), a hydrophilic phase (aqueous), and a homogenizing phase (e.g., alcohol). As alcohol is allowed to evaporate, such as happens after placement of a hydroalcoholic gel on the skin, line AB in the phase diagram is traversed (for reference, composition A in the phase diagram comprises 60% homogenizing phase,

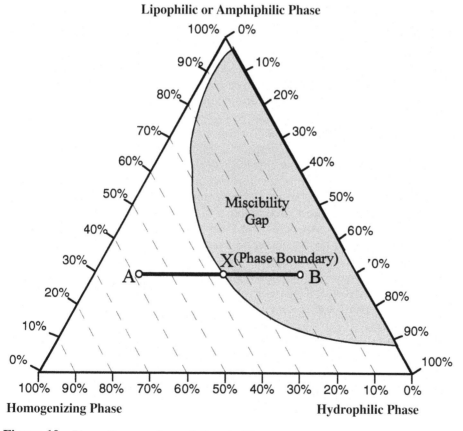

Figure 10 Phase diagram of a partially miscible system.

30% lipophilic phase, and 10% hydrophilic phase). As the homogenizing phase either evaporates or is absorbed, next indicated is composition X, the point of the phase boundary condition. After crossing this point into the *miscibility gap*, phase separation, or phase inversion occurs, usually very rapidly. At this time the migration of solution components proceeds from an area of low concentration to high concentration (an almost pure one of each phase).

Migration of a particle (of a solvent or a solute) from low concentration to high concentration seems contrary to Fick's second law (but recall that Fick's law is only for ideal conditions. The enhancer can also be driven into the stratum corneum according to this physics. When a solute (drug) starts out dissolved in the homogeneous phase (at compositional point A; see Fig. 10), and it subsequently finds itself in the miscibility gap after separation into two phases, it can now be postulated to be in two distinct solutions, supersaturated in one or both. If it prefers one phase thermodynamically, it may (and probably will) not have time to equilibrate into the preferred phase before the phases separate (kinetics overrule). Now, the chemical potential for the drug (and enhancer) realized may be of temporarily very high magnitude. The solute (or permeation enhancer, or both) may now relieve this potential energy by partitioning into the stratum corneum membrane (or by giving off heat, crystallizing, etc.). If the formulation can produce a chemical potential gradient at the moment of interfacial skin contact, and preserve the gradient for the appropriate time fame (assuming the barrier is opened by the appropriate enhancer system), then a formulation can drive high amounts of drug into the stratum corneum. This is also true for concomitantly driving molecules of the enhancer into the stratum corneum, preferably in the same time frame, so that enhancement may be optimal.

This effect may be termed an *energetic epidermal injection*, or *hyperflux*, condition. The effect is largely to be measured empirically, because of the complexities, and each drug must have its formulation optimized for amount of drug delivered and the time course of delivery. Another way to say this is that the escaping tendency (fugacity) of any formulation component (drug or enhancer) will change over time and can be maximized for each component following physical principles (tending over time to always decrease enthalpy and increase entropy).

Figure 11 shows the complexity arising after evaporation of even a very simple emulsion (85). After evaporation of the decane and water, different compositions and complex structures may appear. Prediction of fluxes from such structures becomes very difficult indeed. The structures and forms that appear depend on the balance between thermodynamic and kinetic considerations as time progresses after dosage. Formed structures that are low-energy (low-enthalpy), such as the lamellar liquid crystal depicted in Figure 11 may lower the fugacity of any single subsumed component. On the other hand, if phase-inverted systems appear, then the kinetics may be such that the phase (or the solute, or both) may be injected into the stratum corneum with ease.

Three postulations can be summarized as follows: (1) a metastable state can be achieved by combining solvents of limited miscibility; (2) when the homogenizing solvent evaporates, situations can be produced in which the gradients of solute concentration and gradients in the chemical potential are of different sign; and (3) in this state, the diffusion coefficient, defined by Fick's law, will be negative. Thus, a component can be induced to flow from a region of low concentration to a region of high concentration within the vehicle or within the stratum corneum (101).

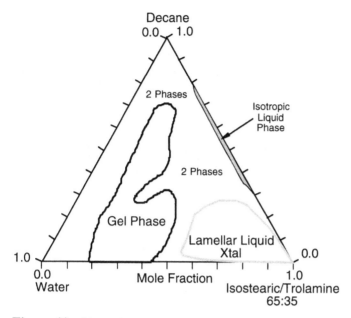

Figure 11 Phase diagram illustrating the complexities of emulsion evaporation.

C. Miscellaneous Chemical Enhancers

In addition to SEPA and Azone, a few other specifically designed enhancing mole-cules have been evaluated. These include 1-[2-(decylthio)ethyl]azacyclopentan-2-one (HPE-101), 4-decyloxazolid-2-one (Dermac SR-38), and dodecyl-*N,N*-dimethyl-amino isopropionate (DDAIP, NexACT 88) (Fig. 12).

1. HPE-101

HPE-101 is believed to have a mechanism of action similar to that of Azone and is also very sensitive to the vehicle of application (86,87). Thus, although the enhancer significantly increased the urinary excretion of indomethacin following topical ap-plication to hairless mice, it was dependent on the application vehicle (87). When

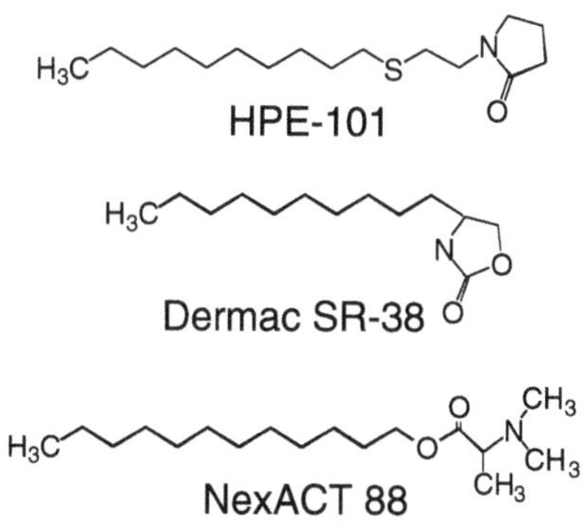

Figure 12 Structures of specifically designed skin penetration enhancers.

the enhancer was applied in solution in polar solvents, such as dipropylene glycol, triethylene glycol, diethylene glycol, glycerin, water, or triethanolamine, enhancement ratios varied between 1.5- and approximately 67-fold. However, when applied in solution in more lipophilic solvents, such as ethanol, isopropanol, oleyl alcohol, isopropyl myristate, or hexylene glycol, no enhancement was observed (Table 2). These data stress the importance of optimization of the delivery vehicle, not only for the drug, but also for the enhancer. Combinations of HPE-101 with cyclodextrins appear to be useful means to improve drug permeation across the skin (88). All of the available data, however, have been obtained using hairless mice or other small laboratory animal skin. Small laboratory animals, especially the hairless mouse, can be uniquely sensitive to skin penetration enhancement and, as yet, the effectiveness of HPE-101 on human skin has not been reported.

2. Dermac SR-38

Dermac SR-38 is one of a series of oxazolidinones, cyclic urethane compounds, evaluated as transdermal enhancers (89). The compound was designed to mimic natural skin lipids (such as ceramides), to be nonirritating, and to be rapidly cleared from the systemic circulation following absorption (90). In animal and human safety studies, Dermac SR-38 demonstrated a good skin tolerance (no observed irritancy

Table 2 HPE-101 Enhancement Ratios for Indomethacin Absorption in Hairless Mouse (Based on Urinary Excretion)—Vehicle (Solvent) Effects

Vehicle	ER[a]
Dipropylene glycol	21.1
Triethylene glycol	67.0
Hexylene glycol	1.54
Diethylene glycol	15.2
1,3-Butylene glycol	9.47
Trimethylene glycol	21.3
Glycerin	8.19
Water	3.09
Oleyl alcohol	0.86
Isopropanol	0.96
Ethanol	0.78
Silicone	2.47
Peppermint oil	1.08
Olive oil	0.78
Isopropyl myristate	0.80
Triethanolamine	5.88

[a]ER, enhancement ratio/urinary excretion of indomethacin when applied topically with HPE-101 divided by urinary excretion of indomethacin when applied topically without HPE-101.
Source: Ref. 87.

or sensitization at levels of 1–10% (w/w); moderate to severe irritation in rabbit at 100%); and a low degree of acute toxicity ($LD_{50(rat\ oral)} > 5.0$ g/kg). The compound was evaluated for its ability to enhance the human skin permeation of diverse drugs from dermal and transdermal delivery systems (Table 3). The data for minoxidil clearly indicate an enhancer concentration-dependent effect for permeation enhancement. Dermac SR-38 enhances the skin retention of both retinoic acid, when applied in Retin A cream, and dihydroxyacetone, when applied in a hydrophilic cream (91).

3. NexACT 88

NexACT 88 (dodecyl-*N,N*-dimethylaminoisopropionate; DDAIP) is one of a series of dimethylamino alkanoates, reported to be biodegradable, which were developed as potential nontoxic skin permeation enhancers (92). Much of the early work was carried out using shed snake skin and, with this model, most of these compounds were equal to, or more active than, Azone (25,92). Studies using human skin indicated that dodecyl-*N,N*-dimethylaminoacetate (DDAA) was a more effective enhancer of absorption of propranolol hydrochloride and sotalol than was Azone (74,93). Structural optimization of the compounds led to the identification of the lead candidate dodecyl-*N,N*-dimethylaminoisopropionate (94,95) that appeared to be more effective than DDAA. Mechanism of action studies indicated that the distribution of DDAIP in stratum corneum lipids was somewhat different from that of DDAA (96), suggesting that other interactions were contributing to the penetration enhancement effect. It is possible that, in addition to its effect on stratum corneum lipids, DDAIP may interact with keratin and potentially increase stratum corneum hydration.

Table 3 Dermac SR-38 Enhancement Ratios for Drug Permeation Across Human Skin

Drug	Vehicle	Dermac SR-38 (wt%)	ER[a]
Lidocaine	Cream	2.5	3.2
Prilocaine	Cream	2.5	4.1
Diclofenac	Gel	2.5	2.3
Hydrocortisone	Cream	1.0	3.0
Indomethacin	Cream	1.0	1.8
Minoxidil	Solution	2.0	2.6
	Solution	5.0	7.3
	Solution	10.0	10.4
Alprazolam	Patch	5.4	1.6
Diltiazem HCl	Cream	3.0	2.4
Isosorbide dinitrate	Patch	8.0	3.0
	Cream	1.4	1.6
Morphine sulfate	Patch	3.0	1.8
Progesterone	Patch	5.0	5.1
Nifedipine	Cream	5.0	2.6

[a]ER, enhancement ratio/in vitro skin permeation of drug when applied with Dermac SR-38 divided by skin permeation of drug when applied without Dermac SR-38.
Source: Ref. 90.

Preliminary animal toxicity studies carried out using DDAA and NexACT 88 indicated that these compounds had low toxicity (92), but may be mildly irritating to rabbit skin at high concentrations. The overall safety and efficacy data suggest that the dimethylamino alkanoates may be useful as transdermal penetration enhancers. More human skin data would be beneficial, and it is important to appreciate that, in common with other skin penetration enhancers, reduction to practice in a clinical situation is a prime requisite for determining efficacy.

4. Others

Other compounds have been identified and have undergone preliminary evaluation as potential skin-penetration enhancers. The data are, however, very limited and the candidate enhancers are mentioned here solely for completeness. The biodegradable fatty acid esters of *N*-(2-hydroxyethyl)-2-pyrrolidone (decyl and oleyl) were synthesized and evaluated for enhancer activity using hairless mouse skin (97). Permeation of hydrocortisone was enhanced twofold. The activity of *n*-pentyl-*N*-acetylprolinate as a skin permeation enhancer has been determined using human skin (98).

III. SUPERSATURATED SYSTEMS

The effect of formulation on drug bioavailability is much greater in topical drug delivery than in any other route of administration. Components of the vehicle, including chemical penetration enhancers, can interact with the drug and the skin to influence both the rate and extent of absorption. However, increasing the number of excipients in the vehicle will inevitably lead to an increase in the potential of the vehicle to induce some form of dermatotoxicity (irritancy, sensitization, etc.). Supersaturated systems provide an enhancer strategy with the capability of low-dose therapeutic efficacy and improved control of percutaneous absorption. Furthermore, these systems can be formulated with a reduced number of excipients and, therefore, may lead to a reduction of dermatotoxic events.

A. Physicochemistry of Supersaturated Systems

1. Supersaturation

James (99) defined a *solution* as a molecular dispersion of a solute in a solvent. Under equilibrium conditions, a solution is capable of a continuous variation in composition within certain limits. At the lower limit the solution is represented by the pure solvent and, at equilibrium, the upper limit occurs when the solvent is incapable of dissolving any more solute and a second phase of undissolved solute is present. A solution in equilibrium with undissolved solute is known as a saturated solution. In saturated solutions the thermodynamic activity of the solute is equal to that of the pure solute, and this value is arbitrarily taken as unity. Thus, supersaturated solutions are those in which the thermodynamic activity of drug in solution is greater than unity. Supersaturated systems may be produced using the dependence of solubility on temperature, pH, and solvent composition, and by utilizing changes in these factors and by other methods, to be described.

2. Crystal Growth and Effects of Additives

Solutions that contain concentrations of solute in excess of the saturated solubility will undergo phase change to form a suspension of solid that is in equilibrium with its saturated solution. This process is known as *recrystallization*. Excess solute in solution (*supersaturation*), however, will not necessarily always cause crystallization to occur. In addition to supersaturation, crystallization requires formation of nuclei of critical size, followed by crystal growth around the nuclei. At high levels of supersaturation solutions are labile and nucleation and crystal growth occur spontaneously. At lower levels of supersaturation a metastable zone may be produced as shown in Figure 13.

Additives are materials that interfere with the processes of nucleation and crystal growth and act to stabilize supersaturated solutions. The mechanism of action of these materials is uncertain, but they may work by inhibition of bulk diffusion, inhibition of crystal surface diffusion, or by poisoning crystal growth sites. Inhibition of bulk diffusion by an increase in bulk viscosity, seems unlikely to be important in topical creams, gels, and ointments, but may be significant in the viscous adhesive layer of transdermal patches and in stratum corneum. The crystal growth model of Cabrera and Frank (100) suggests that surface migration of molecules adsorbed onto a crystal is required before incorporation into the crystal growth sites can occur. Put simply, molecules setting down on a flat surface of a crystal have only a single binding surface and are likely to redissolve. At step-and-kink (corner) sites there are two and three surfaces for bonding; thus, redissolution is much less likely, enabling the molecule to be incorporated into the crystal structure. Thus some additives, for example, polymers, bind to the surface of the crystal to inhibit the surface diffusion of alighting molecules to the step-and-kink sites and allow redissolution to occur. Figure 14 shows the effects of polymer additives on the rate of crystal growth from an eightfold supersaturated solution of hydrocortisone acetate. All polymers studied showed some effect, with the 1% hydroxypropyl methylcellulose (HPMC) system

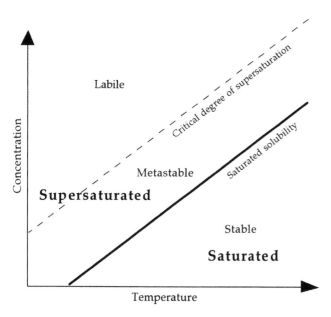

Figure 13   Diagram showing the critical degree of supersaturation and the different stability states of supersaturated systems. (From Ref. 124).

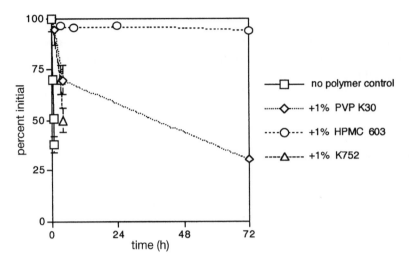

Figure 14 The effects of various polymer additives on the rate of crystal growth from an eightfold supersaturated solution of hydrocortisone acetate. All of the polymers studied showed some effect, with the 1% hydroxypropyl methylcellulose system showing no change over 3 days compared with a 30-min half-life in the no-polymer control. (From Ref. 109.)

showing no change over 3 days compared with the 30-min half-life of the untreated system.

Kondo et al. (101) studied the effect of a wide range of polymers on plasma levels of nifedipine following application of supersaturated solutions to rat skin. All polymer systems generated much higher plasma levels than control systems. For example, polyvinylpyrrolidone systems gave 40-fold increases in plasma levels of nifedipine, and CAP systems demonstrated 75-fold increases. Recently, Ikeda et al. (102) demonstrated superior effects of hydrophobically modified HPMC over HPMC in inhibiting recrystallization of indomethacin in a topical gel. Other authors have also published in this area (103–105). It is clear that for many compounds high levels of supersaturation can be stabilized using polymeric systems. The ability of additives, particularly polymers, to stabilize supersaturated solutions over useful time periods is the key to the practical use of supersaturation in topical drug delivery. Thus, a major area for further research is structure–activity of polymer antinucleant additives.

Other compounds that are similar in structure to the crystal molecule may work by a poisoning mechanism (i.e., partial blocking of step-and-kink sites) to inhibit crystal growth. To our knowledge, however, these have not yet been used in topical drug delivery systems.

B. Supersaturation Solutions to Enhance Penetration

1. Methods to Form Supersaturated Solutions

Supersaturated systems designed to enhance percutaneous penetration have been produced mainly by utilizing changes in the dependence of drug solubility on temperature and solvent composition. Some work has been conducted on supersaturated solutions formed from amorphous materials produced by mechanical processes

(106,107). Earlier reviews that are available on enhancement of percutaneous penetration contain sections on supersaturation (108–112). Only data from solid dispersion systems and cosolvent systems that address key issues on supersaturation will be reviewed here; thus, these methods are described only briefly.

a. Solid (Molecular) Dispersion Systems, Coprecipitates. Solid dispersion systems were first produced by Chiou and Riegelman (113) in an attempt to increase the dissolution rate of poorly water-soluble drugs. Such systems are made either by melting drug and carrier (often a polymer) together or by dissolving them in a common volatile solvent, with subsequent removal of solvent by heat. This process is similar to that currently used to prepare thin-film transdermal patches, and this will be discussed later. Depending on the conditions, dissolution of solid dispersions will often result in generation of supersaturated states. These supersaturated states will often be maintained for a considerable period by the antinucleant and anticrystal growth effects of the polymeric carrier (103,114–117); therefore, it is not surprising that such systems have been investigated for their ability to increase percutaneous absorption (118–121) [see also, Japanese patents: J6 3093-714-A, J6 3093-715-A, J6 3307-818-A, J6 3307-819-A, and J6 3297-320-A].

b. Mixed Cosolvent Systems. Saturation solubility plots often show an exponential increase with cosolvent composition, depending on the relative polarities of the solute and binary cosolvent system (Fig. 15; curve B′B). A basic property of these systems is that by mixing suitable solute–cosolvent solutions, subsaturated, saturated, and supersaturated solutions can be formed. Figure 15 shows schematically that mixing system B (saturated solute in 100% solvent b) with system A (no solute in 100% solvent a) will result in systems C (subsaturated), D (saturated), E and F (both supersaturated), depending on the ratio of A to B. In practice, systems A and B may themselves be mixed, and system B need not necessarily be saturated. The degree of saturation is calculated by dividing the resulting concentration after mixing

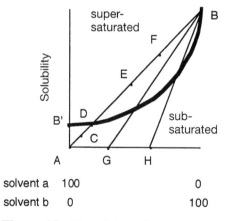

Figure 15   The mixing of system B (saturated solute in 100% solvent b) with system A (no solute in 100% solvent a) will result in systems C (subsaturated), D (saturated), and E and F (both supersaturated), depending on the ratio of A to B used. In practice, systems A and B may themselves be mixed, and system B need not necessarily be saturated. (From Ref. 122.)

with the experimentally determined value of the saturated solubility at that same composition. Thus, supersaturated solutions of known "thermodynamic activity" and drug concentration can be produced readily using a simple-mixing process. This process is similar to the premix process currently used to manufacture topical semi-solids. Because of their versatility to form supersaturated solutions at specific drug concentrations and degrees of saturation, mixed cosolvent systems are being used increasingly as research tools (122–126).

2. Membrane Transport from Supersaturated Solutions

It is clear that high degrees of supersaturation can be stabilized by the use of additives and, theoretically, this should lead to enhancement of membrane transport that is linearly proportional to the degree of saturation. Figure 16 shows a simple physical model (8) that illustrates thermodynamic activity gradients in the skin after topical application of saturated and threefold supersaturated systems. At the outer surface of the stratum corneum activity in the vehicle ($a_{vehicle}$) is equal to activity in the stratum corneum (a_{sc}); thus, following application of a threefold supersaturated solution, the outer layer of the stratum corneum is also threefold supersaturated. A key question, especially if the applied supersaturated state is stabilized with an antinucleant additive, is what happens to the physical state of the drug within the stratum corneum? The two extreme possibilities are shown in Figure 16. In case 1, immediate precipitation of the drug occurs in the outer surface of the stratum corneum, the thermodynamic activity gradient is as shown by the path A–B–C, and there is no significant increase in transport over a saturated system. In case 2, no precipitation occurs. There is a linear drop in thermodynamic activity over the barrier along path A–C, with supersaturation being maintained over two-thirds of the barrier (this depends on the initial degree of supersaturation), and the increase in membrane transport is linearly proportional to the degree of saturation. In the next section membrane transport from saturated and supersaturated systems will be reviewed.

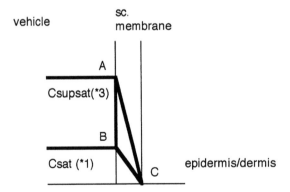

Figure 16   Diagram of a simple physical model (after Higuchi) that illustrates the theoretical thermodynamic activity gradients in the skin after topical application of supersaturated systems. If immediate precipitation of the drug occurs in the outer layer of the stratum corneum, the thermodynamic activity gradient is as shown by the path A–B–C. If no precipitation occurs, there is a linear drop in the thermodynamic activity over the barrier along path A–C.

a. Cellulose Porous Membranes. The transport of hydrocortisone across porous cellulose membranes from supersaturated solutions formed from polyvinyl pyrolidane (PVP) coprecipitates has been studied by several groups (118–120). In these systems PVP worked as both a carrier, to form the coprecipitate dispersion system, and as an additive, to prevent crystal growth. Flux was linearly proportional to the degree of supersaturation up to approximately 12-fold supersaturation (118). A similar experiment (119) demonstrated that at early time points, up to approximately 6 h, flux was broadly proportional to the degree of saturation (14-fold supersaturation generated a 10-times increase in flux). At later times, and at higher degrees of supersaturation, crystallization was seen in the donor phase, and flux no longer increased linearly with degree of supersaturation. Figure 17 (120) shows that for up to sixfold supersaturation of hydrocortisone, the experimental data are a good fit with the theoretical line drawn from the origin through point Cs–Js, the flux from a saturated solution. At higher degrees of supersaturation, rapid crystallization was seen in the donor phase. Merkle (120) also studied transport of benzodiazepines across porous cellulose membranes from supersaturated solutions formed from PVP coprecipitates. Figure 18 demonstrates that for nitrazepam, and up to 10-fold supersaturation, the experimental data are a good fit with the theoretical point Cs–Js. Similar results were reported for clonazepam (14-fold increased activity–flux) and flunitrazepam (5-fold increased activity–flux). Higuchi and Farrar (127) reported flux of digoxin from supersaturated solutions formed from PVP coprecipitates of up to 20 times that from saturated solutions, provided crystallization was prevented by inclusion of a further additive.

In summary, for a variety of compounds, large increases in flux across cellulose porous membranes can be observed with increasing degrees of supersaturation, provided crystallization within the donor vehicle can be prevented by use of additives.

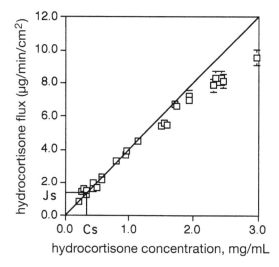

Figure 17 Transport of hydrocortisone–alcohol across cellulose membrane: for up to a sixfold supersaturation, the experimental data are a good fit with the theoretical line drawn from the origin through point Cs–Js, the flux from a saturated solution. At higher degrees of supersaturation, rapid crystallization was seen in the donor. (From Ref. 120.)

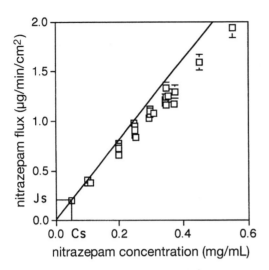

Figure 18 Transport of nitrazepam across cellulose membrane: for up to a tenfold super-saturation, the experimental data are a good fit with the theoretical line drawn from the origin through point Cs–Js, the flux from a saturated solution. (From Ref. 120.)

However, the mechanism of transport through porous membranes is not fully representative of membranes in which transport is dependent on partitioning into and diffusion through the membrane and in which efficient exclusion of additives from the membrane may occur.

b. Silicone Membranes. Silastic (polydimethylsiloxane; silicon rubber) has been widely used as a model membrane for the comparison of relative penetration rates in transport studies (128,129), and has been validated to show response to changes in thermodynamic activity (122,128). Davis and Hadgraft (122) used mixed cosolvent systems to form supersaturated solutions of hydrocortisone acetate, stabilized with 1% hydroxypropylmethyl cellulose as shown in Figure 14. Figure 19 (122) shows that, for systems up to eightfold supersaturated, flux across Silastic is linearly proportional to the degree of saturation. Pellett et al. (124) studied the transport of piroxicam across Silastic from supersaturated solutions in propylene glycol–water, stabilized with hydroxypropylmethyl cellulose. Owing to precipitation problems in the donor phase, an upper limit of four times supersaturation was evaluated and donor solutions were renewed every 12 h. Figure 20 (124) shows that, for systems up to fourfold supersaturated, flux across silastic is linearly proportional to the degree of saturation. Megrab et al. (126) studied the transport of estradiol across Silastic from supersaturated solutions in propylene glycol–water, stabilized with providone (PVP-2). Flux across Silastic from saturated solutions of estradiol increased with increasing propylene glycol content owing to the effect of propylene glycol on increasing estradiol-saturated solubility within the membrane. For this reason, flux from supersaturated systems was normalized by flux from the corresponding saturated vehicle. Figure 21 (126) shows that only at low degrees of supersaturation (up to fourfold) is normalized flux linearly proportional to the degree of saturation. Throughout these studies no apparent instability was seen in the donor phase, even at higher degrees of supersaturation. Thus, the possibility exists for precipitation

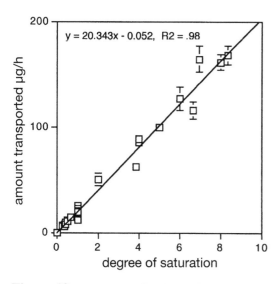

Figure 19 Transport of hydrocortisone acetate across a Silastic membrane: for up to an eightfold supersaturation, flux is linearly proportional to the degree of saturation. (From Ref. 122.)

within the Silastic membrane. Similar results have been obtained with supersaturated solutions of diclofenac (130).

    c. Human Skin In Vitro. In two studies (124,126), comparisons of flux across silastic and human skin were made. Pellett et al. (124) found, as with Silastic, that flux of piroxicam across full-thickness human skin in vitro was linearly proportional to the degree of supersaturation. Figure 22 (126) shows that for up to approximately 12-fold saturation systems, flux of estradiol across epidermal membranes was linearly proportional to the degree of supersaturation. From these data it appears that supersaturated states may be more stable within stratum corneum than in a Silastic membrane.

    d. Synergy Between Supersaturation and Chemical Enhancers. There is anecdotal evidence for synergy between supersaturation and chemical penetration enhancement. For example, the patent literature describes several results that suggest enhanced percutaneous penetration in systems containing a drug in solution, a volatile solvent, a residual component (often a potential lipophilic enhancer), and an antinucleant additive. For example, Kondo (101) described supersaturated formulations of nifedipine based on acetone (volatile) and propylene glycol plus isopropyl myristate (nonvolatile) solvents. The degree of saturation in the residual phase after evaporation was calculated at 12-fold, yet up to 75-fold increases in flux were observed, presumably owing to the known coenhancer effects of propylene glyco–isopropyl myristate mixtures.

    Earlier, it was suggested that the increase in the drug-saturated solubility in the stratum corneum and supersaturated drug concentration in the stratum corneum were independent variables. Megrab et al. (126) measured uptake (solubility) of estradiol in stratum corneum from saturated and supersaturated solutions in various propylene

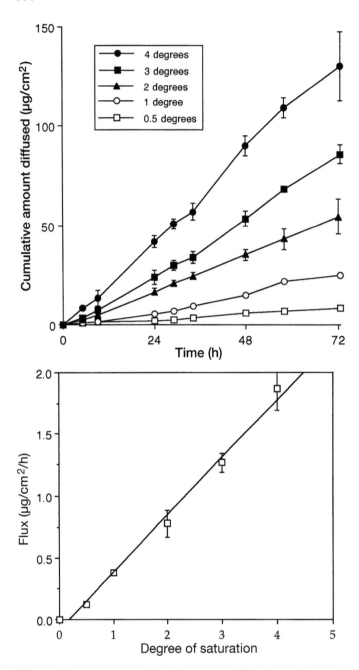

Figure 20 Transport of piroxicam across Silastic membrane: (a) diffusion profiles; (b) flux as a function of degree of supersaturation. For up to fourfold supersaturation the flux across Silastic is linearly proportional to the degree of saturation. (From Ref. 124.)

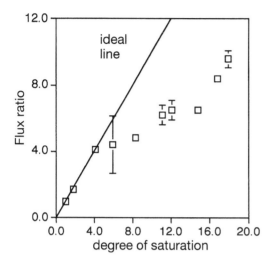

Figure 21 Transport of estradiol across a Silastic membrane: Only at low degrees of supersaturation, up to four, is the flux ratio (flux normalized by flux from the saturated vehicle) linearly proportional to the degree of saturation. Throughout these studies no apparent instability was seen in the donor phase, even at higher degrees of supersaturation, and it is likely that crystallization occurred in the Silastic membrane. (From Ref. 126.)

glycol–water vehicles. As propylene glycol content in saturated solutions increased, so did uptake of estradiol. As degree of supersaturation increased, so also did uptake ratio (normalized for increase in uptake from the corresponding saturated vehicle). Figure 23 shows clearly the synergy between enhancers working to increase drug solubility in the stratum corneum and supersaturation. Pellett et al. (131) studied the

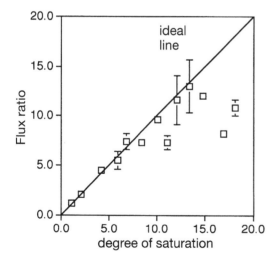

Figure 22 Transport of estradiol across human epidermal membrane. Up to an approximately 12-fold supersaturation, flux across epidermal membranes is linearly proportional to the degree of saturation. (From Ref. 126.)

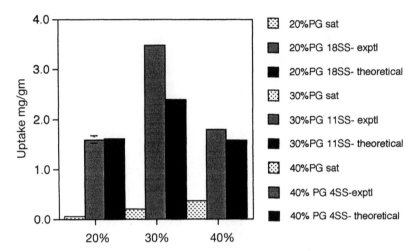

Figure 23 Uptake (solubility) of estradiol into human epidermal membrane from three vehicles (20, 30, and 40% propylene glycol; PG) at saturation and at 18-, 11-, and 4-fold supersaturation, respectively. The theoretical uptake values for the supersaturated systems were obtained by multiplying the uptake from the saturated system by the theoretical degree of supersaturation. The figure shows the potential for synergy between enhancers working to saturated solubility in the stratum corneum and supersaturation. (From Ref. 126.)

synergistic interaction between oleic acid, which is thought to enhance skin permeation by increasing drug diffusivity in the stratum corneum, and supersaturation in human skin in vitro. Penetration from saturated, oleic acid-pretreated skin, sixfold supersaturated, and oleic acid-pretreated skin with sixfold supersaturated flurbiprofen systems, is shown in Figure 24. The results demonstrate the potential for synergy between enhancers working to increase drug diffusivity in the stratum corneum and supersaturation.

e. Conclusions. Provided supersaturated solutions can be stabilized in the donor phase, there is considerable evidence that, as predicted (8), they have the potential to deliver a proportional increase in membrane transport. Given the mechanisms of increased transport (i.e., supersaturation of the membrane), it seems reasonable to suggest that physical stability within the membrane is also important. There is some reassuring evidence from the systematic studies reviewed here, that gross crystallization does not occur within the stratum corneum, and that the stratum corneum, in fact, may be a more superior membrane in this respect than Silastic membrane. The viscosity of the stratum corneum lipids or natural antinucleant agents may play a role. Recently, Pellett et al. (125) have demonstrated that application of supersaturated solutions of piroxicam in vitro to human skin gave rise to linearly proportionally higher drug levels in tape-stripped stratum corneum and viable epidermal–dermal compartments. This is an encouraging result that supports the potential usefulness of supersaturated systems for the improvement of topical therapy. In vivo studies in humans, using techniques such as microdialysis, are now required to confirm these in vitro results. The studies of Megrab and Pellett et al. (126,131), describing synergy between supersaturation and chemical enhancers, generated con-

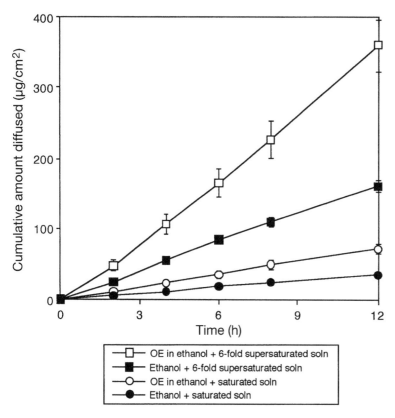

Figure 24 Penetration across human stratum corneum from saturated oleic acid (OE)-enhanced (pretreated skin), sixfold supersaturated, and OE with sixfold supersaturated flurbiprofen systems: the potential for synergy between enhancers working to increase drug diffusivity in the stratum corneum and supersaturation is shown. (From Ref. 131.)

fidence that future developments in this area will provide greatly increased skin penetration and permeation and improve dermal and transdermal therapy.

C. Inadvertent Supersaturation During Manufacture

Potts and Cleary (132) described how the film-coating technology used in the production of nonreservoir transdermal patches could lead to supersaturation of drug within the drug-containing adhesive layer. In the manufacture of these patches the drug and adhesive polymer are dissolved in a solvent that is subsequently evaporated by heat (the process used to produce a solid dispersion system). As has been reviewed here, this may have dramatic and beneficial effects on drug transport (and therapy), at least initially. However, drug release and transport from transdermal patches manufactured by solvent evaporation film-coating techniques may decrease with time if the supersaturated drug recrystallizes (132–134). There is little experimental information in this area. Kubota (135) compared skin permeation of betamethasone benzoate from freshly made silicone adhesive patches, a suspension, and a saturated solution. Permeation from the patch was approximately 2.5 times greater than either cream or saturated control, and the authors suggest supersaturation as one possible mechanism.

Heat-premix processes are commonly used in the manufacture of gels, creams, and ointments. Cooling and mixing a premix of drug in solution with miscible non-solvents (e.g., water) can be expected to result in formation of supersaturated systems, at least initially. Henmi (136) described a heat process that gave rise to a supersaturated indomethacin gel, which subsequently showed reduced release and skin permeation with time.

Potentially large increases in skin penetration and possibly therapeutic effect can occur from inadvertent supersaturation during manufacture. Of special concern is that during the clinical development program, during which freshly made formulations are used, high skin penetration (and efficacy) data may be generated, and this may not be matched by the stored commercial product. This puts a great responsibility on the use of stability and bioequivalence techniques to ensure quality and efficacy on storage. It is important to consider this potential problem in dermatological and transdermal development programs.

Finally, it is also feasible that supersaturated conditions may inadvertently occur during actual therapeutic use of transdermal patch systems. Many transdermal systems are occlusive and will accumulate water as a result of physiological transepidermal water loss. This may produce high activity states for the drug and thereby enhance penetration into and permeation across the stratum corneum.

IV. VESICLES

A. Liposomes

Mezei first suggested that liposomes could be useful drug-carrier systems for the local treatment of skin diseases (137,138). The suggestion was based on drug disposition data obtained following topical application of the steroid triaminolone acetonide incorporated in phospholipid liposomes formulated as lotions or gels. Encapsulation of triaminolone acetonide into liposomes resulted in a vehicle-dependent 4.5- to 4.9-fold increase in the amount of drug recovered from the epidermis. Similarly, topical application of liposomal preparations of hydrocortisone resulted in higher concentrations of the steroid in skin strata, compared with application of the drug in an ointment formulation (139,140). Continuing work by Mezei's group generated further data for other dermatological drugs, such as econazole, minoxidil, and retinoic acid (141–143). Overall, the work of Mezei suggested that application of the dermatological drugs in liposomal form, when compared with the conventional formulations, led to increased drug concentration in the skin and subcutaneous tissues and decreased biodisposition in plasma and remote sites (Table 4).

These encouraging early observations were followed by several confirmatory research and clinical investigations, most notably those of Weiner's group at the University of Michigan (144–148) and Korting's group at Ludwig–Maximilians University in Munich (149–151). Many other studies have indicated the potential of phospholipid liposomes to increase the skin content of topically applied drugs. Touitou et al. (152,153) evaluated the penetration of xanthines (dyphylline and caffeine) in liposomal and nonliposomal formulations and found that delivery to the skin was markedly influenced by the formulation in which the liposomes were applied. Thus,

Table 4 Effect of Liposomal Encapsulation on Drug Disposition Following Topical Application

Site	Triamcinolone acetonide	Econazole	Minoxidil	Retinoic acid
Epidermis	443[a]	872	465	261
Dermis	482	265	451	283
Subcutaneous tissue	72	206	1036	44
Plasma	47	50	55	56
Liver	93	63	76	124
Heart	89	97	85	91

[a]Drug concentration as a percentage of control (the control value was obtained from non-liposomal formulations).
Source: Refs. 137–141.

liposomes containing dyphylline were formulated into a polyethylene glycol ointment, a carbomer gel, and a polyethylene glycol base containing oleic acid and diglycol. The largest skin permeation was found for the latter formulation, and the authors suggested that the slower permeation from the polyethylene glycol ointment may be due to higher skin partitioning and retention of the drug. Enhanced skin retention of caffeine was also observed following application in liposomal preparations (153). Phospholipid liposome encapsulation also enhances skin retention of the topically applied local anesthetics lidocaine and tetracaine (154,155), cyclosporine (156), interferon (157,158), and T_4 endonuclease V (159). Reports on the latter two compounds are particularly interesting, for they indicate the potential usefulness of liposomes in localization of biologically active proteins and peptides within the skin.

Liposomes have also been prepared from lipid mixtures similar in composition to the stratum corneum intercellular lipid (146,160–162). The "skin" lipids (fully described in Chap. 1) usually comprise ceramides, derived from bovine brain; cholesterol, cholesterol sulfate, and a fatty acid. When using the rationale that liposomes formed from skin lipids would interact more favorably with the stratum corneum intercellular lipid regions than phospholipid-based liposomes, Weiner's group (146) tested this hypothesis in the HSV-1 guinea pig model. They demonstrated that skin lipid liposomes containing interferon were more effective at reducing lesion score than were interferon-loaded phospholipid-based liposomes, and they generated increased levels of the drug in the deeper skin layers (157). Fresta and Puglisi (161) evaluated the effect of incorporation of hydrocortisone, triamcinolone, and betamethasone into phospholipid and skin lipid liposomes on in vitro human skin distribution of the steroid, and then compared this with conventional ointment formulation. The data clearly showed that, although the formulation appeared to have no effect on the stratum corneum content of the three steroids, liposomal formulation increased the epidermal and dermal concentrations. Furthermore, skin lipid liposomes increased the concentration of drug in deeper skin strata over that resulting from phospholipid liposome encapsulation. The authors confirmed the in vitro data by demonstrating in vivo a greater blanching effect (vasoconstriction) following application of the steroid in skin lipid liposomes. Finally, the authors investigated the systemic distribution of the steroids following topical application on guinea pigs.

Quite clearly (Table 5) liposomal incorporation reduced systemic disposition when compared with the ointment formulation, although there appeared to be little or no difference between the liposomal formulations.

B. Niosomes

Niosomes are another vesicle system that has been investigated for potential modification of skin permeation (163,164). Niosomes comprise nonionic surfactants, such as polyoxyethylene alkyl ethers, and may be prepared as single or multilamellar vesicles. Surfactants of this type enhance skin permeation (165), and this is likely to play a role in any modification of permeation using these vehicles. The effect of nonionic surfactant vesicles on the skin permeation of estradiol was dependent on the physical state of the niosome (164). Thus, whereas niosomes prepared from polyoxyethylene(3)stearyl ether and existing in the gel state did not increase estradiol permeation, those prepared from polyoxyethylene(3)lauryl ether and polyoxyethylene(10)oleyl ether—both existing as liquid crystalline vesicles—significantly enhanced transport. Further experiments, in which the skin was pretreated with unloaded niosomes indicated that the enhanced transport of estradiol from drug-loaded vesicles was not wholly a result of surfactant-induced penetration enhancement. The authors postulated that niosomes fused at the surface of the stratum corneum, generating high local concentrations of estradiol that resulted in increased thermodynamic activity of the permeant in the upper layers of the stratum corneum.

C. Mode of Action

The precise mode of interaction between lipid vesicles and skin remains unclear. Certainly, despite earlier claims (159), there is considerable doubt about the ability of whole vesicles to permeate intact stratum corneum. Most evidence suggests that vesicles can penetrate the outer cell layers of the stratum corneum (166–168) where desmosomal linkages have become disrupted and, presumably, the keratinocytes are less tightly bound and surrounded by a mixture of intercellular lipid and sebum. However, continuing diffusion of vesicles through the approximately 60-nm intercellular space of the deeper layers of the stratum corneum seems unlikely. Current thinking suggests that lipid vesicles fuse with endogenous lipid, either on the surface or in the outermost layers of the stratum corneum (168,169). The fusion is followed

Table 5 Effect of Liposomal Encapsulation on Triamcinolone Disposition Following Topical Application

Site	Ointment	Phospholipid liposome	Skin lipid liposome
Skin	1.36[a]	1.75	1.87
Subcutaneous tissue	3.73	2.22	2.05
Blood	0.04	0.02	0.02
Liver	0.19	0.13	0.09
Heart	0.09	0.09	0.09

[a]Drug concentration: mg/g tissue for skin, μg/g tissue for all other sites.
Source: Ref. 161.

by structural changes in the deeper layers of the stratum corneum, as evidenced by freeze–fracture electron microscopy and small angle X-ray–scattering techniques. These structural changes are presumed to be the result of intercellular diffusion of vesicle lipid components (not intact vesicles) to the deeper layers, interaction with and disruption of endogenous lipid lamellae. It is simple to postulate that this interaction or disruption of lipid lamellae will lead to an increase in skin permeation rates, but this does not explain the observed increase in skin retention of permeants, as described earlier. Apparent increased skin retention may be an artifact from exogenous lipid depot formation on the skin surface. On the other hand, formations of lipid aggregates, possibly comprising mixtures of endogenous and exogenous lipid (170,171), which have been observed in deeper layers of the stratum corneum (164), may provide a reservoir for topically applied drugs.

D. Transfersomes

Despite the evidence to the contrary, Cevc (172) suggested that it was possible for whole vesicles to cross intact stratum corneum. The basic premise for this hypothesis was the driving force provided by the osmotic gradient between the outer and inner layers of the stratum corneum and the development of specific mixes of lipids to form modified liposomes, termed transfersomes. The requirement for the osmotic gradient to be maintained suggests that transfersomes will not function in occlusive conditions, and careful formulation is necessary. Because of their unique structure (described as a mix of phosphatidylcholine, sodium cholate, and ethanol) (173), transfersomes are reputed to be very flexible vesicles and capable of transporting their contents through the tortuous intercellular route of the stratum corneum. Thus, dermal application of local anesthetics in transfersomes was "nearly as swift in action as an injection" (173), and the application of the corticosteroids, triamcinolone acetonide, dexamethasone, and hydrocortisone, encapsulated in transfersomes, resulted in more reliable site-specificity for the drug (174).

E. Follicular Delivery

Lieb et al. (175,176) proposed that liposomes may be useful for targeting drugs to skin follicles for the treatment of diseases, such as acne and alopecia. Their initial experiments, using the hamster ear pilosebaceous unit, demonstrated that carboxyfluorescein, incorporated into phospholipid liposomes, was more efficiently targeted to follicles than when formulated as a simple aqueous solution, a propylene glycol (5%) solution, or a sodium dodecyl sulfate (0.05%) solution (175). However, most of the carboxyfluorescein was located in the epidermis. In later experiments, application of cimetidine, incorporated in phospholipid and nonionic liposomes, was compared with its application in a 50% alcohol solution (176), and generated data that was similarly equivocal. In this case, although small amounts of drug were located within the pilosebaceous unit, most was located on the surface or within the stratum corneum (determined by tape-stripping). Nonetheless, the data showed that the liposomal formulations were considerably more effective at delivering the drug to the stratum corneum and the follicles than was the alcoholic solution. Interestingly, the phospholipid liposomes delivered approximately twice as much drug to these compartments as the nonionic liposomes.

More recently, this group has demonstrated that nonionic liposomes were capable of delivering the macromolecules cyclosporine and interferon-alfa into pilosebaceous units (177). Three nonionic and one phospholipid liposome were evaluated. The most effective vesicle comprised glyceryl dilaurate (57% by weight), cholesterol (12%), and polyoxyethylene(10)stearyl ether (28%), which facilitated the follicular deposition of both hydrophilic and hydrophobic drugs. The authors speculated that the low-melting point of glyceryl dilaurate (30°C) may have resulted in fluidization of the liposome bilayers following application of the formulation, and this led to partial release of polyoxyethylene(10)stearyl ether (a known skin penetration enhancer) (165), which further led to an enhancement of drug deposition. Given these studies, the authors investigated the liposomal delivery of plasmid DNA to the skin (178). Their results suggested that nonionic–cationic liposomes (comprising glyceryl dilaurate, cholesterol polyoxyethylene(10)stearyl ether, and 1,2-dioleoyloxy-3-(trimethylammonio)-propane) could deliver expression plasmid DNA to perifollicular cells and mediate transient transfection in vivo. These data show promise for future topical gene therapy for a number of dermatological diseases caused by abnormal regulation of soluble cytokines.

Other recent data on liposome-mediated drug delivery to follicular regions has been contradictory. Whereas, Bernard et al. (179) demonstrated that phospholipid liposomes were useful for targeting delivery of an antiandrogen to the sebaceous gland, Tshan et al. (180) found no enhancement of follicular deposition of isotretinoin when applied in liposomal formulation. Overall, the number and variety of constituents from which lipid vesicles have been constructed, the large number of permeants and formulations studied, and the diversity of experimental methods render it difficult to establish any systematic ground rules. It is clear, however, that although there is considerable evidence for improved drug delivery characteristics, much remains to be achieved in this field.

V. CONCLUDING REMARKS

In this chapter several strategies for enhancing skin penetration have been reviewed. Quite clearly, there are many occasions when the ability to increase the rate and extent of dermal and transdermal drug delivery would be beneficial to therapy. This is reflected in the considerable research effort in this area. In recent years, much of this effort has focused on the development of chemical penetration enhancers, the use of supersaturated systems, the use of vesicular systems, and physical methods, such as iontophoresis and ultrasound. Limitations of space have precluded a discussion of physical methods of enhancement. However, further and more extensive information on physical and chemical enhancement of skin penetration can be found (9–11,181–183).

REFERENCES

1. Poulsen BJ, Young E, Coquilla V, Katz M. Effect of vehicle composition on the in vitro release of fluocinolone acetonide and its acetate ester. J Pharm Sci 57:928–993, 1972.
2. Ostrenga J, Steinmetz C, Poulsen BJ. Significance of vehicle composition I: relationship between topical vehicle composition, skin permeability and clinical efficacy. J Pharm Sci 60:1175–1179, 1971.

3. Lippold BC, Schneemann H. The influence of vehicles on the local bioavailability of betamethasone-17-benzoate from solution and suspension-type ointments. Int J Pharm 22:31–43, 1984.

4. Guy RH. Current status and future prospects of transdermal drug delivery. Pharm Res 13:1765–1769, 1996.

5. Vaughan CD. Using solubility parameters in cosmetics formulations. J Soc Cosmet Chem 36:319–333, 1985.

6. Sloan KB, Koch SAM, Siver KG, Flowers FP. Use of solubility parameters of drug and vehicle to predict flux through skin. J Invest Dermatol 87:244–252, 1986.

7. Cooper ER, Merritt BW, Smith RL. Effect of fatty acids and alcohols on the penetration of acyclovir across human skin in vitro. J Pharm Sci 74:688–689, 1985.

8. Higuchi T. Physical chemical analysis of percutaneous absorption process from creams and ointments. J Soc Cosmet Chem 11:85–97, 1960.

9. Walters KA, Hadgraft J, eds. Pharmaceutical Skin Penetration Enhancement. New York: Marcel Dekker, 1993.

10. Smith EW, Maibach HI, eds. Percutaneous Penetration Enhancers. Boca Raton: CRC Press, 1995.

11. Prausnitz MR. The effects of electric current applied to skin: a review for transdermal drug delivery. Adv Drug Deliv Rev 18:395–425, 1996.

12. Beastall JC, Hadgraft J, Washington C. Mechanism of action of Azone as a percutaneous penetration enhancer—lipid bilayer fluidity and transition–temperature effects. Int J Pharm 43:207–213, 1988.

13. Rolland A, Brzokewicz A, Shroot B, Jamoulle JC. Effect of penetration enhancers on the phase-transition of multilamellar liposomes of dipalmitolyphosphatidylcholine—a study by differential scanning calorimetry. Int J Pharm 76:217–224, 1991.

14. Lewis D, Hadgraft J. Mixed monolayers of dipalmitoylphosphatidylcholine with Azone or oleic-acid at the air–water interface. Int J Pharm 65:211–218, 1990.

15. Hoogstraate AJ, Verhoef J, Brussee J, IJzerman AP, Spies F, Boddé HE. Kinetics, ultrastructural aspects and molecular modeling of transdermal peptide flux enhancement by *N*-alkylazacycloheptanones. Int J Pharm 76:37–47, 1991.

16. Schuckler F, Lee G. The influence of Azone on monomolecular films of some stratum corneum lipids. Int J Pharm 70:173–186, 1991.

17. Schuckler F, Bouwstra JA, Gooris GS, Lee G. An X-ray–diffraction study of some model stratum corneum lipids containing Azone and dodecyl-L-pyroglutamate. J Controlled Release 23:27–36, 1993.

18. Engblom J, Engstrom S, Fontell K. The effect of the skin penetration enhancer Azone on fatty acid–sodium soap–water mixtures. J Controlled Release 33:299–305, 1995.

19. Engblom J, Engstrom S. Azone and the formation of reversed monocontinuous and bicontinuous lipid–water phases. Int J Pharm 98:173–179, 1993.

20. Hadgraft J, Peck J, Williams DG, Pugh WJ, Allan G. Mechanisms of action of skin penetration enhancers and retarders—Azone and analogs. Int J Pharm 141:17–25, 1996.

21. Hadgraft J, Walters KA, Wotton PK. Facilitated transport of sodium-salicylate across an artificial lipid-membrane by Azone. J Pharm Pharmacol 37:725–727, 1985.

22. Irwin WJ, Smith JC. Extraction coefficients and facilitated transport—the effect of absorption enhancers. Int J Pharm 76:151–159, 1991.

23. Fuhrman LC, Michniak BB, Behl CR, Malick AW. Effect of novel penetration enhancers on the transdermal delivery of hydrocortisone: an in vitro species comparison. J Controlled Release 45:199–206, 1997.

24. Bhatt PP, Rytting JH, Topp EM. Influence of Azone and lauryl alcohol on the transport of acetaminophen and ibuprofen through shed snake skin. Int J Pharm 72:219–226, 1991.

25. Hirvonen J, Rytting JH, Paronen P, Urtti A. Dodecyl *N,N*-dimethylamino acetate and Azone enhance drug penetration across human, snake, and rabbit skin. Pharm Res 8: 933–937, 1991.

26. Itoh T, Wasinger L, Turunen TM, Rytting JH. Effects of transdermal penetration enhancers on the permeability of shed snakeskin. Pharm Res 9:1168–1172, 1992.

27. Niazy EM. Differences in penetration-enhancing effect of Azone through excised rabbit, rat, hairless mouse, guinea-pig and human skins. Int J Pharm 130:225–230, 1996.

28. Okamoto H, Hashida M, Sezaki H. Effect of 1-alkyl- or 1-alkenylazacycloalkanone derivatives on the penetration of drugs with different lipophilicites through guinea pig skin. J Pharm Sci 80:39–45, 1991.

29. Fincher TK, Yoo SD, Player MR, Sowell JW, Michniak BB. In-vitro evaluation of a series of *N*-dodecanoyl-L-amino acid methyl-esters as dermal penetration enhancers. J Pharm Sci 85:920–923, 1996.

30. Godwin DA, Michniak BB, Player MR, Sowell JW. Transdermal and dermal enhancing activity of pyrrolidinones in hairless mouse skin. Int J Pharm 155:241–250, 1997.

31. Michniak BB, Chapman JM, Seyda KL. Facilitated transport of 2 model steroids by esters and amides of clofibric acid. J Pharm Sci 82:214–219, 1993.

32. Michniak BB, Player MR, Chapman JM, Sowell JW. In vitro evaluation of a series of Azone analogs as dermal penetration enhancers 1. Int J Pharm 91:85–93, 1993.

33. Michniak BB, Player MR, Fuhrman LC, Christensen CA, Chapman JM, Sowell JW. In-vitro evaluation of a series of Azone analogs as dermal penetration enhancers 2. (Thio)amides. Int J Pharm 94:203–210, 1993.

34. Michniak BB, Player MR, Chapman JM, Sowell JW. Azone analogs as penetration enhancers—effect of different vehicles on hydrocortisone acetate skin permeation and retention. J Controlled Release 32:147–154, 1994.

35. Michniak BB, Player MR, Fuhrman LC, Christensen CA, Chapman JM, Sowell JW. In-vitro evaluation of a series of Azone analogs as dermal penetration enhancers 3. Acyclic amides. Int J Pharm 110:231–239, 1994.

36. Michniak BB, Player MR, Godwin DA, Phillips CA, Sowell JW. In-vitro evaluation of a series of Azone analogs as dermal penetration enhancers 4. Amines. Int J Pharm 116:201–209, 1995.

37. Phillips CA, Michniak BB. Topical application of Azone analogs to hairless mouse skin—a histopathological study. Int J Pharm 125:63–71, 1995.

38. Phillips CA, Michniak BB. Transdermal delivery of drugs with differing lipophilicities using Azone analogs as dermal penetration enhancers. J Pharm Sci 84:1427–1433, 1995.

39. Michniak BB, Player MR, Sowell JW. Synthesis and in-vitro transdermal penetration enhancing activity of lactam *N*-acetic acid-esters. J Pharm Sci 85:150–154, 1996.

40. Bezema FR, Marttin E, Roemele PEH, Brussee J, Bodde HE, Degroot HJM. H-2 NMR evidence for dynamic disorder in human skin induced by the penetration enhancer Azone. Spectrochim Acta [Pt A] Mol Biomol Spectr 52:785–791, 1996.

41. Bouwstra JA, Peschier LJC, Brussee J, Bodde HE. Effect of *N*-alkyl-azocycloheptan-2-ones including Azone on the thermal-behavior of human stratum corneum. Int J Pharm 52:47–54, 1989.

42. Bouwstra JA, Devries MA, Gooris GS, Bras W, Brussee J, Ponec M. Thermodynamic and structural aspects of the skin barrier. J Controlled Release 15:209–220, 1991.

43. Bouwstra JA, Gooris GS, Brussee J, Salomonsdevries MA, Bras W. The influence of alkyl-azones on the ordering of the lamellae in human stratum corneum. Int J Pharm 79:141–148, 1992.

44. Quan DY, Maibach HI. An electron-spin-resonance study. I. Effect of Azone on 5-doxyl stearic acid-labeled human stratum corneum. Int J Pharm 104:61–72, 1994.

45. Quan DY, Cooke RA, Maibach HI. An electron-spin-resonance study of human epidermal lipids using 5-doxyl stearic acid. J Controlled Release 36:235–241, 1995.
46. Gay CL, Murphy TM, Hadgraft J, Kellaway IW, Evans JC, Rowlands CC. An electron-spin resonance study of skin penetration enhancers. Int J Pharm 49:39–45, 1989.
47. Groundwater PW, Hadgraft J, Harrison JE, Watkinson AC. The synthesis of perdeuterated Azone (D(35)-1-dodecylhexahydro-2*h*-azepin-2-one). J Label Comp Radiopharm 34:1047–1053, 1994.
48. Harrison JE, Groundwater PW, Brain KR, Hadgraft J. Azone induced fluidity in human stratum corneum—a Fourier–transform infrared spectroscopy investigation using the perdeuterated analog. J Controlled Release 41:283–290, 1996.
49. Ongpipattanakul B, Burnette RR, Potts RO, Francoeur ML. Evidence that oleic acid exists in a separate phase within stratum corneum lipids. Pharm Res 8:350–354, 1991.
50. Bonina FP, Montenegro L. Penetration enhancer effects on in vitro percutaneous absorption of heparin sodium salt. Int J Pharm 82:171–177, 1992.
51. Harrison JE, Watkinson AC, Green DM, Hadgraft J, Brain KR. The relative effect of Azone and Transcutol on permeant diffusivity and solubility in human stratum corneum. Pharm Res 13:542–546, 1996.
52. Diez–Sales O, Watkinson AC, Herraez–Dominguez M, Javaloyes C, Hadgraft J. A mechanistic investigation of the in vitro human skin permeation enhancing effect of Azone. Int J Pharm 129:33–40, 1996.
53. Brain KR, Hadgraft J, Lewis D, Allan G. The influence of Azone on the percutaneous absorption of methotrexate. Int J Pharm 71:R9–R11, 1991.
54. Patil S, Szolarplatzer C, Maibach HI. Effect of topical laurocapram (Azone) on the in vitro percutaneous permeation of sodium lauryl sulfate using human skin. J Invest Dermatol 106:947, 1996.
55. Szolarplatzer C, Patil S, Maibach HI. Effect of topical laurocapram (Azone) on the in vitro percutaneous permeation of sodium lauryl sulfate using human skin. Acta Derm Venereol 76:182–185, 1996.
56. Hadgraft J, Williams DG, Allan G. Azone mechanisms of action and clinical effect. In: Walters KA, Hadgraft J, eds. Pharmaceutical Skin Penetration Enhancement. New York: Marcel Dekker, pp 175–197, 1993.
57. Chatterjee DJ, Li WY, Koda RT. Effect of vehicles and penetration enhancers on the in vitro and in vivo percutaneous absorption of methotrexate and edatrexate through hairless mouse skin. Pharm Res 14:1058–1065, 1997.
58. Gonsho A, Imanidis G, Vogt P, Kern ER, Tsuge H, Su MH, Choi SH, Higuchi WI. Controlled (trans)dermal delivery of an antiviral agent (acyclovir) 1. An in vivo animal model for efficacy evaluation in cutaneous HSV-1 infections. Int J Pharm 65:183–194, 1990.
59. Higo N, Hinz RS, Lau DTW, Benet LZ, Guy RH. Cutaneous metabolism of nitroglycerin in vitro 2. Effects of skin condition and penetration enhancement. Pharm Res 9:303–306, 1992.
60. Hwang GC, Lin AY, Chen W, Sharpe RJ. Development and optimization of a methotrexate topical formulation. Drug Dev Ind Pharm 21:1941–1952, 1995.
61. Ogiso T, Iwaki M, Tsuji S. Percutaneous absorption of bromhexine in rats. Chem Pharm Bull 39:1609–1611, 1991.
62. Swart PJ, Weide WL, Dezeeuw RA. In vitro penetration of the dopamine-D_2 agonist N-0923 with and without Azone. Int J Pharm 87:67–72, 1992.
63. Ogiso T, Iwaki M, Tanino T, Oue H. Percutaneous absorption of pindolol and pharmacokinetic analysis of the plasma concentration. J Pharmacobiodyn 15:347–352, 1992.
64. Ruan LP, Liang BW, Tao JZ, Yin CH. Transdermal absorption of nitrendipine from adhesive patches. J Controlled Release 20:231–236, 1992.

65. Wiechers JW, Drenth BFH, Jonkman JHG, Dezeeuw RA. Percutaneous absorption of triamcinolone acetonide from creams with and without Azone in humans In vivo. Int J Pharm 66:53–62, 1990.

66. Wiechers JW. Absorption, distribution, metabolism, and excretion of the cutaneous penetration enhancer Azone. Pharm Weekblad Sci Ed 12:116–118, 1990.

67. Wiechers JW, Drenth BFH, Adolfsen FAW, Prins L, Dezeeuw RA. Disposition and metabolic profiling of the penetration enhancer Azone 1. In vivo studies—urinary profiles of hamster, rat, monkey, and man. Pharm Res 7:496–499, 1990.

68. Wiechers JW, Drenth BFH, Jonkman JHG, Dezeeuw RA. Percutaneous absorption and elimination of the penetration enhancer Azone in humans. Pharm Res 4:519–523, 1987.

69. Wiechers JW, Drenth BFH, Jonkman JHG, Dezeeuw RA. Percutaneous absorption, metabolism, and elimination of the penetration enhancer Azone in humans after prolonged application under occlusion. Int J Pharm 47:43–49, 1988.

70. Wiechers JW, Drenth BFH, Jonkman JHG, Dezeeuw RA. Percutaneous absorption, metabolic profiling, and excretion of the penetration enhancer Azone after multiple dosing of an Azone-containing triamcinolone acetonide cream in humans. J Pharm Sci 79:111–115, 1990.

71. Wotton PK, Mollgaard B, Hadgraft J, Hoelgaard A. Vehicle effect on topical drug delivery 3. Effect of Azone on the cutaneous permeation of metronidazole and propylene glycol. Int J Pharm 24:19–26, 1985.

72. Watkinson AC, Hadgraft J, Bye A. Aspects of the transdermal delivery of prostaglandins. Int J Pharm 74:229–236, 1991.

73. Ganga S, Ramarao P, Singh J. Effect of Azone on the iontophoretic transdermal delivery of metoprolol tartrate through human epidermis in vitro. J Controlled Release 42:57–64, 1996.

74. Hirvonen J, Kontturi K, Murtomaki L, Paronen P, Urtti A. Transdermal iontophoresis of sotalol and salicylate—the effect of skin charge and penetration enhancers. J Controlled Release 26:109–117, 1993.

75. Kalia YN, Guy RH. Interaction between penetration enhancers and iontophoresis: effect on human skin impedance in vivo. J Controlled Release 44:33–42, 1997.

76. Ueda H, Isshiki R, Ogihara M, Sugibayashi K, Morimoto Y. Combined effect of ultrasound and chemical enhancers on the skin permeation of aminopyrine. Int J Pharm 143:37–45, 1996.

77. Samour CM, Daskalakis S. U.S. Patent 4,861,764, 1989.

78. Gyurik RJ, et al. SEPA penetration enhancement of econazole in human skin. In: Brain KR, James VJ, Walters KA, eds. Prediction of Percutaneous Penetration. Vol. 4b. Cardiff: STS Publishing, pp 124–126, 1996.

79. Gauthier ER, Gyurik RJ, Krauser SF, Pittz EP, Samour CM. SEPA absorption enhancement of polar and non-polar drugs. In: Brain KR, James VJ, Walters KA, eds. Perspectives in Percutaneous Penetration. Vol. 5a. Cardiff: STS Publishing, p 79, 1997.

80. Yoneto K, Ghanem A–H, Higuchi WI, Peck KD, Li SK. Mechanistic studies of the 1-alkyl-2-pyrroloidones as skin permeation enhancers. J Pharm Sci 84:312–317, 1995.

81. Cornwell PA, Barry BW, Bouwstra JA, and Gooris GS. Modes of action of terpene penetration enhancers in human skin; differential scanning calorimetry, small-angle X-ray diffraction and enhancer uptake studies. Int J Pharm 127:9–26, 1996.

82. van Hal DA, Jeremiasse E, Junginger HE, Spies F, Bouwstra JA. Structure of fully hydrated human stratum corneum: a freeze–fracture electron microscope study. J Invest Dermatol 106:89–95, 1996.

83. Engström S, Forslind B, Engblom J. Lipid polymorphism—a key to an understanding of skin penetration. In: Brain KR, James VJ, Walters KA, eds. Prediction of Percutaneous Penetration. Vol. 4b. Cardiff: STS Publishing, pp 163–166, 1996.

84. Strathman H. Production of microporous media by phase inversion processes. In: Lloyd DR, ed. Materials Science of Synthetic Membranes. ACS Symp Ser 269:165–195, 1996.

85. Langlois B, Friberg S. Evaporation from a complex emulsion system. J Soc Cosmet Chem 44:23–34, 1993.

86. Yano T, Higo N, Furukawa K, Tsuji M, Noda K, Otagiri M. Evaluation of a new penetration enhancer 1-[2-(decylthio)ethyl]azacyclopentan-2-one (HPE-101). J Pharmacobiodyn 15:527–533, 1992.

87. Yano T, Higo N, Fukuda K, Tsuji M, Noda K, Otagiri M. Further evaluation of a new penetration enhancer, HPE-101. J Pharm Pharmacol 45:775–778, 1992.

88. Adachi H, Irie T, Uekama K, Manako T, Yano T, Saita M. Combination effects of *O*-carboxymethyl-*O*-ethyl-β-cyclodextrin and penetration enhancer HPE-101 on transdermal delivery of prostaglandin E_1 in hairless mice. Eur J Pharm Sci 1:117–123, 1993.

89. Rajadhyaksha VJ. Oxalodinone penetration enhancing compounds. U.S. Patent 4,960,771, 1990.

90. Pfister WR, Rajadhyaksha VJ. Oxazolidinones: a new class of cyclic urethane transdermal enhancer (CUTE). Proc Int Symp Controlled Release Bioact Mater 24:709–710, 1997.

91. Pfister WR, Rajadhyaksha VJ. Oxazolidinones: a new class of cyclic urethane transdermal enhancer (CUTE). Pharm Res 12:S280, 1995.

92. Wong O, Huntington A, Nishihata T, Rytting JH. New alkyl *N,N*-dialkyl-substituted amino acetates as transdermal penetration enhancers. Pharm Res 6:286–295, 1989.

93. Wongpayapkul L, Chow D. Comparative evaluation of various enhancers on the transdermal permeation of propranolol hydrochloride. Pharm Res 8:S140, 1991.

94. Büyüktimkin S, Büyüktimkin N, Rytting JH. Synthesis and enhancing effect of dodecyl 2-(*N,N*-dimethylamino)-propionate (DDAIP) on the transepidermal delivery of indomethacin, clonidine, and hydrocortisone. Pharm Res 10:1632–1637, 1993.

95. Turunen TM, Büyüktimkin S, Büyüktimkin N, Urtti A, Paronen P, Rytting JH. Enhanced delivery of 5-fluorouracil through shed snake skin by two new transdermal penetration enhancers. Int J Pharm 92:89–95, 1993.

96. Turunen TM, Urtti A, Paronen P, Audus KL, Rytting JH. Effect of some penetration enhancers on epithelial membrane lipid domains: evidence from fluorescence spectroscopy studies. Pharm Res 11:288–294, 1994.

97. Lambert WJ, Kudla RJ, Holland JM, Curry JT. A biodegradable transdermal penetration enhancer based on *N*-(2-hydroxyethyl)-2-pyrrolidone I. Synthesis and characterization. Int J Pharm 95:181–192, 1993.

98. Harris WT, Tenjarla SN, Holbrook JM, Smith J, Mead C, Entrekin J. *N*-Pentyl *N*-acetylprolinate. A new skin penetration enhancer. J Pharm Sci 84:640–642, 1995.

99. James KC. Solubility and Related Properties. New York: Marcel Dekker, 1986.

100. Frank FC. The influence of dislocations on crystal growth. Discuss Faraday Soc 5:48–54, 1949.

101. Kondo S, Yamanaka C, Sugimoto I. Enhancement of transdermal delivery by superfluous thermodynamic potential III. Percutaneous absorption of nifedipine in rats. J Pharmacobiodyn 10:743–749, 1987.

102. Ikeda K, Saitoh I, Oguma T, Takagishi Y. Effect of hydrophobically modified hydroxypropylmethyl cellulose on recrystallization from supersaturated solutions of indomethacin. Chem Pharm Bull 42:2320–2326, 1994.

103. Hasagawa A, Taguchi M, Suzuki R, Miyata T, Nakagawa H, Sugimoto I. Supersaturation mechanisms of drugs from solid dispersions with enteric coating agents. Chem Pharm Bull 36:4941–4950, 1988.

104. Pearson A. Inhibition of a growth in a model pharmaceutical semisolid. PhD dissertation, Strathclyde University, England No BRD-97084.

105. Yeoh TY. Use of polymer additives to inhibit phenytoin nucleation and crystal growth from supersaturated aqueous solutions. PhD dissertation, Purdue University, West Lafayette, IN, No. DA9513096.

106. Morita M, Horita S. Effect of crystallinity on the percutaneous absorption of corticosteroids II. Chemical activity and biological activity. Chem Pharm Bull 33:2091–2097, 1985.

107. Szeman J, Ueda H, Szejtli J, Fenyvesi W, Watanabe Y, Machida Y, Nagai T. Enhanced percutaneous absorption of homogenised tolnaftate/β–cyclodextrin polymer ground mixture. Drug Design Deliv 1:325–332, 1987.

108. Lippold BC. How to optimize drug penetration through the skin. Pharm Acta Helv 67: 294–300, 1992.

109. Davis AF, Hadgraft J. Supersaturated solutions as topical drug delivery systems. In: Walters KA, Hadgraft J, eds. Pharmaceutical Skin Penetration Enhancement. New York: Marcel Dekker, pp 243–267, 1993.

110. Fischer W. Trends in transdermal drug delivery. In: Junginger HE, ed. Pharmaceutical Technology: Drug Targeting and Delivery. London: Ellis Horwood, pp 190–202, 1992.

111. Zatz JL. Enhancing skin penetration of actives with the vehicle. Cosmet Toiletr 109: 27–36, 1994.

112. Pefile S, Smith EW. Transdermal drug delivery vehicle design and formulation. S Afr J Sci 93:147–150, 1997.

113. Chiou WL, Riegelmann S. Pharmaceutical applications of solid dispersion systems. J Pharm Sci 60:1281–1301, 1971.

114. Hasegawa A, Nakagawa H, Sugimoto I. Application of solid dispersions of spray–dried nifedipine with enteric-coating agents to prepare a sustained-release dosage form. Chem Pharm Bull 33:1615–1619, 1985.

115. Hasegawa A, Kawamura H, Nakagawa H, Sugimoto I. Physical properties of solid dispersions of poorly water soluble drugs with enteric coating agents. Chem Pharm Bull 33:3429–3435, 1985.

116. Corrigan OI, Farvar MA, Higuchi WI. Drug membrane transport enhancement using high energy drug polvinylpyrrolidone (PVP) coprecipitates. Int J Pharm 5:229–238, 1980.

117. Merkle HP. Drug polvinylpyrrolidone co-precipitates: kinetics of drug release and formation of supersaturated solutions. In: Digenis GA, Ansell J, eds. Proceedings International Symposium Povidone. Lexington: University of Kentucky, pp 202–216, 1983.

118. Norman FH. In: Roche EB, ed. Design of Biopharmaceutical Properties Through Prodrugs and Analogs, Alpha Series. pp 198, 1997.

119. Corrigan OI. Drug–polvinylpyrrolidone coprecipitates. In: Smith EW, Maibach HI, eds. Percutaneous Penetration Enhancers. Boca Raton: CRC Press, pp 221–232, 1995.

120. Merkle HP. Transdermal delivery systems. Methods Findings Exp Clin Pharmacol 11: 135–153, 1989.

121. Dittgen M, Bombor R. Process for the production of a pharmaceutical preparation. German patent DD 217 989-A1, 1985.

122. Davis AF, Hadgraft J. Effect of supersaturation on membrane transport: 1. Hydrocortisone acetate. Int J Pharm 76:1–8, 1991.

123. Pellet MA, Davis AF, Hadgraft J. Effect of supersaturation on membrane transport: 2. Piroxicam. Int J Pharm 111:1–6, 1994.

124. Pellet MA, Castellano S, Hadgraft J, Davis AF. The penetration of supersaturated solutions of piroxicam across silicone membrane and human skin in vitro. J Controlled Release 46:205–214, 1997.

125. Pellet MA, Roberts MS, Hadgraft J. Supersaturated solutions evaluated with an in vivo stratum corneum tape stripping technique. Int J Pharm 151:91–98, 1997.

126. Megrab NA, Williams AC, Barry BW. Oestradiol permeation through human skin and Silastic membrane: effects of propylene glycol and supersaturation. J Controlled Release 36:277–294, 1995.

127. Higuchi WI, Farrar MA. Drug membrane transport enhancement using high energy drug povidone coprecipitates. In: Digenis GA, Ansell J, eds. Proceedings International Symposium Povidone, Lexington: University of Kentucky, pp 71, 1983.

128. Flynn GL, Smith RW. Membrane diffusion III. Influence of solvent composition and permeant solubility on membrane transport. J Pharm Sci 61:61–66, 1972.

129. Tanaka S, Takanashima Y, Murayama H, Tsuchiya S. Studies on drug release from ointments: V. Release of hydrocortisone butyrate from topical dosage forms to silicone rubber. Int J Pharm 27:29–38, 1985.

130. Pellett MA. Unpublished observations.

131. Pellett MA, Watkinson AC, Brain KR, Hadgraft J. Diffusion of flurbiprofen across human stratum corneum using synergistic methods of enhancement. In: Brain KR, James VJ, Walters KA, eds. Perspectives in Percutaneous Penetration. Vol. 5a. Cardiff: STS Publishing, p 86, 1997.

132. Potts RO, Cleary GW. Transdermal delivery: useful paradigms. J Drug Targeting 3:247–251, 1995.

133. Ma X, Taw J, Chiang C–M. Inhibition of crystallization of steroid drugs in transdermal patches. Proc Int Symp Controlled Release Bioact Mater 22:712–713, 1995.

134. Ma X, Taw J, Chiang C–M. Control of drug crystallization in transdermal matrix system. Int J Pharm 143:115–119, 1996.

135. Kubota K, Sznitowska M, Maibach H. Percutaneous permeation of betamethasone 17-valerate from different vehicles. Int J Pharm 96:105–110, 1993.

136. Henmi T, Fujii M, Kikuchi K, Matsumoto M. Application of an oily gel formed by hydrogenated soybean phospholipids as a percutaneous absorption-type enhancer. Chem Pharm Bull 42:651–655, 1994.

137. Mezei M, Gulasekharam V. Liposomes—a selective drug delivery system for the topical route of administration: lotion dosage forms. Life Sci 26:1473–1477, 1980.

138. Mezei M, Gulasekharam V. Liposomes—a selective drug delivery system for the topical route of administration: a gel dosage form. J Pharm Pharmacol 34:473–474, 1982.

139. Wohlrab W, Lasch J. Penetration kinetics of liposomal hydrocortisone in human skin. Dermatologica 174:18–22, 1987.

140. Kim MK, Chung SJ, Lee MH, Cho AR, Shim CK. Targeted and sustained delivery of hydrocortisone to normal and stratum corneum-removed skin without enhanced skin absorption using a liposome gel. J Controlled Release 46:243–251, 1997.

141. Mezei M. Liposomes as a skin delivery system. In: Breimer DD, Speiser P, eds. Topics in Pharmaceutical Sciences. Amsterdam: Elsevier, pp 345–358, 1985.

142. Mezei M. Liposomal drug delivery system for the topical route of administration. In: Tipnis HP, ed. Controlled Release Dosage Forms. Bombay: Bombay College of Pharmacy, pp 47–57, 1988.

143. Foong WC, Harsanyi BB, Mezei M. Biodisposition and histological evaluation of topically applied retinoic acid in liposomal, cream and gel dosage form. In: Hanin I, Pepeu G, eds. Phospholipids—Biochemical, Pharmaceutical and Analytical Considerations. New York: Plenum, pp 279–282, 1990.

144. Ganesan MG, Weiner ND, Flynn GL, Ho NFH. Influence of liposomal drug entrapment on percutaneous absorption. Int J Pharm 20:139–154, 1984.

145. Ho NFH, Ganesan MG, Weiner ND, Flynn GL. Mechanisms of topical delivery of liposomally entrapped drugs. J Controlled Release 2:61–65, 1985.

146. Egbaria K, Weiner N. Liposomes as a topical drug delivery system. Adv Drug Deliv Rev 5:287–300, 1990.

147. Egbaria K, Weiner N. Topical delivery of liposomally encapsulated ingredients evaluated by in vitro diffusion studies. In: Braun–Falco O, Korting HC, Maibach HI, eds. Liposome Dermatics. Berlin: Springer-Verlag, pp 172–181, 1992.

148. du Plessis J, Ramachandran C, Weiner N, Müller DG. The influence of particle size of liposomes on the deposition of drug into skin. Int J Pharm 103:277–282, 1994.

149. Korting HC, Zienicke H, Schäfer–Korting M, Braun–Falco O. Liposome encapsulation improves efficacy of betamethasone dipropionate in atopic eczema, but not in psoriasis vulgaris. Eur J Clin Pharmacol 29:349–352, 1991.

150. Schäfer–Korting M, Korting HC, Ponce–Pöschl E. Liposomal tretinoin for uncomplicated acne vulgaris. Clin Invest 72:1086–1091, 1994.

151. Schmid MH, Korting HC. Therapeutic progress with topical liposome drugs for skin disease. Adv Drug Deliv Rev 18:335–342, 1996.

152. Touitou E, Shaco–Ezra N, Dayan N, Jushynski M, Rafaeloff R, Azoury R. Dyphylline liposomes for delivery to the skin. J Pharm Sci 81:131–134, 1992.

153. Touitou E, Levi–Schaffer F, Dayan N, Alhaique F, Riccieri F. Modulation of caffeine skin delivery by carrier design: liposomes versus permeation enhancers. Int J Pharm 103:131–136, 1994.

154. Foldvari M, Gesztes A, Mezei M. Dermal drug delivery by liposome encapsulation: clinical and electron microscopic studies. J Microencap 7:479–489, 1990.

155. Foldvari M. In vitro cutaneous and percutaneous delivery and in vivo efficacy of tetracaine from liposomal and conventional vehicles. Pharm Res 11:1593–1598, 1994.

156. Egbaria K, Ramachandran C, Weiner N. Topical delivery of cyclosporin: evaluation of various formulations using in vitro diffusion studies in hairless mouse skin. Skin Pharmacol 3:21–28, 1990.

157. Egbaria K, Ramachandran C, Kittayanond D, Weiner N. Topical delivery of liposomal IFN. Antimicrob Agents Chemother 34:107–110, 1990.

158. Short SM, Rubas W, Paasch BD, Mrsny RJ. Transport of biologically active interferon-gamma across human skin in vitro. Pharm Res 12:1140–1145, 1995.

159. Yarosh D, Bucana C, Cox P, Alas L, Kibitel J, Kripke M. Localization of liposomes containing a DNA repair enzyme in murine skin. J Invest Dermatol 103:461–468, 1994.

160. William A, Wertz PW, Landmann L, Downing DT. Preparation of liposomes from stratum corneum lipids. J Invest Dermatol 87:582–584, 1986.

161. Fresta M, Puglisi G. Corticosteroid dermal delivery with skin–lipid liposomes. J Controlled Release 44:141–151, 1997.

162. Gray GM, White RJ. Epidermal lipid liposomes: a novel nonphospholipid membrane system. Biochem Soc Trans 7:1129–1131, 1979.

163. Hofland HEJ, van der Geest R, Boddé HE, Junginger HE, Bouwstra JA. Estradiol permeation from nonionic surfactant vesicles through human stratum corneum in vitro. Pharm Res 11:659–664, 1994.

164. Schreier H, Bouwstra J. Liposomes and niosomes as topical drug carriers: dermal and transdermal drug delivery. J Controlled Release 30:1–15, 1994.

165. French EJ, Pouton CW, Walters KA. Mechanisms and prediction of nonionic surfactant effects on skin permeability. In: Walters KA, Hadgraft J, eds. Pharmaceutical Skin Penetration Enhancement. New York: Marcel Dekker, pp 113–143, 1993.

166. Lasch J, Laub R, Wohlrab W. How deep do intact liposomes penetrate into human skin? J Controlled Release 18:55–58, 1991.

167. Schubert R, Joos M, Deicher M, Magerle R, Lasch J. Destabilization of egg lecithin liposomes on the skin after topical application measured by perturbed $\gamma\gamma$ angular correlation spectroscopy (PAC) with ^{111}ln. Biochim Biophys Acta 1150:162–164, 1993.

168. Hofland H, Bouwstra JA, Boddé HE, Spies F, Junginger HE. Interactions between liposomes and human stratum corneum in vitro: freeze fracture electron microscopial

visualization and small angle X-ray scattering studies. Br J Dermatol 132:853–866, 1995.

169. Bouwstra JA, Hofland HEJ, Spies F, Gooris GS, Junginger HE. Changes in the structure of the human stratum corneum induced by liposomes. In: Braun–Falco O, Korting HC, Maibach HI, eds. Liposome Dermatics. Berlin: Springer-Verlag, pp 121–136, 1992.
170. Blume A, Jansen M, Ghyczy M, Gareiss J. Interation of phospholipid liposomes with lipid model mixtures for stratum corneum. Int J Pharm 99:219–228, 1993.
171. Korting HC, Stolz W, Schmid MH, Maierhofer G. Interaction of liposomes with human epidermis reconstructed in vitro. Br J Dermatol 132:571–579, 1995.
172. Cevc G, Blume G. Lipid vesicles penetrate into intact skin owing to the transdermal osmotic gradients and hydration force. Biochim Biophys Acta 1104:226–232, 1991.
173. Planas ME, Gonzalez P, Rodriguez L, Sanchez S, Cevc G. Noninvasive percutaneous induction of topical analgesia by a new type of drug carrier, and prolongation of local pain insensitivity by anesthetic liposomes. Anesth Analg 75:615–621, 1992.
174. Cevc G, Blume G, Schätzlein. A. Transferosomes-mediated transepidermal delivery improves the regio-specificity and biological activity of corticosteroids in vivo. J Controlled Release 45:211–226, 1997.
175. Lieb LM, Ramachandran C, Egbaria K, Weiner N. Topical delivery enhancement with multilamellar liposomes into pilosebaceous units: I. In vitro evaluation using fluorescent techniques with the hamster ear model. J Invest Dermatol 99:108–113, 1992.
176. Lieb LM, Flynn G, Weiner N. Follicular (pilosebaceous unit) deposition and pharmacological behavior of cimetidine as a function of formulation. Pharm Res 11:1419–1423, 1994.
177. Niemiec SM, Ramachandran C, Weiner N. Influence of nonionic liposomal composition on topical delivery of peptide drugs into pilosebaceous units: an in vivo study using the hamster ear model. Pharm Res 12:1184–1188, 1995.
178. Niemiec SM, Latta JM, Ramachandran C, Weiner N, Roessler BJ. Perifollicular transgenic expression of human interleukin-1 receptor antagonist protein following topical application of novel liposome–plasmid DNA formulations in vivo. J Pharm Sci 86: 701–708, 1997.
179. Bernard E, Dubois J–L, Wepierre J. Importance of sebaceous glands in cutaneous penetration of an antiandrogen: target effect of liposomes. J Pharm Sci 86:573–578, 1997.
180. Tshan T, Steffen H, Supersaxo A. Sebaceous gland deposition of isotretinoin after topical application: an in vitro study using human facial skin. Skin Pharmacol 10:126–134, 1997.
181. Hsieh DS, ed. Drug Permeation Enhancement—Theory and Applications. New York: Marcel Dekker, 1994.
182. Braun–Falco O, Korting HC, Maibach HI, eds. Liposome Dermatics. Berlin: Springer-Verlag, 1992.
183. Lai PM, Roberts MS. Iontophoresis. In: Roberts MS, Walters KA, eds. Dermal Absorption and Toxicity Assessment. New York: Marcel Dekker, pp 371–414, 1998.

7

Dermatological Formulation and Transdermal Systems

KENNETH A. WALTERS and KEITH R. BRAIN

An-eX Analytical Services Ltd., Cardiff, Wales

I. INTRODUCTION

Over the past few decades there have been many advances in our understanding of the physicochemical properties of both formulation systems and their ingredients. These have led to the ability to develop physically, chemically, and biologically stable products. There has also been a significant increase in our knowledge of the properties of skin and the processes that control skin permeation. The ground rules for skin permeation were laid down by Scheuplein and Blank in the late 1960s and early 1970s (1), and these have been updated on a reasonably regular basis (2–8). We have learned, for example, that the permeation of compounds across intact skin is controlled fundamentally by the stratum corneum, and it is the chemical composition and morphology of this layer that usually determines the rate and extent of absorption (9,10). Similarly, we have discovered how to modify this barrier, by chemical or physical means and, thereby, alter the rate of diffusion of many permeating molecules (11,12).

A basic deficiency, however, in the application of our understanding of the barrier properties of the skin to dermatological and transdermal therapy is that this knowledge has largely been generated by investigations on normal, rather than pathological, skin. The relevance of such information to diseased skin, for which permeation characteristics are probably significantly altered, has yet to be fully established. There is some data on transport across skin that has been artificially damaged (13–18), and limited information on permeation through diseased skin has been obtained in the clinic (19–22).

In modern-day pharmaceutical practice, therapeutic compounds are applied to the skin for dermatological (within the skin), local (regional), and for transdermal (systemic) delivery. Whatever the target site or organ, it is usually a prerequisite that the drug crosses the outermost layer of the skin, the stratum corneum. However, a major function of the stratum corneum is to provide a protective barrier to the ingress of xenobiotics and to control the rate of water loss from the body: Evolution has generated a robust and durable barrier that fulfills its biological function throughout an individual's lifetime. A basic, yet thorough, understanding of the structure and transport properties of this membrane is essential to the rational development of topical dosage forms, and this has been provided elsewhere in this volume.

II. DRUG CANDIDATE SELECTION

Although it may appear to be a simple task to select lead compounds for pharmaceutical product development, based on therapeutic rationale and compound safety and efficacy, the practicalities of this procedure are somewhat more complex. For the most part, therapeutic efficacy is dependent on the ability of a compound to cross biological barriers, travel to the target site, and interact with specific receptors. However, as pointed out and excellently reviewed (3), it is often more appropriate in dermatological therapy to select compounds based on their inability to breach relevant biological barriers. Because the site of action may be the skin surface, the stratum corneum, the viable epidermis, the appendages, the dermis, or the local subcutaneous tissues, the rules of candidate selection will vary. For the purposes of this discussion it will be assumed that the therapeutic rationale for dermal drug delivery has been established and that a series of compounds with appropriate pharmacological activity have been identified. It will also be assumed that each compound within the series possesses equivalent chemical and physical stability. In other words, drug candidate selection need only be based on the ability to deliver the compound to its site of action.

Earlier chapters in this volume have taught that the primary requirement for a compound to penetrate the skin is the ability to leave the delivery system and enter the stratum corneum. Furthermore, this characteristic is dependent on the stratum corneum–vehicle partition coefficient of the compound, for which the octanol–water partition coefficient is often used as a surrogate. Whereas it is immediately apparent that a high value for this parameter will favor delivery into the stratum corneum, it will not favor movement into the more hydrophilic regions of the viable epidermis. Furthermore, the rate of diffusion through the stratum corneum and lower layers of the skin is linked to the molecular volume of the permeant. It is evident that a compound with a high octanol–water partition coefficient and a relatively high molecular volume will possess a high affinity for the stratum corneum (i.e., be substantive to the stratum corneum). This principle is used extensively in the design of sunscreen agents, for which it is not uncommon to add a medium-length or branched-chain alkyl to the UV-absorbing molecule to increase residence time in the skin and reduce systemic uptake.

The use of the skin as a route of delivery into the systemic circulation was neither commercially nor scientifically exploited until the 1950s, when ointments containing agents such as nitroglycerin and salicylates were developed. Angina could be controlled for several hours by applying an ointment containing 2% nitroglycerin (23). Similarly, topical salicylates could be absorbed through the skin into arthritic joints. More recently, nonsteroidal anti-inflammatory agents, such as ibuprofen and ketoprofen, and hormonal steroids, such as estradiol and testosterone, have been developed and marketed in semisolid preparations. A major problem with transdermal semisolid preparations, however, is that of control. Drug concentrations in plasma or duration of action, are not reliably predictable for several reasons, including the amount and area of application and dosage frequency.

The specific advantages of transdermal therapy have been fully discussed elsewhere (24). Briefly, transdermal devices are easy to apply, can remain in place for up to 7 days (depending on the system), and are easily removed following, or during, therapy. The reduced-dosing frequency, and the production of controllable and sustained plasma levels, tend to minimize the risk of undesirable side effects sometimes observed after oral delivery. The avoidance of extensive hepatic first-pass metabolism is a further advantage. The major limitation to transdermal drug delivery is the intrinsic barrier property of the skin. Although marketed patch-type transdermal delivery systems are available for only a limited number of drugs (e.g., scopolamine, nitroglycerin, clonidine, estradiol, fentanyl, testosterone, and nicotine), several other candidates are at various stages of development. Many of the drugs under investigation do not intrinsically possess any significant ability to cross the skin; therefore, ways must be found to improve their transdermal delivery. This could be achieved by the use of prodrugs designed such that they are more rapidly absorbed than the parent compound, yet are metabolized to the active species before receptor site occupancy (25,26). Physical methods, such as iontophoresis (27), electroporation (28), and sonophoresis (29) have proved experimentally useful for increasing the skin permeation of several compounds. Alternatively, the barrier may be modulated using thermodynamic strategies, or chemically modified to reduce diffusive resistance by the use of penetration enhancers, both of which are discussed in Chapter 6. Such developmental strategies will increase the number of candidate drugs for transdermal delivery in the future.

III. PREFORMULATION AND FORMULATION

Preformulation encompasses those studies that should be carried out before the commencement of formulation development. The primary goal of the preformulation process is to permit the rational development of stable, safe, efficacious dosage forms, and it is concerned mainly with the characterization of the physicochemical properties of the drug substance. At the preformulation stage, the final route of drug administration is usually undecided; therefore, any protocols must be able to cover all required aspects. The preformulation study has several distinct phases (Table 1). A detailed description of all of the studies that form part of the preformulation stage are given elsewhere (30) and will not be considered here. The only important aspect of preformulation that is specific to dermatological and transdermal formulation concerns drug delivery characteristics. These have been fully discussed elsewhere in this volume (see Chapters 4 and 5).

A. Formulation of Dermatological Products

The selection of formulation type for dermatological products is usually influenced by the nature of the skin lesion and the opinion of the medical practitioner. Kitson and Maddin (31) elegantly stated: "It is idle to pretend that the therapy for skin

Table 1   Preformulation Tasks Listed in
Approximate Chronologic Order

Preformulation Task
1. General description of the compound
2. Calorimetry
3. Polymorphism
4. Hygroscopicity
5. Analytical development
6. Intrinsic stability
7. Solubility and partitioning characteristics
8. Drug delivery characteristics

diseases, as currently practiced, has its origins in science." To this day a practicing dermatologist would prefer to apply a "wet" formulation (ranging from simple tap-water to complex emulsion formulations, with or without drug) to a wet lesion and a "dry" formulation (e.g., petrolatum) to a dry lesion. The preparation of such formulations as poultices and pastes is extemporaneous, and it is unlikely that the industrial pharmaceutical formulator will be required to develop products of this type. Solutions and powders lack staying power (retention time) on the skin and can afford only transient relief. In modern-day pharmaceutical practice, semisolid formulations are the preferred vehicles for dermatological therapy because they remain in situ and deliver the drug over extended time periods. In most cases, therefore, the developed formulation will be an ointment, emulsion, or gel. Typical constituents for these types of formulations are shown in Table 2.

1. Ointments

An *ointment* is classified as any semisolid containing fatty material and intended for external application (*United States Pharmacopeia; USP*). There are four types of ointment base, and these are listed in the *USP* as hydrocarbon base, absorption base, water-removable base, and water-soluble base. Only the hydrocarbon bases are completely anhydrous. The anhydrous hydrocarbon bases, which contain straight or branched hydrocarbons with chain lengths ranging from C_{16} to C_{30}, which may also contain cyclic alkanes, are used principally in nonmedicated form. A typical formulation contains fluid hydrocarbons (mineral oils and liquid paraffins) mixed with a longer alkyl chain, higher-melting point, hydrocarbons (white and yellow soft paraffin and petroleum jelly). The difference between white and yellow soft paraffin is simply that the white version has been bleached. Hard paraffin and microcrystalline waxes are similar to the soft paraffins, except that they contain no liquid components. These anhydrous mixtures tend to produce formulations that are greasy and unpleasant to use, but the addition of solid components, such as microcrystalline cellulose, can reduce the greasiness. Improved skin feel can also be attained by the incorporation of silicone materials, such as polydimethylsiloxane oil or dimethicones. Silicones are often used in barrier formulations that are designed to protect the skin against water-soluble irritants.

Although the nonmedicated anhydrous ointments are extremely useful as emollients, their value as topical drug delivery systems is limited by the relative insolu-

Table 2 Constituents of Semisolid Formulations

Function	Sample ingredients	
Polymeric thickeners	Gums	Acrylic acids
	Acacia	Carbomers
	Alginates	Polycarbophil
	Carageenan	Colloidal solids
	Chitosan	Silica
	Collagen	Clays
	Tragacanth	Microcrystalline cellulose
	Xanthan	Hydrogels
	Celluloses	Polyvinyl alcohol
	Sodium carboxymethyl	Polyvinylpyrrolidone
	Hydroxyethyl	Thermoreversible polymers
	Hydroxypropyl	Poloxamers
	Hydroxypropylmethyl	
Oil phase	Mineral oil	Isopropyl myristate
	White soft paraffin	Isopropyl palmitate
	Yellow soft paraffin	Castor oil
	Beeswax	Canola oil
	Stearyl alcohol	Cottonseed oil
	Cetyl alcohol	Jojoba oil
	Cetostearyl alcohol	Arachis (Peanut) oil
	Stearic acid	Lanolin (and derivatives)
	Oleic acid	Silicone oils
Surfactants	Nonionic	Anionic
	Sorbitan esters	Sodium dodecyl sulfate
	Polysorbates	Cationic
	Polyoxyethylene alkyl ethers	Cetrimide
	Polyoxyethylene alkyl esters	Benzalkonium chloride
	Polyoxyethylene aryl ethers	
	Glycerol esters	
	Cholesterol	
Solvents	Polar	Polyethylene glycols
	Water	Propylene carbonate
	Propylene glycol	Triacetin
	Glycerol	Nonpolar
	Sorbitol	Isopropyl alcohol
	Ethanol	Medium-chain triglycerides
	Industrial methylated spirit	
Preservatives	Antimicrobial	Antioxidants
	Benzalkonium chloride	α-Tocopherol
	Benzoic acid	Ascorbic acid
	Benzyl alcohol	Ascorbyl palmitate
	Bronopol	Butylated hydroxyanisole
	Chlorhexidine	Butylated hydroxytoluene
	Chlorocresol	Sodium ascorbate
	Imidazolidinyl urea	Sodium metabisulfite
	Paraben esters	Chelating agents
	Phenol	Citric acid
	Phenoxyethanol	Edetic acid
	Potassium sorbate	
	Sorbic acid	
pH adjusters	Diethanolamine	Sodium hydroxide
	Lactic acid	Sodium phosphate
	Monoethanolamine	Triethanolamine

bility of many drugs in hydrocarbons and silicone oils. It is possible to increase drug solubility within a formulation by incorporation of hydrocarbon-miscible solvents, such as isopropylmyristate or propylene glycol, into the ointment. Although increasing the solubility of a drug within a formulation may often decrease the release rate, it does not necessarily decrease the therapeutic effect. It is well accepted that simple determination of release rates from formulations may not be predictive of drug bioavailability. For example, when formulated in a simple white petrolatum–mineral oil ointment, the release rate of betamethasone dipropionate was considerably higher than when the drug was formulated at the same concentration (0.05%) in an augmented, and more clinically effective, ointment that contained propylene glycol (Fig. 1) (32). It is also important to appreciate that various grades of petrolatum are commercially available, and that the physical properties of these materials will vary depending on the source and refining process. Even slight variations in physical properties of the constituents of an ointment may have substantial effects on drug release behavior (33).

The preparation of ointment formulations may appear to be a simple matter of heating all of the constituents to a temperature higher than the melting point of all

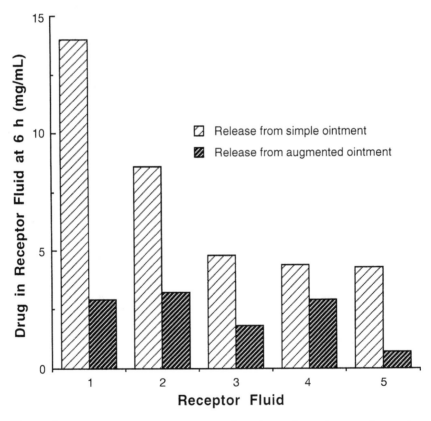

Figure 1 Release rates of betamethasone dipropionate from ointments into various receptor fluids. (1) 5% hexane in acetonitrile; (2) octanol; (3) acetonitrile; (4) 60% acetonitrile in water; (5) 95% ethanol. (From Ref. 32.)

of the excipients and then cooling with constant mixing. The reality, however, is that the process is somewhat more complex and requires careful control over various parameters, particularly the cooling rate. Rapid cooling, for example, creates stiffer formulations in which there are numerous small crystallites, whereas a slow-cooling rate results in the formation of fewer, but larger, crystallites and a more fluid product. Further information on temperature effects and ointment phase behavior are available (34–36).

2. Gels

The common characteristic of all gels is that they contain continuous structures that provide solid-like properties (4). Depending on their constituents, gels may be clear or opaque, and be polar, hydroalcoholic, or nonpolar. The simplest gels comprise water thickened with either natural gums (e.g., tragacanth, guar, or xanthan), semi-synthetic materials (e.g., methylcellulose, carboxymethylcellulose, or hydroxyethyl-cellulose), synthetic materials (e.g., carbomer–carboxyvinyl polymer), or clays (e.g., silicates or hectorite). Gel viscosity is generally a function of the amount and molecular weight of the added thickener.

There are a variety of semisynthetic celluloses in use as thickeners in gel formulations. These include methylcellulose (MC), carboxymethylcellulose (CMC), hydroxyethylcellulose (HEC), hydroxypropylcellulose (HPC), and hydroxypropyl-methylcellulose (HPMC). These celluloses can be obtained in diverse molecular weight grades, and the higher molecular weight compounds are used at 1–5% (w/w). In the development of prototype gel formulations, it is useful to evaluate a variety of different types of cellulose. For example, if clarity of the gel is a major requirement, HPMC is preferable to MC. It is also important to appreciate that some celluloses may exhibit specific incompatibilities with other potential formulation ingredients. For example, HEC is incompatible with several salts, and MC and HPC are incompatible with parabens. This latter incompatibility limits the choice of preservative for gel formulations that are based on MC and HPC. Finally, the presence of oxidative materials (e.g., peroxides, or other ingredients containing peroxide residues) in formulations gelled with celluloses should be avoided because oxidative degradation of the polymer chains may cause a rapid decrease in formulation viscosity (37).

As the branched-chain polysaccharide gums, such as tragacanth, pectin, carrageenan, and guar, are of naturally occurring plant origin, they can have widely varying physical properties, depending on their source. They are usually incorporated into formulations at concentrations between 0.5 and 10%, contingent on the required viscosity. Viscosity may be enhanced synergistically by the addition of inorganic suspending agents, such as magnesium aluminum silicate. Tragacanth, a mixture of water-insoluble and water-soluble polysaccharides, is negatively charged in aqueous solution and, therefore, incompatible with many preservatives when formulated at a pH of 7 or higher. Similarly, xanthan gum, which is produced by bacterial fermentation, is incompatible with some preservatives. Alginic acid is a hydrophilic colloidal carbohydrate obtained from seaweed and the sodium salt, sodium alginate, is used at 5–10% as a gelling agent. Film gels may be obtained by incorporation of small amounts of soluble calcium salts (e.g., tartrate on citrate). Many gums are ineffective in hydroalcoholic gels containing more than 5% alcohol.

The natural clay thickeners (e.g., bentonite and magnesium aluminium silicate) are useful for thickening aqueous gels containing cosolvents, such as ethanol, iso-propanol, glycerin, and propylene glycol. These materials possess a lamellar structure that can be extensively hydrated. The flat surfaces of bentonite are negatively charged, whereas the edges are positively charged. These clays swell in the presence of water because of hydration of the cations and electrostatic repulsion between the negatively charged faces. Thixotropic gels form at high concentrations at which the clay particles combine in a flocculated structure in which the edge of one particle is attracted to the face of another. The rheological properties of these clay dispersions, therefore, are particularly sensitive to the presence of salts. Bentonite, a native colloidal hydrated aluminium silicate (mainly montmorillonite), can precipitate under acidic conditions, and formulations must be at pH 6 or higher. A synthetic clay (colloidal silicon dioxide) is also useful for thickening both aqueous and nonpolar gels. The concentration of clay usually required to thicken formulations is 2–10%.

By far the most extensively employed gelling agents in the pharmaceutical and cosmetic industries are the carboxyvinyl polymers known as carbomers. These are synthetic high molecular weight polymers of acrylic acid, cross-linked with either allylsucrose or allyl ethers of pentaerythritol. Pharmaceutical grades of these carbomers are available (e.g., Carbopol 981NF; B. F. Goodrich Performance Materials). In the dry state, a carbomer molecule is tightly coiled, but when dispersed in water the molecule begins to hydrate and partially uncoil, exposing free acidic moieties. To attain maximum thickening effect the carbomer molecule must be fully uncoiled, and this can be achieved by one of two mechanisms (Fig. 2). The most common method is to convert the acidic molecule to a salt, by the addition of an appropriate neutralizing agent. For formulations containing aqueous or polar solvent, carbomer gellation can be induced by the addition of simple inorganic bases, such as sodium or potassium hydroxide. Less polar or nonpolar solvent systems may be neutralized with amines, such as triethanolamine or diethanolamine, or a number of alternative amine bases (e.g., diisopropanolamine, aminomethyl propanol, tetrahydroxypropyl ethylenediamine, and tromethamine) may be employed. Neutralization ionizes the carbomer molecule, generating negative charges along the polymer backbone, and the resultant electrostatic repulsion creates an extended three-dimensional structure. Care must be taken not to under- or overneutralize the formulation, as this will result in viscosity or thixotropic changes (38). Overneutralization will reduce viscosity, as the excess base cations screen the carboxy groups and reduce electrostatic repulsion. Hydrated molecules of carbomer may also be uncoiled in aqueous systems by the addition of 10–20% of hydroxyl donors, such as a nonionic surfactant or a polyol, that are able to hydrogen-bond with the polymer. Maximum thickening will not be as instantaneous using this mechanism, as it is with base neutralization, and may take several hours. Heating accelerates the process, but the system should not be heated to more than 70°C. Because they are synthetic, carbomer bases vary little from lot-to-lot, although batch-to-batch differences in mean molecular weight may result in variations in the rheological characteristics of aqueous dispersions (39).

3. Emulsions

The most common emulsions used in dermatological therapy are creams. These are two-phase preparations in which one phase (the dispersed or internal phase) is finely dispersed in the other (the continuous or external phase). The dispersed phase can

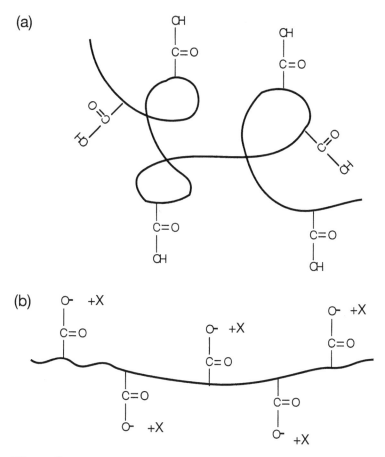

Figure 2 A tightly coiled carbomer molecule will (a) hydrate and swell when dispersed in water. (b) The molecule will completely uncoil to achieve maximum thickening when it is converted from the acid form to the salt form on neutralization.

have either a hydrophobic-based (oil-in-water creams; O/W), or be aqueous based (water-in-oil creams; W/O). Whether a cream is O/W or W/O depends on the properties of the system used to stabilize the interface between the phases. Because there are two incompatible phases in close conjunction, the physical stability of creams is always tenuous, although it may be maximized by the judicious selection of an appropriate emulsion-stabilizing system. In most pharmaceutical emulsions, the stabilizing systems comprise either surfactants (ionic or nonionic), polymers (nonionic polymers, polyelectrolytes, or biopolymers), or mixtures of these. The most commonly used surfactant systems are sodium alkyl sulfates (anionic), alkylammonium halides (cationic), and polyoxyethylene alkyl ethers or polysorbates (nonionic). These are often used alone, or in conjunction with nonionic polymerics, such as polyvinyl alcohol or poloxamer block copolymers, or polyelectrolytes, such as polyacrylic–polymethacrylic acids.

The physicochemical principles underlying emulsion formulation and stabilization are extremely complex and will not be covered in depth here. The interested reader is referred to the volume edited by Sjöblom (40). Briefly, an *emulsion* is formed when two immiscible liquids (usually, oil and water) are mechanically agitated. When this occurs in the absence of any form of interfacial stabilization during agitation, both liquids will form droplets that rapidly flocculate and coalesce into two phases on standing. *Flocculation* is the term used to describe the close accumulation of two or more droplets of dispersed phase without loss of the interfacial film, and it is largely the result of van der Waals attraction. The flocculated droplets may then coalesce into one large droplet, with the loss of the interfacial film. In practice, there is a brief period when one of the phases becomes the continuous phase because the droplets of this liquid coalesce more rapidly than the droplets of the other. Physical stability of an emulsion is determined by the ability of an additive to counteract the van der Waals attractions, thereby reducing flocculation and coalescence of the dispersed phase. This may be achieved in two ways: an increase in the viscosity of the continuous phase, which will reduce the rate of droplet movement, or the establishment of an energy barrier between the droplets, or both. Although increasing the viscosity of the continuous phase will reduce the rate at which droplets flocculate, in pharmaceutical shelf-life terms, a stable system can be generated in this way only if the continuous phase is gelled and the droplet diameter is smaller than 0.1 μm.

In pharmaceutical emulsions it is more common to develop stability using the energy barrier technique and to complement this stabilization, if necessary, by increasing the viscosity of the continuous phase. The basis of the energy barrier is that droplets experience repulsion when they approach each other. Repulsion can be generated either electrostatically, by the establishment of an electric double layer on the droplet surface, or sterically, by adsorbed nonionic surfactant or polymeric material. Electrostatic repulsion is provided by ionic surfactants that, when adsorbed at the oil–water interface, orient such that the polar ionic group enters the water. Some of the surfactant counterion (e.g., the sodium ion of sodium dodecyl sulfate) will separate from the surface and form a diffuse cloud surrounding the droplet. This diffuse cloud, together with the surface charge from the surfactant, forms the electric double layer, and electrostatic repulsion occurs when two similarly charged droplets approach each other. For obvious reasons, this method of emulsion stabilization is appropriate only for O/W formulations. In addition, it is important to appreciate that emulsions stabilized by electrostatic repulsion are extremely sensitive to the presence of additional electrolytes, which will disrupt the electrical double layer.

Steric repulsion may be produced using nonionic surfactants or polymers, such as polyvinyl alcohol or poloxamers. The specific distribution of the polyethoxylated nonionic surfactants and block copolymers (Fig. 3a and b) results in the formation of a thick hydrophilic shell of polyoxyethylene chains around the droplet. Repulsion is then afforded by both mixing interaction (osmotic repulsion) and entropic interaction (volume restriction), the latter as a result of a loss of configurational entropy of the polyoxyethylene chains when there is significant overlap. For polymeric materials without definitive hydrophobic and hydrophilic regions, the adsorption energy is critical to generation of steric repulsion. The adsorption energy must not be so low that there is no polymer adsorption, nor so high that there is complete polymer adsorption to the droplet. In either of these cases, there will be none of the loops, or tails, (see Fig. 3c) that are essential to steric repulsion. Polymeric steric repulsion is usually achieved using block copolymers, such as poloxamers, which consist of

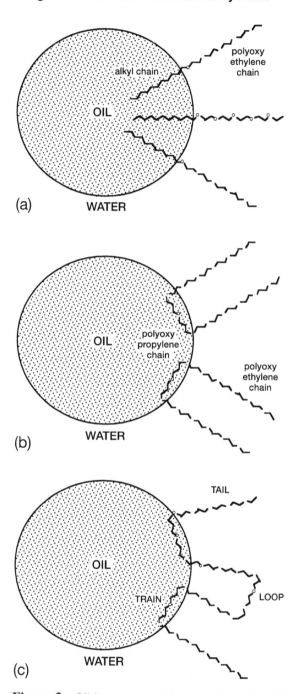

Figure 3 Oil-in-water emulsions may be stabilized by (a) nonionic surfactants, (b) poloxamer block copolymers, or (c) polymeric materials. The hydrophilic chains produce repulsion by mixing interaction (*osmotic*) or volume restriction (*entropic*).

linked polyoxyethylene and polyoxypropylene chains. More recently, polyacrylic acid polymers linked to hydrophobic chains (Pemulen; B. F. Goodrich) have been used as primary emulsification systems in O/W formulations and are listed in the *USP* as carbomer 1342. These materials form very stable emulsions because the polyacrylic acid chain, anchored to the oil droplet by the alkyl methacrylate moieties, considerably increases the surface charge on the oil droplet, forming a strong electrical barrier at the interface. Furthermore, emulsion stability is enhanced by an increase in the viscosity of the continuous phase.

Usually, for O/W emulsions, and always, for W/O emulsions, it is necessary to select an emulsification system based on surfactants. Although ionic surfactants can be used only for O/W emulsions, nonionic surfactants may be used for both O/W and W/O formulations. Although, at first glance, the choice of surfactant system appears limitless (there are hundreds to select from), there are some basic guidelines to aid the formulator. As the first priority, use of pharmaceutically approved surfactants (or those having a Drug Master File in place with the FDA) will save a considerable amount of regulatory justification. Raw material suppliers will frequently provide guidelines, although these will obviously be biased toward the use of their products. However, it should be appreciated that suppliers have considerable experience in the applications and uses of their products, and they are a very useful resource.

An approximate guide to emulsion formulation is provided by the hydrophilic–lipophilic balance (HLB) system that generates an arbitrary number (usually between 0 and 20) that is assigned to a particular surfactant. The HLB value of a polyoxyethylene-based nonionic surfactant may be derived from:

$$HLB = \frac{\text{mol\% hydrophobic group}}{5}$$

and the HLB of a polyhydric alcohol fatty acid ester (e.g., glyceryl monostearate) may be derived from

$$HLB = 20 \left(\frac{1 - S}{A} \right)$$

where S is the saponification number of the ester and A the acid number of the fatty acid. When it is not possible to obtain a saponification number (e.g., lanolin derivatives), the HLB can be calculated from

$$HLB = \frac{(E + P)}{5}$$

where E is the weight percent (wt%) of the polyoxyethylene chain and P the wt% of the polyhydric alcohol group in the molecule.

From the foregoing equations it is apparent that hydrophilic surfactants have high HLB values, and lipophilic surfactants have low HLB values. It is generally recognized that surfactants with HLB values between 4 and 6 are W/O emulsifiers and those with HLB values between 8 and 18 are O/W emulsifiers. It is also generally recognized, although poorly understood, that mixtures of surfactants create more stable emulsions than the individual surfactants. The overall HLB of a surfactant mixture (HLB$_M$) can be calculated from

$$\text{HLB}_M = f\text{HLB}_A + (1 - f)\text{HLB}_B$$

where f is the weight fraction of surfactant A. The required emulsifier HLB values for several oils and waxes are given in Table 3. Importantly, the HLB system can be used only as an approximation in emulsion design, and stability of an emulsion cannot be guaranteed by the use of an emulsifier mix with an appropriate HLB value. For example, creaming of an emulsion, a typical physical stability problem, is much more dependent on the viscosity of the continuous phase than the characteristics of the interfacial film.

Mixtures of surfactants create more stable emulsions than the individual surfactants. A reasonable and coherent explanation for this is given by the gel network theory (41,42). Briefly, this theory relates the consistencies and stabilities of O/W creams to the presence or absence of viscoelastic gel networks in the continuous phase. These networks form when there is an amount of mixed emulsifier, in excess of that required to stabilize the interfacial film, that can interact with the aqueous continuous phase. In its simplest form, a cream consists of oil, water, and mixed emulsifier. Official emulsifying waxes may be cationic (cationic emulsifying wax BPC; a mixture of cetostearyl alcohol and cetyltrimethylammonium bromide, 9:1), anionic (emulsifying wax BP; a mixture of cetostearyl alcohol and sodium dodecyl sulfate, 9:1) or nonionic (nonionic emulsifying wax BPC; a mixture of cetostearyl alcohol and cetomacrogol, 4:1). The theory dictates that when the cream is formulated it is composed of at least four phases (Fig. 4):

1. Bulk water
2. A dispersed oil phase
3. A crystalline hydrate
4. A crystalline gel phase composed of bilayers of surfactant and fatty alcohol separated by layers of interlamellar-fixed water

An examination of ternary systems (containing emulsifying wax and water, but no oil phase) by X-ray diffraction indicated that, for all emulsifying systems, addition of water caused swelling of the interlamellar spaces. Ionic emulsifying systems possess a greater capacity to swell than nonionic systems. Swelling in ionic systems is an electrostatic phenomenon, whereas in nonionic systems, it is due to hydration of the polyoxyethylene chains and is limited by the length of this chain. The gel network

Table 3 HLB Values for Several Oils and Waxes

Constituent	Emulsion type	
	O/W	W/O
Liquid paraffin	12	4
Hard paraffin	10	4
Stearic acid	16	—
Beeswax	12	5
Castor oil	14	—
Cottonseed oil	9	—

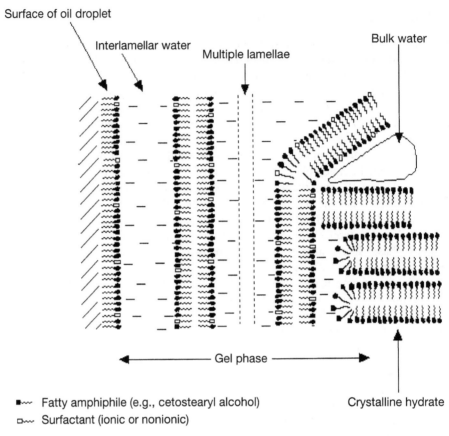

Surface of oil droplet

Interlamellar water

Multiple lamellae

Bulk water

Gel phase

■⌁ Fatty amphiphile (e.g., cetostearyl alcohol)
□⌁ Surfactant (ionic or nonionic)

Crystalline hydrate

Figure 4 The gel network theory suggests that when a cream is formulated it is composed of four phases: bulk water, a dispersed oil phase, a crystalline hydrate, and a crystalline gel phase composed of bilayers of surfactant and fatty alcohol separated by layers of interlamellar-fixed water. (Courtesy of Dr. G. M. Eccleston.)

theory also offers an explanation for the observation that nonionic O/W creams thicken on storage. This change is related to additional gelation in the continuous phase as a result of slow hydration of the polyoxyethylene chains of the surfactant, which reduces the amount of free water in the formulation.

A relatively stable emulsion formulation may be prepared from a simple four-component mixture: oil, water, surfactant, and fatty amphiphile. In practice, however, things are never that straightforward. In addition to the four principle components, a pharmaceutical emulsion formulation will also contain a drug, and is likely to contain a cosolvent for the drug, a viscosity enhancer, a microbiological preservative system, a pH adjusting–stabilizing buffer, and an antioxidant system. All of these additional components are required so that the formulation is capable of delivering the correct amount of drug to the therapeutic application site from a formulation that is free from microbial contamination and is essentially physically unchanged from the day of manufacture.

B. Preservation of Semisolid Formulations

All pharmaceutical semisolid formulations that are not sterilized unit-dose products can support the growth of microorganisms. Preservatives are ingredients that prevent or retard microbial growth and protect formulations from spoilage. The use of preservatives is required to prevent product damage caused by microorganisms during manufacture, storage, and inadvertent contamination by the patient during use. Similarly, preservatives serve to protect consumers from possible infection from contaminated products. The characteristics of an ideal preservative system are shown in Table 4. No single preservative meets all of these characteristics for all formulations (43), and it is often necessary to use a system containing a combination of individual preservatives. It is also important to appreciate that preservatives are intrinsically toxic materials so that a balance must be achieved between antimicrobial efficacy and dermal toxicity.

The most commonly used preservatives in pharmaceutical products are the parabens (alkyl esters of *p*-hydroxybenzoic acid, such as methyl and propyl paraben). These compounds are highly effective against both gram-positive bacteria and fungi at low concentrations (e.g., 0.1–0.3% parabens combinations provide effective preservation of most emulsions). Because of their widespread use, the toxicological profiles of the parabens have been extensively researched, and the safety in use of the lower esters (methyl, ethyl, propyl, and butyl) has been established (45). Other preservatives widely used in topical pharmaceutical formulations include benzoic acid, sorbic acid, benzyl alcohol, phenoxyethanol, chlorocresol, benzalkonium chloride, and cetrimide. Each has particular advantages and disadvantages, which makes combination preservatives particularly effective. For example, although methyl paraben is highly active against gram-positive bacteria and moderately active against yeasts and molds, it is only weakly active against gram-negative bacteria. However, a combination of methyl paraben with phenoxyethanol provides a preservative system that is also highly active against gram-negative species.

The acid preservatives (benzoic and sorbic acids) are active only as free acids and formulations containing these preservatives must be buffered to acid pH values (pH < 5). A list of pharmaceutical preservatives useful in topical formulations is given in Table 5, together with their microbiological and physicochemical properties. More information on preservatives and preservative systems may be found in the *British Pharmaceutical Codex* (The Pharmaceutical Press) and in the excellent text by Orth (43).

Table 4 Characteristics for an Ideal Preservative

1. Effective at low concentrations against a wide spectrum of microbes
2. Soluble in the formulation at the required concentration
3. Nontoxic and nonsensitizing to the consumer at in-use concentrations
4. Compatible with other formulation components
5. No physical effect on formulation characteristics
6. Stable over a wide range of pH and temperature
7. Inexpensive

Source: Ref. 44.

Table 5 Microbiological and Physicochemical Properties of Selected Preservatives

Preservative	Antimicrobial activity[a]				In-use Conc (%)	pH range[b]	O/W[c]
	Gram+	Gram−	Molds	Yeasts			
Benzoic acid	1	2	3	3	0.1	2–5	3–6
Sorbic acid	2	2	2	1	0.2	<6.5	3.5
Phenoxyethanol	2	1	3	3	1.0	Wide	—
Methyl paraben	1	3	2	2	0.4–0.8	3.0–9.5	7.5
Propyl paraben	1	3	2	2	0.4–0.8	3.0–9.5	80
Butyl paraben	1	3	2	2	0.4–0.8	3.0–9.5	280
Chlorocresol	1	2	3	3	0.1	<8.5	117–190
Benzalkonium Cl	1	2	3	2	0.01–0.25	4–10	<1
Cetrimide	1	2	3	2	0.01–0.1	4–10	<1

[a]1, highly active; 2, moderately active; 3, weakly active.
[b]Optimal pH range for activity.
[c]Oil–water partition coefficient.

Interestingly, despite their widespread use and excellent safety profile, there is increasing interest in the potential systemic exposure to preservatives following application in pharmaceutical and cosmetic products. Skin penetration data are not available in the literature for many preservatives, and much of the publicly available data has been obtained under conditions inappropriate for risk assessment. By far the most available data concern the parabens.

Although most studies on skin permeation of parabens have evaluated one or two of the homologous series, recently Dal Pozzo and Pastori [46] reported on the *in vitro* human skin permeation of six parabens (methyl, ethyl, propyl, butyl, hexyl, and octyl esters). The permeants were applied to abdominal epidermal membranes, mounted in diffusion chambers, either as solid compounds (deposited in acetone) or as saturated solutions in various vehicles, including three typical emulsion formulations (two O/W and one W/O). Following application of the unformulated pure substance, maximum flux decreased with increasing lipophilicity, from 65.0 μg/cm^2 h^{-1} for methyl paraben to 13.7 μg/cm^2 h^{-1} for hexyl paraben. Similarly, when applied as saturated aqueous solutions, the maximum flux decreased with increasing lipophilicity. In the latter case, however, when flux was normalized to the vehicle concentration of the permeant, the permeability coefficient increased with increasing lipophilicity. Addition of 50% propylene glycol or 20% polyethylene glycol 400 to the aqueous solutions did not alter the profile of permeation, although the permeability coefficients were somewhat reduced. On the other hand, when the parabens were dissolved in liquid paraffin, the relation between permeability coefficient and lipophilicity appeared parabolic, maximum flux occurred for the butyl ester, and the highest permeability coefficient occurred for the propyl ester. These results are consistent with theory and demonstrate that the application vehicle can significantly affect skin permeability characteristics of compounds.

In emulsion systems the existence of two distinct phases (oil and water) results in distribution of the parabens according to their physicochemical characteristics, and this can be influenced by the presence of other ingredients, such as cosolvents and surfactants. Furthermore, these excipients may also affect skin barrier properties. It was observed that permeation of the parabens from two O/W emulsions was higher than expected, based on the data from simple vehicles, and that permeation from the O/W emulsions was higher than that from the W/O emulsion (46). These results were rationalized on the basis of permeant release from the formulation, and it was assumed that the external lipid phase of the W/O emulsion retained parabens.

It is difficult, however, to relate this data to conditions of actual consumer exposure because, in the DalPozzo and Pastori study (46), the formulations were applied at infinite dose (which effectively generated occlusive conditions) and at an artificially high permeant concentration (0.7% w/w). When low finite doses are applied to the skin surface, the vehicle is continually changing. Whether the vehicle is O/W or W/O, the water content will be released by the shear forces generated by application to the skin, most of the water will evaporate, and this should tend to reduce the skin penetration of these preservatives.

It is theoretically possible to reduce skin absorption of the parabens by formulation manipulation. This may be achieved by altering the distribution pattern of the preservative within the formulation or by complexation. Care must be taken, however, to ensure that any formulation modification does not interfere with the antimicrobial activity of the preservative system (47). One such modification involved the complexation of methyl paraben with 2-hydroxypropyl-β-cyclodextrin (48). Although the aqueous solubility of methyl paraben was increased considerably in the presence of the cyclodextrin, the percutaneous penetration of the preservative through hairless mouse skin in vivo over 24 h was reduced by 66%. Even if this system clearly had benefit in terms of potential reduction in systemic absorption of the preservative, there was no indication of any possible alteration in preservative efficacy. Incorporation of butyl paraben into liposomes prepared from phosphatidyl-choline–cholesterol–dicetyl–phosphate had no effect on preservative efficacy, although the antimicrobial effect was proportional to the free, and not the total, concentration of the preservative (49). A further study by these authors demonstrated that incorporation of butyl paraben in some liposomal systems had little effect on the permeation of the preservative across guinea pig skin in vivo or in vitro (50), although incorporation of increasing amounts of lipid into the liposome tended to decrease the percutaneous absorption.

Although there is considerable evidence that parabens can penetrate the skin, permeation and systemic availability of intact compounds are likely to be considerably reduced by transcutaneous and systemic metabolism. Furthermore, because these preservatives are present at concentrations of 0.1–0.2% w/w in topical pharmaceutical formulations, in use, dermal exposure to these compounds is relatively low. In the cosmetic industry there is a trend toward preservative-free and self-preserving formulations (51). However, before taking this route, the pharmaceutical formulator must consider the potential implications on efficacy and safety of the product.

C. Drug Release from Semisolid Formulations and SUPAC-SS

Determination of the ability of a semisolid formulation to release a drug, and the pattern and rate of this release, are important aspects of formulation development and optimization. However, it is also important to appreciate that the data generated must not be over interpreted. Release studies normally involve measurement of drug diffusion out of a mass of formulation into a receiving medium that is separated from the formulation by a synthetic membrane (52,53). A detailed study of the data from this type of system can generate invaluable data concerning the physical state of the drug in the formulation. For example, examination of the early experimental models and their refined updates that were derived to describe drug release from semisolids, reveals that release patterns are dependent on whether the drug is present as a solution or suspension within the formulation (54–56). These subtle differences, together with variations in the rate of release, may be used to determine such parameters as drug diffusivity within the formulation matrix, the particle size of suspended drug, and the absolute solubility of a drug within a complex formulation (Fig. 5) (57). However, it is generally agreed that drug release-rate data cannot predict skin permeation or bioavailability. Application of a formulation to the skin is an ephemeral situation. The formulation will undergo considerable shearing forces, solvents will evaporate, and excipients may interact with the skin and modulate bioavailability. Nonetheless, release-rate determinations are important for purposes other than formulation development and characterization.

The FDA has issued a guidance document (SUPAC-SS Nonsterile Semisolid Dosage Forms, U.S. Department of Health and Human Services, Food and Drug

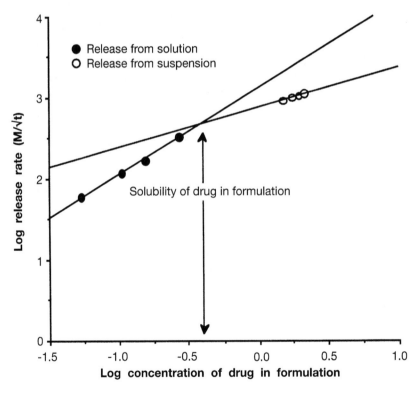

Figure 5   Illustration of the use of release rates from semisolid preparations to determine drug solubility within a formulation. Data shows the release rates of benzocaine from propylene glycol/water gels as a function of drug concentration in the formulation. (From Ref. 85.)

Administration, Center for Drug Evaluation and Research, May 1997) that recommends the use of in vitro drug-release testing in the scale-up and postapproval changes for semisolids (SUPAC-SS). The FDA intends to promote the use of this test as a quality assurance tool to monitor minor differences in formulation composition or changes in manufacturing sites, but not yet as a routine batch-to-batch quality control test. Thus, although the agency is suggesting in vitro release rate data for level 2 and level 3 changes in formulation components and composition, such data are not required for a level 1 change. In the former case, the in vitro release rate of the new or modified formulation should be compared with a recent batch of the original formulation, and the 90% confidence limit should fall within the limits of 75–133%. Similarly, in vitro release testing is suggested for level 2 changes in manufacturing equipment, processes, and scale-up, and level 3 changes in manufacturing site. Recently, the use of in vitro testing as a quality assurance tool has been questioned, especially for a hydrophilic formulation containing the highly water-soluble drug, ammonium lactate (58), for which the method was insufficiently specific to differentiate between small differences in drug loading or minor compositional and processing changes.

D. Formulation of Transdermal Products

This section concentrates on those aspects of development specific to transdermal delivery systems.

1. Pharmacokinetic Modeling

To assess the feasibility of the transdermal route for a specific therapeutic agent, several pharmacokinetic factors, including rate of absorption and elimination, must be considered. Guy and Hadgraft (59) developed an early kinetic model for topical drug delivery. Various rate constants were described including release rate from the device (first- or zero-order); back diffusion from the stratum corneum to the patch (usually insignificant); diffusion rates of the compound across the stratum corneum and viable epidermis; reverse rate constant (enabling prediction of the stratum corneum–viable epidermis partition coefficient) and the plasma clearance rate constant. This model has been used to predict the plasma concentration of several transdermally administered drugs and has shown remarkable similarities between predicted and actual plasma concentration profiles, suggesting that the model is useful for evaluating the feasibility of potential transdermal drug delivery candidates.

Although this model provides predictive quantitative data based mainly on physicochemical and pharmacokinetic parameters, other aspects of the percutaneous absorption process are more difficult to predict. Of most concern is the potential for metabolic degradation of the drug, but only limited data are available on quantitative aspects of cutaneous metabolism. Modeling is possible, by estimating metabolic rate constants and varying the possible residence time of the drug in the skin. More lipophilic drugs, which have a relatively long residence time in the skin, will be more exposed to cutaneous metabolic enzymes and be less bioavailable than their hydrophilic counterparts.

More recent models for prediction of skin permeation and pharmacokinetics have become increasingly elaborate and take into account such parameters as the hydrogen-bonding capability of the permeant and potential vehicle effects (60).

2. Biopharmaceutical Considerations

A fundamental consideration in the development of transdermal therapeutic systems is whether the dermal delivery route can provide the requisite bioavailability for drug effectiveness. This is ultimately determined by the skin-penetration rate of the drug, the potential for metabolism during permeation across the skin, and the biological half-life of the drug. Penetration rates may be modified, if necessary, by the use of penetration enhancers, but drug metabolism and plasma clearance cannot be influenced by any simple means. Although prediction of skin penetration and bioavailability of drugs from transdermal therapeutic systems has been reasonably accurate, there is no doubt that testing of formulated patches in vitro and in vivo will continue to be the most accurate means of evaluating their usefulness.

A wide variety of experimental approaches have been developed for in vitro drug permeability determination through skin, and guidelines have been established to rationalize this aspect of pharmaceutical development (see Chapter 5). In the early stages of product development, skin penetration rates from prototype vehicles and patches are usually determined in vitro using simple diffusion cells and skin from a variety of animals. Although, the use of in vitro systems provides little quantitative information on the transcutaneous metabolism of candidate drugs, a major advantage is that experimental conditions can be controlled precisely so that the only variables are in the prototype formulations. In the latter stages of product development, when quantitative skin permeation data is required, human skin should be the membrane of choice for use in in vitro systems. Although methods are available to improve the sensitivity of in vitro skin penetration measurements (61), it is essential, at this stage, to ensure that account is taken of the inherent variability in human skin permeation.

Factorial design and artificial neural networks have been used in the optimization of transdermal drug delivery formulations using in vitro skin permeation techniques (62–64). For example, Kandimalla et al. (63) optimized a vehicle for the transdermal delivery of melatonin using the response surface method and artificial neural networks. Briefly, three solvents (water, ethanol, and propylene glycol) were examined either as single solvents or binary and ternary mixtures. Measurements of skin flux, lag time, and solubility were made for ten vehicles and compared with values predicted from both a response surface generated from a quartic model and an artificial neural network employing a two-layered back-propagation network with all ten design points in the hidden layer. Predictability of flux using both statistical techniques was good (Table 6), suggesting that such models may be useful in preliminary formulation optimization.

A major drawback of transdermal delivery systems is the potential for localized irritant and allergic cutaneous reactions. At the earlier stages of formulation development, it is, therefore, important to evaluate both drugs and excipients for their potential to cause irritation and sensitization (see Chapters 10 and 11). This is true for all transdermal systems, but especially for those that may stay in place for prolonged periods. The degree of primary and chronic irritation, and the potential to cause contact allergy, photoirritation, and photoallergy should be determined. Normally, the drug and excipients are initially separately evaluated for contact irritation and sensitization in animal models before evaluation in human subjects. It must, however, be emphasized that animal data are often not predictive of the human situation. Evaluation of skin irritation and delayed contact hypersensitivity should

Table 6 Experimental Versus Predicted Flux[a] of Melatonin

Application vehicle[b]	Experimental (μg/cm^2 h^{-1})[c]	ANN Prediction (μg/cm^2 h^{-1})	RSM Prediction (μg/cm^2 h^{-1})
W:E:P (20:60:20)	11.32 $\pm$ 0.86	12.17	12.73
W:E (40:60)	10.89 $\pm$ 1.36	11.83	11.76
W:P (40:60)	7.54 $\pm$ 1.39	6.75	7.50

[a]Flux was predicted on the basis of artificial neural networks (ANN) or response surface methods (RSM).
[b]W, water; E, ethanol; P, propylene glycol.
[c]Data are means $\pm$ standard deviation ($n = 3$) and represent flux through rat dorsal skin.
Source: Ref. 63.

always be carried out using the final and complete formulation in human volunteers. Fortunately, most of the observed skin reactions to transdermal systems are transient and mild and disappear within hours of patch removal.

3. Design Considerations

All patch-type transdermal delivery systems developed to date can be described by three basic design principles: drug in adhesive, drug in matrix (usually polymeric), and drug in reservoir (Fig. 6). In the latter the reservoir is separated from the skin by a rate-controlling membrane. Although there are many differences in the design of transdermal delivery systems, several features are common to all systems including the release liner, the pressure-sensitive adhesive and the backing layer, all of which must be compatible for a successful product. For example, if a system is designed in such a way that the drug is intimately mixed with adhesive, or diffuses from a reservoir through the adhesive, the potential for interaction between drug and adhesive, which can lead to either a reduction of adhesive effectiveness, or the formation of a new chemical species, must be fully assessed. Similarly, residual monomers, catalysts, plasticizers, and resins may react to give new chemical species. Additionally, the excipients, including enhancers, or their reaction products, may interfere with adhesive systems. Incompatibilities between the adhesive system and other formulation excipients, although undesirable, may not necessarily be impeding and designs in which the adhesive is remote from the drug delivery area of the system may be developed (see Fig. 6d). There are three critical considerations in the selection of a particular system: adhesion to skin, compatibility with skin, and physical or chemical stability of total formulation and components.

All devices are secured to the skin by a skin-compatible pressure-sensitive adhesive. These adhesives, usually based on silicones, acrylates, or polyisobutylene, can be evaluated by shear-testing and assessment of rheological parameters. Standard rheological tests include creep compliance (which measures the ability of the adhesive to flow into surface irregularities), elastic index (which determines the extent of stretch or deformation as a function of load and time) and recovery following deformation. Skin-adhesion performance is based on several properties, such as initial and long-term adhesion, lift, and residue. The adhesive must be soft enough to ensure initial adhesion, yet have sufficient cohesive strength to remove cleanly, leaving no residue. Because premature lift will interfere with drug delivery, the cohesive and

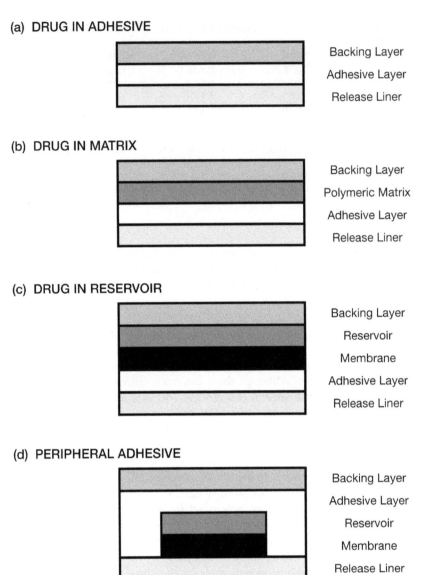

Figure 6 Typical transdermal drug delivery system designs.

adhesive properties must be carefully balanced and maintained over the period of intended application. This can be evaluated only by wear-testing, in which a placebo patch is applied to skin.

Skin adhesion is affected by shape, conformability, and occlusivity and round patches tend to be more secure than those of sharply angled geometry. If the patch is able to conform to the skin contours it resists lifting and buckling with movement. The presence of water may affect adhesive properties; therefore, the occlusivity of the system must be taken into consideration. Occlusion for prolonged periods can lead to excessive hydration and problems associated with microbial growth that may increase the possibility of irritation or sensitization to the various system components.

The backing material and release liner can be fabricated from a variety of materials, including polyvinylchloride, polyethylene, polypropylene, ethylene vinyl acetate, and aluminium foil. The principal requirement is that they are impervious to the drug and other formulation excipients. The most useful backing materials are those that conform with the skin and provide sufficient resistance to transepidermal water loss to allow some hydration of the stratum corneum, yet maintain a healthy subpatch environment. The release liner must be easily separated from the adhesive layer without lifting off any of the adhesive to which it is bound. Liners are usually films or coated papers, and silicone release coatings are used with acrylate- and rubber-based adhesive systems, whereas fluorocarbon coatings are used with silicone adhesives (65).

4. Drug and Enhancer Incorporation

The three principal methods for incorporating the active species into a transdermal system have led to the loose classification of patches as membrane, matrix, or drug-in-adhesive types. It is, however, quite possible to combine the main types of patch; for example, by placing a membrane over a matrix, or using a drug-in-adhesive in combination with a membrane–matrix device to deliver an initial bolus dose.

Membrane patches contain a delivery rate-controlling membrane between the drug reservoir and the skin. Microporous membranes, which control drug flux by the size and tortuosity of pores in the membrane, or dense polymeric membranes, through which the drug permeates by dissolution and diffusion, may be used. Several materials can be used as rate-controlling membranes (e.g., ethylene–vinyl acetate copolymers, silicones, high-density polyethylene, polyester elastomers, and poly-acrylonitrile). Ideally, the membrane should be permeable only to the drug and enhancer (if present) and should retain other formulation excipients. Membranes have also been designed such that they allow differential permeation of an enhancer and drug (66–68). This type of membrane, sometimes designated as a one-way membrane, is useful when the drug is present in the adhesive and the enhancer is formulated in a reservoir.

Asymmetric polymeric membranes have also been evaluated for use in transdermal delivery systems (69). Asymmetric poly(4-methyl-1-pentene) membranes, fabricated using a dry–wet inversion method, were used to control the delivery of nitroglycerin. The release rates of nitroglycerin were strongly influenced by the nature and amount of the nonsolvent (butanol) used, together with the solvent (cyclohexane), in the casting process. This is, perhaps, not surprising, as increasing the amount of nonsolvent increases the porosity of the cast membrane. The concept of fine-tuning delivery of a drug through a given membrane by subtle adjustment of the porosity creates some exciting new possibilities in transdermal technology (70).

In all marketed membrane-controlled transdermal systems, the rate-controlling membrane is fabricated from synthetic polymeric materials. Thacharodi and Rao (71) evaluated the potential of two biopolymers (fetal calf skin collagen and chitosan) in membrane systems for delivery of nifedipine. Chitosan (deacetylated chitin) is a widely distributed major constituent of the shells of marine shellfish. It was concluded that the permeability of both biopolymers could be readily adjusted by altering the fabrication method or cross-linking and, because these polymers were biocompatible, they were more suitable for use as rate-controlling membranes in transdermal systems.

A variety of materials can be used in the drug reservoir, ranging from simple formulations (such as mineral oil), to complex formulations (such as aqueous–alcoholic solutions and gels, with or without various cosolvents, and polymeric materials). A definite requirement for a reservoir system is that it can permit zero-order release of the drug over the delivery period. Essentially, this requires that the reservoir material is saturated with the drug over the period of product application, which can be achieved by formulating the drug as a suspension.

The second type of transdermal system is the matrix design, in which the drug is uniformly dispersed in a polymeric matrix, through which it diffuses to the skin surface. Here, the polymeric matrix, which may comprise silicone elastomers, polyurethanes, polyvinyl alcohol, polyvinylpyrrolidones, and such, may be considered the drug reservoir. Several steps are involved in the drug delivery process: principally dissociation of drug molecules from the crystal lattice, solubilization of the drug in the polymer matrix, and diffusion of drug molecules through the matrix to the surface of the skin. Many variables may affect the dissolution and diffusion rates, making it particularly difficult, but not impossible, to predict release rates from experimental or prototype formulations (72). For a drug to be released from a polymeric matrix under zero-order kinetics, the drug must be maintained at saturation in the fluid phase of the matrix, and the diffusion rate of the drug within the matrix must be much greater than the diffusion rate in the skin.

Several methods can be used to alter the release rate of a drug or an enhancer from a polymeric matrix, and some of these are illustrated by a study on release of several drugs from silicone matrices (73). Silicone medical-grade elastomers (polydimethylsiloxanes) are flexible, lipophilic polymers, with excellent compatibility with biological tissues, that can be coformulated with hydrophilic excipients, such as glycerol, and inert fillers, such as titanium dioxide, to alter release kinetics. Increasing the amount of glycerol in the matrix increased the release rate of indomethacin, propranolol, testosterone, and progesterone, whereas incorporation of inert fillers (titanium dioxide or barium sulfate) tended to reduce the release rate. Hydrophilic drug-release rates from polydimethylsiloxane matrices were also increased by up to three orders of magnitude using polydimethylsiloxane–polyethylene oxide graft copolymers (74). These data demonstrate that release rates can be modulated to achieve a desired profile by simple formulation modification.

Perhaps the simplest form of transdermal drug delivery device, which is now most commonly employed, is the drug-in-adhesive system. This involves formulating the drug, and enhancer if present, in an adhesive mixture that is subsequently coated onto a backing membrane, such as a polyester film, to produce an adhesive tape. This simplicity is, however, deceptive and several factors, involving potential interaction between drug or enhancer and the adhesive, need to be considered. These can involve chemical interactions resulting in interference with adhesive performance, breakdown of the active species, or formation of new chemical entities. Additionally, the physicochemical characteristics of the drug and adhesive system may provide very different release rates for hydrophilic and hydrophobic drugs: for example, silicone adhesives are typically lipophilic, which limits solubility of hydrophilic drugs within the adhesive matrix.

Incorporation of other excipients, such as skin permeation enhancers, into a drug-in-adhesive system may alter drug release rates and adhesive properties. For example, addition of 1% urea to a polyacrylate pressure-sensitive adhesive resulted in loss of adhesion and skin contact could not be maintained over the required period (75). A strategy to reduce the influence of drug and enhancer on adhesive properties was to design a system in which there was no contact between these constituents and the adhesive, by limiting the adhesive to a boundary laminate that surrounded a drug–enhancer-releasing layer. One disadvantage of this type of system, however, is that the drug–enhancer-releasing layer may not remain in intimate contact with the skin. If high levels of liquid skin penetration enhancers are incorporated into drug-in-adhesive transdermal patches, there is likely to be a loss in cohesiveness, resulting in patch slipping and skin residues following patch removal. Cohesive strength, can be increased by high levels of cross-linking in acrylate adhesives, but this may alter both long-term bonding and drug release rates. These problems may be overcome by use of grafted copolymer adhesives, such as ARcare ETA Adhesive Systems (76), for which reinforcement is achieved mainly through phase separation of the side chain within the continuous polymer network. A variety of side chains are available and enhancer concentrations up to 30% can be incorporated without seriously affecting the adhesive properties. This work has been limited to fatty acid ester type enhancers and application to other enhancer types remains to be established. It may also be possible to maintain adhesive properties in the presence of skin penetration enhancers by using different molecular weight blends of acrylic copolymers (77,78).

When enhancers are incorporated into transdermal systems it is important to appreciate that it is a fundamental requirement that the enhancer, as well as the drug, is released by the adhesive. Furthermore, it is probable that the presence of the enhancer may increase ski permeation of other formulation excipients and that this may have an influence on local toxicity. Much remains to be done in the field of enhancer incorporation into transdermal drug delivery systems and it is encouraging to observe the increasing efforts of adhesive manufacturers in this sphere.

5. System Manufacture and Testing

The manufacturing processes for reservoir, matrix, and drug-in-adhesive transdermal systems are, to a large extent, similar. All involve the following stages: preparing the drug; mixing the drug (with other excipients and penetration enhancers, if required) with the reservoir, matrix, or adhesive; casting into films and drying (or molding and curing); laminating with other structural components (e.g., backing layer, rate-controlling membrane, and release liner); die-cutting; and finally, packaging. Casting and lamination are the most critical steps in the manufacturing process: tensions and pressures must be carefully controlled to provide a wrinkle-free laminate that ensures reproducible adhesive-coating thickness and uniform drug content (79).

As with all controlled-release delivery systems, final product checks include content uniformity, release-rate determination, and physical testing. Content-uniformity evaluation involves removing a random sample of patches from a batch and assaying the amount of drug present. Of the several methods available for determining drug release rates from controlled-release formulations, the U.S. Pharmaceutical Manufacturers Association (PMA) Committee (80) has recommended three: the "paddle over disk" (which is identical with the *USP* paddle dissolution apparatus, except that the transdermal system is attached to a disk or cell resting at the bottom of the vessel that contains medium at 32°C); the "cylinder-modified *USP* basket" (which is similar to the *USP* basket method, except that the system is attached to the surface of a hollow cylinder immersed in medium at 32°C), and the "reciprocating disk" (in which patches attached to holders are oscillated in small volumes of medium, allowing the apparatus to be useful for systems delivering low concentrations of drug). Researchers at the FDA have developed a modified paddle procedure (essentially the "paddle over disk" method) for determining drug release from transdermal systems (81). One problem with the original method was the mode of maintaining the patch in position in the dissolution beaker, and a device to improve and maintain placement of the patch was subsequently suggested (82). Although the industry is moving rapidly in method development, much remains to be established in the field of pharmacopeal standards for transdermal drug delivery systems.

IV. REVIEW OF CLINICAL USE OF TRANSDERMAL SYSTEMS

Despite the barriers to systemic delivery of drugs through intact human skin, there have been several successful transdermal products. The remainder of this chapter describes recent clinical studies on several of these products, including those delivering nicotine, glyceryl trinitrate, estrogens, androgens, fentanyl, clonidine, and nonsteroidal anti-inflammatory agents (NSAIDs).

A. Nicotine

1. Smoking Cessation

Data on efficacy, safety, and pharmacoeconomics of transdermal (TD) nicotine therapy for smoking cessation up to 1994 were reviewed (86). TD nicotine more than doubled success rates of smoking cessation in motivated subjects smoking 10–15, or more, cigarettes a day. Application site reactions (erythema or burning ≤16%, transient itch ≤50%) caused discontinuation in 10%, or fewer of subjects. Sleep disturbance due to nocturnal nicotine absorption occurred in less than 13% of subjects. Attempts to develop a nicotine transdermal system (TDS), with reduced adverse skin reactions, used in vitro human cadaver skin permeation to demonstrate a permeation profile comparable with those from Habitrol and Nicoderm, and a clinical study compared systemic bioavailability and pharmacokinetic profiles (87). Immediate effects of TD nicotine on sleep architecture, snoring, and sleep-ordered breathing in 20 nonsmoking subjects with a history of habitual snoring receiving placebo or a nicotine TD system delivering 11 mg/24 h were reported (88). Patches were applied at 6 PM and removed after 12 h. Mean nicotine level was nondetectable with placebo and 7.8 ± 2.3 ng/mL with active therapy. Nicotine significantly decreased total sleep time by 33 min, sleep efficiency from 89.7 to 83.5%, and rapid eye movement (REM) sleep from 18.8 to 15.1%. Although sleep index was unchanged, mean snoring intensity decreased by 1.1 dB (p = 0.01) with nicotine. Nausea (50%) and vomiting (20%) were predominant side effects.

Concurrent administration of the nicotine antagonist mecamylamine and nicotine TD was evaluated (89) in a trial with 48 healthy smokers. Nicotine TD (6–8 weeks) plus oral mecamylamine (2.5–5 mg twice a day for 5 weeks) was compared with nicotine TD plus placebo. Mecamylamine treatment began 2 weeks before smoking cessation. At 7 weeks with mecamylamine, the continuous abstinence rate tripled (50%) and point abstinence doubled (58%) compared with that for placebo. Continuous abstinence was threefold higher for mecamylamine at 6 months and ninefold higher at 12 months. It was concluded that agonist–antagonist therapy may substantially improve smoking cessation. The pharmacokinetic and pharmacodynamic interactions between TD mecamylamine and intravenous nicotine were investigated (90). Cigarette smokers received TD mecamylamine (6 mg/24 h) and placebo patches for 7 days each. On day 5, subjects received a combined infusion of deuterium-labeled nicotine and cotinine; the disposition kinetics of nicotine and cotinine, and cardiovascular and plasma catecholamine responses to nicotine were measured. Fifty percent of subjects were studied under alkaline urine conditions and 50% under acidic urine conditions. Steady-state plasma mecamylamine concentrations were double (12.2 vs. 6.3 ng/mL), consistent with lower renal clearance (2.1 vs. 5.8 mL/min kg^{-1}) during alkaline compared with acidic urine conditions. Mecamylamine did not significantly affect clearance of nicotine or cotinine, but did significantly reduce the volume of distribution and inhibited the epinephrine-releasing effects of nicotine. It was concluded that mecamylamine had little effect on nicotine clearance and was not expected to affect steady-state levels during TD nicotine dosing. Reduction of the volume of distribution of nicotine by mecamylamine suggested that part of the antagonism of nicotinic CNS effects by mecamylamine may be due to a pharmacokinetic interaction, probably decreased binding to nicotine receptors, or transport of nicotine into the brain, and may decrease potential adverse cardiovascular effects of coadministered nicotine.

Stable isotope-labeled nicotine was infused simultaneously with TD nicotine to determine absolute bioavailability and absorption kinetics from TD systems of different designs (91). Rapid intravenous infusion of nicotine slowed absorption of TD nicotine, probably by constricting dermal blood vessels, thereby limiting percutaneous absorption. A study evaluated safety and tolerability of a 44-mg/day dose for smoking cessation in subjects who smoked 20 or more, cigarettes per day (92). Smokers received 44 mg/day of TD nicotine for 4 weeks, followed by 4 weeks at 22 mg/day. Thirty-eight of forty patients (95%) completed the initial 4 weeks, and 36 (90%) completed the entire study. Confirmed smoking cessation rates were 65% (4 weeks) and 55% (8 weeks), and self-reported smoking cessation at 3 months was 50%. Sleep complaints were reported by 33% of subjects during the 44-mg phase. It was concluded that 44 mg/24 h nicotine patch therapy in heavy smokers was safe, tolerable, and without significant adverse events. Efficacy and safety of 22- and 44-mg doses of TD nicotine paired with minimal, individual, or group counseling to improve smoking cessation rates were compared in an 8-week clinical trial using daily cigarette smokers (15 or fewer cigarettes per day for ≥ 1 year) and random assignment to dose and counseling conditions (93). Four weeks of 22- or 44-mg transdermal nicotine followed by 4 weeks of dosage reduction (2 weeks of 22 mg followed by 2 weeks of 11 mg). Smoking cessation rates for TD doses and levels of counseling did not differ significantly 8 or 26 weeks postquitting date. In those receiving minimal contact, 44 mg produced greater abstinence (68%) at 4 weeks than 22 mg (45%). Participants with minimal-contact adjuvant treatment were less abstinent at 4 weeks than those receiving individual or group counseling (56 vs. 67%). Although a 44-mg dose decreased desire to smoke more than a 22-mg dose, this effect was unrelated to success in quitting smoking. The 44-mg dose produced significantly higher frequencies of nausea (28%), vomiting (10%), and erythema, with edema at the patch site (30%) than did the 22-mg dose (10, 2, and 13%, respectively). It was concluded that there was no general, sustained, benefit of initiating TD nicotine therapy with a 44-mg dose or providing intense adjuvant cessation treatment.

The effects of using nicotine gum with nicotine TD were evaluated in healthy subjects (374) at their work-setting in a 1-year trial (94): 149 subjects received nicotine TD plus nicotine gum; 150 nicotine TD plus placebo gum; and 75 placebo TD plus placebo gum. Treatment duration was 12 weeks with a 16-h 15-mg TDS, then 10 mg (6 weeks) and 5 mg (8 weeks). Gum use was unrestricted during the first 6 months, with recommendation to use at least four pieces a day. It was concluded that adding nicotine gum use to nicotine TD use in subjects smoking 10, or more, cigarettes a day increased abstinence rates. Efficacy of TD nicotine as an adjunct to advice and support in patients attending hospital was investigated (95) in 234 in- and outpatients with smoking-related respiratory or cardiovascular disease (ages 18–75 years) advised to stop smoking. Advice was reinforced by a Smoking Cessation Counselor initially and at 2, 4, 8, and 12 weeks, supplemented by a 24-h TDS in adjusted doses over the study period. Patients, who no longer smoked at 12 weeks, were followed up at 26 and 52 weeks and then self-reported abstinence was validated. Of patients receiving TD nicotine 24/115 (21%) were verified as non-smokers at 12, 26, and 52 weeks, compared with 17 of 119 (14%) of the placebo group. Cessation was related to increasing age and lower Fagerstrom score. Minor skin reactions and nausea were more frequent in the nicotine group (47 vs. 34% and 12 vs. 3%,, respectively), but severe skin reactions were rare (about 5%).

Nicotine concentrations in gastric juice, saliva, and plasma were monitored after Nicorette TD systems (15 mg/16 h) were applied to seven healthy volunteers (96). Nicotine concentration was highest in gastric juice, followed by saliva then plasma which suggested ion-trapping of nicotine base in acidic gastric juice.

The incremental cost-effectiveness (based on physician time and retail cost of nicotine TD and benefits, based on quality-adjusted life years; QALYs), of addition of nicotine TD to smoking cessation counseling was investigated in men and women smokers (25–69 years of age) receiving primary care (97). TD use produced one additional lifetime quitter at a cost of 7,332 dollars. Incremental cost-effectiveness of nicotine TD by age group ranged from 4,390 to 10,943 dollars per QALY for men and from 4,955 to 6,983 dollars per QALY for women. A clinical strategy involving limiting prescription renewals to patients successfully abstaining for the first 2 weeks improved cost-effectiveness by 25%. These data provided support for routine use of nicotine TD as an adjunct to smoking cessation counseling and for health insurance coverage of TD nicotine therapy. Metanalysis estimated cost-effectiveness of nicotine TD as an adjunct to brief physician counseling during routine office visits (98). Depending on age, average costs per year of life saved ranged from 965 to 1,585 dollars for men and from 1,634 to 2,360 dollars for women. Incremental costs per year of life saved ranged from 1,796 to 2,949 dollars for men and from 3,040 to 4,391 dollars for women. It was concluded that the nicotine TD is cost-effective and less costly per year of life saved than other widely accepted medical practices and that physicians and third-party payers should recommend the nicotine patch to patients who wish to stop smoking.

Assessment of the use and effectiveness of free nicotine TD among Medicaid and uninsured smokers concluded that barriers of access to effective treatment for smoking cessation need to be eliminated (99). The majority (90%) of participants found the TD system useful, 14% were abstinent for 6 months or more, and there was no support for inappropriate use. The cost-effectiveness, for the National Health Service (NHS) in the United Kingdom, of allowing general practitioners to prescribe TD nicotine patches for up to 12 weeks was evaluated (100) using data from a randomized, placebo-controlled efficacy trial of nicotine TD and a survey of associated resource use in 30 general practitioner's (GP) surgeries in England. The health benefit of TD nicotine treatment was calculated in the number of life years that would be saved by stopping smoking at various ages, and used an abstinence–contingent treatment model to calculate incremental cost per life year saved by GP counseling with nicotine-patch treatment over GP counseling alone. If the NHS allowed prescription of nicotine TD for up to 12 weeks, the incremental cost per life year saved would be 398 pounds per person younger than 35 years; 345 pounds for ages 35–44 years; 432 pounds for ages 45–54 years; and 785 pounds for ages 55–65 years.

Cardiovascular effects of smoking, effects of nicotine without tobacco smoke, and available data on cardiovascular risk during nicotine replacement therapy were reviewed (101). Although nicotine gum and TD are approved for over-the-counter sale, and smokers with cardiovascular disease are advised to seek physician counseling before using nicotine products, information on product safety in such patients is not readily available. Nicotine can contribute to cardiovascular disease, by hemodynamic consequences of sympathetic neural stimulation and systemic catecholamine release, but there are many other potential cardiovascular toxins in cigarette smoke. Doses of nicotine obtained by cigarette smoking usually exceed those delivered TD, and cardiovascular effects of nicotine are generally more intense when delivered rapidly by cigarette smoking, rather than more slowly by TD nicotine or nicotine gum. As the dose–cardiovascular response for nicotine is flat, the effects of cigarette smoking in conjunction with replacement therapy are similar to those of cigarette smoking alone. Although cigarette smoking increases blood coagulability, a major cardiovascular risk factor, TD nicotine does not appear to.

Wide variations in levels of nicotine (and cotinine) have been observed after nasal and transdermal delivery. Sources of individual variability in nicotine and cotinine plasma levels after use of replacement systems or cigarette smoking were evaluated (102). Cigarette smokers received four treatments of 5-days duration each, including (a) cigarette smoking (16 cigarettes a day); (b) TD nicotine (15 mg a day); (c) nicotine nasal spray, (24×1-mg doses a day); (d) placebo nicotine nasal spray (24 doses a day). Disposition kinetics were determined by infusion of deuterium-labeled nicotine and cotinine. There was considerable individual variation in daily dose of nicotine absorbed (nasal spray $\times$ 5, TD $\times$ 2–3) and in plasma nicotine and cotinine levels. Plasma nicotine levels were determined predominantly by nicotine clearance, whereas cotinine levels were determined most strongly by nicotine dose and, to a lesser extent, by clearance of cotinine and fractional conversion of nicotine to cotinine. To compensate for individual differences in clearance, individualization of dosing based on therapeutic monitoring and comparison with nicotine or cotinine levels during cigarette smoking before treatment may be required to optimize nicotine therapy.

Determinants of variability and the utility of baseline plasma concentrations as predictors of concentrations during TD treatment was evaluated using data from smoking cessation ($n = 466$) and pharmacokinetic studies ($n = 12$) (103). Plasma concentrations of nicotine and cotinine were highly variable. Indirect estimates of plasma clearance (baseline plasma concentration per number of cigarettes per day) together with other factors accounted for 33% or less, variability during TD treatment in the smoking cessation study. In contrast, 75–99% was accounted for by direct measurements of plasma clearances and systemic doses of nicotine in the pharma-cokinetic study. It was concluded that plasma concentrations of nicotine and cotinine during TD nicotine treatment are poorly predicted by clinical history or baseline plasma concentrations.

The frequency of adverse effects associated with use of TD nicotine was es-timated by metanalysis of data from 47 reports of 35 clinical trials (104). Few adverse cardiovascular outcomes were reported, and no excess of these outcomes was de-tected among patients assigned to TD use. Incidence of minor adverse effects (es-pecially sleep disturbances, nausea or vomiting, localized skin irritation, and respi-ratory symptoms) was clearly elevated among TD groups. Incidence of nausea or vomiting appeared lowest when the patch dose was tapered.

Application of urine and serum nicotine and cotinine excretion rates for as-sessment of nicotine replacement in light, moderate, and heavy smokers undergoing TD therapy was investigated (105,106). Subjects were stratified as light (10–15 cig-arettes per day), moderate (16–30 cigarettes per day), or heavy (more than 30 cig-arettes per day) smokers and assigned to a daily 24-h–TD system delivering a dose of 0, 11, 22, or 44 mg. Steady-state urinary excretion rates of nicotine and cotinine were attained in 2 and 3 days, respectively, at all doses, independent of smoking rate. Significant underreplacement occurred with the 11-mg/day dose, particularly in moderate and heavy smokers (<50%), and at 22 mg/day, nicotine replacement was still less than 100% in most subjects. Only a dose of 44 mg/day provided mean replacement exceeding 100%, regardless of baseline smoking rate. Steady-state plasma concentrations of nicotine and cotinine were attained in 1 and 3 days, re-spectively, at all doses, independent of baseline smoking rate.

Individualization of dosage is desirable and plasma cotinine levels at steady state (>3 days of therapy) can be used to calculate percentage replacement using baseline levels. Attempts were made to identify variables associated with long-term smoking cessation following hospitalization (107). Patients were assigned to (a) min-imal care (MC; brief physician-delivered motivational message); (b) counseling plus active nicotine patch (CAP; motivational message, 6-week supply of nicotine TD, and extended bedside and telephone counseling); and (c) counseling plus placebo patch (CPP). At 6 months, abstinence rates were 4.9, 6.5, and 9.7% for MC, CPP, and CAP treatments, respectively, but not significantly different. In another trial (108) 308 smokers were randomly allocated to (a) 3-g dextrose tablets and 15-mg nicotine TD; (b) dextrose and placebo patch; (c) placebo tablets and nicotine TD; (d) placebo tablets and placebo patch. The proportion of smokers abstinent in each group was 49% (dextrose plus active patch); 44% (dextrose plus placebo patch); 36% (placebo tablet plus active patch); and 30% (placebo tablet plus placebo patch). The difference between dextrose and placebo tablets (13%) was significant, but that between active and placebo patches (6%) was not.

Efficacy of using nicotine TD for five months in combination with a nicotine nasal spray for 1 year was evaluated (109) in 237 smokers (22 to 66-years old) receiving either nicotine TD (5 months) with nicotine nasal spray for 1 year ($n = 118$) or nicotine TD with placebo spray ($n = 119$). TD treatment comprised 15-mg nicotine (months 1–3), 10-mg (month 4), and 5-mg (month 5). The combination of use of nicotine TD for 5 months with a nicotine nasal spray for 1 year was a more effective method of smoking cessation than use of a patch only. Sustained abstinence rates for the patch and nasal-spray group and the patch-only group were 51 versus 35% at 6 weeks, 37 versus 25% at 3 months, 31 versus 16% at 6 months, 27 versus 11% at 12 months, and 16 versus 9% at 6 years.

Long-term smoking-cessation efficacy of varying doses of TD nicotine was evaluated 4–5 years after quitting among patients enrolled in the Transdermal Nicotine Study Group investigation (110). Self-reported continuous quit rate for patients assigned 21 mg (20.2%) was significantly higher than that for patients assigned 14 mg (10.4%), 7 mg (11.8%), or placebo (7.4%). Relapse rates among treatment conditions were similar to those 1-year after cessation.

Plasma cotinine replacement levels of 56 outpatient smokers using a 21-mg/day TDS (Nicoderm CQ) were reported (111). Cotinine replacement was 35–232% (mean 107%; median 90.5%). Baseline cotinine level, previous quitting attempts, gender, and Fagerstrom Tolerance Questionnaire scores were significantly correlated with the percentage of cotinine replacement. Baseline cotinine level plus gender was the most powerful predictor combination. Predictors and timing of adverse experiences during TD nicotine therapy were investigated (112). Intervention consisted of brief behavioral counseling, a booklet containing smoking cessation advice, instructions on patch use, and a 12-week course of decreasing TD nicotine. Most adverse experiences were mild. Sleep problems occurred in 48% and usually started on the day of smoking cessation. Application site reactions occurred in 34%, most frequently after 6 days of therapy. Significant predictors of sleep problems were female gender and successfully quitting smoking. Predictors of application site reactions were psoriasis or eczema, other skin conditions, age younger than 40 years, female gender, and trade or university education level. Substantially increased nicotine intake during therapy, compared with baseline smoking, occurred in 8% who smoked concurrently, and 4% of those who did not. It was concluded that sleep disturbance during therapy was primarily associated with tobacco withdrawal, rather than excess nicotine from TD treatment. Comparison between nicotine gum, TD, nasal spray, and inhaler was made with assessments at quit date and after 1, 4, and 12 weeks (113). Products did not differ in their effects on withdrawal discomfort, urge to smoke, or rate of abstinence. Continuous validated 12-week abstinence rates were 20% (gum), 21% (TD), 24% (spray), and 24% (inhaler). Compliance with recommended treatment use was high for TD, low for gum, and very low for spray and inhaler.

A quantitative gas chromatography–mass spectrometry (GC–MS) method for simultaneous determination of total and free *trans*-3'-hydroxycotinine (THOC) and cotinine (COT) in human urine during nicotine transdermal therapy was developed (114). Results from six consecutive 24-h urine collections in 71 subjects, who used daily TD nicotine doses of 11, 22, and 44 mg, showed that free THOC was 76% of total THOC, and free COT was 48% of total COT. The effect of 24-h nicotine patches in smoking cessation was evaluated among over-the-counter customers in Denmark

(115). A total of 522 customers who smoked ten or more cigarettes per day were randomized to either nicotine or placebo patches; 24-h patches were offered for a 3-month period. Those smoking 20 or more cigarettes per day started with patches of 21 mg/day. Those smoking less started with 14-mg/day patches, and all participants were gradually reduced to 7-mg/day patches. There was a significant increase in smoking cessation rates but, after 8 weeks of follow-up, only among smokers who used 21-mg/day patches. No significant differences in smoking cessation rates were seen among smokers who started with low-dose nicotine or placebo patches. The Collaborative European Antismoking Evaluation (CEASE) was a multicenter, randomized, double-blind placebo-controlled smoking cessation study, the objectives of which were to determine whether higher dosage and longer duration of nicotine patch therapy increased the success rate (116). Thirty-six chest clinics enrolled 3575 smokers, who were allocated placebo or either standard- or higher-dose nicotine TD (15 and 25 mg daily) each given for 8 or 22 weeks, with adjunctive moderately intensive support. The 12-month sustained success rates were: 25-mg TD for 22 weeks, 15.4%; 25-mg TD for 8 weeks, 15.9%; 15-mg TD for 22 weeks, 13.7%; 15-mg TD for 8 weeks, 11.7%; and placebo, 9.9%.

Nicotine plasma levels and safety of nicotine TD in smokers undergoing situations suspected to result in increased nicotine plasma levels were assessed (117). Effects of increasing nicotine intake through sequential administration of nicotine TD (day 2), TD followed by consumption of nicotine gum (day 3), and TD followed by gum consumption and cigarette smoking (day 4) were examined; nicotine plasma levels increased transiently after addition of each nicotine source. Mean AUCs (0–24 h) for nicotine were 453 ± 120 ng h^{-1} mL^{-1} (day 2); 489 ± 143 ng h^{-1} mL^{-1} (day 3); and 485 ± 143 ng h^{-1} mL^{-1} (day 4). A second study evaluated effects of physical exercise on kinetics and safety of two types of nicotine transdermal device. Mean delivered dose was higher with Nicoderm than Habitrol, and the products were not bioequivalent. During a 20-min–exercise period, nicotine plasma levels increased by $13 \pm 9\%$ for Nicoderm and $30 \pm 20\%$ for Habitrol. After exercise, subjects taking Habitrol had a higher incidence of adverse events compared with baseline values, but safety profiles remained acceptable. It was concluded that both superimposed nicotine sources and physical exertion resulted in short-lived plasma nicotine elevations and temporarily increased nicotine pharmacodynamic parameters, but without increased risk.

Short-term effects of TD nicotine replacement in pregnancy were examined (118). After customary smoking cessation efforts had failed, six prenatal patients (28 to 37-weeks gestation) who smoked one to two packs per day were admitted for a period of 21 h. Maternal and fetal assessments, including vital signs, biophysical profile, and electronic fetal monitoring, amniotic fluid index, and umbilical artery Doppler examinations were made, and salivary levels of cotinine and nicotine levels were determined. There were no measurable differences in fetal or maternal well-being. During TD use salivary nicotine levels increased to 19.0 ± 13.5 $\mu g/L$ at 480 min (consistent with levels in nonpregnant adults).

Surprisingly, salivary cotinine concentrations were much lower (~ 50 $\mu g/L$) than those in smoking nonpregnant adults and varied little over the period that the patch was worn. Weight changes in subjects receiving variable doses of TD nicotine replacement were assessed in 70 subjects receiving placebo or to 11-, 22-, or 44-mg/day doses of TD nicotine and 1 week inpatient treatment with outpatient follow-up through 1 year (119). The study included 1 week of intensive inpatient treatment with active TD therapy for a further 7 weeks. Counseling sessions were provided weekly during patch therapy, with long-term follow-up at 3, 6, 9, and 12 months. Forty-two subjects were confirmed as nonsmokers at all weekly visits during TD therapy, and their 8-week weight change from baseline (3.0 ± 2.0 kg) was negatively correlated with the percentage of cotinine replacement ($r = -0.38$, p = 0.012) and positively correlated with baseline weight and age. Men had higher 8-week weight gain (4.0 ± 1.8 kg) than women (2.1 ± 1.7 kg). This suggested that higher replacement levels of nicotine may delay postcessation weight gain in both men and women, but did not identify predictive factors.

TD nicotine use, nicotine and cotinine levels, and fetal effects were investigated in pregnant cigarette smokers aged 18 years or older, whose fetuses were beyond 24-weeks gestational age (120). Serial measurements of mother and fetus were made at baseline while the mother was smoking, while abstaining from smoking, and while using TD nicotine therapy for 4 days in the hospital. Nonpregnant women smokers of similar age were used as comparators. No evidence of fetal compromise was seen while nicotine patch therapy was administered. Morning serum cotinine levels were significantly higher in nonpregnant than in pregnant subjects, but afternoon levels were not significantly different. Steady-state urinary levels of nicotine and cotinine were also not significantly different between pregnant versus nonpregnant patients. On inpatient days 2, 3, and 4 for women not smoking, but wearing nicotine TD, the morning fetal heart rates were significantly reduced relative to baseline when subjects were smoking.

Abuse liability and dependence potential of nicotine gum, TD, spray, and inhaler were compared in 504 male and female smokers (121). No significant differences between products in terms of satisfaction or subjective dependence, except at week 15 when no patch users rated themselves as dependent. Continued use of nicotine replacement at week 15 was related to rate of delivery of nicotine: 2% for patch, 7% for gum and inhaler, and 10% for spray. Cessation of nicotine replacement between weeks 12 and 15 was not accompanied by withdrawal discomfort or increased frequency of urges to smoke. It was concluded that abuse liability was low for all products.

2. Treatment of Ulcerative Colitis

Ulcerative colitis is largely a disease of nonsmokers and anecdotal reports have suggested that smoking and nicotine may improve symptoms (122). Patients with active ulcerative colitis were treated with either nicotine TD or placebo patches for 6 weeks. All patients had been taking mesalamine, and some were also receiving low doses of glucocorticoids. These medications were continued during the study. Incremental doses of nicotine were used and most patients tolerated 15–25 mg/24 h: 17 of 35 patients in the nicotine group had complete remissions, compared with 9 of 37 patients in the placebo group. Patients in the nicotine group had greater improvement in global clinical and histological grades of colitis, lower stool frequency, less abdominal pain, and less fecal urgency. More of the nicotine group had minor side effects (23 vs. 11 in placebo group), and withdrawals owing to ineffective therapy were more common in the placebo group (3 vs. 8).

The value of TD nicotine for maintenance of remission was studied in 80 patients with ulcerative colitis in remission, using either TD nicotine or placebo patches for 6 months (123). Incremental doses were given for 3 weeks to achieve a maintenance dose (most tolerated 15 mg for 16 h daily). All patients were taking mesalamine at study entry, but this was stopped when maintenance nicotine doses were achieved. Twenty-two patients in the nicotine group were prematurely withdrawn from the study, 14 because of relapse and 8 for other reasons, including side effects and protocol violations. In the placebo group, 20 patients were withdrawn prematurely, 17 owing to relapse and 3 for other reasons. Among patients using 15-mg–nicotine patches, serum nicotine and cotinine concentrations were lower than expected, which may have reflected poor compliance. Side effects were reported by 35 patients, 21 in the nicotine group and 14 in the placebo group. TD nicotine alone was no better than placebo in maintaining remission of ulcerative colitis, and early withdrawal because of side effects was more common in the nicotine group.

Nicotine alone was compared with prednisolone in 61 patients with active ulcerative colitis treated with either nicotine TD or 15 mg of prednisolone for 6 weeks (124). Incremental nicotine doses were given for the first 9 days. Of the 43 patients who completed the trial, 6 of 19 in the nicotine group achieved full sigmoidoscopic remission, compared with 14 of 24 with prednisolone. In those who completed this study, nicotine alone appeared to be of only very modest benefit in acute colitis.

Use of TD nicotine in mildly to moderately active ulcerative colitis was investigated (125) in 64 nonsmoking patients with mildly to moderately active ulcerative colitis despite the use of medication. These were stratified (on the basis of smoking history, extent of disease, and concomitant therapy) and assigned to daily treatment with TD nicotine ($n = 31$) at highest-tolerated dose (11 mg for 1 week and then ≤ 22 mg for 3 weeks) or placebo ($n = 33$). At 4 weeks, 39% of those who received nicotine showed clinical improvement compared with 9% who received placebo. Four patients receiving nicotine discontinued therapy because of side effects. At week 4, the nicotine group had trough serum concentrations of 12.3 ± 8.4 ng/mL (nicotine) and 192 ± 95 ng/mL (cotinine). It was concluded that transdermal nicotine at ≤ 22 mg/d for 4 weeks was effective in controlling the clinical manifestations of mildly to moderately active ulcerative colitis.

A pilot trial of nicotine TD as an alternative to corticosteroids in ulcerative colitis was reported (126). In ten patients with mild-to-moderate clinical relapses of ulcerative colitis during mesalamine treatment and with a previous history of poorly tolerated steroids, TD nicotine (15 mg daily) was added for 4 weeks. Clinical remission was achieved in seven patients and persisted for up to 3 months after nicotine withdrawal.

A second study (127) investigated long-term effects. Patients with mild-to-moderate clinical relapses of left-sided ulcerative colitis during maintenance treatment with mesalamine were allocated additional treatment with either TD nicotine or prednisone for 5 weeks. The first consecutive 15 patients in each group with clinical and endoscopic signs of remission were followed-up for 6 months, while continuing mesalamine maintenance treatment. Relapses of active colitis were observed in 20% of patients formerly treated with nicotine and 60% of patients in the prednisone group, and relapses occurred earlier in the latter group. As patients with mild-to-moderate active colitis treated with mesalamine plus TD nicotine appeared to suffer fewer relapses than patients treated with mesalamine plus oral prednisone a long-term follow-up was carried out (128). Thirty patients with remission of distal colitis were monitored for 12 months and relapsed patients retreated in a crossover manner. Recurrences were observed in 14 of 15 patients initially treated with steroids and in 7 of 15 subjects who received TD nicotine.

3. Other Indications

Individuals with major depression have a high frequency of cigarette smoking, and TD nicotine can produce short-term improvement in mood. The effects of nicotine patches (17.5 mg) on 12 nonmedicated outpatients with major depression were studied over 4 continuous days (129). Two patients dropped out of the study because of nausea and vomiting. There was significant improvement in depression after day 2 of TD nicotine and patients relapsed 3 or 4 days after the final nicotine dose. Although nicotine TDS produced short-term improvement of depression, with minor side effects, nicotine TD was not recommended for clinical use in depression because of the high health risks of nicotine. It was concluded that analogues might be developed that can improve depression without major risks.

The therapeutic response to nicotine TD was investigated in patients with Tourette's syndrome (130). Twenty patients (17 children and adolescents, 3 adults) were studied following application of two patches (2 × 7 mg/24 h). There was a broad range in individual response, but each patch application produced a significant reduction in the Yale Global Tic Severity Scale scores, for an average duration of approximately 1–2 weeks. This suggested that TD nicotine could be an effective adjunct to neuroleptic therapy of Tourette's syndrome. Nicotine gum and nicotine TD were used to reduce motor and vocal tics of children (age 8 years or older; weight ≥25 kg), adolescents, and adults (131). Reduction of tics was seen during chewing of nicotine gum, but improvement lasted no longer than 1 h after chewing. With nicotine TD, motor and vocal tics were reduced 45% over baseline in 85% of 35 subjects within 30 min to 3 h after patch application. Relief of symptoms with a single 7-mg patch, left on the skin for 24 h, persisted for variable periods up to 120 days. Application of a second patch for 24 h when symptoms returned resulted in similar reduction in tic severity and frequency, which persisted an average of 13 ± 3 days.

Short-term nicotine injections have improved attentional performance in patients with Alzheimer's disease (AD), but little is known about prolonged effects of nicotine. A study evaluated clinical and neuropsychological effects of extended TD nicotine application in AD subjects over a 4-week period (132). Patients were treated with nicotine TD (Nicotrol) for 16 h/day at the following doses: 5 mg/day (week 1), 10 mg/day (weeks 2 and 3), and 5 mg/day (week 4). Nicotine significantly improved attentional performance, with a significant reduction in errors of omission, which continued throughout nicotine administration, and variability of reaction time for correct responses was also significantly reduced. Nicotine did not improve performance on other tests measuring motor and memory function.

4. Poisoning with Nicotine Patches

To evaluate potential adverse effects from inadvertent exposure, three marketed TD nicotine products: Nicoderm (drug reservoir and rate-controlling membrane); Nicotinell (nicotine solution dispersed in cotton gauze between layers of adhesive); and Niconil (nicotine in gel matrix), were administered topically and orally to dogs (133). Topical nicotine doses were 1–2 mg/kg 24 h^{-1} for all products, with plasma concentrations 43 ng/mL, or less. Of 12 topical exposures (with Nicotinell and Niconil) 2 were associated with clinical signs (excess salivation or emesis). Oral doses (2.8 mg/kg [one patch] to 13.4 mg/kg [two patches] over 25–57 h), were two- to ninefold higher than the oral doses reported to produce severe toxicity in children, and the higher dose was within the known lethal range for dogs. Oral dosing of Nicotinell and Niconil (two patches per dog) produced vomiting in 2 of 12 exposures. No clinical signs were observed with either topical or oral dosing of Nicoderm. Characteristics and outcomes of U.S. poisoning cases, involving dermal human application of multiple nicotine TD systems, were evaluated (134). Nine cases of dermal exposure to 2–20 nicotine TD systems resulted from intentional misuse or suicide attempts and included concomitant exposure to other drugs in 7 of 9 cases. Mean age was 45 years, and 7 of 9 patients were female. All suffered medical complications, including seizures, other CNS changes, cardiovascular effects, and respiratory failure, but plasma nicotine–cotinine levels did not correlate with the severity of illness. Eight patients were hospitalized, but all recovered.

B. Glyceryl Trinitrate and Related Drugs

Measurement of plasma concentrations of glyceryl trinitrate (GTN) is very difficult owing to the unusual pharmacokinetics, with very rapid disappearance from plasma, and large intraindividual and interindividual variations (135). There is extensive first-pass hepatic extraction after oral administration and plasma levels are often undetectable. With controlled-release TD GTN systems, plasma concentrations can be maintained over 24 h, but with fluctuations and important intra- and interindividual variability. After administration of GTN by any route, glyceryl dinitrates and mononitrates are found in plasma. Bioavailability and main pharmacokinetic parameters of the GTN metabolites 1,2- and 1,3-glyceryl dinitrate (1,2-GDN and 1,3-GDN) were determined following application of two types of GTN TDS in healthy volunteers (136). For 1,2-GDN, AUCs(0-T_{last}) were 23.77 h ng^{-1} mL^{-1} (patch A) and 27.83 h ng^{-1} mL^{-1} (patch B). Peak plasma levels were 2.45 ng/mL (at 6.4 h) and 2.93 ng/mL (at 8.3 h), respectively. For 1,3-GDN AUCs(0-T_{last}) were 3.32 h ng^{-1} mL^{-1} (patch A) and 3.81 h, ng^{-1} mL^{-1} (patch B). Peak plasma levels were 0.35 ng/mL (at 6.4 h) and 0.41 ng/mL (at 7.9 h), respectively. Statistical comparison showed bioequivalence between these TD systems for the metabolites investigated. Typical side effects observed after nitrate therapy also occurred.

The efficacy of lisinopril, TD GTN, and their combination in improving survival and ventricular function after acute myocardial infarction (AMI) was assessed (137). A total of 19,394 patients were randomized from 200 coronary care units in Italy and assigned 6 weeks of oral lisinopril (5-mg initial dose and then 10 mg daily), or open control, as well as nitrates (intravenous for the first 24 h followed by TD GTN 10 mg/day) or open control. Lisinopril, started within 24 h after AMI symptoms began, produced significant reductions in overall mortality and in the combined outcome measure of mortality and severe ventricular dysfunction. Administration of TD GTN did not show any independent effect on the same outcome measures. Systematic combined administration of lisinopril and GTN produced significant reductions in overall mortality and in the combined endpoint. The effects of a GTN TDS on platelet aggregation was examined in eight normal volunteers (138). A significant effect of TD GTN on platelet aggregation was demonstrated in the presence and absence of iloprost, but the clinical significance of the antiplatelet effect of TD GTN remained unknown.

The anti-inflammatory and analgesic effects of TD GTN was studied in 21 patients with mild to moderate leg varicose veins who underwent vein sclerotherapy in both legs (139). The vein in one leg was treated every 8 h with GTN and compared with a placebo ointment applied to the vein of the other leg. Inflammation signs were observed in all cases 15 min after first application. Intensity of inflammation signs were 26% in GTN-treated veins and 61.5% in placebo-treated veins. One hour later only 63% of cases in the GTN group, but all cases in the placebo group, showed signs of thrombophlebitis. All veins in the GTN group were free of signs of thrombophlebitis in fewer than 48 h, whereas, of the placebo group, 45% required more than 48 h.

Intermittent TD GTN therapy with a 10- to 12-h–patch-free period each day has documented clinical benefits. The antianginal and anti-ischemic effects of three dose levels of TD GTN applied for 12 h daily for 30 days and the development of tolerance and rebound were assessed (140). There was a significant increase in treadmill walking time to moderate angina in each GTN patch group, compared with placebo, at time points up to 12 h throughout the 30-day period. Secondary efficacy parameters supported the primary efficacy results,and there was no evidence of tolerance or rebound. Transdermal GTN is widely used to treat angina pectoris, but development of tolerance is a major problem (141). The effects of short (5 h) and prolonged (3 days) exposure to transdermal GTN patches on the development of tolerance in terms of hemodynamics and vascular reactivity in the conscious rabbit were, therefore, investigated. It was concluded that in the rabbit, prolonged exposure to clinical GTN patches caused hemodynamic compensation and baroreflex resetting, but no evidence of vascular reactivity tolerance.

The efficacy of adding transdermal GTN or oral *N*-acetylcysteine, or both, to conventional medical therapy was examined (142) in a trial of 200 patients with unstable angina, followed-up for 4 months. Death, myocardial infarction, or refractory angina requiring revascularization occurred in 31% of patients receiving GTN, 42% of those receiving *N*-acetylcysteine, 13% of those receiving GTN plus *N*-acetylcysteine, and 39% of those receiving placebo. There was higher probability of no treatment failure when receiving both GTN and *N*-acetylcysteine than with placebo, *N*-acetylcysteine, or GTN alone. However, combination of GTN and *N*-acetylcysteine was associated with a high incidence of side effects (35%), mainly intolerable headache.

The relation between tolerance development, counterregulatory responses, and arterial vasodilating effects were studied in 20 patients with stable angina pectoris who were exercise tested before, after 2 h, and 24 h of nitrate patch treatment (143). Effects observed after 2 h of treatment on exercise duration, ST-segment depression, blood pressure, and heart rate were usually lost by 24 h, although effects on arterial pulse curves persisted after 24 h, with a mean change from baseline of 29%, compared with 33% at 2 h. After 24 h, a significant decrease in hematocrit and an increase in body weight were observed. Hematocrit changes correlated with loss of clinical efficacy. It was concluded that clinical nitrate tolerance may be observed despite maintenance of arterial vasodilating effects, and that tolerance is more related to plasma volume expansion as a counterregulatory mechanism.

A subsequent study (144) examined whether GTN TDS treatment for 24 h could induce local cutaneous changes that impaired drug delivery and clinical efficacy. Twenty angina patients were exercise-tested after 2 and 24 h of treatment and 2 h after device renewal. The TDS was either renewed at a new skin location or on the previous application site. Clinical efficacy, the effect seen on plethysmography, and GTN plasma concentrations, all tended to increase after TDS renewal, regardless of the application site indicating that cutaneous changes of clinical importance were not demonstrated. The effect of GTN in patients with chronic heart failure (CHF) treated with angiotensin-converting enzyme (ACE) inhibition was evaluated (145). High-dose (50–100 mg) TD GTN or placebo were given daily for 12 h in 29 patients with CHF. Exercise time (4 h after patch application) showed progressive improvement during GTN administration. GTN decreased left ventricular end-diastolic and end-systolic dimensions and augmented LV fractional shortening.

Clinical efficacy of, and patient tolerance to, sustained-release GTN TDS in the treatment of Raynaud's phenomenon was reported (146). Patients had primary Raynaud's disease or Raynaud's phenomenon secondary to systemic sclerosis. GTN TD (0.2 mg/h) was effective in reducing both number and severity of Raynaud's attacks, but headaches led to withdrawal of 8 patients and occurred in about 80% of remaining patients. A clinical study in 20 patients with shoulder pain syndrome caused by supraspinatus tendinitis was conducted to determine whether TD GTN has analgesic action in this condition (147). One 5-mg GTN (Nitroplast) patch (or placebo patch) was applied per day over 3 days in the most painful area. Follow-up showed a significant decrease in intensity of pain at 24 and 48 h in the GTN group, but no change in the placebo group. It was concluded that GTN could be a useful approach to management of this common condition and other tendon musculoskeletal disorders.

The 24-h effect of TD GTN on splanchnic hemodynamics in nine patients with biopsy-proved liver cirrhosis was evaluated (148). GTN tape, capable of releasing 15 mg of drug in 24 h, was applied to chest skin at 7 AM of the second day. After GTN application, the mean portal blood velocity and flow significantly decreased by 18 and 22%, whereas superior mesenteric artery velocity decreased and resistance indices increased. This indicated that GTN, from a transdermal long-acting system, significantly influenced portal hemodynamics in liver cirrhosis, and the use of GTN was proposed for long-term clinical studies to test efficacy in preventing gastrointestinal bleeding.

The role of TD GTN, as a source of exogenous nitric oxide, in the management of primary dysmenorrhea was investigated (149) in a multinational study. Eighty-eight patients from six countries were evaluated during three menstrual cycles while receiving GTN (0.1 mg/h) or placebo patches. The data indicated that TD GTN, as a source of exogenous nitric oxide, was useful as a modulator of uterine contractility and represented a new and mechanistically different therapeutic alternative for management of primary dysmenorrhea.

The maternal and fetal cardiovascular effects of TD GTN compared with ritodrine for acute tocolysis (150) were studied in 60 women in preterm labor. At doses required for acute tocolysis, TD GTN had minimal effects on maternal pulse, blood pressure (BP), or fetal heart rate, and significantly fewer adverse cardiovascular effects than intravenous ritodrine. It was concluded that TD GTN may be a safer treatment for women in preterm labor.

The efficacy of TD GTN and intravenous ritodrine as tocolytics was also eval-uated in an international study (151). A total of 245 women with preterm labor and intact membranes between 24- and 36-weeks gestation were randomized to TD GTN (10 to 20 mg patch) or intravenous ritodrine. GTN and ritodrine prolonged gestation by 74% to 37 weeks. There was no significant difference in the proportion of women receiving GTN or ritodrine who delivered within the specified days from study entry or weeks of gestation, and no serious maternal side effects were reported for either. It was concluded that there was no overall difference between GTN and ritodrine in the acute tocolysis of preterm labor, but there was a suggested advantage of GTN over ritodrine in reducing preterm delivery rate. Maternal side effect profile and treatment discontinuation rates were fewer for GTN, suggesting it was the safer alternative.

The nitric oxide (NO) donor morpholinosydnonmine has been reported to in-hibit insulin release in isolated pancreatic islets; accordingly, the effect of TD GTN, an alternative NO donor, on glucose-stimulated insulin release was studied in healthy, young, male volunteers (152). Oral glucose tolerance tests were performed in the presence of placebo or TD GTN (~0.4 mg/h of GTN) in the same patients, with a 2-week intertest interval. Glucose-stimulated maximum increases in plasma insulin immunoreactivity were 36.3 ± 5 and 78.8 ± 6.1 mU/mL in the presence of active and placebo patches, respectively, although both fasting and postload blood glucose levels were equal. Active patches significantly decreased blood pressure with a mar-ginal increase in heart rate. It was concluded that inhibition of glucose-stimulated insulin release by TD GTN without causing hyperglycemia may be a novel com-ponent of the antianginal action mechanism of nitrates.

Benzoxazinones are a potent new class of organic nitrates used in cardiovas-cular therapy that have a coronary vascular selectivity greater than that of GTN and isosorbide dinitrate. The ability of these new derivatives to reach therapeutic steady-state plasma concentrations after TD administration was investigated in vitro using human skin (153). Two members of this class: sinitrodil (ITF 296) and ITF 1129 were compared with GTN, isosorbide dinitrate, and nicorandil at two concentrations (0.08% w/v and saturated solutions). Sinitrodil was considered a good candidate for transdermal administration.

C. Estrogens

Numerous clinical studies have compared the effects of TD and alternative delivery strategies for steroid hormones on a range of factors and, also, comparisons have been made between reservoir and matrix type TDS, as well as topical gels. The most common use of TD steroids is in hormone replacement therapy (HRT) in women. The reduction in estrogen production in menopause may cause hot flashes, sweating, mood and sleep disturbances, fatigue, and urogenital dysfunction. The effectiveness of estrogen-based HRT in ameliorating these symptoms, and in preventing long-term effects, such as osteoporosis, is well established (154). Comparative trials indicated that 625 μg of oral, conjugated estrogens, 20 μg oral ethinyl estradiol and 50 μg TD estradiol had equivalent efficacy in relief of mild-to-moderate menopausal symp-toms and prevention of bone mineral loss. Concomitant progestogen therapy is usu-ally included, if the uterus is intact, to protect against endometrial hyperplasia and carcinoma. Addition of progestogen maintains and may enhance bone-conserving effects of estrogen, and continuous regimens appear to reduce incidence of irregular menses. Adverse reactions are predominantly local skin irritation with TDS (14%) and systemic effects common to most forms of HRT, including breast tenderness, flushing, headache, and irregular bleeding, occurring in 2% or fewer of patients. Cost–benefit and cost–effectiveness of HRT in treatment of menopausal symptoms suggested that conjugated estrogens and TD estradiol compared well with alternative therapies, such as veralipride and Chinese medicines.

A combined TDS estradiol–norethisterone system was designed to deliver both estradiol and norethisterone at a constant rate for up to 4 days (155). TD norethisterone did not appear to alter the potentially beneficial effects of TD estradiol on total cholesterol, low-density lipoprotein (LDL) or triglyceride levels, or metabolic parameters of bone resorption or vaginal cytology. Protection from the effects of unopposed estradiol was achieved by sequential treatment with TD estradiol–norethisterone for 2 weeks of each 28-day cycle, and most patients experienced a regular vaginal bleeding pattern with this regimen. Menopausal symptoms were improved to a similar extent during estradiol-only and combined estradiol–norethisterone phases. The system was well accepted in clinical trials and generally well tolerated, the most common adverse effect being local irritation. Replacement therapy in 12 amenorrheic adolescents with gonadal dysgenesis treated with TD estradiol 100 μg (Estraderm TTS-100) twice weekly for 3 weeks, plus medroxyprogesterone acetate (MPA; Provera) for the last 11 days, following an interval of 1 week, was reported (156). No significant changes were recorded in FSH, LH, estradiol-17β, and PRL serum levels, but significant decreases of TC values and atheromatic indices 1 (TC/HDL) and 2 (LDL/HDL), significant increase in apolipoproteins-A$_1$, and beneficial effect on bone mass were seen at the end of treatment.

The efficacy and acceptability of a TD norethisterone device was assessed for 6 months in 18 patients of confirmed menopausal status (157). Therapy was continuous application of Estraderm TTS 50 for 28 days, with additional application of 2-norethisterone acetate patches for the last 12 days, repeated for six cycles. There was significant improvement in hot flashes and sweating, and withdrawal vaginal bleeding was established at regular intervals in 9 of 15 who completed 6 months of therapy. Histological examination of endometrial biopsies showed secretory activity in 56% of samples after 6 months, but no evidence of hyperplasia, premalignant, or malignant changes in the remaining biopsies. Efficacy and overall acceptability of 100- and 200-μg twice-weekly doses of Estraderm TTS in the treatment of severe PMS were compared (158).

Women with severe PMS received Estraderm TTS continuously with either esterone 10 mg or MPA 5 mg, from day 17 to day 26 of each cycle. There was no difference in change in total-ESA$_{max}$ between the Estraderm 100-μg and 200-μg groups, but there was a greater dropout rate and incidence of side effects attributed to estrogen in the higher-dosage group. Mean estradiol levels were 300 (100 μg) and 573 (200 μg) pmol/L, and 100 μg suppressed midluteal progesterone from a mean of 35.5 to 3.4. It was concluded that the lower-dose therapy was as effective in reducing symptom levels in severe PMS and was better tolerated.

Efficacy and tolerability of Menorest 50 was compared with Estraderm TTS 50 in treatment of postmenopausal symptoms in 205 women with moderate to severe vasomotor symptoms (159). After a 4-week treatment-free period, each woman received a cyclic regimen (25 days of a 4-week cycle) of Menorest 50 (matrix-type) or Estraderm TTS 50 (reservoir-type) twice weekly for 12 weeks. Oral progestin was also given for 10 days each cycle. Significant reduction in hot flashes was observed in each group compared with baseline. There were no significant differences in mean plasma estradiol levels and mean estradiol/estrone ratio (>1.0) in both groups after 10 weeks. Menorest 50 showed better local tolerability than Estraderm TTS 50.

Effects of daily intrauterine release of 20 μg of levonorgestrel by an intrauterine device on climacteric symptoms, bleeding pattern, and endometrial histological features in postmenopausal women receiving transdermal estrogen replacement therapy was evaluated in 40 parous postmenopausal women over a period of 1 year (160). Twenty women receiving a continuous TD daily dose of 50 μg of estradiol had a levonorgestrel-releasing intrauterine contraceptive device inserted, whereas the control group ($n = 20$) received a continuous oral dose of 2 mg of estradiol valerate and 1 mg of norethisterone acetate daily. Both treatment regimens effectively relieved climacteric symptoms.

The effect of TD estrogen replacement therapy on lipoprotein (Lpα) and other plasma lipoproteins was studied (161) in 30 women who had undergone a total abdominal hysterectomy and bilateral salpingo-oophorectomy for benign gynecological conditions treated with 1.5 mg of 17β-estradiol gel applied daily for 12 consecutive months. Plasma lipoproteins were measured before treatment and at 6- and 12-month intervals. There was significant reduction in Lpα levels during the first 6 months of treatment, with median values falling from 7.87 to 6.16 mg/dL, but during the second 6 months, median concentration increased to 9.38 mg/dL. Significant reductions in apoprotein A-I, apoprotein B, HDL-C, and HDL(3)-C were present after 6 months, but at study completion these values were no different than those at baseline. By avoiding the "first-pass" effect, this method of delivery did not appear to produce the sustained changes in lipoproteins seen with oral treatment. After menopause the hemostatic balance shifts toward a latent hypercoagulable state and the effects on hemostasis of HRT with TD estradiol and oral sequential MPA were evaluated (162). The balance between procoagulant factors and inhibitors were studied in 255 women in physiological menopause for 1–5 years, allocated to 1 year of treatment with cyclic TD E2 (50 μg/day for 21 days) plus MPA (10 mg/day from days 10 to 21), continuous TD E2 (50 μg/day for 28 days) plus MPA (10 mg/day from days 14 to 25), or placebo. Continuous treatment gave significantly lower final values of fibrinogen, factor VII, antithrombin III, protein S, and heparin cofactor II than placebo.

Effects of HRT on bone mineral density (BMD) and disease activity in postmenopausal women with rheumatoid arthritis (RA) were investigated in 62 patients with RA, 22 taking placebo and 40 receiving HRT (TD estradiol patches twice weekly for 48 weeks plus norithisterone tablets when clinically indicated) (163). Fifty-nine percent of placebo and 78% of HRT groups completed 48 weeks. At entry, BMD values in the lumbar spine and femoral neck were similar to those in matched controls, whereas at the distal radius, BMD was significantly reduced to about 50% of control values. In the HRT group, spinal BMD increased significantly by +0.94% at 48 weeks, but BMD at femoral neck and distal radius did not change in either group. In the HRT group, there was significant improvement in well-being and articular index. Because older women often experience side effects with conventional HRT, a low-dose preparation (Estraderm 25) was compared with conventional HRT (Estraderm 50) in patients with bone loss (164). A total of 196 women were studied over 1 or 2 years, with 80 reaching 3 years of treatment. In the lumbar spine, BMD increased maximally in year 1 in all groups, and the gain was maintained after 3 years. Only 3.9% of patients were nonresponders at this site after 3 years. Mean changes after 3 years were 8.1 ± 6.8% for Estraderm 25 and 9.0 ± 8.3% for Estraderm 50. At the femoral neck, 10.4% of patients were nonresponders after 3 years, and changes were significant only in Estraderm 25 in women older than 67 years, and Estraderm 50 in those younger than 67 years. BMD change over 3 years at the lumbar spine and femoral neck correlated with menopausal age. Use of Estraderm 50 was not associated with a greater response of bone mass and there was no evidence of increasing BMD response as estradiol dosage per kilogram of body weight increased. It was reported that oral and transdermal 17β-estradiol provided similar benefits in clinical studies (165). The lowest effective doses were 0.625 mg/day for conjugated estrogens, 2 mg/day for oral 17β-estradiol, 1.5 μg/day for 17β-estradiol gel, and 50 g/day of 17β-estradiol TDS.

Decrease in incidence of osteoporotic fractures was achieved only when the duration of HRT exceeded 7 years. The responses of various biochemical markers for bone turnover to TD estradiol were measured in 11 postmenopausal women over 24 weeks (166) and compared with the within-subject variability of markers in 11 untreated healthy postmenopausal women. Mean decrease in markers of bone formation ranged from 19% for procollagen type I C-terminal propeptide to 40% for procollagen type I N-terminal propeptide (PINP). The mean decrease in markers of bone resorption ranged from 10% for tartrate-resistant acid phosphatase (TRAP), to 67% for C-terminal cross-linked telopeptide. The ability to detect a response differed between markers and was not dependent on the magnitude of response to therapy. The highest number of responders were found using PINP (9 of 11) and osteocalcin (9 of 11), and free deoxypyridinoline (8 of 11) and total deoxypyridinoline (7 of 11). Lumbar spine BMD defined four patients as responders.

The comparative effects on BMD in routine clinical practice use of tibolone and estrogen (unopposed or combined with cyclic progestogen) in postmenopausal women who had not previously received estrogen or other menopausal therapy were assessed (167). BMD was measured in the spine and hip at 12-month intervals over 3 years in 82 postmenopausal women referred for climacteric therapy. Thirty-five women received tibolone, 24 TD estradiol alone, and 12 conjugated equine estrogens together with cyclic progestogen; 11 received no therapy other than calcium. Spinal BMD increased significantly in those taking tibolone over 3 years. In those receiving conjugated equine estrogens and cyclic progestogen, spinal BMD also increased significantly over years 1 and 2, but not year 3. Although spinal BMD rose over 3 years in women treated with TD estradiol alone, this was not significant. No significant change in BMD of spine or hip was observed in the control group.

A significant difference in increase of spinal BMD between treatment groups was observed at 2 years in favor of those taking tibolone or conjugated equine estrogens, compared with TD estradiol. The most common side effect and reason for discontinuation with Norplant use is bleeding disturbance and, therefore, a 6-week application of a patch releasing 100 μg/day estradiol was investigated as a method of reducing this problem (168). Of 98 Norplant users, 34 had normal, and 64 abnormal, bleeding patterns. Estradiol (33) or placebo (31) TDS were randomly used to treat patients with abnormal bleeding. Although there was clinical improvement in the estradiol group, this was not significant.

Changes in serum lipoproteins, apoproteins, and coagulation factors, induced in postmenopausal women treated by the Gynaderm TDS (designed to deliver 50 μg of 17β-estradiol per day) were studied over 6 months in 53 hysterectomized, healthy, postmenopausal women (169). One patch was applied twice weekly. There were no significant changes in levels of total cholesterol, triglycerides, HDL, or LDL, a significant rise in apolipoprotein A-I (apo A-I) level at 3 months was not sustained after 6 months, and there was a significant drop in apo A-II level after 6 months. Changes in apo B and Lpα were not significant. There were significant falls in levels of antithrombin III and protein S, and a significant rise in factor VII. Changes in levels of fibrinogen and protein C were not significant. TD estradiol administration caused minimal changes in lipoprotein metabolism and the statistically significant changes in the thrombophilia profile parallel those observed with oral HRT.

A clinical study evaluated the effect of HRT on plasma lipoproteins and Lpα profile in 42 menopausal women with primary hypercholesterolemia (total cholesterol >240 mg/dL) (170). Patients were randomly assigned to (a) TD estradiol, 50 μg + medroxyprogesterone, 10 mg/day for 12 days; (b) conjugated equine estrogens, 0.625 mg/day + MPA 10 mg/day for 12 days; (c) no treatment. Total cholesterol and LDL cholesterol significantly decreased after 6 months in both treated groups in comparison with untreated women, but HDL cholesterol and triglycerides showed minimal changes.

Comparison was made between TD estradiol (0.05 mg/day) and oral norethisterone acetate (2.5 mg/day) administered for 12 days every 2 or 3 months to patients whose menopause had begun at least 4 years earlier (171). Study duration was two long cycles in each group within 7–10 months. Efficacy (group E/NA = 94/92%), and systemic tolerability (95/97%) were good and continuation of spaced-out treatment accepted by 88%(E) and 87%(NA). Major skin reactions occurred in 7%(E) and 4%(NA). Progestin-associated withdrawal bleedings occurred in 61% of patients; mean duration 4.3 $\pm$ 1.9/4.8 $\pm$ 1.6 days, and breakthrough bleeding requiring sonographic or histological work-up in 8%(E) and 13%(NA).

Effects of a HRT regimen of continuous estrogen and interrupted progestogen, administered transdermally, on the endometria of 15 healthy postmenopausal women, and the pattern of bleeding and relief of menopausal symptoms were investigated in a volunteer pilot study of up to 6-months duration involving weekly application of an estrogen-only TDS releasing 50 μg estradiol per day interspersed with a combined estrogen and progestogen TDS releasing 50 μg estradiol and 250 μg norethisterone acetate per day for 3 days (172). Transvaginal ultrasound measurements of endometrial thickness and endometrial biopsies were performed in month 3 at the end of both the estrogen-only phase of treatment and combined estrogen–progestogen phase. Treatment provided relief of hot flashes and, by month 6, 71% women who completed treatment had no vaginal bleeding. No endometrial hyperplasia or atypical changes were observed in biopsies, and ultrasound measurements demonstrated a thin endometrium. Reduced immunostaining for Ki67 was observed in endometrium from the combined phase of treatment compared with estrogen-only phase, consistent with progestogenic-antagonism of proliferation. Exposure to progestogen did not suppress steroid receptors, as similar immunostaining was observed in both treatment phases.

The tolerability, adhesion, and efficacy of the matrix-type estradiol TDS, Oesclim 50, were compared with those of Estraderm TTS 50 (a reservoir-type system) in a multicenter clinical trial (173). The patches were applied twice weekly for 24 days of each 28-day cycle, over 4 cycles. Oral progestogen was taken by nonhysterectomized patients for the last 12 days of estrogen therapy in each cycle. In the Oesclim 50 group, 4.2% of applications caused a local skin reaction, compared with 9.5% in the Estraderm TTS 50 group. Both were well tolerated, although 7 patients in the Oesclim 50 group, and 12 in the Estraderm TTS 50 group, discontinued owing to adverse events. There was no significant difference between the percentages of patients with signs of hyperestrogenism (20.3% in Oesclim and 20.0% in Estraderm TTS 50 group). Adhesion was significantly better for Oesclim 50 (6.0% detached) than for Estraderm TTS 50 (11.3% detached). Greater adhesion of Oesclim 50 was particularly apparent, with three times fewer Oesclim 50 systems becoming detached during a shower or bath. Each treatment produced significant and comparable improvements in vasomotor symptoms, other menopausal symptoms, and gynecological assessments. A near-maximal effect on vasomotor symptoms was observed after approximately 1 month of treatment, and this was maintained for the entire treatment period.

The local skin tolerability of Oesclim 50 was also compared with that of Estraderm TTS 50 (174). In the first study, the modified Draize–Shelanski–Jordan method of sensitization was used to compare cutaneous tolerability of repeated applications of patches in 24 healthy postmenopausal women. This indicated no sensitizing potential or induction of allergic reactions. The second study was a multicenter clinical trial involving 283 healthy menopausal women over 4 months. In this study, 4.2% of applications in the Oesclim group provoked reactions, compared with 9.5% in the Estraderm group, and 25.9% treated with Oesclim and 39.9% receiving Estraderm experienced one or more reactions. Redness and itching were the most frequent reactions in both groups. Durations of reactions were significantly shorter in the Oesclim group, with higher percentage of durations of less than 1 h and lower percentage of durations less than 48 h. No reactions in the Oesclim group led to premature removal, compared with 11 in the Estraderm group. One patient in the Oesclim group discontinued treatment because of an application site reaction, but seven in the Estraderm group discontinued.

A comparison between efficacy and safety of two sizes of Lyrelle (matrix type) and Estraderm TTS 50 (reservoir type) TDS was made in 394 hysterectomized postmenopausal women in a multicenter trial (175). A significant decrease in mean number of hot flashes per day was observed in all groups from the end of cycle 1, reaching 90% at the end of cycle 7. There was no significant difference between Lyrelle 50 and Estraderm at any time point for any parameter, although between-group differences for Lyrelle 80 and Estraderm occurred in cycles 1–3 in favor of Lyrelle 80. A similar effect on blood lipid levels was observed in all groups.

The efficacy, bleeding patterns, and safety of continuous TD and sequential TD progestogen therapy were compared with those of oral progestogen therapy in postmenopausal women receiving TD estrogen (176). In a 1-year (13 treatment periods, 28 days each), study, 774 postmenopausal women received 50 μg/day of continuous TD estradiol with either continuous or sequential TD norethisterone acetate (NETA) in daily doses of 170 or 350 μg in a single TDS or sequential oral progestogen (1 mg norethisterone [NET] or 20 mg dydrogesterone per day). The average number of hot flashes per day decreased from prestudy by over 90%, and this reduction was unaffected by different progestogen regimes. With sequential progestogen, the bleeding incidence and number of bleeding days did not change over the course of the study, but were lower in the low-dose TD progestogen group. With continuous progestogen, the incidence of bleeding decreased in both low- and high-dose groups, from 35 and 45% in treatment period 1, to 25 and 15%, respectively, at the end of treatment. Adverse event incidence was similar in both groups, with 23–36% reporting events possibly or probably related to HRT (excluding vaginal bleeding). Lipoprotein-α was reduced in all but the oral progestogen group. It was concluded that continuous and sequential TD estrogen–progestogen treatments with estradiol–NETA are effective and safe alternatives to continuous TD estrogen and oral sequential progestogen for treatment of menopausal symptoms. Continuous TD therapy with estradiol–NETA may be more acceptable for most patients (i.e., those who wish to avoid monthly bleeds), whereas the sequential regimen may be preferable when monthly bleeding may be appropriate.

The safety and efficacy of TD estrogen replacement therapy in liver-transplanted menopausal women was investigated (177). Thirty-two menopausal women who had undergone liver transplantation at least 6 months earlier, received TD estradiol replacement therapy in combination with progestin (Estracomb Ciba, 50 μg/24 h, 250 μg/24 h) if the uterus was intact, or estradiol alone (Estraderm Ciba, 50 μg). Liver function and hemostatic parameters were measured at 0, 3, and 6 months and gynecological transvaginal ultrasound (TVS) performed at 0 and 6 months. Efficacy of hormonal treatment was assessed from serum concentrations of estradiol, estrone, FSH, LH, and SHBG, by measuring endometrial thickness with TVS and recording changes in subjective climacteric symptoms at 0 and 6 months. Safety was assessed by measuring liver enzyme activity, liver synthesis functions, and coagulation factors. Therapy did not impair any liver parameters measured, no thrombotic effect was detected, and hormonal effects of the regimen were verifiable biochemically, clinically, and by TVS.

Effects of continuous TD estradiol, with or without sequential oral MPA, on serum lipids and lipoproteins in menopausal women were investigated in 62 healthy menopausal women (178). Group A included 38 hysterectomized women treated with continuous TD estradiol only (50 μg daily). Group B included 24 menopausal women, with an intact uterus, treated with TD estradiol (50 μg daily) and MPA (10 mg daily for first 12 days of each calendar month). Serum lipids and lipoproteins were reviewed after 6 months. In group A there was a small reduction in total cholesterol (-5.5%) and slight lowering in LDL-cholesterol (-5.7%). In group B, there were no significant changes in total cholesterol and LDL-cholesterol. HDL-cholesterol levels did not change significantly with unopposed TD estradiol or additional sequential MPA. Serum triglyceride concentrations decreased significantly in both groups (-13.9 and -13.4%, respectively). Serum lipid changes did not differ between groups. A multicenter trial (179) compared incidence of amenorrhea in 54 postmenopausal women (mean age, 54.9 $\pm$ 0.6 years) who underwent six 4-week cycles of continuous HRT combining progestin–nomegestrol acetate 2.5 mg/day with one of three estrogens: percutaneous 17β-estradiol gel (1.5 mg/day, group G), TD 17β-estradiol patch (50 μg/day, group P), or oral estradiol valerate (2 mg/day, group O). The rate of amenorrhea varied significantly according to type of estrogen preparation (calculated cycle-by-cycle, rates were 67–83% [group G], 25–56% [group P], and 53–61% [group O]). Overall rates of persistent amenorrhea were not different between groups for cycles 1 through 3, but for cycles 4 through 6, significantly more women in groups G and O (67 and 46%, respectively) experienced amenorrhea than did those in group P (12%). Amenorrhea rates for the entire six-cycle period were 78% for group G, 48% for group P, and 60% for group O, although these differences were not statistically significant. Differences in rates could not be attributed to endometrial atrophy, because endometrial thickness did not differ significantly among groups. Calculated as a function of the number of women included in the trial, the percentage of amenorrheic women was highest with group G, although findings were similar for group O. Two 11-week, placebo-controlled studies (180) compared the Climara 7-day matrix patch (at two dose levels) with 625 μg/day oral conjugated equine estrogen, found that both the 50- and 100-μg/day estradiol patches had a positive effect on climacteric symptoms. Tolerance was good and similar for both. Studies of skin irritation and adhesion revealed that the 7-day patch was well tolerated and that, although irritation was similar to that associated with Estraderm, adhesion was superior. Absorption of estradiol was higher and more consistent from buttock than abdomen, suggesting that choice of application site may require further investigation.

Efficacy, safety, and tolerability of an estradiol gel (1.0 mg of estradiol daily; Divigel/Sandrena) in HRT of postmenopausal women were compared with those of an estradiol TDS (50 μg/24 h estradiol, Estraderm TTS) over 12 months with 120 postmenopausal women (181). Dydrogesterone tablets (Terolut), 10-mg daily for the first 12 days of every month, were used as the progestogen component of therapy. Twenty-five women without HRT served as reference group for BMD measurements. Both treatment regimens were equally effective in alleviating climacteric symptoms, preserving BMD, and were equally safe. A trend toward heavier bleeding was detected in patients treated with the estradiol TD. A nonsignificant decrease of total cholesterol and triglyceride, but no change in high-density lipoprotein cholesterol was observed in both groups.

Acceptability of treatment was higher in the gel (96.4%) than patch group (90.7%). Only two (3.3%) women using the gel complained of skin irritation, whereas 28 patients (46.7%) using the patch reported this effect. Two doses of TD estradiol gel (Divigel/Sandrena) plus oral sequential MPA were compared with oral estradiol valerate plus oral sequential MPA (Divina/Dilena) in postmenopausal women with climacteric complaints or already using HRT in a 2-year comparative study (182). Groups received either (a) 1 g of gel containing 1 mg estradiol for 3 months plus 20 mg of oral MPA during last 14 days; (b) 2 g of gel containing 2 mg of estradiol for 21 days plus 10 mg oral MPA during the last 14 days; (c) 2 mg of estradiol valerate tablets for 3 weeks plus 10 mg of oral MPA during the last 10 days. With each preparation, climacteric complaints were significantly reduced, good bleeding control was obtained, BMD was maintained, and bone turnover was reduced. Lipid parameters showed no unfavorable changes. Continuation rates were similar in all groups, with 74% of patients completing the first year, and 94% of patients who elected to continue completing the second year. Tolerability of gel was good, with only 1.7% of patients discontinuing because of skin irritation.

Estradiol and estrone concentrations and bioavailability were compared after a single dose and at a steady state during oral estradiol valerate, TD estradiol gel, and TD estradiol TDS treatments (183). In study A, 12 healthy postmenopausal women received 1.5 mg of estradiol as a TD gel or a 2-mg estradiol valerate tablet daily for 14 days. In study B, 15 postmenopausal women were treated for 18 days with 1.5-

mg estradiol gel or a TD system releasing estradiol 50 μg/24 h (replaced every 72 h). Tablet and transdermal gel yielded similar serum estradiol profiles with a peak concentration 4–5 h after administration. The TDS gave relatively stable estradiol levels during the mid-third of the wearing time, whereas much lower levels were observed at the beginning and end. There was no difference in fluctuation of peak-and-trough estradiol levels between gel (56 or 67%) and tablet (54%), but fluctuation was greater with the TDS (89%). Bioavailability of estradiol from the gel was 61%, compared with the tablet, and 109%, compared with the TDS. Gel was not bioequivalent with tablet or TD and individual dose adjustments may be needed when changing administration forms.

The effect of HRT on the conjunctiva in postmenopausal women was the subject of a clinical study in 11 postmenopausal women receiving TD estradiol or TD estradiol plus MPA for 4 months (184). Significant increases in serum estradiol levels, vaginal maturation value, and mild cytological maturation changes in conjunctival epithelium were observed. The changes were significant and the data support the view that HRT induces cytological maturation changes in conjunctival epithelium in postmenopausal women.

A case of sensitization to estrogen was reported in a patient receiving Estraderm (185). A 40-year-old woman suffered from cyclic skin disorders and at each menses developed pruritus and erythematous papulovesicular lesions over the members and trunk. Prick and patch tests with alcoholic solutions of estrone alone and serum tests for antiethinyl–estradiol antibodies and antiprogesterone antibodies were positive. In cases of progesterone sensitization, treatment of choice is estrogen inhibition of ovulation, whereas for estrogen sensitization, antiestrogen treatment appears more effective. Bilateral ovariectomy may be required in difficult cases.

The history, current clinical practice, choice of methods, and number of prescriptions and sales of HRT in the United States were reviewed (186). The percentage of women currently utilizing HRT was greater in women aged 40–60 (35%), but fell with ages older than 65 (15%), and declined further in women older than 80 (7%). News media and physicians were the largest source of information on HRT, and obstetricians and gynecologists were predominant prescribers. Women who spontaneously developed menopause early and younger women undergoing castration were more likely to take TD estrogen, but 86% of U.S. prescriptions were for oral estrogens (with conjugated equine estrogens [70%] the market leaders). Of the prescriptions for women with a uterus 50% were for combined continuous estrogen and progestogen, whereas 42% contained cyclic estrogen and progestogen. Although use of HRT by postmenopausal women in the United States increased, the percentage of current users remained lower than anticipated, in spite of widespread media and educational efforts of benefits. TD estrogens were used more commonly in women in the early postmenopausal period, whereas in women with a uterus, most U.S. physicians prescribed combined estrogen plus progestogen, but used oral, rather than TD, estrogen.

Bioavailability, pharmacokinetics, and tolerability of two matrix transdermal delivery systems providing 50 μg/24 h of estradiol were compared in 20 healthy postmenopausal women (187). Menorest (3–4 days suggested use) and Climara (7 days suggested use) were compared at steady-state in two 14-day treatment periods separated by a 4-week washout, with plasma estradiol monitored in the second week of each treatment. No differences between treatments relative to AUC, $C_{(max)}$, $C_{(min)}$,

$C_{(average)}$, or fluctuations in plasma estradiol. $T_{(max)}$ was significantly shorter for Menorest than for Climara, and $C_{(max)}$ and $C_{(min)}$ were significantly higher for the second Menorest patch than for the first. Three cases of erythema with Menorest and a total of 21 skin reactions in 15 subjects with Climara were reported. Systemic tolerability was similar between treatments with 8 estrogen-related adverse events in 8 subjects with Menorest and 13 events in 10 subjects with Climara. It was concluded that although the bioavailability of estradiol from these TDS was similar, the products were not bioequivalent because $T_{(max)}$ was significantly shorter for Menorest than for Climara. A comparison of continuous combined TD delivery of estradiol–norethindrone acetate and estradiol alone for menopause was carried out to determine whether a continuous estradiol–norethindrone acetate TD delivery system reduced incidence of endometrial hyperplasia in postmenopausal women more than that of TD estradiol alone (188). A total of 625 postmenopausal women were assigned to one of four treatments: TD estradiol 50 μg/day, or TD estradiol–norethindrone acetate, with 50-μg estradiol and 140, 250, or 400 μg/day of norethindrone acetate. Endometrial hyperplasia was found in 37.9% in the estradiol-alone group versus 0.8%, 1%, and 1.1% in the estradiol–norethindrone acetate 50–140, 50–250, and 50–400 groups, respectively. Uterine bleeding was less frequent in the estradiol–norethindrone acetate 50–140 group. The estradiol–norethindrone acetate combination TDS showed skin tolerance comparable with that of estradiol alone.

Bioavailability of two 100-μg daily 17β-estradiol TDS (once-a-week matrix patch and twice-a-week reservoir patch) was compared in healthy postmenopausal women (189) in a two-period, crossover study with two 8-day treatment periods separated by a minimum 7-day washout. Subjects were assigned to either (a) matrix patch applied to abdomen and worn for 7 consecutive days, or (b) reservoir patch applied to abdomen and worn for 4 days, followed immediately by a second reservoir patch worn for 3 days. Three-hours after patch application serum estradiol levels were significantly higher than levels at time of patch application. After 12 h, mean serum estradiol level in women with matrix patches was 98.20 ± 44.97 pg/mL, significantly higher than in women with the reservoir patch (62.20 ± 16.21 pg/mL). An AUC (0–168 h) with the matrix patch was also higher than for reservoir patch.

Left ventricular heart function and its response to long-term estrogen replacement therapy was assessed in 30 postmenopausal women, 20 of whom had modest-to-severe hot flashes and 10 of whom had never had them (190). Continuous TD estradiol was given to women with surgically induced menopause, and a combination of TD estradiol and sequential MPA to those with spontaneous menopause. Although HRT significantly improved heart function in healthy postmenopausal women, there appeared to be some minor differences in response between those with flashes and nonflashers.

Effects of TD estradiol on serum triglycerides in menopausal women with preexisting mild-to-moderate hypertriglyceridemia were evaluated (191). Forty-four women (posthysterectomy and maintained on 50-μg unopposed estradiol for 6 months) were divided into those with normal baseline triglyceride concentrations (0.4–2 mmol/L) and those with raised baseline readings (>2–4 mmol/L). Significant reductions in serum triglyceride concentrations occurred in both groups (−9.6 and −17%, respectively). TD estradiol therapy may be a useful treatment option in menopausal women with preexisting hypertirglyceridemia.

A summary of the tolerability and safety of Oesclim, which was developed with the objective of providing improved local skin tolerability and adhesion, while minimizing hyperestrogenic effects, has been published (192). In a comparative clinical trial, Oesclim resulted in fewer than half as many application site reactions as Estraderm TTS (4.3 vs. 9.5%), and the duration of reactions was significantly lower in the Oesclim group. Oesclim was well tolerated in all clinical trials and reported to have a estrogen-specific tolerability comparable to Estraderm TTS.

Low-dose Oesclim (25 μg/day) was associated with reduction in hyperestrogenic side effects compared with higher doses. In a study of long-term Oesclim therapy, 79% of patients wished to continue therapy after 1 year, and in a follow-up study, 79.8% wished to continue at the end of 3 years.

Efficacy and safety of three dosages of Oesclim, delivering 0.025, 0.050, or 0.100 mg 17β-estradiol per 24 h, in treatment of moderate to severe vasomotor symptoms was evaluated in a multicenter trial (193). A total of 196 highly symptomatic menopausal women received 12 weeks of continuous unopposed treatment with one of the three dosages of Oesclim or a matching placebo patch. Reduction in frequency of moderate-to-severe vasomotor symptoms was statistically significant compared with placebo from week 2 onward in the Oesclim 50 and 100 groups, and from week 3 onward in the Oesclim 25 group. Symptom severity was also reduced. Estrogen-related adverse events were less frequent in the Oesclim 25 group.

Significant differences in estradiol bioavailability were reported from two similarly labeled estradiol matrix TDS (Alora and Evorel) (194). The fluctuation index produced by Evorel was significantly higher than that with Alora (135 vs. 76%) and the estradiol baseline-corrected AUC was significantly lower for Evorel than Alora (1870.6 vs. 2871.8 pg h^{-1} mL^{-1}). Efficacy, safety and compliance with Climara 50 and Climara 100 were evaluated in 100 women (195). Reductions in weekly frequency of hot flashes were 58.6% in the Climara 50 group and 72.1% in the Climara 100 group. Of 64 patients with sleep disturbances, 51 reported some improvement, 10 had no advantage, and 3 were worse. Incidence of temporary minor side effects was dose-related (38 vs. 70%).

A multicenter study assessed the efficacy, safety, and tolerability of a low-dose (0.0375 mg/day) estradiol matrix TDS for the treatment of moderate-to-severe post-menopausal hot flashes in healthy women (196). Estradiol matrix or matching placebo patches were administered over three 4-week–treatment cycles to 257 patients (130 estradiol, 127 placebo). Assessments of treatment effectiveness significantly favored the estradiol patch over placebo. This lowest available dose estradiol TD provided significant relief from moderate-to-severe postmenopausal hot flashes and was well tolerated. A 3-year study enrolled 277 early postmenopausal women to examine the efficacy of a matrix 17β-estradiol TDS, at three dosages (25, 50, and 75 μg/day) combined with sequential oral dydrogesterone 20 mg/day, in preventing bone loss (197). At 2 years, difference from placebo in percentage change from baseline of L1-4 lumbar spine BMD was 4.7 $\pm$ 0.7% (25 μg/day), 7.3 $\pm$ 0.7% (50 μg/day), and 8.7 $\pm$ 0.7% (75 μg/day). There were also significant increases in femoral neck, trochanter, and total hip BMD with all doses of estradiol, compared with placebo. Most patients receiving estradiol also had a significant gain (>2.08%) in lumbar spine bone mass and clinically significant and dose-related decreases in total serum osteocalcin, serum bone alkaline phosphatase, and urinary C-telopeptide, with all three markers of bone turnover returning to premenopausal levels.

The effects of four doses (0.025, 0.05, 0.06, and 0.1 mg/day) of a 7-day TD 17β-estradiol delivery system on bone loss in postmenopausal women were evaluated in a multicenter study (198). At 24 months, doses of 0.025, 0.05, 0.06, and 0.1 mg/day resulted in mean increases in BMD of the lumbar spine of 2.37, 4.09, 3.28, and 4.70%, respectively, and increased BMD of the total hip by 0.26, 2.85, 3.05, and 2.03%, respectively. All increases were significantly greater than placebo. Consistent and significant improvements in biochemical markers of bone turnover were also noted in all treatment groups.

Two estradiol TDS that released 25 or 37.5 μg/day were compared with a placebo patch on 156 patients in natural or surgical menopause suffering from at least five hot flashes per day, treated continuously for 12 weeks, without progestin opposition (199). "Responders" (patients with fewer than three hot flashes per day at end of treatment), were 82 and 90% with 25 or 37.5 μg/day, respectively, both significantly more than placebo (44%).

Efficacy and tolerability of a matrix patch delivering estradiol at doses of 0.05 and 0.10 mg/day (Estraderm MX 50, 100) in treatment of moderate to severe postmenopausal symptoms was compared (200). A total of 254 postmenopausal women received 0.10, 0.05 mg, or placebo for 12 weeks continuously. TDS were applied twice weekly to the buttocks with each patient wearing two patches simultaneously. Patches containing 0.10 and 0.05 mg estradiol were superior to placebo in reducing hot flashes per 24 h after 4, 8, and 12 weeks of treatment. For all other efficacy parameters studied, both dosage strengths were superior to placebo at all time points. It was concluded that this matrix patch offered an effective and well-tolerated dosage form and may be particularly suitable for women who experience local sensitivity to alcohol-containing systems.

The effect of the administration route and cigarette smoking on plasma estrogen levels during HRT was evaluated in 14 healthy postmenopausal women (6 smokers and 8 nonsmokers) (201). All patients randomly received cyclic therapy with estradiol and norethisterone orally or TD, each for 6 months. Plasma levels of estrone, estradiol, and estrone sulfate, all were 40–70% lower in smokers than nonsmokers when HRT was given orally. Oral dosing caused higher extradiol/estradiol sulfate and estrone/estradiol sulfate ratios compared with TD therapy in smokers (40.2 vs. 7.0; and 3.2 vs. 0.8, respectively).

Pharmacokinetics of Fem7 (an estradiol matrix-type TDS, applied once weekly) was investigated in 36 healthy postmenopausal women at doses of 25, 50, 75, and 100 μg/24 h (202). Maximum plasma estradiol and estrone concentrations occurred 14–20 h after patch application, remained within the therapeutic range until removal, and returning to baseline within 12 h. Plasma estradiol concentrations increased in a dose-dependent manner for all dose levels, and plasma estrone increased for the three highest doses. Treatment was well tolerated at all dose levels and no severe adverse reactions were reported.

The efficacy of two strengths of TD estradiol matrix with daily oral doses of conjugated equine estrogens in reducing the frequency of moderate-to-severe hot flashes in postmenopausal women was evaluated (203). An estradiol TDS (Alora 0.05 or 0.1 mg/day) administered twice weekly or oral doses of conjugated equine estrogens (CEE 0.625 or 1.25 mg) administered daily were given to 321 highly symptomatic postmenopausal women for 12 weeks. Results indicated no significant differences at any time point in mean frequency or mean percentage reduction in frequency of moderate-to-severe hot flashes between patients given Alora 0.1 mg/day or CEE 1.25 mg/day, or between the Alora 0.05 mg/day and CEE 0.625 mg/day groups by week 12. There were no serious or unexpected adverse events with the TDS, and local skin tolerability was excellent. Other estrogenic effects were comparable between TD and oral administration groups, except for lower incidence of bleeding in women receiving the lower transdermal dose.

The effect of estradiol TD in postmenopausal women with confirmed pollakiuria and urinary incontinence was investigated in ten women using Estraderm TTS 2 mg for 8 weeks (204). In seven cases, severity of urinary incontinence was "very effective" in three cases, "improved" in two, "slightly improved" in one, and "no change" in one.

Postmenopausal women (especially those older than 60 years) prefer HRT that avoids cyclical uterine bleeding and continuous combined HRT regimens were primarily introduced to avoid bleeding and increase compliance. In a multicenter study, 136 women at least 2 years postmenopausal, with mild-to-moderate menopausal symptoms, received either Estragest TTS 0.125/25 (delivering 0.125 mg norethisterone acetate [NETA] and 25 μg estradiol per day) or placebo for 6 months (205). After 4, 12, and 24 weeks the Kupperman index was significantly lower in the Estragest group, and the severity of vaginal dryness and dyspareunia at 12 and 24 weeks was also reduced. The proportion of superficial cells increased significantly in the Estragest, but not the placebo group. The percentage of patients reporting amenorrhea with Estragest ranged from 80 (month 2) to 87% (month 6). In a second multicenter study lasting 1 year, 441 postmenopausal women received one of three continuous combined HRT regimens: group A, Estragest TTS 0.125/25; group B, TDS delivering estradiol 50 μg and NETA 0.25 mg/day, group C, oral tablets containing 2 mg estradiol and NETA 1 mg/day. During treatment cycles 4–6, amenorrhea was achieved in 73% (group A), 47% (group B), and 66% (group C). During treatment cycles 10–12, proportions increased to 86% (A), 65% (B), and 79% (C). Bleeding patterns in groups A and C were not significantly different, but superior to those in group B. It was concluded that Estragest TTS 0.125/25 was effective in treatment of mild-to-moderate menopausal symptoms and urogenital complaints, induced a high rate of amenorrhea and provided good endometrial protection.

Effects of three commonly prescribed estrogen replacement therapies (oral conjugated equine estrogens [CEE; $n = 37$], oral micronized estradiol [ME; $n = 25$], and TD estradiol [TE; $n = 24$]) on the concentrations of blood sex hormone-binding globulin (SHBG), estradiol, and estrone were studied (206). Increases in SHBG concentrations were 100, 45, and 12% for subjects receiving CEE, ME, and TE regimens, respectively. Decreases in the percentage estradiol not bound to protein and increases in the percentage of estradiol bound to SHBG correlated with therapy-mediated changes in concentrations of this protein.

Systemic bioavailability and plasma profiles of 17β-estradiol after application of three matrix patches: Menorest, Tradelia, and Estraderm MX, claiming to deliver 50 μg/day were evaluated (207). All patches were each worn randomly by 21 postmenopausal women volunteers over 96 h, separated by an at least a 7-day washout period. T_{max} (32 h) was the only pharmacokinetic parameter identical for all patches. Menorest produced the highest estradiol bioavailability, judged by the $AUC_{(0-96\ h)} = 3967.8 \pm 1651.8$ pg/mL h^{-1}, $C_{(average)} = 41.3 \pm 21.3$ pg/mL, $C_{(min)} = 36.8 \pm 8.6$ pg/mL. Tradelia ($AUC_{(0-96\ h)} = 3737.9 \pm 1637.6$ pg/mL h^{-1}, $C_{(average)} = 38.9 \pm 17.0$ pg/

mL, and $C_{(min)}$ = 33.8 ± 26.7 pg/mL) was not significantly less than Menorest. Estraderm MX showed the lowest estradiol profiles (AUC$_{(0-96 h)}$ = 3192.1 ± 1646.0 pg/mL h^{-1}, $C_{(max)}$ = 38.9 ± 25.1 pg/mL, $C_{(average)}$ = 33.2 ± 17.1 pg/mL). Menorest showed the smallest fluctuation over the entire test period, similar to Estraderm MX, whereas Tradelia showed the highest fluctuation and the highest $C_{(max)}$ = 48.0 ± 20.3 pg/mL. When estradiol baseline levels, before patch application, were individually subtracted from the subsequent estradiol level, Estraderm MX was not bioequivalent to Menorest. A circadian curve pattern of estradiol plasma level was observed for all patches, and in the evening, higher plasma levels were always detected. Individual comparison of AUC$_{(0-96 h)}$ for each patch showed large interindividual variability (2000–8000 pg/mL h^{-1}) for all patches, but relatively small individual variability. Women with high estradiol bioavailability (high-responders) maintained high bioavailability with all patches, and women identified as low- and medium-responders remained the same, regardless of the applied patch. Side effects were approximately equal in all patches, with a maximum after 72 h.

Two long-term multicenter studies compared the efficacy on climacteric symptoms of a new active matrix estradiol TDS (CAS 50-28-2) with a reference reservoir patch (both releasing 50 μg/day) (208). One group received the matrix patch and the other the reservoir patch in 4-week cycles, with twice-weekly application of patches for 3 weeks, followed by 1-week washout. Progestin opposition was with MPA; 5 mg/day orally in the last 11 days of patch application in a German study and with 10 mg/day in the last 12 days of patch application in an Italian study. Each study was divided into two parts: (a) with three 4-week cycles and (b) for ten 4-week cycles. In the German study both patches quickly relieved climacteric symptoms during the first 3 weeks of application, as shown by rapid decrease of the Kupperman Index. At the end of part 1, 91% (matrix) and 96% (reservoir) group reported relief from climacteric symptoms; at the end of part 2, these were 98% and 95%, respectively. Both patches were systemically fairly well tolerated and only 4.5 (matrix) and 3.9% (reservoir) discontinued owing to adverse reactions. Relative to local skin reactions, the matrix patch was significantly better tolerated and adhesion was better. In the Italian study both patches relieved climacteric symptoms during the first 3 weeks of application. At the end of part 1 both patches relieved 95% of patients and at the end of part 2, 100% of patients were relieved. Patches were systemically equally fairly well tolerated with premature discontinuations for systemic adverse drug reactions in 5.0% (matrix) and 3.9% (reservoir) groups. As in the German study, matrix patches were significantly better tolerated.

The effect of short-term HRT with estradiol and norethisterone on the pharmacokinetics of phenazone was investigated in ten women at least 6 weeks after ovariectomy (209). Each patient received TD estradiol (4 mg, every 3 days) for a period of 18 days, and subsequently oral norethisterone (5 mg/day) for 10 days, with an interval of 1 day. Pharmacokinetic studies were performed before estradiol administration, on the last day of estradiol treatment and on the last day of norethisterone administration. Short-term administration of estradiol did not modify the pharmacokinetics of phenazone.

Percutaneous absorption of progesterone in postmenopausal women treated by application of progesterone cream to the skin was evaluated in six postmenopausal women over a 4-week period (210). Transdermal estradiol, 0.05 mg, was applied 2 days before first application of progesterone (30 mg/d) and continued throughout the study, with patches changed twice weekly. Progesterone cream was applied once a day for 2 weeks. On days 15–29 progesterone cream was applied twice daily (60 mg/d). Serum 17β-estradiol and progesterone were measured over 24 h on day 1 and at weekly intervals for the study duration. Individual serum 17β-estradiol concentrations ranged from 40 to 64 pg/mL, but intraindividual concentrations remained constant. Serum progesterone concentrations were 1.6–3.3 ng/mL. After 2 weeks of percutaneous dosing, progesterone concentrations were sustained for at least 8 h and were consistent within an individual. An increase in progesterone concentration occurred after 4 weeks, compared with 2 weeks. Individually, a significant correlation was seen between absorption of 17β-estradiol and progesterone.

Serum and urinary hormone levels following short- and long-term administration of two regimens of progesterone cream in postmenopausal women were evaluated (211) in a multiple-dose study using 24 healthy postmenopausal women. Subjects were allocated to progesterone cream, 40 mg daily, or 20 mg twice daily, for 42 days. Serum progesterone was measured on days 1 and 42 before the morning dose, and at 2, 4, 6, 12, and 24 h after the morning dose. Serum FSH, estradiol, testosterone, and urinary pregnanediol-3-glucuronide were also measured on days 1 and 42. The mean progesterone concentration rose at each sampling time between days 1 and 42 and there was evidence of a rise in pregnanediol-3-glucuronide over the study course. There were no changes in FSH, estradiol, or testosterone and no differences were detected between the regimens.

The use of TD progesterone cream for vasomotor symptoms and postmenopausal bone loss was also investigated (212). One hundred two healthy women, within 5 years of menopause, were assigned to TD progesterone cream or placebo. Subjects were instructed to apply 0.25 teaspoon of cream (containing 20 mg progesterone or placebo) to the skin daily. Subjects received daily multivitamins and 1200 mg of calcium; symptoms were reviewed ever 4 months. In the treatment group, 69% and 55% in the placebo group initially complained of vasomotor symptoms. Improvement or resolution of these symptoms was noted in 83% of treatment subjects and 19% of placebo subjects. However, the number of women showing a BMD gain of more than 1.2% did not differ significantly.

In addition to the widespread use of HRT TD, steroid delivery has also been investigated as a means of contraception. A once-a-week (monophasic) contraceptive TDS was designed to simultaneously deliver a low-dose combination of levonorgestrel (LNG) and 17β-estradiol (E2) for fertility regulation in females (213). In vitro permeation studies using human cadaver skin indicated 6.0 $\pm$ 0.9 μg/day cm^{-2} of LNG and 2.9 $\pm$ 0.5 μg/day cm^{-2} of E2 could be delivered. A 7-day dermal toxicity study on six rabbits indicated minimal potential to cause skin irritation, and histopathological examination revealed only mild-to-moderate inflammation. A phase 1 bioavailability–dose proportionality clinical study, consisting of pretreatment, treatment, and posttreatment cycles, was conducted on fertile Chinese women. During the pretreatment cycle, 48 subjects were given placebo patches to study wearability (including skin irritation and adhesion tests). During the treatment cycle, each subject in the test groups received weekly application of 1 (A), 2 (B), or 3 (C) 10-cm^2 patches, and group D received daily 150 μg of LNG and 35 μg of ethynyl estradiol, orally. The wearability study indicated patches were very well accepted. Residual assay of used TDS indicated delivery of LNG and E2 at rates of approximately 5.0 μg/cm^2 day^{-1} and 4.0 μg/cm^2, day^{-1}, respectively, during the treatment cycle. Ra-

dioimmunoassay (RIA) of serum samples demonstrated therapeutically effective serum concentration of LNG and serum profiles of progesterone, luteinizing hormone
(LH), and follicle-stimulating hormone (FHS) also indicated that ovulation inhibition
occurred in most of the subjects wearing TDS. No subject became pregnant, and
posttreatment hormonal profiles indicated that most returned to their normal menstrual cycle on termination of patch use. A second evaluation of TD delivery of
steroids for contraception has been reported (214). A TDS changed weekly and delivering both E(2) and LNG at daily dosages of 3.8 ± 0.8 and 2.9 ± 0.7 $\mu g/cm^2$
day^{-1}, respectively, showed ovulation suppression. An alternative progestin (ST
1435) penetrated skin when formulated in acetylated lanolin or an hydroalcoholic
gel and produced ovulation suppression at a dose of 2 mg/day in a small number of
cycles. For contraceptive purposes TD systems must be perfectly adhesive, well
tolerated locally, and achieve nearly 100% efficacy. These targets are very challeng-
ing, although the potential advantages are so high that the concept deserves further
development.

A pilot clinical study was carried out to evaluate the cognitive and neuroendocrine response to estrogen administration for postmenopausal women with Alzheimer's disease (AD) (215). Twelve women with probable AD of mild-to-moderate
severity completed the study. During an 8-week treatment period, 6 received 0.05
mg/day dosage of 17β-estradiol by a patch and the remainder a placebo patch. Significant effects of estrogen treatment were observed on attention and verbal memory.
In women treated with estrogen, verbal memory enhancement was positively correlated with estradiol plasma levels and negatively correlated with plasma concentrations of insulin-like growth factor-binding protein-3 (IGFBP-3). This suggested that
estrogen replacement may enhance cognition for postmenopausal women with AD.

A trial assessed the effectiveness and tolerability of transdermal estrogen in
men with hot flashes after hormonal therapy for prostate cancer (216). Twelve men
with moderate to severe hot flashes received either low-dose (0.05 mg) or high-dose
(0.10 mg) estrogen patches applied twice weekly for 4 weeks. After a 4-week washout, each patient received the alternative dose for 4 weeks. There was a significant
reduction in overall severity of hot flashes seen with both low- and high-dose estrogen patches, but a significant reduction in daily frequency of hot flashes was seen
only at the high dose. Eighty-three percent reported either mild, moderate, or major
improvement in symptoms with either low- or high-dose patch. Mild, painless breast
swelling or nipple tenderness was noted in 17 and 42% of the men treated with low-
and high-dose estrogen, respectively. FSH levels decreased significantly at both
doses. Estradiol levels increased from 12.1 to 16.4 (low-dose) and 26.9 (high-dose)
pg/mL, but there was no significant change in serum testosterone or LH levels.

D. Androgens

An important aim in treating male hypogonadism is restoration of physiological
concentrations of testosterone and metabolites. New methods for testosterone delivery that have provided increased options for men requiring hormonal replacement
therapy have been reviewed (217). Intramuscular administration of testosterone is
associated with early, high serum levels followed by a gradual decline over the
dosing interval. The TDS now available as alternatives include Testoderm (applied
to the scrotum) and Androderm (applied to nonscrotal skin). Most patients achieved

normal serum testosterone levels with circadian variation and normal estradiol levels. Serum LH levels generally decreased, but not to suppressed levels, and Testoderm use leads to an increase in plasma dihydrotestosterone (DHT). Clinical response in mood, energy level, and sexual function were improved with both systems and were generally comparable with intramuscular injection. There were no clinically significant changes in laboratory parameters, including prostatic specific antigen (PSA), and prostate size did not increase above normal. Skin reactions were common and may require discontinuation of therapy. Patients with inadequate scrotal size may not achieve satisfactory results with Testoderm. Although patches are more expensive than intramuscular (IM) injections, they require less frequent office visits and both transscrotal and transdermal systems offer a good alternative for hypogonadal men who do not desire fertility during the treatment period.

Scrotal testosterone patches can produce normal serum levels mimicking diurnal variations. This was followed in hypogonadal men treated transdermally for up to 10 years (218). Eleven men (age 35.9 ± 9.8 years) at start of study were treated with transscrotal patches (Testoderm) because of primary ($n = 4$) or secondary ($n = 7$) hypogonadism. Clinical examinations were performed every 3 months during the first 5 years and every 6 months thereafter. On daily application of one patch, testosterone levels rose from 5.3 ± 1.3 to 16.7 ± 2.6 nmol/L at month 3 and remained in the normal range throughout treatment. Serum DHT rose from 1.3 ± 0.4 to 3.9 ± 1.4 nmol/L and estradiol from 52.3 ± 9.3 to 71.3 ± 9.6 pmol/L and remained stable. Patients reported no local side effects apart from occasional itching. No relevant changes occurred in clinical chemistry and hemoglobin and erythrocyte counts remained normal. Bone density increased slightly from 113.6 ± 5.4 to 129.7 ± 9.3 mg/cm^3. In the nine patients who were younger than 50 years prostate volumes showed a small, but insignificant, increase from 16.8 ± 1.5 to 18.8 ± 2.1 mL during therapy. In two older patients, prostate volume remained constant or decreased slightly during therapy. Prostate-specific antigen levels were constantly low in all patients.

The possibility of immediate adverse effects of short-term testosterone administration to older men with low bioavailable testosterone, especially on the symptoms of benign prostate hyperplasia, was investigated (219). A 9-week intervention with either intramuscular testosterone enanthate (200 mg every 3 weeks), TD testosterone (two 2.5-mg patches per day), or neither, was followed by a 9-week observation period. Twenty-seven men (age 74 ± 3 years) with no medical conditions known to affect bone turnover and with total testosterone levels less than 350 ng/dL or bioavailable testosterone levels less than 128 ng/dL were included. All men receiving testosterone treatment increased levels above their own baseline, but only six of nine men receiving TD testosterone achieved bioavailable testosterone levels in the normal range for young men. No side effects were reported using intramuscular delivery, but five of nine men using TD testosterone developed a rash.

As part of a phase III multicenter study, pharmacokinetics and metabolism of a permeation-enhanced testosterone transdermal system and the influence of the application site were investigated in 34 hypogonadal men (21–65 years of age) (220). After an 8-week androgen washout period, two patches were applied to the back for 24 h. Serum concentrations of total testosterone (T), bioavailable testosterone (BT), DHT, and estradiol [E(2)] increased from hypogonadal levels into normal physiological ranges and declined to baseline levels within 24 h after system removal. Peak concentrations occurred about 8 h after application for T and BT and at 13 h for DHT and E(2). Estimated half-lives were: T, 1.29 ± 0.71 h; BT, 1.21 ± 0.75 h; DHT, 2.83 ± 0.97 h; and E(2), 3.53 ± 1.93 h. The influence of application site was evaluated by applying two patches for 24 h to the abdomen, back, chest, shin, thigh, or upper arm. Hormone profiles were qualitatively similar at each site, but C_{ss} values were significantly different. Based on BT levels, rank ordering of sites was back > thigh > upper arm > abdomen > chest > shin. DHT/T and E(2)/T ratios showed negligible site variation.

Further investigations of hormone levels, pharmacokinetics, clinical response, and safety of a permeation-enhanced testosterone transdermal system (TD) in the treatment of hypogonadal men were reported (221). This was a multicenter study with four consecutive periods: period I (3 weeks), evaluation of current androgen therapy (primarily testosterone enanthate injections (mean dose 229 mg; mean interval 26 d); period II (8 weeks), androgen washout; period III (3–4 weeks), single-dose pharmacokinetic studies of TD systems; period IV (12 months), efficacy, safety, and steady-state pharmacokinetic evaluation of TD systems (5 mg/day nominal delivery rate of testosterone). Thirty-seven hypogonadal men 21–65 years old enrolled, 34 entered periods III and IV and 29 (9 primary, 20 secondary hypogonadism) completed the study. Four patients withdrew because of adverse events. Measurements included morning serum levels of total testosterone (T); bioavailable testosterone (BT), DHT, and E(2) levels; circadian pattern of T profiles and 24-h time-averaged T level; LH levels in patients with primary hypogonadism; and reduction of hypogonadal symptoms. Safety assessments included skin tolerability, prostate parameters, lipid profile, and systemic parameters. Twelve months of TD therapy normalized morning serum T levels in 93% of patients, and produced more than 80% normalization of BT, DHT, and E(2) levels. The TD system mimicked the circadian variation in T levels seen in healthy young men and normalized 24-h time–average T levels in 86% of patients. LH was suppressed in eight of nine men with primary hypogonadism, and normalized in five of these. Subjective symptoms of hypogonadism, including decreased libido and fatigue, showed improvement after 2–4 weeks of treatment in most patients. Most adverse events were local skin reactions. Prostate assessments showed a lower prostate-specific antigen level during TD therapy compared with IM injections (0.66 vs. 1.00 μg/L), but prostate size did not differ significantly between the treatment regimens.

Weight loss associated with human immunodeficiency virus (HIV) infection is multifactorial in its pathogenesis, but it was speculated that hypogonadism contributed to depletion of lean tissue and muscle dysfunction (222). Effects of testosterone replacement, using Androderm, on lean body mass, body weight, muscle strength, health-related quality of life, and HIV disease markers were evaluated. Testosterone replacement in HIV-infected men with low testosterone levels was considered safe and associated with a 1.35-kg gain in lean body mass, a significantly greater reduction in fat mass than achieved with placebo treatment, and an increased red cell count and improvement in role limitation owing to emotional problems. It was concluded that further studies were required to assess whether such supplementation can produce clinically meaningful changes in muscle function and disease outcome in HIV-infected men.

Markedly decreased serum androgen levels occur in women with acquired immunodeficiency syndrome (AIDS) and may be a contributing factor to the wasting syndrome. A pilot study of the effects of androgen replacement therapy was conducted to determine efficacy in terms of change in serum testosterone, safety parameters, and tolerability, and to investigate testosterone effects on weight, body composition, quality of life, and functional indexes (223). Fifty-three ambulatory women with AIDS wasting syndrome, free of new opportunistic infection within 6 weeks of study initiation and with serum levels of free testosterone less than normal reference range, were enrolled. Subjects weighed 92 ± 2% of ideal body weight, and had lost 17 ± 1% of their maximum weight. Subjects received two placebo patches (PP), one active and one placebo patch (AP), or two active patches (AA) applied twice weekly to the abdomen for 12 weeks. Nominal delivery rates were 150 and 300 μg/day, respectively, for AP and AA groups. Serum free testosterone levels increased significantly from 1.2 ± 0.2 to 5.9 ± 0.8 pg/mL (AP) and from 1.9 ± 0.4 to 12.4 ± 1.6 pg/mL (AA). Testosterone administration was generally well tolerated locally and systemically, with no adverse trends in hirsutism scores, lipid profiles, or liver function tests. Improved social functioning and pain score were observed in AP- versus PP-treated patients. These data suggested that testosterone administration may improve the status of women with AIDS wasting.

The literature on androgen replacement for erectile dysfunction was evaluated by metanalysis (224). Study inclusion criteria were testosterone given as the only therapy for erectile dysfunction and a clearly stated definition of response for evaluating treatment. Sixteen of 73 articles published between 1966 and 1998 were included. Overall response rate was 57%, and patients with primary versus secondary testicular failure had a response rate of 64% versus 44%. Intramuscular and oral methods of delivery were equivalent (response rates 51.3 and 53.2%, respectively) but response to TD therapy was significantly different (80.9%).

It is acknowledged that women may experience symptoms secondary to androgen deficiency, and there is substantial evidence that prudent androgen replacement can be effective in relieving both physical and psychological symptoms of androgen insufficiency (225). Testosterone replacement for women is now available in a variety of formulations. It appears to be safe, with the caveat that doses are restricted to the "therapeutic" window for androgen replacement in women, such that the beneficial effects on well-being and quality of life are achieved without incurring undesirable virilizing side effects.

For treatment of adult hypogonadal men, nightly 24-h application of the Androderm testosterone TDS (5 mg/day) has been demonstrated to be effective by a series of clinical pharmacokinetic studies (226). For treatment of adolescent males, physiological replacement can be approximated by modifying the dose and duration of Androderm application to mimic patterns of nocturnal testosterone secretion observed during puberty. A clinical audit was reported on the acceptability and efficacy as a treatment for hypogonadism of the first transdermal testosterone therapy available in the United Kingdom (Andropatch), compared with existing androgen replacement options (227). Serum testosterone and questionnaire data on treatment efficacy, side effects, therapy preference, sexual dysfunction, and partner's attitudes to therapy were obtained from 50 hypogonadal men prescribed long-term testosterone replacement. Eighty percent returned analyzable questionnaires and, of these, 84% experienced adverse effects with TD therapy, usually dermatological problems. Twenty-two percent elected to continue with TD therapy, 72% returned to depot, and 5% to oral therapy. The reservoir patches were judged to be too large, uncomfortable, visually obtrusive and noisy; thus, the pharmacokinetic advantages were largely out-weighed by low patient acceptability. The pharmacokinetics of three testosterone-containing TDS were evaluated in healthy male volunteers (228). Type 1 and 2 were nonscrotal membrane patches differing in adhesive type. Six subjects were treated with low-dose testosterone type 1, high-dose testosterone type 1, and low-dose tes-tosterone type 2. To eliminate the influence of endogenous serum testosterone, se-cretion was suppressed by the GnRH antagonist cetrorelix. Physiological testosterone levels were achieved during the 24-h application period. Maximal serum levels were achieved after 4 h with both types, and both enabled a physiological circadian profile to be achieved.

Effects of TD testosterone replacement therapy using a permeation-enhanced system on plasma lipolytic enzymes (hepatic and lipoprotein lipase), LDL, and HDL subfraction concentrations were assessed (229). Ten patients with primary testicular failure were started on Testoderm therapy and evaluated before and after 3 months of treatment. Serum testosterone level increased to within the normal range in all subjects, whereas serum DHT increased to supranormal values. Plasma hepatic lipase (HL) activity increased after testosterone replacement (24.7 ± 7.5 vs. 29.2 ± 8.3 μmol free fatty acid released per hour) and the increase in HL correlated with the increase in DHT. No significant change was seen in the HDL2 subfraction but HDL3 decreased after treatment (0.93 ± 0.17 vs. 0.79 ± 0.14 mmol/L). Whether these changes adversely influence the cardiovascular risk in the long term remains to be determined.

Local skin reactions at application site are common adverse events with trans-dermal testosterone and an open-label, controlled pilot study evaluated whether top-ical pretreatment with triamcinolone acetonide, 0.1% cream, reduced the incidence or severity of chronic skin irritation in health volunteers (230). At all assessment points, more subjects had lower cumulative scores with pretreatment than without pretreatment.

E. Fentanyl

Transdermal fentanyl has been used widely in the United States since it was approved in 1990 (231). The first clinical report of TD fentanyl therapy in cancer pain involved five patients. Pain relief was established with intravenous fentanyl and a TDS se-lected to deliver the same hourly dose while the intravenous infusion was tapered over 6 h. The TDS was changed every 24 h for a total of 3–156 days. The initial study demonstrated steady-state plasma levels were linearly related to the fentanyl dose. A multicenter trial was conducted in 39 patients. The TD fentanyl dose was established from a conversion table based on the dose of oral immediate-release morphine required to control pain. The fentanyl patches were changed every 72 h and immediate-release morphine used on an as-required basis for incidental pain. This trial further demonstrated that patients could be converted from oral morphine to an equianalgesic dose of transdermal fentanyl and that pain relief could be main-tained for a lengthy time period on an outpatient basis.

The analgesic, pharmacokinetic, and clinical respiratory effects of 72-h appli-cation of two TD fentanyl patch sizes in patients undergoing abdominal hysterectomy were evaluated (232). Fentanyl TDS, releasing 50 μg/h (TF-50) or 75 μg/h (TF-75) fentanyl or placebo patches were applied to 120 women 2 h before abdominal hysterectomy under general anesthesia. All patients had postoperative access to supple-mental morphine using patient-controlled pumps. VAS pain scores, supplementary analgesia, fentanyl plasma concentration, continuous hemoglobin saturation, respi-ratory pattern, and adverse effects data were collected. Visual analogue scale (VAS) pain scores and supplemental morphine use significantly decreased in the TF-75 group in the postanesthesia care unit, and for both TF-50 and TF-75 groups for 8–48 h postoperatively. Between 5 and 36 h, the TD fentanyl groups had significantly increased abnormal respiratory patterns, including apneic episodes and episodes of slow respiratory rate, and a significantly increased requirement for oxygen supple-mentation. Nine patients in the TD fentanyl groups were withdrawn because of severe respiratory depression, but none in the placebo group. Although fentanyl plasma concentration were higher in the TF-75 than in TF-50 group, differences were not significant. Fentanyl plasma concentration decreased significantly 48 h after patch application. Although good analgesia was the result of this combination therapy, it was associated with a high incidence of respiratory depression requiring intensive monitoring, oxygen supplementation, patch removal in about 11% of patients, and opioid reversal with naloxone in about 8% of patients.

It is generally recommended that patients should be titrated with a short-acting narcotic to control their cancer pain before they are converted to a fentanyl TDS. However, immediate TD fentanyl therapy in patients with uncontrolled cancer pain with direct titration of fentanyl TDS according to clinical necessity on a day-to-day basis, with the availability of morphine solution for rescue medication has been evaluated (233). The major objective was simplification of therapy. On average, sufficient pain control was reached within 48 h of the start of TD fentanyl, and VAS values at all follow-up times were significantly lower than pretreatment values. Mean fentanyl doses were 70 μg/h (week 1), 98 μg/h (week 2), 107 μg/h (week 3) and 116 μg/h (week 4). Mean morphine doses as rescue medication steadily decreased from 11 mg/day in week 1 to 3 mg/day in week 4 of treatment, although differences were not significant. It was concluded that titration with a short-acting narcotic before conversion to TD fentanyl was not necessary, provided the patients were well monitored.

A comparison of TD fentanyl with placebo for postoperative analgesia in or-thopedic surgery was reported (234). Forty adult patients had general anesthesia with propofol, isoflurane in nitrous oxide, and oxygen, and small bolus of alfentanil or sufentanil. Preoperatively, one group received TD fentanyl (75 μg/h) for 72 h, and the second group a placebo patch. Morphine was given postoperatively as necessary. Eleven patients receiving fentanyl needed morphine, compared with 19 in the placebo group, and mean morphine dose was significantly lower in the fentanyl group. One fentanyl group patient had decreased oxygen saturation and intense sedation, neces-sitating administration of naloxone. The mean maximum plasma fentanyl concentra-tion was 1.63 ng/mL.

An alternative fentanyl TDS was evaluated (235). Male adult surgical patients received 650 or 750 μg of fentanyl intravenously as part of the induction of anes-thesia, and plasma fentanyl concentrations were measured over the following 24 h. On the first postoperative day a fentanyl TD was placed on the upper torso for 24 h and then removed. Plasma fentanyl concentrations were measured for 72 h after device application and each patient's clearance and unit disposition function were determined. During the 72 h after application of the TDS, the amount of fentanyl absorbed and the absorption rate were also determined. Residual fentanyl in the transdermal device was measured to determine absolute bioavailability. Of 14 subjects receiving TD fentanyl, 3 had clinically significant fentanyl toxicity, mandating early removal of the TDS. In the remaining subjects the fentanyl concentration from 12 to 24 h varied over a 20-fold range (0.34–6.75 ng/mL). In subjects wearing the device for 24 h, terminal half-life of fentanyl after device removal was 16 h. Bioavailability of transdermally administered fentanyl was 63 ± 35% and rate of fentanyl absorption from 12 to 2 h in subjects still wearing the device ranged from 10 to 230 μg/h. In two subjects, the rate within the first 6 h briefly exceeded 300 μg/h and both demonstrated fentanyl toxicity, requiring early device removal. It was concluded that the Cygnus transdermal fentanyl device produced more highly variable plasma fentanyl concentrations than those reported for the Duragesic device, which is contraindicated for postoperative analgesia.

The use of transdermal fentanyl was evaluated in ten patients aged 9–16 years with sickle cell pain crisis (236) who received TD fentanyl at 25 ($n = 7$) or 50 ($n = 3$) μg/h and morphine use was >2.5 mg/h. Average TD fentanyl dose was 0.77 ± 0.37 μg/kg h^{-1} on day 1 and 1.17 ± 0.46 μg/kg h^{-1} on day 2. The fentanyl concentration at 24 and 48 h was 0.60 ± 0.31 and 1.18 ± 0.44 ng/mL, respectively. There was a significant relation between dose and fentanyl concentration and no difference in any clinical-monitoring parameter between day 1 and day 2, although seven of ten patients reported subjective improvement in pain control. No adverse effects were noted, but it was concluded that improved understanding of dose–effect relations for TD fentanyl in children and adolescents was necessary before adequate pain control could be achieved by this route.

At the start of TD fentanyl treatment, depot accumulation of the drug within skin results in a significant delay (17–48 h) before achievement of maximum plasma concentration (237). However, concomitant use of short-acting morphine maintained pain relief during the titration period, and supplementary medication decreased with duration of TD fentanyl treatment. Patient preference for TD fentanyl was indicated by the number of patient requests for continued use at the end of the study (up to 95%). Although postoperative TD fentanyl is contraindicated, supplementary patient-controlled analgesia was significantly reduced in patients receiving 75 μg/h TD fentanyl, compared with placebo. Some patients with previously uncontrolled pain became completely pain-free. The most common adverse events during TD fentanyl therapy included vomiting, nausea, and constipation. Most serious adverse event was hypoventilation, which occurred more frequently in postoperative (4%) than in cancer patients (2%). In surgical patients, fentanyl-associated respiratory events usually occurred within 24 h of patch application, although there were isolated reports of late onset (up to 36 h).

Respiratory depression was observed in a 2-year-old boy unintentionally exposed to a fentanyl TDS (238). He was found unresponsive after sleeping in bed with his grandmother and, after intubation and ventilation, a fentanyl TDS was discovered on his back. Removal of the device and treatment with naloxone resolved symptoms. This was the first reported case of secondary exposure to a fentanyl patch causing clinically significant respiratory depression in the pediatric population, and it emphasized a new hazard.

A multicenter evaluation of TD fentanyl for cancer pain relief in 53 patients who required 45 mg or more of oral morphine daily was reported (239). After 1-

week stabilization on oral morphine, patients were transferred to an appropriate dose of TD fentanyl (25, 50, 75, or 100 μg/h) administered by patch every 3 days. TD fentanyl was titrated to pain relief and patients were followed for up to 3 months. Mean duration of TD fentanyl use was 58 $\pm$ 32 days. Mean daily morphine dose during the last 2 days of stabilization was 189 $\pm$ 20 mg, and mean initial fentanyl patch dose was 58 $\pm$ 6 μg/h. The mean daily morphine dose taken as required for breakthrough pain at study completion was 35 mg. Mean final fentanyl dosage at study completion was 169 $\pm$ 29 μg/h. Pain relief was rated as good or excellent by 82% of patients during the treatment period and 63% preferred TD fentanyl to their last analgesic therapy. Side effects related to the patch were nausea (13%), vomiting (8%), skin rash (8%), and drowsiness (4%). Thirty percent of patients reported adverse experiences related to the patch, and 17% had to be discontinued.

Pain-related treatment satisfaction, patient-perceived side effects, functioning, and well-being in patients with advanced cancer receiving TD fentanyl (Duragesic) or sustained-release oral morphine (MS Contin or Oramorph SR) were compared (240). Patients were more satisfied with TD fentanyl than sustained-release oral morphine. Patients receiving TD fentanyl were more satisfied overall with their pain medication and also reported a significantly lower frequency and effect of a pain medication's side effects. Measures of pain intensity, sleep adequacy, and symptoms demonstrated no significant differences between treatment groups.

Forty-eight patients with noncancer neuropathic pain, who had participated in a randomized controlled trial with intravenous fentanyl infusions, next received prolonged TD fentanyl (241). Eighteen patients stopped prematurely owing to inadequate pain relief, side effects, or both. Pain relief among the remaining 30 patients completing the 12-week–dose titration protocol was substantial in 13 and moderate in 5, and quality of life improved by 23%. Psychological dependence or induction of depression was not observed, and tolerance emerged in only 1 patient. There was a significant positive correlation between pain relief obtained with intravenous and prolonged TD fentanyl. It was concluded that long-term TD fentanyl may be effective in noncancer neuropathic pain without clinically significant management problems. Testing with intravenous fentanyl may assist in selecting neuropathic pain patients who can benefit from treatment by the transdermal route.

Disposition of fentanyl in dogs after intravenous and TD administration was investigated (242). Each of six clinically normal beagles received intravenous fentanyl (50 μg/kg body weight) and TD fentanyl (50 μg/h). TD fentanyl produced average steady-state concentrations of 1.6 ng/mL. Actual delivery rate of fentanyl was 27.5–99.6% of the theoretical rate. Mean elimination half-life after patch removal was 1.39 h. It was concluded that TD fentanyl had clinically useful potential as an analgesic in dogs. Plasma fentanyl concentrations were obtained in six intact, mixed-breed adult dogs (two males, four females), weighing 19.9 $\pm$ 3.4 kg, after application of three sizes of fentanyl TDS delivering 50, 75, or 100 μg/h (243). Results are summarized in Table 7.

TD fentanyl was evaluated in children with cancer pain (244) and measures of analgesia, side effects, and skin changes obtained for a minimum of two doses (6 treatment days). Treatment was well tolerated and 10 of 11 patients continued the treatment. Time to peak plasma concentration ranged from 18 to more than 66 h in patients using a 25-μg/h patch. Compared with published data from adults, mean

Table 7 Comparison Between Three Transdermal Fentanyl Patches in Dogs

Fentanyl delivery (μg/h)	Plasma fentanyl[a] (ng/mL)	Total AUC (ng/h mL^{-1})	Elimination half-life (app) (h)
50	0.7 ± 0.2	46 ± 12.2	3.6 ± 1.2
75	1.4 ± 0.5	101.2 ± 41.4	3.4 ± 2.7
100	1.2 ± 0.5	80.4 ± 38.3	2.5 ± 2.0

[a]Levels between 24 and 72 h.
Source: Ref. 243.

clearance and volume of distribution of transdermal fentanyl were equal, but variability was less.

It has been claimed that fentanyl patches are less suitable for patients who are elderly or terminally ill and dying. However, a retrospective survey of 205 cancer patients who died within hospital-based home care in Norrkoping, Sweden between January 1997 and June 1998 reported that 34 patients used fentanyl TD, and in 30 it was possible to evaluate analgesic efficacy. Estimated efficacy was good (93%) or moderate (88%) in patients younger or older than 65 years of age (245).

Constipation and use of laxatives were investigated in patients with chronic cancer pain treated with oral morphine and TD fentanyl (246). Of 46 patients, treated with slow-release morphine, 30–1000 mg/day for 6 days, 39 were switched to TD fentanyl (0.6–9.6 mg/day) with a conversion ratio of 100:1. Median fentanyl doses increased from 1.2 to 3.0 mg/day throughout the 30-day TD treatment period. Mean pain intensity decreased slightly after conversion, although the number of patients with breakthrough pain or who requiring immediate-release morphine as a rescue medication was higher with TD fentanyl. The number of patients with bowel movements did not change after the opioid switch, but the number of patients taking laxatives was significantly reduced from 78–87% of patients per treatment day (morphine) to 22–48% (fentanyl). Constipation is an almost universal side-effect of prolonged opioid analgesia, and resulting discomfort can be more severe than the pain itself, leading to reduction of analgesic use and consequently increased pain (247). Because of higher lipophilicity, fentanyl penetrates the blood–brain barrier more easily and lower doses are possible; thus, comparatively less opioid is available in the gastrointestinal tract to block local receptors. The cost of treating constipation among terminally ill cancer patients receiving TD fentanyl and 12-h sustained-release morphine in Ontario, Canada was examined (248). Cost of managing constipation in a patient receiving TD fentanyl and 12-h sustained-release morphine was estimated at 31.77 and 52.76 dollars, respectively, during the first 2 weeks of treatment. When the acquisition costs of opioids are included, the two-week cost of managing a patient with TD fentanyl and 12-h sustained-release morphine was 123.24 and 119.70 dollars, respectively. It was concluded that acquisition costs alone should not dictate treatment, and care should always be tailored to the needs and preferences of individual patients.

The issues involved in switching opioids to transdermal fentanyl in a clinical setting have been evaluated (249). Case records of patients treated with TD fentanyl were retrospectively examined and conversion ratios calculated. Opioid therapy was switched to TD fentanyl during inpatient treatment for 53 patients and during out-patient treatment for 11 patients. Before conversion patients were treated with slow-release morphine (48%), immediate-release morphine (17%), buprenorphine (11%), tramadol (11%), levomethadone (5%), tilidine/naloxone (5%), and piritramid (3%). Reasons for opioid rotation included inadequate pain relief (33%), patients' wish to reduce oral medication (20%), gastrointestinal side effects (31%), vomiting (13%), constipation (19%), and dysphagia (27%). Reduction of side effects was reported by 10 of 19 patients. In 12 of 21 patients, when medication was switched because of inadequate pain relief, reduction in pain intensity was reported. It was concluded that conversion to TD therapy may readjust the balance between opioid analgesia and side effects.

F. Anti-inflammatories

A multicenter study of comparative efficacy, tolerability, and acceptability of a flurbiprofen local-action transcutaneous (LAT) patch (40 mg b.d.) and piroxicam gel (3 cm, 0.5% q.d.s), was conducted in general practice in the United Kingdom in 137 men and women with soft-tissue rheumatism of the shoulder or elbow tag, epicondylitis, tendinitis, bursitis, or adhesive capsulitis (250). Patients received one therapy for 4 days before crossing over for a further 4 days, followed by 6 days of their preferred therapy. Assessment of severity of pain, tenderness, and overall clinical condition was carried out at baseline and at 4, 8, and 14 days. There was a significant reduction in severity of pain in favor of flurbiprofen LAT (42 vs. 26%). Eligible dataset analysis revealed significant differences, in favor of flurbiprofen LAT, in severity of lesion tenderness and overall change in clinical condition. Superior efficacy was also indicated at the end of the crossover phase when 69% chose to continue treatment with flurbiprofen LAT. There were also significant differences in favor of flurbiprofen LAT in assessments for night pain, quality of sleep, and patients' overall opinion of treatment.

G. Clonidine

Pharmacokinetic and pharmacodynamic properties and safety, of a TD clonidine system (M-5041T) were evaluated after single and repeated applications (251). In the single-application study, one patch delivering 4, 6, or 8 mg was applied for 3 days to eight healthy subjects. In a repeated-application study, A (0–72 h), B (72–144 h), and C (144–216 h) TD systems delivering 6 mg were applied in seven healthy subjects. In the single-application study, plasma clonidine level, C_{max} and AUC increased in a dose-dependent manner, but not significantly. The blood pressure (BP)-lowering effect of 8 mg was greater than that of 4 and 6 mg. Adverse effects were reported, but did not cause withdrawal. In the repeated-application study, plasma clonidine increased up to 48 h after application of patch A, and remained stable until removal of patch C. C_{max} and AUC did not differ significantly. During an active period, BP decreased significantly during treatment, but BP at midnight did not change significantly. Mild erythema and systemic adverse effects were reported. In a second study (252), TD clonidine (M-5041T) was compared with oral clonidine (Catapres TTS). One TDS containing 6 mg of clonidine was applied on the right chest for 3 days or one tablet of Catapres TTS (0.075 mg) was given orally every 12 h for 3 days in eight healthy subjects. Plasma clonidine concentration increased gradually after application of TDS and decreased gradually after removal, whereas it increased rapidly then decreased rapidly after each dose of Catapres TTS. Elimination half-life of clonidine after patch removal was significantly greater than after the final dose of Catapres TTS. There was no significant difference in C_{max}, AUC, or BP-lowering between the treatments. Adverse symptoms occurred more frequently during Catapres TTS therapy and were observed when plasma clonidine concentration was relatively high.

Clinical pharmacological evaluation of a clonidine TDS was undertaken in a variety of phase I volunteer studies to assess potential effects of site of application, pharmacokinetic linearity, and absolute bioavailability (253). Delivery of clonidine from Catapres TTS was influenced by application site. The order of permeability was chest > upper arm > upper thigh. Pharmacokinetics of clonidine following TD delivery were linear with slight nonlinearity between devices of differing area. Absolute bioavailability of clonidine from Catapres TTS was approximately 60% and calculated in vivo absorption rate (4.32 ± 1.68 μg/h) was in good agreement with claimed performance (0.1 mg/day).

Premedication with oral and TD clonidine in postoperative sympatholysis was investigated in 61 patients undergoing elective major noncardiac surgery (254). The treatment group were premedicated with a clonidine TDS (0.2 mg/d), applied the night before surgery and left in place for 72 h, and were given 0.3-mg oral clonidine 60–90 min before surgery. Clonidine reduced enflurane requirements, intraoperative tachycardia, and myocardial ischemia (1 of 28 clonidine patients vs. 5 of 24 placebo). However, postoperatively, the heart rate decreased for only the first 5 h, and incidence of postoperative myocardial ischemia (6 of 28 clonidine vs. 5 of 26 placebo) did not differ. Clonidine significantly reduced plasma levels of epinephrine and norepinephrine on the first postoperative morning. It was suggested that larger doses of clonidine should be investigated for their ability to decrease postoperative tachycardia and myocardial ischemia. Surgical trauma induces diffuse sympathoadrenal activation that contributes to perioperative cardiovascular complications in high-risk patients. Regional anesthetic and analgesic techniques can attenuate this "stress response" and reduce rates of adverse perioperative events, but their postoperative use is logistically difficult and expensive. Hence, use of transdermal clonidine to blunt the stress response throughout the perioperative period was evaluated (255). Forty patients scheduled for major upper abdominal surgery were entered in a clinical trial. Patients received either clonidine (0.2 mg orally plus clonidine TTS-3 patch the evening before surgery plus 0.3 mg orally on call to operating room) or matched oral and transdermal placebo. Preoperative transdermal (plus oral) clonidine administration resulted in therapeutic plasma clonidine concentrations over the perioperative period (1.54 ± 0.07 μg/mL) and reduced preoperative epinephrine and norepinephrine levels by 65%. Plasma catecholamines increased in both groups after surgery, but were markedly lower over the postoperative period in patients receiving clonidine (who also had reduced frequency of postoperative hypertension).

A study was performed to determine if a lower than previously reported oral–transdermal clonidine regimen could reduce postoperative morphine requirements without producing systemic side effects (256). Twenty-nine healthy females undergoing elective abdominal hysterectomy received preoperative oral clonidine, 4–6 μg/kg plus a 7 cm^2 clonidine TDS (0.2 mg/24 h), or a placebo tablet and patch. Low-dose clonidine had no potentiating effect on morphine analgesia and postoperative morphine use, VAS pain scores and morphine plasma levels were similar between the groups. The clonidine group experienced a significantly higher incidence of intraoperative and postoperative hypotension and bradycardia than the control group, but no differences were noted in incidence of nonhemodynamic side effects.

Clinical and pharmacological indications suggested reduction of noradrenergic tone occurs in cluster headache, during both active and remission periods, but that sharp fluctuations of the sympathetic system may trigger the attacks (257). It was postulated that continuous administration of low-dose clonidine could be beneficial in the active phase by antagonizing variations in noradrenergic tone. After a run-in week, TD clonidine (5–7.5 mg) was administered for 1 week to 13 patients suffering from episodic or chronic cluster headache. During clonidine treatment, mean weekly frequency of attacks reduced from 17.7 ± 7.0 to 8.7 ± 6.6, pain intensity of attacks (VAS) from 98.0 ± 7.2 to 41.1 ± 36.1 mm, and duration from 59.3 ± 21.9 to 34.3 ± 24.6 min. This strongly suggested that TD clonidine may be effective in the preventive treatment of cluster headache.

To test a previous clinical observation that approximately 25% of patients with painful diabetic neuropathy appeared responsive to clonidine, a formal clinical trial of TD clonidine was conducted (258). In stage 1, 41 patients completed a randomized, three-period crossover comparison of TD clonidine (titrated from 0.1 to 0.3 mg/day) to placebo patches. Twelve apparent responders from stage 1 were entered into the second stage, consisting of an additional four double-blind, randomized, 1-week treatment periods with TD clonidine and placebo. Stage 1 showed that, for the total group, pain intensity differed little between clonidine and placebo. In stage 2, however, the 12 apparent responders from stage 1 had 20% less pain with clonidine than placebo (95% CI: 4–35% pain reduction), confirming that their pain was responsive to clonidine. Post hoc analysis suggested patients describing their pain as sharp and shooting were more likely to respond to clonidine. These results support the hypothesis that there is a subset of patients with painful diabetic neuropathy who benefit from systemic clonidine administration and also demonstrate the value of an enriched enrollment technique in analgesic trials.

Efficacy and tolerability of TD clonidine in inner-city African-American and Hispanic-American patients with essential hypertension was evaluated in multiple community-based primary care centers in a 12-week trial (259). Patients with diastolic BP higher than 90 mmHg were given TD clonidine (0.1 or 0.2 mg daily). The drug was titrated after 1 month if diastolic BP exceeded 90 mmHg. At 12 weeks, change in blood pressure, adverse effects, and patient satisfaction were assessed. Transdermal clonidine significantly lowered BP by 15.7/12.8 ± 18.1/9.6 mmHg, and heart rate by 3 ± 9 beats per minute. There were no differences in BP reduction according to race, ethnicity, gender, or age. Most common adverse effects were pruritus or discomfort at the patch site, dizziness, dry mouth, and fatigue (11% discontinued owing to adverse effects). Sixty-seven percent of patients reported that TD clonidine was more convenient than oral therapy.

H. Miscellaneous

Intraocular pressure (IOP)-lowering effects of a TD system containing pilocarpine that was designed to avoid common side effects in glaucoma treatment with conventional eye drops were reported (260). Two patches, each containing 30 mg of pilocarpine or placebo, were applied to the supraclavicular skin of 24 patients. IOP was recorded before (22.7 ± 5.8 mmHg) and at +12, 16, and 20 h after application. Pilocarpine TD did not significantly reduce IOP although there were detectable plasma levels of pilocarpine at 12 (2.9 ng/mL) and 20 h (1.3 ng/mL) after administration.

The therapeutic effects of selective cholinergic replacement using oral xanomeline, an M_1/M_4 receptor agonist, were assessed in a multicenter study of 343 patients with Alzheimer's disease (261). Improvement in ADAS-Cog provided clinical evidence of involvement of the M_1 muscarinic receptor in cognition, and the favorable effects of xanomeline on disturbing behavior suggested a novel approach for treatment of noncognitive symptoms. Although adverse effects (mainly gastrointestinal) associated with the oral formulation appeared to limit its use, a large-scale study investigating the safety and efficacy of transdermal xanomeline is underway.

The effects of TD scopolamine on pulmonary function, symptoms, and bronchial hyperresponsiveness to methacholine were reported (262). The first study evaluated therapeutic effects of a single scopolamine TD system (Scopoderm TTS, 2.5 cm^2) on forced expiratory volume in 1 s (FEV_1), reversibility, peak expiratory flow (PEF), and symptoms in ten patients with reversible airways disease. The drug was adequately taken up into the systemic circulation, but no significant clinical effects, nor correlations between scopolamine levels and outcome parameters, were observed. Because of the possibility of subtherapeutic doses, a second study with two scopolamine TD systems was carried out in ten patients with bronchial hyperresponsiveness to methacholine. Blood and urine concentrations of free scopolamine were doubled compared with the first study, but there were still no significant effects on FEV_1, PEF, symptoms, and bronchial hyperresponsiveness, although most of the patients now reported adverse side effects. It was concluded that TD administration of scopolamine was not clinically useful in asthma and chronic obstructive pulmonary disease.

Efficiency of TD scopolamine in prophylaxis of postoperative nausea and vomiting (PONV) after otoplasty was evaluated in a post hoc assessment of data (263). Of 50 otoplasty patients 25 received a scopolamine patch before general anesthesia. The placebo group received intravenous atropine (10 μg/kg) during induction. Scopolamine-treated patients suffered more from moderate preoperative bradycardia (8 of 25) than atropine-treated patients (1 of 25). After unilateral otoplasty, no scopolamine-treated patients, but 50% of atropine-treated patients, suffered from PONV. After bilateral otoplasty, respective incidences were 39 and 81%. After unilateral otoplasty no patient needed droperidol, but after bilateral otoplasty, 12 of 19 atropine-treated and 4 of 18 scopolamine-treated patients needed droperidol. It was concluded that TD scopolamine offers effective prophylaxis against PONV, but does not protect from bradycardia in otoplasty. The effects of TD scopolamine (1.5 mg) and oral cyclizine (50 mg) on postural sway, optokinetic nystagmus, and circularvection in humans were investigated (264). Neither scopolamine nor cyclizine (at doses used for relief of motion sickness) had a significant suppressive effect on these aspects of visual–vestibular interaction.

A pilot clinical trial of TD administration of selegiline in HIV-positive patients was performed to obtain preliminary data on safety, tolerability, and effect on HIV-associated cognitive impairment (265). Both selegiline and placebo were well tolerated and improvements favoring the selegiline group were suggested on single tests of verbal memory and motor–psychomotor performance.

TD patches of diltiazem hydrochloride were formulated using ethyl cellulose and polyvinylpyrrolidone. Pharmacodynamic and pharmacokinetic performance of diltiazem hydrochloride after TD administration was compared with that of oral administration in male New Zealand albino rabbits (266). Pharmacokinetic parameters were significantly different between transdermal and oral administration. Terminal elimination half-life of TD-delivered diltiazem was similar to that of oral administration. Skin irritation studies indicated no recognizable changes after patch removal. It was concluded that relative bioavailability of diltiazem hydrochloride and therapeutic activity were fivefold higher after TD than after oral administration. Transdermal histamine has been successfully used for amelioration of symptoms of both relapsing–remitting and progressive multiple sclerosis (267). Of the 55 patients using TD histamine cream 67% had improvements in one or more areas (extremity strength, balance, bladder control, fatigue, activities of daily living, and cognitive functioning) sustained for up to 3 months, and 33% had improvements in three or more areas.

The occlusive properties of a range of hydrocolloid patches on the penetration of triamcinolone acetonide, and hydration, were assessed in vivo using visual assessment (268). There was a significant difference in rates of hydration of patches containing either NaCMC 39%, or pectin 39%, although changes in hydrocolloid composition did not significantly alter the blanching response.

The feasibility of use of water-activated, pH-controlled silicone reservoir devices for TD administration was investigated using timolol maleate (269). Two timolol patches were applied to the arm of 12 volunteers for 81 h and absorption compared with that from an oral tablet formulation (Blocanol). In vivo plasma levels were also compared with those predicted by kinetic simulations. Both steady-state timolol concentrations in plasma and duration could be controlled with water-activated, pH-controlled patches, although considerable, interindividual variability in TD absorption occurred owing to the high fractional skin control in timolol delivery. Timolol patches were well tolerated, skin irritation was mild, and after removal of the patches, skin changes were practically reversed in 24 h. Preclinical tests were also reported (270) including release of timolol from patches and timolol permeation across human cadaver skin.

Systemic effects, local tolerance, and effectiveness on the penis of topical gel formulations containing alprostadil (prostaglandin E_1) plus 5% of the penetration enhancer SEPA versus SEPA alone (placebo) were evaluated in men with erectile dysfunction (271). Application of prostaglandin E_1 gel correlated positively with erectile response (67–75% of patients had an erection compared with 17% of controls.

A soft, stretchy adhesive patch (10 $\times$ 14 cm) containing 5% lidocaine (700 mg) has been used for topical treatment of pain associated with postherpetic neuralgia (272). Systemic absorption was reported as minimal in healthy subjects and patients with postherpetic neuralgia (3% of dose). In clinical trials (12 h–24 days), treatment resulted in a significant reduction in pain intensity and increased pain relief compared with vehicle patch.

Transdermal drug delivery systems are increasingly popular, yet few data exist on emergency medical outcomes after exposures. A retrospective study (273) using data collected through a Regional Poison Information System identified 61 cases over a 5-year period. Routes were dermal (48), oral (10), combined oral and dermal (1), parenteral use of gel residue (1), and combined oral and parenteral (1). Most exposures (72%) were managed by home telephone consultation only. Eleven of 17 patients (18%) evaluated in health care facilities were admitted, including 8 (13%) to intensive care units. Hospital admission correlated statistically with clonidine, fentanyl, oral exposure, and drug abuse. Clonidine exposure also correlated statistically with intensive care admission. One fatality was recorded; all other patients recovered uneventfully.

V. CONCLUDING REMARKS

In this chapter we have described some of the considerations that are important in the design and development of pharmaceutical products intended for application to the skin. In addition, we have described some recent clinical observations of transdermal therapeutic systems. Space limitations dictated that we could not provide an exhaustive review of all factors, and the reader will appreciate that preformulation, scale-up, and safety are not covered. However, these aspects are fully covered either elsewhere in this volume or in some of the recent excellent and fully recommended texts (7,8,83). We have attempted to share our knowledge of the formulation types that are applied to the skin, together with their properties, problems, and clinical usefulness. We hope that our comments will provide the novice formulator with some insights borne of experience, and the experienced formulator with some novel insights in the field of dermatological and transdermal products development and their use.

REFERENCES

1. Scheuplein RJ, Blank IH. Permeability of the skin. Physiol Rev 1971; 51:702–747.
2. Idson B. Percutaneous absorption. J Pharm Sci 1975; 64:901–924.
3. Flynn GL. (1996). Cutaneous and transdermal delivery: processes and systems of delivery. In: Banker GS, Rhodes CT, eds. Modern Pharmaceutics, 3rd ed. New York: Marcel Dekker. 1996; 239–298.
4. Barry BW. Dermatological Formulatins, Percutaneous Absorption. New York: Marcel Dekker. 1983.
5. Shah VP, Maibach HI, eds. Topical Drug Bioavailability, Bioequivalence, and Penetration. New York: Plenum Press. 1993.
6. Schaefer H, Redelmeir TE. Skin Barrier, Principles of Percutaneous Absorption. Basel: Karger, 1996.
7. Roberts MS, Walters KA, eds. Dermal Absorption and Toxicity Assessment. New York: Marcel Dekker. 1998.
8. Bronaugh RL, Maibach HI, eds. Percutaneous Absorption, Drugs, Cosmetics, Mechanisms, Methodology. 3rd ed. New York: Marcel Dekker. 1999.
9. Elias PM. Lipids and the epidermal permeability barrier. Arch Dermatol Res 1981; 270:95–117.
10. Raykar PV, Fung MC, Anderson BD. The role of protein and lipid domains in the uptake of solutes by human stratum corneum. Pharm Res 1988; 5:140–150.
11. Walters KA, Hadgraft J, eds. Pharmaceutical Skin Penetration Enhancement. New York: Marcel Dekker, 1993.

12. Smith EW, Maibach HI, eds. Percutaneous Penetration Enhancers. Boca Raton: CRC Press, 1995.

13. Behl CR, Flynn GL, Barrett M, Walters KA, Linn EE, Mohamed Z, Kurihara T, Ho NFH, Higuchi WI, Pierson CL. Permeability of thermally damaged skin II: immediate influences of branding at 60°C on hairless mouse skin permeability. Burns 1981; 7: 389–399.

14. Flynn GL, Behl CR, Walters KA, Gatmaitan OG, Wittkowsky A, Kurihara T, Ho NFH, Higuchi WI, Pierson CL. Permeability of thermally damaged skin III: influence of scalding temperature on mass transfer of water and n-alkanols across hairless mouse skin. Burns 1982; 8:47–58.

15. Bronaugh RL, Stewart RF. Methods for in vitro percutaneous absorption studies. V. Permeation through damaged skin. J Pharm Sci 1985; 74:1062–1066.

16. Scott RC, Dugard PH, Doss AW. Permeability of abnormal rat skin. J Invest Dermatol 1986; 86:201–207.

17. Scott RC. In vitro absorption through damaged skin. In: Bronaugh RL, Maibach HI, eds. In Vitro Percutaneous Absorption, Principles, Fundamentals and Applications. Boca Radon: CRC Press. 1991; 129–135.

18. Bond JR, Barry BW. Damaging effect of acetone on the permeability barrier of hairless mouse skin compared with that of human skin. Int J Pharm 1988; 41:91–93.

19. van der Valk PGM, Nater JP, Bleumink E. Vulnerability of the skin to surfactants in different groups of eczema patients and controls as measured by water vapour loss. Clin Exp Dermatol 1985; 10:98–103.

20. Turpeinen M, Salo OP, Leisti S. Effect of percutaneous absorption of hydrocortisone on adrenocortical responsiveness in infants with severe skin disease. Br J Dermatol 1986; 115:475–484.

21. Turpeinen M. Absorption of hydrocortisone from the skin reservoir in atopic dermatitis. Br J Dermatol 1991; 124:358–360.

22. Fartasch M, Diepgen TL. The barrier function in atopic dry skin. Acta Derm Venereol 1992; 176(suppl):26–31.

23. Reichek N, Goldstein RE, Redwood DR. Sustained effects of nitroglycerin ointment in patients with angina pectoris. Circulation 1974; 50:348–352.

24. Cleary GW. Transdermal delivery systems: a medical rationale. In: Shah VP, Maibach HI, eds. Topical Drug Bioavailability, Bioequivalence, and Penetration. New York: Plenum Press, 1993; 17–68.

25. Anderson BD. Prodrugs and their topical use, In: Shah VP, Maibach HI, eds. Topical Drug Bioavailability, Bioequivalence, and Penetration. New York: Plenum Press. 1993; 69–89.

26. Ahmed S, Imai T, Otagiri M. Stereoselective hydrolysis and penetration of propranolol prodrugs: in vitro evaluation using hairless mouse skin. J Pharm Sci 1995; 84:877–883.

27. Lai PM, Roberts MS. Iontophoresis. In: Roberts MS, Walters KA, eds. Dermal Absorption and Toxicity Assessment. New York: Marcel Dekker. 1998; 371–414.

28. Prausnitz MR. The effects of electric current applied to skin: a review for transdermal drug delivery. Adv Drug Deliv Rev 1996; 18:395–425.

29. McElnay JC, Benson HAE, Hadgraft J, Murphy TM. (1993). The use of ultrasound in skin penetration enhancement. In: Walters KA, Hadgraft J, eds. Pharmaceutical Skin Penetration Enhancement. New York: Marcel Dekker. 1993; 293–309.

30. Steele GW. 2001. Preformulation. In: Gibson M, ed. Pharmaceutical Preformulation and Formulation. Interpharm Press. Interpharm Book.

31. Kitson N, Maddin S. Drugs used for skin diseases. In: Roberts MS, Walters KA, eds. Dermal Absorption and Toxicity Assessment. New York: Marcel Dekker. 1998; 313–326.

32. Zatz JL, Varsano J, Shah VP. In vitro release of betamethasone dipropionate from petrolatum-based ointments. Pharm Dev Technol 1996; 1:293–298.

33. Kneczke M, Landersjö L, Lundgren P, Führer C. In vitro release of salicylic acid from two different qualities of white petrolatum. Acta Pharm Suec 1986; 23:193–204.

34. Osborne DW. Phase behavior characterization of propylene glycol, white petrolatum, surfactant ointments. Drug Dev Ind Pharm 1992; 18:1883–1894.

35. Osborne DW. Phase behavior characterization of ointments containing lanolin or a lanolin substitute. Drug Dev Ind Pharm 1993; 19:1283–1302.

36. Pena LE, Lee BL, Stearn JF. Structural rheology of a model ointment. Pharm Res 1994; 11:875–881.

37. Dahl T, He G–X, Samuels G. Effect of hydrogen peroxide on the viscosity of a hydroxyethylcellulose-based gel. 1998; Pharm Res 15:1137–1140.

38. Planas MD, Rodriguez FG, Dominguez MH. The influence of neutralizer concentration on the rheological behaviour of a 0.1% Carbopol hydrogel. Pharmazie 1992; 47:351–355.

39. Pérez–Marcos B, Martínez–Pacheco R, Gómez–Amoza, Souto C, Concheiro A, Rowe RC. Interlot variability of carbomer 934. Int J Pharm 1993; 100:207–212.

40. Sjöblom J, ed. Emulsions and Emulsion Stability. New York: Marcel Dekker. 1996.

41. Eccleston GM. Multiple phase oil-in-water emulsions. J Soc Cosmet Chem 1990; 41:1–22.

42. Eccleston GM. Functions of mixed emulsifiers and emulsifying waxes in dermatological lotions and creams. Colloids Surfaces 1997; 123:169–182.

43. Orth DS. Handbook of Cosmetic Microbiology. New York: Marcel Dekker, 1993.

44. Takruri H, Anger CB. (1989). Preservation of dispersed systems. In: Lieberman HA, Rieger MM, Banker GS, eds. Pharmaceutical Dosage Forms: Disperse Systems. Vol. 2. New York: Marcel Dekker. 1989; 73–114.

45. Cosmetic Ingredient Review. Final report on the safety of methyl, ethyl, propyl and butyl paraben. JACT 1984; 3(5).

46. Dal Pozzo A, Pastori N. Percutaneous absorption of parabens from cosmetic formulations. Int J Cosmet Sci 1996; 18:57–66.

47. Kurup TRR, Wan LSC, Chan LW. Interaction of preservatives with macromolecules. Part II. Cellulose derivatives. Pharm Acta Helv 1995; 70:187–193.

48. Tanaka M, Iwata Y, Kouzuki Y, Taniguchi K, Matsuda H, Arima H, Tsuchiya S. Effect of 2-hydroxypropyl-β-cyclodextrin on percutaneous absorption of methyl paraben. J Pharm Pharmacol 1995; 47:897–900.

49. Komatsu H, Higaki K, Okamoto H, Miyagawa K, Hashida M, Sezaki H. Preservative activity and in vivo percutaneous penetration of butyl paraben entrapped in liposomes. Chem Pharm Bull 1986; 34:3415–3422.

50. Komatsu H, Okamoto H, Miyagawa K, Hashida M, Sezaki H. Percutaneous absorption of butyl paraben from liposomes in vitro. Chem Pharm Bull 1986; 34:3423–3430.

51. Kabara JJ, Orth DS, eds. Preservative-Free and Self-Preserving Cosmetics and Drugs —Principles and Practice. New York: Marcel Dekker. 1997.

52. Shah VP, Elkins J, Hanus J, Noorizadeh C, Skelly JP. In vitro release of hydrocortisone from topical preparations and automated procedure. Pharm Res 1991; 8:55–59.

53. Chattaraj SC, Kanfer I. Release of acyclovir from semisolid dosage forms: a semiautomated procedure using a simple Plexiglass flow-through cell. Int J Pharm 1995; 125:215–222.

54. Higuch T. Physical chemical analysis of percutaneous absorption process from creams and ointments. J Soc Cosmet Chem 1960; 11:85–97.

55. Higuchi T. Rate of release of medicaments from ointment bases containing drugs in suspension. J Pharm Sci 1961; 50:874–875.

56. Bunge AL. Release rates from topical formulations containing drugs in suspension. J Control Release 1998; 52:141–148.

57. Flynn GL, Caetano PA, Pillai RS, Abriola LM, Amidon GE. Recent perspectives in percutaneous penetration: in vitro drug release and its relationship to drug delivery. In: Brain KR, Walters KA, eds. Perspectives in Percutaneous Penetration. Vol. 6b. Cardiff, Wales: STS Publishing. 1999 (in press).

58. Kril MB, Parab PV, Genier SE, DiNunzio JE, Alessi D. Potential problems encountered with SUPAC-SS and the in vitro release testing of ammonium lactate cream. Pharm Technol 1999; March:164–174.

59. Guy RH, Hadgraft J. The prediction of plasma levels of drugs following transdermal application. J Control Release 1985; 1:177–182.

60. Pugh WJ, Hadgraft J, Roberts MS. Physicochemical determinants of stratum corneum permeation. In: Roberts MS, Walters KA, eds. Dermal Absorption and Toxicity Assessment. New York: Marcel Dekker. 1998; 245–268.

61. Kasting GB, Filloon TG, Francis WR, Meredith MP. Improving the sensitivity of in vitro skin penetration experiments. Pharm Res 1994; 11:1747–1754.

62. Giannakou SA, Dallas PP, Rekkas DM, Choulis NH. Development and in vitro evaluation of nimodipine transdermal formulations using factorial design. Pharm Dev Technol 1998; 3:517–525.

63. Kandimalla KK, Kanikkannan N, Singh M. Optimization of a vehicle mixture for the transdermal delivery of melatonin using artificial neural networks and response surface method. J Control Release 1999; 61:71—82.

64. Takayama K, Takahara J, Fujikawa M, Ichikawa H, Nagai T. Formula optimization based on artificial neural networks in transdermal drug delivery. J Control Release 1999; 62:161–170.

65. Marecki NM. Design considerations in transdermal drug delivery systems. Proceedings 10th Pharmaceutical Technology Conference. 1987; 311–318.

66. Yuk SH, Lee SJ, Okano T, Berner B, Kim SW. One-way membrane for transdermal drug delivery systems. I. Membrane preparation and characterization. Int J Pharm 1991; 77:221–229.

67. Yuk SH, Lee SJ, Okano T, Berner B, Kim SW. One-way membrane for transdermal drug delivery systems. II. System optimization. Int J Pharm 1991; 77:231–237.

68. Okabe H, Suzuki E, Saitoh T, Takayama K, Nagai T. Development of novel transdermal system containing d-limonene and ethanol as absorption enhancers. J Control Release 1994; 32:243–247.

69. Wang D–M, Lin F–C, Chen L–Y, Lai J–Y. Application of asymmetric TPX membranes to transdermal delivery of nitroglycerin. J Control Release 1998; 50:187–195.

70. Lai J–Y, Lin F–C, Wu T–T, Wang D–M. On the formation of macrovoids in PMMA membranes. J Membr Sci 1999; 155:31–43.

71. Thacharodi D, Rao KP. Rate-controlling biopolymer membranes as transdermal delivery systems for nifedipine: development and in vitro evaluations. Biomaterials 1996; 17:1307–1311.

72. Iordanskii AL, Feldstein MM, Markin VS, Hadgraft J, Plate NA. Modeling of the drug delivery from a hydrophilic transdermal therapeutic system across polymer membrane. Eur J Pharm Biopharm 2000; 49:287–293.

73. Pfister WR, Sheeran MA, Watters DE, Sweet RP, Walters P. Methods for altering release of progesterone, testosterone, propranolol, and indomethacin from silicone matrices: effects of cosolvents and inert fillers. Proc Int Symp Control Release Bioact Mater 1987; 14:223–224.

74. Ulman KL, Gornowicz GA, Larson KR, Lee C–L. Drug permeability of modified silicone polymers. I. Silicone-organic block copolymers. J Control Release 1989; 10: 251–260.

75. Hille T. Technological aspects of penetration enhancers in transdermal systems. In: Walters KA, Hadgraft J, eds. Pharmaceutical Skin Penetration Enhancement. New York: Marcel Dekker. 1993; 335–343.

76. Adhesives Research, Inc., Transdermal Technical Brief, Enhancer-Tolerant Adhesives.

77. Ko CU, Wilking SL, Birdsall J. Pressure sensitive adhesive property optimizations for the transdermal drug delivery systems. Pharm Res 1995; 12:S-143.

78. te Hennepe HJC. Solution acrylic polymers for medical applications. National Starch & Chemical B.V. 1993.

79. Jenkins AW. Developing the Fematrix transdermal patch. Pharm J 1995; 255:179–181.

80. PMA Committee Report. Transdermal Drug Delivery Systems. Pharmacop Forum 1986; 12:1798–1807.

81. Shah VP, Tymes NW, Yamamoto LA, Skelly JP. In vitro dissolution profile of transdermal nitroglycerin patches using paddle method. Int J Pharm 1986; 32:243–250.

82. Man M, Chang C, Lee PH, Broman CT, Cleary GW. New improved paddle method for determining the in vitro drug release profiles of transdermal delivery systems. J Control Release 1993; 27:59–68.

83. Roy SD. Preformulation aspects of transdermal drug delivery systems, In: Ghosh TK, Pfister WR, Sum SI, eds. Transdermal and Topical Drug Delivery Systems. Buffalo Grove: Interpharm Press, 1997; 139–166.

84. Gibson M, ed. Pharmaceutical Preformulation and Formulation. Englewood, IHS Health Group, 2001.

85. Caetano PA, Flynn GL, Farinha AR, Toscano CF, Campos RC. The in vitro release test as a means to obtain the solubility and diffusivity of drugs in semisolids. Proc Int Symp Control Release Bioact Mater 1999; 26:375–376.

86. Gourlay S. The pros and cons of transdermal nicotine therapy. Med J Aust 1994; 160: 152.

87. Lin SS, et al. Transdermal nicotine delivery systems—multiinstitutional cooperative bioequivalence studies. Drug Dev Ind Pharm 1993; 19:2765–2793.

88. Davila DG, et al. Acute effects of transdermal nicotine on sleep architecture, snoring, and sleep-disordered breathing in nonsmokers. Am J Respir Crit Care Med 1994; 150: 469–474.

89. Rose JE, et al. Mecamylamine combined with nicotine skin patch facilitates smoking cessation beyond nicotine patch treatment alone. Clin Pharmacol Ther 1994; 56:86–99.

90. Zevin S, Jacob P III, Benowitz NI. Nicotine–mecamylamine interactions. Clin Pharmacol Ther 2000; 68:58–66.

91. Benowitz NL. Clinical-pharmacology of transdermal nicotine. Eur J Pharm Biopharm 1995; 41:168–174.

92. Fredrickson PA, et al. High-dose transdermal nicotine therapy for heavy smokers—safety, tolerability and measurement of nicotine and cotinine levels. Psychopharmacology 1995; 122:215–222.

93. Jorenby DE, et al. Varying nicotine patch dose and type of smoking cessation counseling. JAMA 1995; 274:1347–1352.

94. Kornitzer M, et al. Combined use of nicotine patch and gum in smoking cessation—a placebo-controlled clinical-trial. Prev Med 1995; 24:41–47.

95. Campbell IA, Prescott RJ, TjederBurton SM. Transdermal nicotine plus support in patients attending hospital with smoking-related diseases: a placebo-controlled study. Respir Med 1996; 90:47–51.

96. Lindell G, Lunell E, Graffner H. Transdermally administered nicotine accumulates in gastric juice. Eur J Clin Pharmacol 1996; 51:315–318.

97. Fiscella K, Franks P. Cost-effectiveness of the transdermal nicotine patch as an adjunct to physicians' smoking cessation counseling. JAMA 1996; 275:1247–1251.

98. Wasley MA, et al. The cost-effectiveness of the nicotine transdermal patch for smoking cessation. Prev Med 1997; 26:264–270.

99. Jaen CR, et al. Patterns of use of a free nicotine patch program for Medicaid and uninsured patients. J Natl Med Assoc 1997; 89:325–328.

100. Stapleton JA, Lowin A, Russell MAH. Prescription of transdermal nicotine patches for smoking cessation in general practice: evaluation of cost-effectiveness. Lancet 1999; 354:210–215.

101. Benowitz NL, Gourlay SG. Cardiovascular toxicity of nicotine: implications for nicotine replacement therapy. J Am Coll Cardiol 1997; 29:1422–1431.

102. Benowitz NL, Zevin S, Jacob P. Sources of variability in nicotine and cotinine levels with use of nicotine nasal spray, transdermal nicotine, and cigarette smoking. Br J Clin Pharmacol 1997; 43:259–267.

103. Gourlay SG, et al. Determinants of plasma concentrations of nicotine and cotinine during cigarette smoking and transdermal nicotine treatment. Eur J Clin Pharmacol 1997; 51:407–414.

104. Greenland S, Satterfield MH, Lanes SF. A meta-analysis to assess the incidence of adverse effects associated with the transdermal nicotine patch. Drug Safety 1998; 18: 297–308.

105. Lawson GM, et al. Application of urine nicotine and cotinine excretion rates to assessment of nicotine replacement in light, moderate, and heavy smokers undergoing transdermal therapy. J Clin Pharmacol 1998; 38:510–516.

106. Lawson GM, et al. Application of serum nicotine and plasma cotinine concentrations to assessment of nicotine replacement in light, moderate, and heavy smokers undergoing transdermal therapy. J Clin Pharmacol 1998; 38:502–509.

107. Lewis SF, et al. Transdermal nicotine replacement for hospitalized patients: a randomized clinical trial. Prev Med 1998; 27:296–303.

108. West R, Willis N. Double-blind placebo controlled trial of dextrose tablets and nicotine patch in smoking cessation. Psychopharmacology 1998; 136:201–204.

109. Blondal T, et al. Nicotine nasal spray with nicotine patch for smoking cessation: randomised trial with six year follow up. Br Med J 1999; 318:285–289.

110. Daughton DM, et al. The smoking cessation efficacy of varying doses of nicotine patch delivery systems 4 to 5 years post-quit day. Prev Med 1999; 28:113–118.

111. Gariti P, et al. Cotinine replacement levels for a 21 mg/day transdermal nicotine patch in an outpatient treatment setting. Drug Alcohol Depend 1999; 54:111–116.

112. Gourlay SG, et al. Predictors and timing of adverse experiences during transdermal nicotine therapy. Drug Safety 1999; 20:545–555.

113. Hajek P, et al. Randomized comparative trial of nicotine polacrilex, a transdermal patch, nasal spray, and an inhaler. Arch Intern Med 1999; 159:2033–2038.

114. Ji AJ, et al. A new gas chromatography–mass spectrometry method for simultaneous determination of total and free *trans*-3′-hydroxycotinine and cotinine in the urine of subjects receiving transdermal nicotine. Clin Chem 1999; 45:85–91.

115. Sonderskov J, et al. Nicotine patches in smoking cessation: a randomized trial among over-the-counter customers in Denmark. Ugeskr Laeger 1999; 161:593–597.

116. Tonnesen P, et al. Higher dosage nicotine patches increase one-year smoking cessation rates: results from the European CEASE trial. Eur Respir J 1999; 13:238–246.

117. Homsy W, et al. Plasma levels of nicotine and safety of smokers wearing transdermal delivery systems during multiple simultaneous intake of nicotine and during exercise. J Clin Pharmacol 1997; 37:728–736.

118. Wright LN, et al. Transdermal nicotine replacement in pregnancy: maternal pharmacokinetics and fetal effects. Am J Obstet Gynecol 1997; 176:1090–1094.

119. Dale LC, et al. Weight change after smoking cessation using variable doses of transdermal nicotine replacement. J Gen Intern Med 1998; 13:9–15.

120. Ogburn PL, Jr, et al. Nicotine patch use in pregnant smokers: nicotine and cotinine levels and fetal effects. Am J Obstet Gynecol 1999; 181:736–743.

121. West R, et al. A comparison of the abuse liability and dependence potential of nicotine patch, gum, spray and inhaler. Psychopharmacology 2000; 149:198–202.

122. Pullan RD, et al. Transdermal nicotine for active ulcerative-colitis. N Engl J Med 1994; 330:811–815.

123. Thomas GAO, et al. Transdermal nicotine as maintenance therapy for ulcerative-colitis. N Engl J Med 1995; 332:988–992.

124. Thomas GAO, et al. Transdermal nicotine compared with oral prednisolone therapy for active ulcerative colitis. Eur J Gastroenterol Hepatol 1996; 8:769–776.

125. Sandborn WJ, et al. Transdermal nicotine for mildly to moderately active ulcerative colitis—a randomized, double-blind, placebo-controlled trial. Ann Intern Med 1997; 126:364.

126. Guslandi M, Tittobello A. Pilot trial of nicotine patches as an alternative to corticosteroids in ulcerative colitis. J Gastroenterol 1996; 31:627–629.

127. Guslandi M, Tittobello A. Outcome of ulcerative colitis after treatment with transdermal nicotine. Eur J Gastroenterol Hepatol 1998; 10:513–515.

128. Guslandi M. Long-term effects of a single course of nicotine treatment in acute ulcerative colitis: remission maintenance in a 12-month follow-up study. Int J Colorectal Dis 1999; 14:261–262.

129. SalinPascual RJ, et al. Antidepressant effect of transdermal nicotine patches in nonsmoking patients with major depression. J Clin Psychiatry 1996; 57:387–389.

130. Shytle RD, et al. Transdermal nicotine for Tourette's syndrome. Drug Dev Res 1996; 38:290–298.

131. Silver AA, Shytle RD, Sanberg PR. Clinical experience with transdermal nicotine patch in Tourette syndrome. CNS Spectr 1999; 4(2):68–76.

132. White HK, Levin ED. Four-week nicotine skin patch treatment effects on cognitive performance in Alzheimer's disease. Psychopharmacology 1999; 143:158–165.

133. Matsushima D, Prevo ME, Gorsline J. Absorption and adverse-effects following topical and oral-administration of 3 transdermal nicotine products to dogs. J Pharm Sci 1995; 84:365–369.

134. Woolf A, et al. Self-poisoning among adults using multiple transdermal nicotine patches. J Toxicol Clin Toxicol 1996; 34:691–698.

135. Bogaert MG. Clinical pharmacokinetics of nitrates. Cardiovasc Drugs Ther 1994; 8: 693–699.

136. Hutt V, et al. Bioequivalence evaluation of the metabolites 1,2-glyceryl and 1,3-glyceryl dinitrate of 2 different glyceryl trinitrate patches after 12-h usage in healthy-volunteers. ArzneimitteForschung 1994; 44:1317–1321.

137. Devita C, et al. Gissi-3—effects of lisinopril and transdermal glyceryl trinitrate singly and together on 6-week mortality and ventricular function after acute myocardial infarction. Lancet 1994; 343:1115–1122.

138. Andrews R, et al. Inhibition of platelet-aggregation by transdermal glyceryl trinitrate. Br Heart J 1994; 72:575–579.

139. Berrazueta JR, et al. Local transdermal glyceryl trinitrate has an antiinflammatory action on thrombophlebitis induced by sclerosis of leg varicose veins. Angiology 1994; 45:347–351.

140. Parker JO, et al. Intermittent transdermal nitroglycerin therapy in angina-pectoris—clinically effective without tolerance or rebound. Circulation 1995; 91:1368–1374.

141. Serone AP, Angus JA, Wright CE. Baroreffex resetting but no vascular tolerance in response to transdermal glyceryl trinitrate in conscious rabbits. Br J Pharmacol 1996; 118:93–104.

142. Ardissino D, et al. Effect of transdermal nitroglycerin or *N*-acetylcysteine, or both, in the long-term treatment of unstable angina pectoris. J Am Coll Cardiol 1997; 29:941–947.

143. Klemsdal TO, Mundal HH, Gjesdal K. Mechanisms of tolerance during treatment with nitroglycerin patches for 24 h. Eur J Clin Pharmacol 1996; 51:227–230.

144. Klemsdal TO, et al. Transdermal nitroglycerin: clinical and pharmacokinetic consequences of renewing the patch and the application site. Eur J Clin Pharmacol 1997; 52:379–381.

145. Elkayam U, et al. Double-blind, placebo-controlled study to evaluate the effect of organic nitrates in patients with chronic heart failure treated with angiotensin-converting enzyme inhibition. Circulation 1999; 99:2652–2657.

146. Teh LS, et al. Sustained-release transdermal glyceryl trinitrate patches as a treatment for primary and secondary raynauds-phenomenon. Br J Rheumatol 1995; 34:636–641.

147. Berrazueta JR, et al. Successful treatment of shoulder pain syndrome due to supraspinatus tendinitis with transdermal nitroglycerin. A double blind study. Pain 1996; 66: 63–67.

148. Zoli M, et al. Transdermal nitroglycerin in cirrhosis. A 24-hour echo-Doppler study of splanchnic hemodynamics. J Hepatol 1996; 25:498–503.

149. Moya RA, et al. Transdermal glyceryl trinitrate in the management of primary dysmenorrhea. Int J Gynecol Obstet 2000; 69:113–118.

150. Black RS, et al. Maternal and fetal cardiovascular effects of transdermal glyceryl trinitrate and intravenous ritodrine. Obstet Gynecol 1999; 94:572–576.

151. Lees CC, et al. Glyceryl trinitrate and ritodrine in tocolysis: an international multicenter randomized study. Obstet Gynecol 1999; 94:403–408.

152. Kovacs P, et al. Effect of transdermal nitroglycerin on glucose-stimulated insulin release in healthy male volunteers. Eur J Clin Invest 2000; 30:41–44.

153. Minghetti P, et al. In vitro skin permeation of sinitrodil, a member of a new class of nitrovasodilator drugs. Eur J Pharm Sci 1999; 7:231–236.

154. Whittington R, Faulds D. Hormone replacement therapy. 1. Pharmacoeconomic appraisal of its therapeutic use in menopausal symptoms and urogenital estrogen deficiency. Pharmacoeconomics 1994; 5:419–445.

155. Wiseman LR, McTavish D. Transdermal estradiol norethisterone—a review of its pharmacological properties and clinical use in postmenopausal women. Drugs Aging 1994; 4:238–256.

156. Creatsas G, et al. Transdermal estradiol plus oral medroxyprogesterone acetate replacement therapy in primary amenorrheic adolescents—clinical, hormonal and metabolic aspects. Maturitas 1994; 18:105–114.

157. Normantaylor JQ, et al. Hormone replacement therapy by the transdermal administration of estradiol and norethisterone. Maturitas 1994; 18:221–228.

158. Smith RNJ, et al. A randomized comparison over 8 months of 100 mu-g and 200 mu-g twice weekly doses of transdermal estradiol in the treatment of severe premenstrual-syndrome. Br J Obstet Gynaecol 1995; 102:475–484.

159. Pornel B. Efficacy and safety of Menorest in two positive-controlled studies. Eur J Obstet Gynecol Reprod Biol 1996; 64(SS):S35–S37.

160. Raudaskoski TH, et al. Transdermal estrogen with a levonorgestrel-releasing intrauterine-device for climacteric complaints—clinical and endometrial responses. Am J Obstet Gynecol 1995; 172(1 pt1):114–119.

161. Haines CJ, et al. The effect of percutaneous oestrogen replacement therapy on Lp(a) and other lipoproteins. Maturitas 1995; 22:219–225.

162. Crosignani PG, Cortellaro M, Boschetti C. Effects on haemostasis of hormone replacement therapy with transdermal estradiol and oral sequential medroxyprogesterone ac-

etate: a 1-year, double-blind, placebo-controlled study. Thrombo Haemost 1996; 75: 476–480.

163. Macdonald AG, et al. Effects of hormone replacement therapy in rheumatoid-arthritis —a double-blind placebo-controlled study. Ann Rheum Dis 1994; 53:54–57.

164. Evans SF, Davie MWJ. Low and conventional dose transdermal oestradiol are equally effective at preventing bone loss in spine and femur at all post-menopausal ages. Clin Endocrinol 1996; 44:79–84.

165. Roux C. Estrogen therapy in postmenopausal osteoporosis what we know and what we don't. Rev Rhum 1997; 64:402–409.

166. Hannon R, et al. Response of biochemical markers of bone turnover to hormone replacement therapy: impact of biological variability. J Bone Miner Res 1998; 13:1124–1133.

167. Prelevic GM, et al. Comparative effects on bone mineral density of tibolone, transdermal estrogen and oral estrogen/progestogen therapy in postmenopausal women. Gynecol Endocrinol 1996; 10:413–420.

168. Boonkasemsanti W, et al. The effect of transdermal oestradiol on bleeding pattern, hormonal profiles and sex steroid receptor distribution in the endometrium of Norplant users. Hum Reprod 1996; 11(S2):115–123.

169. Habiba M, et al. Thrombophilia and lipid profile in post-menopausal women using a new transdermal oestradiol patch. Eur J Obstet Gynecol Reprod Biol 1996; 66:165–168.

170. Perrone G, et al. Effect of oral and transdermal hormone replacement therapy on lipid profile and Lp(a) level in menopausal women with hypercholesterolemia. Int J Fertil Menopausal Stud 1996; 41:509–515.

171. Rabe T, et al. Spacing-out of progestin—efficacy, tolerability and compliance of two regimens for hormonal replacement in the late postmenopause. Gynecol Endocrinol 1997; 11:383–392.

172. Cameron ST, et al. Continuous transdermal oestrogen and interrupted progestogen as a novel bleed-free regimen of hormone replacement therapy for postmenopausal women. Br J Obstet Gynaecol 1997; 104:1184–1190.

173. Rozenbaum H, et al. Comparison of two estradiol transdermal systems (Oesclim 50 and Estraderm TTS 50). 2. Local skin tolerability. Maturitas 1996; 25:175–185.

174. Rozenbaum H, et al. Comparison of two estradiol transdermal systems (Oesclim 50 and Estraderm TTS 50). 1. Tolerability, adhesion and efficacy. Maturitas 1996; 25:161–173.

175. AlAzzawi F, et al. A randomised study to compare the efficacy and safety of a new 17 beta-oestradiol transdermal matrix patch with Estraderm TTS 50 in hysterectomised postmenopausal women. Br J Clin Prac 1997; 51:20.

176. Rozenberg S, Ylikorkala O, Arrenbrecht S. Comparison of continuous and sequential transdermal progestogen with sequential oral progestogen in postmenopausal women using continuous transdermal estrogen: vasomotor symptoms, bleeding patterns, and serum lipids. Int J Fertil Womens Med 1997; 42(S2):376–387.

177. Appelberg J et al. Safety and efficacy of transdermal estradiol replacement therapy in postmenopausal liver transplanted women—a preliminary report. Acta Obstet Gynecol Scand 1998; 77:660–664.

178. Bhathena RK, et al. The influence of transdermal oestradiol replacement therapy and medroxyprogesterone acetate on serum lipids and lipoproteins. Br J Clin Pharmacol 1998; 45:170–172.

179. Blanc B, et al. Continuous hormone replacement therapy for menopause combining nomegestrol acetate and gel, patch, or oral estrogen: a comparison of amenorrhea rates. Clin Ther 1998; 20:901–912.

180. Notelovitz M. Clinical experience with a 7-day estrogen patch: principles and practice. Gynecol Endocrinol 1998; 12:249–258.
181. Hirvonen E, et al. Effects of transdermal oestrogen therapy in postmenopausal women: a comparative study of an oestradiol gel and an oestradiol delivering patch. Br J Obstet Gynaecol 1997; 104(S16):26–31.
182. Hirvonen E, et al. Transdermal oestradiol gel in the treatment of the climacterium: a comparison with oral therapy. Br J Obstet Gynaecol 1997; 104(S16):19–25.
183. Jarvinen A, Nykanen S, Paasiniemi L. Absorption and bioavailability of oestradiol from a gel, a patch and a tablet. Maturitas 1999; 32:103–113.
184. Vavilis D, et al. The effect of transdermal estradiol on the conjunctiva in postmenopausal women. Eur J Obstet Gynecol Reprod Biol 1997; 72:93–96.
185. Coustou D, et al. Estrogen dermatitis. Ann Dermatol Venereol 1998; 125:505–508.
186. Carr BR. HRT management: the American experience. Eur J Obstet Gynecol Reprod Biol 1996; 64(SS):S17–S20.
187. Andersson TLG, et al. Bioavailability of estradiol from two matrix transdermal delivery systems: menorest and Climara. Maturitas 2000; 34:57–64.
188. Archer DF, et al. A randomized comparison of continuous combined transdermal delivery of estradiol–norethindrone acetate and estradiol alone for menopause. Obstet Gynecol 1999; 94:498–503.
189. Baracat E, et al. Comparative bioavailability study of two 100-μg daily 17-beta-estradiol transdermal delivery systems: once-a-week matrix patch and twice-a-week reservoir patch in healthy postmenopausal women. Curr Ther Res Clin Exp 1999; 60:129–137.
190. Beljic T, et al. Effect of estrogen replacement therapy on cardiac function in postmenopausal women with and without flushes. Gynecol Endocrinol 1999; 13:104–112.
191. Bhathena RK, Anklesaria BS, Ganatra AM. The treatment of hypertriglyceridaemia in menopausal women with transdermal oestradiol therapy. Br J Obstet Gynaecol 1999; 106:980–982.
192. Brackman F. Oesclim: summary of tolerability and safety. Maturitas 1999; 33(suppl 1):83–88.
193. Utian WH, et al. Efficacy and safety of low, standard, and high dosages of an estradiol transdermal system (Esclim) compared with placebo on vasomotor symptoms in highly symptomatic menopausal patients. Am J Obstet Gynecol 1999; 181:71–79.
194. Buch AB, et al. Significant differences in estradiol bioavailability from two similarly labelled estradiol matrix transdermal systems. Climacteric 1999; 2:248–253.
195. Caserta R, et al. Efficacy and safety of a weak transdermal estradiol drug delivery system: a pharmacokinetic comparison of two different dosage of formulations. G Ital Ostet Ginecol 2000; 22:279–283.
196. Cohen L, et al. Low-dose 17-beta estradiol matrix transdermal system in the treatment of moderate-to-severe hot flushes in postmenopausal women. Curr Ther Res Clin Exp 1999; 60:534–547.
197. Cooper C, et al. Matrix delivery transdermal 17beta-estradiol for the prevention of bone loss in postmenopausal women. Osteopor Int 1999; 9:358–366.
198. Weiss SR, et al. A randomized controlled trial of four doses of transdermal estradiol for preventing postmenopausal bone loss. Obstet Gynecol 1999; 94:330–336.
199. De Aloysio D, et al. Efficacy on climacteric symptoms and safety of low dose estradiol transdermal matrix patches—a randomized, double-blind placebo-controlled study. Arzneimittelforschung 2000; 50:293–300.
200. De Vrijer B, et al. Efficacy and tolerability of a new estradiol delivering matrix patch (Estraderm MX) in postmenopausal women. Maturitas 2000; 34:47–55.
201. Geisler J, et al. Plasma oestrogen fractions in postmenopausal women receiving hormone replacement therapy: influence of route of administration and cigarette smoking. J Endocrinol 1999; 162:265–270.

202. Geyer D, et al. Pharmacokinetics of Fem7, a once-weekly, transdermal oestrogen replacement system in healthy, postmenopausal women. Gynecol Obstet Invest 1999; 48: 1–6.

203. Good WR, et al. Comparison of Alora estradiol matrix transdermal delivery system with oral conjugated equine estrogen therapy in relieving menopausal symptoms. Climacteric 1999; 2:29–36.

204. Matsumoto S, et al. A study of the clinical effect of estradiol transdermal therapeutic system alone on pollakisuria and urinary incontinence in postmenopausal woman. Jpn J Urol 2000; 91:501–505.

205. Mattsson L–A. Clinical experience with continuous combined transdermal hormone replacement therapy. J Menopause 1999; 6:25–29.

206. Nachtigall LE, et al. Serum estradiol-binding profiles in postmenopausal women undergoing three common estrogen replacement therapies: associations with sex hormone-binding globulin, estradiol, and estrone levels. Menopause 2000; 7:243–250.

207. Rohr UD, Nauert C, Stehle B. 17beta-Estradiol delivered by three different matrix patches 50 μg/day. A three way cross-over study in 21 postmenopausal women. Maturitas 1999; 33:45–58.

208. Rovati LC, et al. Efficacy on climacteric symptoms of a new estradiol transdermal patch with active matrix in comparison with a reference reservoir patch: two long-term randomized multicenter parallel-group studies. Arzneimittelforschung 1999; 49:933–943.

209. Wojcicki J, et al. Effects of estradiol and norethisterone replacement therapy on the pharmacokinetics of phenazone. Med Sci Mon 1999; 5:754–757.

210. Burry KA, et al. Percutaneous absorption of progesterone in postmenopausal women treated with transdermal estrogen. Am J Obstet Gynecol 1999; 180:1504–1511.

211. Carey BJ, et al. A study to evaluate serum and urinary hormone levels following short and long term administration of two regimens of progesterone cream in postmenopausal women. Br J Obstet Gynaecol 2000; 107:722–726.

212. Leonetti HB, Longo S, Anasti JN. Transdermal progesterone cream for vasomotor symptoms and postmenopausal bone loss. Obstet Gynecol 1999; 225–228.

213. Chien TY, et al. Transdermal contraceptive delivery system—preclinical development and clinical-assessment. Drug Dev Ind Pharm 1994; 20:633–664.

214. Sitrukware R. Transdermal application of steroid-hormones for contraception. J Steroid Biochem Mol Biol 1995; 53:247–251.

215. Asthana S, et al. Cognitive and neuroendocrine response to transdermal estrogen in postmenopausal women with Alzheimer's disease: results of a placebo-controlled, double-blind, pilot study. Psychoneuroendocrinology 1999; 24:657–677.

216. Gerber GS, et al. Transdermal estrogen in the treatment of hot flushes in men with prostate cancer. Urology 2000; 55:97–101.

217. Cofrancesco J, Dobs AS. Transdermal testosterone delivery systems. Endocrinologist 1996; 6:207–213.

218. Behre HM, et al. Long-term substitution therapy of hypogonadal men with transscrotal testosterone over 7–10 years. Clin Endocrinol 1999; 50:629–635.

219. Kenny AM, Prestwood KM, Raisz LG. Short-term effects of intramuscular and transdermal testosterone on bone turnover, prostate symptoms, cholesterol, and hematocrit in men over age 70 with low testosterone levels. Endocr Res 2000; 26:153–168.

220. Meikle AW, et al. Pharmaockinetics and metabolism of a permeation-enhanced testosterone transdermal system in hypogonadal men: influence of application site—a clinical research center study. J Clin Endocrinol Metab 1996; 81:1832–1840.

221. Arver S, et al. Long-term efficacy and safety of a permeation-enhanced testosterone transdermal system in hypogonadal men. Clin Endocrinol 1997; 47:727–737.

222. Bhasin S, et al. Effects of testosterone replacement with a nongenital, transdermal system, Androderm, in human immunodeficiency virus-infected men with low testosterone levels. J Clin Endocrinol Metab 1998; 83:3155–3162.

223. Miller K, et al. Transdermal testosterone administration in women with acquired immunodeficiency syndrome wasting: a pilot study. J Clin Endocrinol Metab 1998; 83: 2717–2725.

224. Jain P, Rademaker AW, McVary KT. Testosterone supplementation for erectile dysfunction: results of a meta-analysis. J Urol 2000; 164:371–375.

225. Davis SR. The therapeutic use of androgens in women. J Steroid Biochem Mol Biol 1999; 69:177–184.

226. Mazer NA. New clinical applications of transdermal testosterone delivery in men and women. J Control Release 2000; 65:303–315.

227. Parker S, Armitage M. Experience with transdermal testosterone replacement therapy for hypogonadal men. Clin Endocrinol 1999; 50:57–62.

228. Rolf C, et al. Pharmacokinetics of new testosterone transdermal therapeutic systems in gonadotropin-releasing hormone antagonist-suppressed normal men. Exp Clin Endocrinol Diabetes 1999; 107:63–69.

229. Tan KCB, Shiu SWM, Kung AWC. Alterations in hepatic lipase and lipoprotein subfractions with transdermal testosterone replacement therapy. Clin Endocrinol 1999; 51: 765–769.

230. Wilson DE, et al. Use of topical corticosteroid pretreatment to reduce the incidence and severity of skin reactions associated with testosterone transdermal therapy. Clin Ther 1998; 20:299–306.

231. Simmonds MA. Transdermal fentanyl—clinical development in the United States. Anticancer Drugs 1995; 6(S3):35–38.

232. Sandler AN. Transdermal fentanyl for pain relief—guidelines for appropriate use. CNS Drugs 1997; 7:442–451.

233. Korte W, Morant R. Transdermal fentanyl in uncontrolled cancer pain—titration on a day-to-day basis as a procedure for safe and effective dose-finding—a pilot-study in 20 patients. Supp Care Cancer 1994; 2:123–127.

234. Vanbastelaere M, Rolly G, Abdullah NM. Postoperative analgesia and plasma levels after transdermal fentanyl for orthopedic surgery—double-blind comparison with placebo. J Clin Anesth 1995; 7:26–30.

235. Fiset P, et al. Biopharmaceutics of a new transdermal fentanyl device. Anesthesiology 1995; 83:459–469.

236. Christensen ML, et al. Transdermal fentanyl administration in children and adolescents with sickle cell pain crisis. J Pediatr Hematol Oncol 1996; 18:372–376.

237. Jeal W, Benfield P. Transdermal fentanyl—a review of its pharmacological properties and therapeutic efficacy in pain control. Drugs 1997; 53:109–138.

238. Hardwick WE, King WD, Palmisano PA. Respiratory depression in a child unintentionally exposed to transdermal fentanyl patch. South Med J 1997; 90:962–964.

239. Sloan PA, Moulin DE, Hays H. A clinical evaluation of transdermal therapeutic system fentanyl for the treatment of cancer pain. J Pain Sympt Manage 1998; 16:102–111.

240. Payne R, et al. Quality of life and cancer pain: satisfaction and side effects with transdermal fentanyl versus oral morphone. J Clin Oncol 1998; 16:1588–1593.

241. Dellemijn PLI, vanDuijn H, Vanneste JAL. Prolonged treatment with transdermal fentanyl in neuropathic pain. J Pain Sympt Manage 1998; 16:220–229.

242. Kyles AE, Papich M, Hardie EM. Disposition of transdermally administered fentanyl in dogs. Am J Vet Res 1996; 57:715–719.

243. Egger CM, et al. Comparison of plasma fentanyl concentrations by using three transdermal fentanyl patch sizes in dogs. Vet Surg 1998; 27:159–166.

244. Collins JJ, et al. Transdermal fentanyl in children with cancer pain: feasibility, tolerability, and pharmacokinetic correlates. J Pediatr 1999; 134:319–323.

245. Jakobsson M, Strang P. Fentanyl patches for the treatment of pain in dying cancer patients. Anticancer Res 1999; 19:4441–4442.

246. Radbruch L et al. Constipation and the use of laxatives: a comparison between transdermal fentanyl and oral morphine. Palliat Med 2000; 14:111–119.

247. Haazen L, et al. The constipation-inducing potential of morphine and transdermal fentanyl. Eur J Pain 1999; 3(suppl A):9–15.

248. Flynn TN, Guest JF. Cost of managing constipation among terminally ill cancer patients treated with transdermal fentanyl and 12 hour sustained-release morphine in Canada. J Med Econ 1999; 2:15–32.

249. Elsner F, et al. Switching opioids to transdermal fentanyl in a clinical setting. Schmerz 1999; 13:273–278.

250. Ritchie LD. A clinical evaluation of flurbiprofen LAT and piroxicam gel: a multicentre study in general practice. Clin Rheumatol 1996; 15:243–247.

251. Fujimura A, et al. Pharmacokinetics and pharmacodynamics of a new transdermal clonidine, M-5041t, in healthy-subjects. J Clin Pharmacol 1993; 33:1192–1200.

252. Fujimura A, et al. Comparison of the pharmacokinetics, pharmacodynamics, and safety of oral (Catapres) and transdermal (M-5041t) Clonidine in healthy subjects. J Clin Pharmacol 1994; 34:260–265.

253. Toon S, Phase-I pharmacokinetic assessment of the clonidine transdermal therapeutic system. Eur J Pharm Biopharm 1995; 41:184–188.

254. Ellis JE, et al. Premedication with oral and transdermal clonidine provides safe and efficacious postoperative sympatholysis. Anesth Analg 1994; 79:1133–1140.

255. Dorman T, et al. Effects of clonidine on prolonged postoperative sympathetic response. Crit Care Med 1997; 25:1147–1152.

256. Owen MD, et al. Postoperative analgesia using a low-dose, oral–transdermal clonidine combination: lack of clinical efficacy. J Clin Anesth 1997; 9:8–14.

257. Dandrea G, et al. Efficacy of transdermal clonidine in short-term treatment of cluster headache—a pilot-study. Cephalalgia 1995; 15:430–433.

258. Byassmith MG, et al. Transdermal clonidine compared to placebo in painful diabetic neuropathy using a 2-stage enriched enrollment design. Pain 1995; 60:267–274.

259. Dias VC, et al. Clinical experience with transdermal clonidine in African-American and Hispanic-American patients with hypertension: evaluation from a 12-week prospective, open-label clinical trial in community-based clinics. Am J Ther 1999; 6:19–24.

260. Dinslage S, et al. A new transdermal delivery system for pilocarpine in glaucoma treatment. Ger J Ophthalmol 1996; 5:275–280.

261. Bodick NC, et al. The selective muscarinic agonist xanomeline improves both the cognitive deficits and behavioral symptoms of Alzheimer disease. Alzheimer Dis Assoc Disord 1997; 11(S4):S16–S22.

262. Douma WR, et al. Effects of transdermal scopolamine on pulmonary function, symptoms and bronchial hyperresponsiveness to methacholine. Eur J Pharm Sci 1997; 5: 327–334.

263. Honkavaara P, Pyykko I. Effects of atropine and scopolamine on bradycardia and emetic symptoms in otoplasty. Laryngoscope 1999; 109:108–112.

264. Gowans J, et al. The effects of scopolamine and cyclizine on visual–vestibular interaction in humans. J Vestib Res Equilibr Orient 2000; 10:87–92.

265. Sacktor N, et al. Transdermal selegiline in HIV-associated cognitive impairment: pilot, placebo-controlled study. Neurology 2000; 54:233–235.

266. Rama RP, Diwan PV. Comparative in-vivo evaluation of diltiazem hydrochloride following oral and transdermal administration in rabbits. Indian J Pharmacol 1999; 31: 294–298.

267. Gillson G, et al. Transdermal histamine in multiple sclerosis: part one—clinical experience. Altern Med Rev 1999; 4:424–428.
268. Martin GP, et al. The influence of hydrocolloid patch composition on the bioavailability of triamcinolone acetonide in humans. Drug Dev Ind Pharm, 2000; 26:35–43.
269. Sutinen R, et al. Water-activated, pH-controlled patch in transdermal administration of timolol II. Drug absorption and skin irritation. Eur J Pharm Sci 2000; 11:25–31.
270. Sutinen R, Paronen P, Urtti A. Water-activated, pH-controlled patch in transdermal administration of timolol I. Preclinical tests. Eur J Pharm Sci 2000; 11:19–24.
271. McVary KT, et al. Topical prostaglandin E_1 SEPA gel for the treatment of erectile dysfunction. J Urol 1999; 162:726–731.
272. Comer AM, Lamb HM. Lidocaine patch 5%. Drugs 59:245–249.
273. Roberge RJ, Krenzelok EP, Mrvos R. Transdermal drug delivery system exposure outcomes. J Emerg Med 18:147–151.

8

Bioavailability and Bioequivalence of Dermatological Formulations

CHRISTIAN SURBER

Kantonsspital Basel, Basel, Switzerland

ADRIAN F. DAVIS

GlaxoSmithKline Consumer Healthcare, Weybridge, Kent, England

I. INTRODUCTION

Official guidelines and current textbooks define *bioavailability* as the rate and extent to which the active ingredient or therapeutic moiety is absorbed from the drug product and becomes available at the site of action. From this, *bioequivalence* is defined as the absence of a significant difference in bioavailability between *pharmaceutical equivalents* or *pharmaceutical alternatives*, the latter being dosage forms in which the chemical form, dosage form type, or strength of the therapeutic moiety, differ.

The *absolute bioavailability* is defined as the extent (here, the rate is usually considered unimportant) to which the active ingredient or therapeutic moiety becomes available in the organism from a dosage form in comparison with an intravenously administered standard form, which is taken to be 100% bioavailable. The *relative bioavailability* is defined as the rate and extent to which the active ingredient or therapeutic moiety becomes available in the organism from a dosage form, compared with a standard form administered by the same route.

After the early description of the physiological availability of vitamins in 1945 (1), erratic reports in the early 1960s (2–5) provided the first indications of the potential for bioavailability and bioequivalence problems with multisource drug products. Variations in absorption of active ingredients from different formulations were soon recognized as a potential health hazard when episodes of drug toxicity were reported following changes to the excipients in the pharmaceutical delivery system or changes in manufacturing procedures (6). Major advances in analytical technology in the late 1960s and early 1970s led to further in-depth investigations involving the bioavailability of marketed drug products, and reports of bioinequivalence included a series of drugs such as digoxin (7), phenytoin (8), chloramphenicol (9), and many others. Although early legislation covering quality, safety, and efficacy did not recognize or directly address the issues of bioavailability and bioequivalence, in the last two decades, the bioavailability and bioequivalance of drug products have emerged as important national and international regulatory and scientific issues.

With the growth of the worldwide generic pharmaceutical industry, bioequivalence has added another dimension to the issue of quality of drug products. To be interchangeable with the innovator drug product, a generic drug product must be not only pharmaceutically equivalent or alternative, but also bioequivalent. Thus, bioequivalence plays an important role in assuring the therapeutic quality of the multisource drug formulations that compose most drug products.

The literature contains many examples of how the composition and manufacture of the finished dosage form can alter the effectiveness of the drug. This is particularly true with topical therapy, for which vehicles have profound effects on percutaneous absorption and may also cause vehicle-related local effects. Product efficacy and safety can depend on how much of the drug is ultimately absorbed from its formulation and how rapidly this occurs. Thus, the two considerations—the extent of absorption of the drug from its formulation, and the rate at which it is absorbed —form the basis of bioavailability and bioequivalence testing and are the predictors of therapeutic performance and therapeutic equivalence.

This chapter, which is divided into four main sections, will first discuss current concepts of bioavailability and bioequivalence of dermal and, to a minor extent, transdermal dosage forms. Against this background, the following two sections will discuss related information on model systems to measure bioavailability and on factors influencing the rate and extent of absorption from topical products. In a final section, bioequivalence, therapeutic equivalence, and control of therapy is discussed.

II. BIOAVAILABILITY AND BIOEQUIVALENCE OF TOPICAL DOSAGE FORMS

Application of the term *bioavailability* to topical dosage forms requires, first, a careful definition of the term *topical dosage form* and, second, a specific adaptation of the general definition of bioavailability to the special case of topical dosage form.

Topical (Greek: *topos* = local) dosage forms include ophthalmic, nasal, and otic preparations; mouth, throat, and pulmonary preparations; gastrointestinal, and anorectal preparations; urogenital preparations; and dermatological preparations. These preparations act at target sites that are identical or close to the site of application and can be external (e.g., skin and eyes) or internal (e.g., lungs and stomach). Instillation into a joint or an intradermal or subcutaneous injection may also be considered—following the Greek root—topical or local applications. The motive for topical or local delivery is the direct application of drug to the target site to maximize efficacy, while minimizing systemic absorption, to improve safety. This chapter considers only delivery systems that are placed against the skin.

Delivery systems that are placed against the skin deliver drugs to (a) the local tissue immediately beneath the application site, (b) deeper regions in the vicinity of, but still somewhat remote from, the application site, and (c) the systemic circulation. The objective of the latter is to mediate pharmacological changes at a site totally removed from the application site. The following comprehensive definitions of the different modes of delivery to the skin have been proposed by Flynn and Weiner (10).

A. Modes of Delivery to the Skin

1. Topical Delivery

Topical delivery can be defined as the application of a drug-containing formulation to the skin to directly treat cutaneous disorders (e.g., acne) or the cutaneous manifestations of a general disease (e.g., psoriasis), with the intent of confining the pharmacological or other effect of the drug to the surface of the skin or within the skin. Although systemic absorption may be unavoidable, it is always unwelcome. Semisolid formulations, in all their diversity, dominate the systems for topical delivery, but foams, sprays, medicated powders, solution, and even medicated adhesive systems are in use.

2. Regional Delivery

Regional delivery, in contrast, involves the application of a drug to the skin for the purpose of treating diseases or alleviating disease symptoms in deep tissues beneath the application. Here, the intent is to effect or accent pharmacological actions of the drug within musculature, vasculature, joints, and other, beneath and around the site of application. A selectivity of action over that achieved by systemic administration is sought. Regional activity requires percutaneous absorption and deposition, one is depending on backleakage of drug from the venous drainage of the application site. At best, backdiffusion would be an inefficient process; consequently, substantial systemic uptake, although unwelcome, is unavoidable. Nevertheless, regional concentrations are thought to be higher than can be achieved by systemic administration at the same total body exposure to the drug. The focusing of drugs into tissues in this manner has been difficult to prove unequivocally, and thus considerable scepticism exists concerning the validity of regional therapy. Regional delivery is accomplished with traditional ointments and creams as well as large adhesive patches, plasters, poultices and cataplasms.

3. Transdermal Delivery

Transdermal delivery involves the application of a drug to the skin to treat systemic disease and is aimed at achieving systemically active levels of the drug. Although such traditional dosage forms as ointments can be employed in this kind of therapy (e.g., nitroglycerin ointments), adhesive systems of precisely defined size arc the rule. Here, percutaneous absorption with appreciable systemic drug accumulation is absolutely essential. Ideally, there would be no local accumulation of drug, but such accumulation is unavoidable. The drug is forced through the relatively small diffusional window defined by the contact area of the patch. Consequently, high and potentially irritating or sensitizing concentrations of a drug in the viable tissues underlying the patch are preordained by the nature of the delivery process.

B. Sampled Matrix and Site of Action

For agents that are incorporated into transdermal delivery systems to treat disease of systemic origin, the straightforward concept and measurement of bioavailability, as used for oral drug administration, can be applied. As defined previously, bioavailability is expected to be correlated with the concentration–time profile of drug in the "biophase," which includes the site of drug action. When one considers, for example, bioavailability from an oral dosage form, it is generally accepted that drug concentration in the blood (e.g., areas under the blood or plasma concentration–time [AUC] profile) will adequately reflect the time course of drug presence within the biophase; therefore, this can be used to evaluate bioavailability. Thus, two drug products, including transdermal delivery systems, are considered to be bioequivalent if they yield comparable bioavailability-based plasma concentration–time profiles when administered to the same individuals under similar dosage conditions.

As an aside, the assumption that is then made, is that equivalent pharmacokinetics indicate equivalent therapeutics. Or, more precisely, that observed differences found in pharmacokinetic parameters between formulations under controlled study conditions and in a group of selected healthy volunteers are predictive of negligible differences in performance in the clinical setting. There has been relatively little discussion on this key assumption (11–13), which is addressed in a later section of this chapter considering issues particular to topical and regional delivery.

Application of the term bioavailability to topical and regional delivery preparations where the drug is intended to act locally (e.g., skin or muscle) is more complex. Determination of the relative bioavailability raises the first question: Is it relative to the same drug administered by a different route, usually the oral route, or is it relative to the same drug delivered from a standard topical or regional delivery preparation? Both comparisons may be important depending on the route of administration by which efficacy and safety of the drug has been established.

The second question is: From which matrix should the drug be sampled: stratum corneum, epidermis, dermis, subcutaneous tissue, appendages, muscle, or blood? Although it is the norm for oral and transdermal dosage forms, the determination of drug concentration in the blood cannot often be employed routinely with topical formulations because circulating levels of drug are usually too low to be analyzed by conventional techniques. Additionally, it can be argued that the relevance of any serum (plasma) concentration–time curve of a topical agent is questionable, because the curve reflects the amount of drug after the active moiety has left the site of action.

According to the definition of bioavailability, drug determinations are intended to reflect the rate and extent to which the active ingredient or active moiety becomes available at the site of action. A series of methods are available to quantify or localize drugs in the skin and muscle (14–17). However, even with the most sophisticated methods currently available a straightforward answer is not found. Usually, several bioanalytical techniques are often combined to deliver a more comprehesive picture of the local situation. Development of new and improvement of existing bioanalytical techniques to measure drugs in the skin and muscle is an active area of research that, similar to the effect of the introduction of techniques to measure drug levels in blood in the 1960–1970s, will transform understanding and lead to improvements in biopharmaceutical and therapeutic quality of topical agents.

Another complicating circumstance is that our current knowledge of the target site of many dermatological diseases is still limited. In viral, bacterial, or fungal infections of the skin or soft tissue, both the germ and the localization of the germ are identified: the target site is known. However for most diseases that require local treatment, neither the target site nor the cause of the diseases, which might give some insight into the site of action, are known.

As a consequence, the bioavailability of a topically applied drug or bioequivalence of two topical dosage forms may be determined by several means, specifically, (a) pharmaceutical parameters, (b) biopharmaceutical parameters, or (c) therapeutic parameters, or a combination of these.

C. The Different Equivalencies

1. Pharmaceutical Equivalence

Pharmaceutical parameters describe the influence of a dosage form on a drug. This, for example, may be the stability of a drug or the release rate of a drug from the dosage form. Some regulatory agencies, therefore, have defined *pharmaceutical equivalence* as medicinal products that contain the same amount of the same active substance(s) in the same or similar dosage forms and that meet the same or comparable pharmaceutical standards. The dilemma is, for example, that two transdermal delivery systems, containing different amounts of the same drug, that are delivered to the body with the same kinetics under identical conditions (i.e., they are bioequivalent), are by definition, pharmaceutically inequivalent. This issue becomes even more complex when semisolid dosage forms, in all their diversity, are compared. This leads to the definition of *pharmaceutical alternatives* that are medicinal products that contain the same therapeutic moiety, but differ in the salt or ester of that moiety, or in dosage form type or strength.

2. Biopharmaceutical Equivalence

Pharmacokinetic parameters describe the activity of the body on the course of disposition of the drug. Pharmacokinetics denotes characterization of drug concentration C, in a biological matrix as a function of time. Concentration in the sampled matrix, most often plasma, is denoted $Cm(t)$, emphasizing the time dependence of the parameter. Concentration at the site of drug action (effector–receptor) C_e, may be estimated by mathematical modeling. Pharmacokinetic parameters lead to the definitions of bioavailability and bioequivalence as described. Bioequivalence tests for pharmaceutical products (tablets, capsules, and transdermal delivery patches) that are based on pharmacokinetic parameters are largely standardized and corresponding statistical evaluation methods and acceptance criteria are defined (18,19). For dosage forms intended for topical and regional dermatological treatment, determination of relevant pharmacokinetic parameters and their acceptance criteria is still being debated (19).

It may happen that a well-designed generic product appears to be biopharmaceutically superior to an old innovator product. This case applies particularly to products for topical and regional dermatological treatment, as current standards of absolute bioavailability are often in the range of 1–5% of dose applied, such that there is considerable scope for increase in bioavailability. The continuous improvement of an innovator product by its manufacturer may also lead to an augmented, suprabioavailable, product. In such a case, the competent generic or innovator manufacturer should not be punished by the verdict bioinequivalent. Regulatory authorities intend to make biopharmaceutical improvements worthwhile. Adaption of the strength, however, may be necessary to make the dosage unit bioequivalent to existing ones. Furthermore, the suprabioavailable product may become a new reference product, indicating that marketing a product is a process that requires continuous improvement of the biopharmaceutical quality (20).

Pharmacodynamic parameters describe the activity of the drug on biological processes that may be only indirectly and poorly related to the therapeutic effect. Measurement of effects on a physiological or pathophysiological process as a function of time after administration of two different products may serve as the basis for bioequivalence assessment, in a manner analogous to that based on pharmacokinetic data. Effect parameters do not necessarily avoid problems inherent in use of concentration parameters for assessment of bioavailability. Effect data, although perceived as closely allied to efficacy, will usually be less relevant than drug concentration data. This is because efficacy and pharmacological concentration–effect relations are likely to be different, yet are related by drug concentration. Furthermore, effect measurement may be so imprecise that statistical criteria can be met only with large numbers of subjects. However, the effects of various drug classes are amenable to accurate, precise, and reproducible quantification (e.g., the miotic response of the pupil elicited by chlorpromazine). For dermatological products, only the vasoconstrictor test has so far played an important role (21–23).

Pharmacodynamics can be an alternative to pharmacokinetics in bioavailability assessment, or an adjunct to substantiate the relevance of pharmacokinetic measures. Understanding the concentration–effect relation is important to predict the limitations of either pharmacokinetic or pharmacodynamic assessments. Thus, pharmacokinetics and pharmacodynamics are complementary approaches. Furthermore, integrated pharmacodynamic–pharmacokinetic methods can provide a comprehensive description of the relations among dose, concentration, effect, and time. Recent guidelines on bioequivalence testing of topical corticosteroids exemplify this integrated approach (21). As with bioequivalence tests that are based solely on pharmacokinetics, assumptions are needed (e.g., that single-dose assessment is predictive of multidose treatment [steady state], and that studies with healthy subjects are predictive of therapeutic response in patients).

3. Therapeutic Equivalence

Some regulatory agencies consider a medicinal product *therapeutically equivalent* with another product if it contains the same active substance or therapeutic moiety and, when administered to the same individuals, shows the same efficacy and toxicity as that product, for which efficacy and safety have been established. The safety requirements exclude medicinal products with essentially identical therapeutic outcomes, but with different therapeutic moieties.

For some oral drugs, the concentration–effect relation is well known, and pharmacotherapeutic experience with the active substance is extensive and well described. In these cases, therapeutic studies may not be necessary. For topical drug products these requirements are generally not met. If therapeutic studies are necessary, they should prove therapeutic (clinical) efficacy (the term *efficacy* is restricted to the desired therapeutic endpoint: what is the physician trying to achieve for the patient?)

in the dosage scheme recommended for the indications claimed, and they should compare frequency and severity of side effects with a standard treatment or with a comparably approved drug product.

D. Statistical Considerations

The standard experimental design for bioequivalence comparisons between two oral drugs is a crossover design that permits comparisons with far greater precision than a parallel group study with the same number of subjects, or equally precise comparisons using fewer subjects.

For topical drug formulations, an experimental design that corresponds to a crossover is a bilateral-paired comparison, in which both formulations are tested in the same subject at the same time, on contralateral body sites. With simultaneous testing of topical preparations, it is possible to test more than two sites on the same volunteer or patient at the same time. By doing so, it is also possible to test more than two formulations simultaneously. Alternatively, the formulations being studied could each be applied at multiple sites on the same subject. This within-subject replication would further reduce statistical error, with the consequence of greater precision of results or a reduced sample size requirement. Advantages of the bilateral-paired comparison design relative to crossover designs for oral products may be partly offset by the apparently greater variability in absorption of topical drugs in comparison with oral drugs, (19). A bioequivalence study of topical agents, therefore, may require a larger sample size than a study of oral agents, with their less variable rates.

When bioequivalence studies are performed for oral drug formulations, there is an explicit attempt to select healthy volunteers within a narrow age range to minimize interindividual variability to focus on variation between formulations. The issue of whether observations in such a group can be generalized to patients in whom the drug might be used has not been fully resolved. Similarly, for topical preparations, it is not clear whether the results of tests carried out on healthy skin can be generalized to patients with diseased skin. This issue of the interaction between the formulations and the absorption site has received little study (24).

In conducting statistical tests of bioequivalence, it was the practice, until the early 1980s, to analyze the results using a standard statistical method, such as a *t*-test or analysis of variance (ANOVA). These methods are designed to test the alternative hypothesis of a difference between the agents against the null hypothesis of no difference. These methods are not applicable to bioequivalence studies and do not validly test equivalence (25–27); they do not address the study hypothesis that such studies should address, which is that two formulations are bioequivalent. Hauck and Anderson (26) reasoned as follows: The test for formulations in the analysis of variance is specifically a test of the null hypothesis that the average bioavailabilities are equal, against the alternative that they differ. If the study is sufficiently sensitive, it is possible to detect significant differences of little clinical meaning. Alternatively, if the study is sufficiently insensitive, large differences will not be detected; the evidence will not be sufficiently strong to accept the alternative hypothesis. They believed that when one has a hypothesis to demonstrate, in this case the hypothesis of bioequivalence, the logic of hypothesis testing requires that the hypothesis be demonstrated to be the alternative hypothesis. They proposed a new decision rule,

starting by interchanging the null hypothesis and the alternative hypothesis in connection with preset lower and upper bioequivalence limits, θ_1 and θ_2, respectively.
For the multiplicative model, with $0 < \theta_1 < 1 < \theta_2$, the hypotheses are as follows:

Null hypothesis (H_0):   There is no bioequivalence (bioinequivalence). H_{01}:
                                  $\mu_{\text{test}}/\mu_{\text{ref}} \leq \theta_1$ or H_{02}: $\mu_{\text{test}}/\mu_{\text{ref}} \geq \theta_1$.
Alternative hypothesis (H_1):   There is bioequivalence. H_1: $\theta_1 < \mu_{\text{test}}/\mu_{\text{ref}} < \theta_2$.

For the additive model the corresponding hypotheses have to be formulated in
terms of differences (28). Hypothesis reversal implies that α and β, the type I error
and the type II error, respectively, are also interchanged. The type I error (α) now
is the consumer risk of incorrectly concluding bioinequivalence. Consequently, statistical power ($1 - \beta$) is now the probability of correctly concluding bioequivalence.

There is considerable experience with the design and analysis of bioequivalence
studies for oral formulations (29). Principles and practices for oral products may be
applied to topical products. In particular, the paired comparison study for topical
products is a direct analogue of the crossover study that is standard for oral products.
Advancements in statistical methods over the last decade do not seem to have become generally adopted in the topical arena.

III. MODEL SYSTEMS

A. In Vitro Model Systems

In the last decades much effort has been directed toward the development of in vitro
model systems for the biopharmaceutical evaluation of topical formulations. Academia, industry, and regulatory authorities have developed active cooperation to stan-
dardize these model systems (30). Because of the diversity of the in vitro model
systems available, a classification of the models is useful, as it is helpful to draw a
clear distinction between two main model groups: the drug-release models, and the
drug-absorption models. These models can be used, respectively, for quality assurance and for formulation development of topical dermatological drug products.

To study the quality of topical drug products, the release rate of the drug from
the formulation should not be influenced by the matrix (membrane–liquid receptor)
into which the drug diffuses. A plot of the amount of drug released from a semisolid
formulation per unit area ($\mu g/cm^2$) against the square root of time ($t^{1/2}$) may yield a
straight line, the slope of which represents the release rate (31,32). This release rate
is formulation-specific and can be used to monitor batch-to-batch uniformity.

In formulation development it is useful to have in vitro models that will predict
the relative effect of formulation changes on transport across membranes. In contrast
to drug-release models, the drug-absorption models should give linear plots with
time, indicating that transport is controlled by the membrane. By using artificial
membranes, the drug-absorption model is of use only in predicting drug–vehicle
interactions (i.e., the thermodynamic activity of the drug in the formulation). Models
with artificial membranes are not predictive for vehicle–membrane interactions (e.g.,
effects of chemical enhancers).

Figure 1 gives an overview of the in vitro model systems.

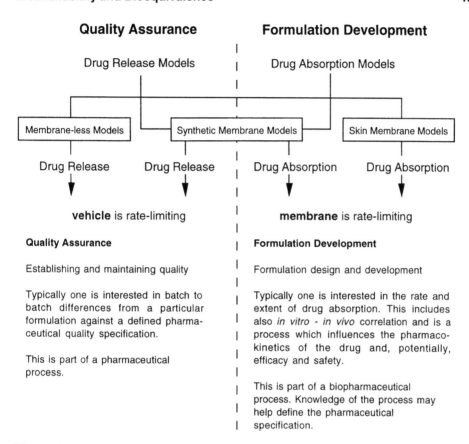

Figure 1 Classification of in vitro models for the evaluation of topical formulations. (Expanded from Ref. 33.)

1. Drug-Release Models

A simple, reliable, reproducible, relevant, and generally accepted in vitro method to assess release of drug from topical dermatological products would be highly valuable for the same reasons that such methods have proved valuable in the development and batch-to-batch quality control of solid oral dosage forms. For this reason, the Food and Drug Administration (FDA), Division of Bioequivalence in the Office of Generic Drugs, and the Division of Anti-Infective Drug Products in the Office of Drug Evaluation II, recommends in an Interim Guidance (34) the development of in vitro release methods for topical dermatological products to accompany in vivo pharmacodynamic and pharmacokinetic data. This guidance (July 1, 1992, final guidance June, 1995) provides detailed recommendations to pharmaceutical sponsors on methods to document bioequivalence of topical corticosteroids.

The two major categories of topical products, transdermal patches and dermatological (topical) products, pose different issues for the development of an in vitro release method.

a. Drug-Release Methods for Transdermal Patches. At present, the *United States Pharmacopeia* (*USP*) has identified three different apparatus–procedures for determining the release rate of transdermal patches (35). These procedures are complex and not universally applicable to the different types of patches presently on the market. Therefore, they have developed a simple method that can be employed to determine in vitro release profiles of all marketed transdermal patches. The method employs a watchglass–patch–Teflon mesh sandwich assembly and the paddle method (*USP* apparatus 2) to determine the release rate of the product, as a quality control procedure to assure batch-to-batch uniformity. The method is applicable to all brands, shapes, and strengths of nitroglycerin patches (36) and gives a release profile that is comparable with that obtained with other more complicated and cumbersome compendial methods (37). The watchglass–patch–Teflon sandwich method has also been applicable to the release test method for scopolamine, clonidine, and estradiol patches. It also yields release data comparable with those obtained from other methods (38).

    b.   Drug-Release Methods for Semisolid Formulations.   Among the first experiments performed on the release of drugs from ointments were those developed by Reddish (39) and Christoff (40). The Reddish procedure involved the measurement of the antiseptic properties of ointments containing various drugs. The experiments were performed by applying ointment to agar that had been inoculated with *Staphylococcus aureus*. After incubation, the agar plates were assessed to determine the presence of a zone of inhibition. Thus, a crude measure of the ointment's ability to liberate a drug was achieved. A variation of this method was introduced (41) in which the minimum time required for an ointment to cause inhibition of bacterial growth was used as an index of ointment efficacy.

    Two procedures, one in which the drug product is placed in direct contact with the receptor phase (42–45), and the other in which the drug product and the receptor medium are separated by a membrane (46–52), have been used. In the first case, the possibility of product mixing and sloughing into the receptor media exists, making the release rate determination difficult.

    Recently, researchers at the FDA have developed a simple method to determine the release of corticosteroids from cream, ointment, and lotion preparations. The method uses a commercially available static diffusion cell system and a synthetic cellulose acetate membrane. The use of a commercially available synthetic membrane obviates the problems associated with the preparation and variability of skin membranes. In a recent study (32), hydrocortisone cream was placed in the donor chamber on the synthetic membrane and the aqueous buffer was used as the receptor medium. The release profile of hydrocortisone was determined over a 6-h interval and plotted as the amount released versus $t^{1/2}$. Hydrocortisone from different batches of the same manufacturer showed similar release rates, but these differed from those of hydrocortisone from another manufacturer (53). When the procedure was used with creams containing sparingly soluble corticosteroids, use of a hydroalcoholic medium as a receptor phase was essential to increase the drug solubility to maintain sink conditions. Betamethasone valerate cream from two manufacturers showed significantly different release rate profiles under such experimental conditions (54). When employing a hydroalcoholic medium as a receptor phase, there is a question of back-diffusion of alcoholic medium through the synthetic membrane and alteration of the integrity of the cream preparation. With betamethasone valerate cream, for which ethanol–water (6:4) was used as the receptor phase, mere traces of alcohol were detected in the cream (donor chamber) at the end of a 6-h study. Microscopic examination of the cream before, and after, the in vitro release experiment showed no differences (55). Thus, the concentration of ethanol had a negligible effect on cream integrity (test criteria not presented!). Hydrocortisone ointment showed a much slower rate of release when compared with the cream (32).

   c. *Conclusion.* The objective in development of drug-release models is to minimize the effect of the membrane on in vitro release. Systems without membranes, which are feasible only if the polarity of the vehicle and the receptor fluid are very different, are not widely used. Today, systems with membranes are more common. The membranes used may vary from relatively wide-mesh mechanical supports (56), to microporous membranes (57). Corbo et al. (57) and Shah et al. (32,53–55) provide some excellent examples of the development and validation of these methods. Guy and Hadgraft (58) have argued the case for flexibility in the design of these models, especially in the composition of the receptor phase, to allow the test to be optimized for drugs of different physicochemistry, especially aqueous solubility. Most methods use the static diffusion cell system (Franz-type cell). This cell has been automated for ease of use and to improve reproducibility (32,59). Very recently the "Enhancer Cell," a modification of the *USP* tablet-dissolution apparatus, has been evaluated for in vitro release testing of topical products (60).

   The importance of measuring the in vitro release rates of drugs from topical preparations and the usefulness of drug release data is further discussed elsewhere (33,61–64).

2. Drug Absorption Models

Drug absorption models are useful in the design and development of formulations. Some of these methods are described in more detail in Chapter 5.

   a. *Synthetic (Artificial) Membrane Models.* Synthetic membrane models do not measure rates of transport equivalent to human skin, nor do they predict the interaction of vehicle components to modify skin permeability (e.g., effects of chemical enhancers). Silastic (polydimethylsiloxane; silicon rubber) has been widely used as a hydrophobic membrane to model diffusion relevant to transport across human skin. For example, Flynn and Smith (65) found a linear relation between the percentage saturation, thus, thermodynamic activity, of p-aminoacetophenone and its steady-state transport across Silastic. Also, Tanaka et al. (66) measured the transport of hydrocortisone butyrate propionate from various vehicles under open conditions. Transport was directly related to drug solubility; hence, the thermodynamic activity, in the vehicle. Because Silastic models may show the $t^{1/2}$ plots, validation of the in vitro model is required to ensure that transport is rate-controlled by thermodynamic activity only. Davis and Hadgraft (67) have described the development and validation of such a method for use in design of supersaturated solutions of hydrocortisone acetate. Hadgraft and Ridout (68) developed novel artificial lipid membranes to mimic the stratum corneum barrier. However, transport rates are approximately 100 times less than across stratum corneum. Other systems, which may be suitable only for use in the design of formulations to optimize thermodynamic activity, include transport across cell-cultured epidermal membranes (69,70). However, these systems may, more readily than Silastic, also respond to effects of vehicle components on the barrier, which may complicate interpretation of the results.

b. Skin Membrane Models. Skin membrane models, both human and animal skin, are widely used in the development of topical dermatological formulations. The consensus of the FDA/AAPS-sponsored *Workshop on Principles and Practices of In Vitro Percutaneous Penetration Studies* (30) was that human skin is the preferred membrane, although the value of studies on animal skin in the formulation development stage was recognized. Human skin is notorious for its high level of barrier variability (71).

3. In Vitro–In Vivo Correlations

Several authors have described the correlation of in vitro release or drug absorption studies with pharmacological effects, such as the onset time of nicotinate erythema (72–75) and vasoconstriction of corticosteroids (54,66,76,77). Furthermore, the permeability of artificial membranes was also correlated with pharmacokinetic parameters, such as the maximum plasma concentration and the amount of drug excreted in urine for salicylic acid (47), indomethacin (78), and nitroglycerin (79).

In vitro release will be influenced by several factors, including bulk viscosity and microviscosity of the vehicle, polarity of the vehicle relative to the receptor phase, hydrodynamic pressure and, also, the thermodynamic activity of the drug in the vehicle. Of these factors only the latter affects permeation in vivo. Thus, and considering only simple vehicles without enhancers, in vitro-release studies have the potential only to predict in vivo permeation when thermodynamic activity dominates the release process. Figure 2 gives an example in which release rates correlate with the amount of drug that penetrated into the stratum corneum and to a pharmacological response (vasoconstriction) (77).

Also, as shown in Figure 3, Shah et al. (54) found a similar correlation with betamethasone valerate creams. However, no correlation was found in an earlier

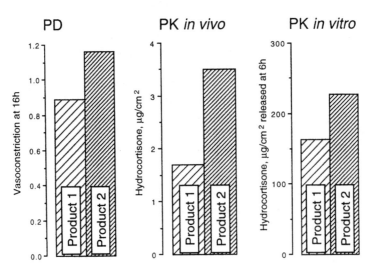

Figure 2 Correlation between pharmacodynamic (PD; vasoconstriction at 16 h), pharmacokinetic (PK; in vivo; $\mu g/cm^2$ in the stratum corneum at 24 h), and in vitro release characteristics (PK; in vitro mg/cm^2 released in 6 h) of two hydrocortisone creams. (From Ref. 77.)

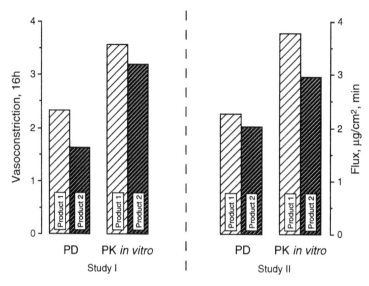

Figure 3 Correlation between pharmacodynamic response (PD; vasoconstriction at 16 h) and in vitro release rate into ethanol/water (6/4) (PK; in vitro mg/cm^2 released in 6 h) for betamethasone valerate cream. (From Ref. 54.)

study on hydrocortisone in which an ointment showed a much slower rate of release compared with cream (32). It is likely that the microviscosity of the ointment caused diffusion through the ointment to rate-limit in vitro release.

Less attention has been given to the correlation of in vitro synthetic membrane absorption models to in vivo absorption, although by definition, these should be better correlated with in vivo absorption in that they respond to drug thermodynamic activity within the vehicle. However, both in vitro release and synthetic membrane absorption models will fail to predict the effect of chemical enhancers on permeation in vivo.

Of all of the in vitro methods, transport across human skin has the highest potential to be predictive of in vivo permeation in humans. How useful the prediction is has been addressed by Franz (80). He compared the permeability of 12 organic compounds across excised skin with that previously found in vivo (81). Franz found a rank order in vitro−in vivo correlation that is reasonable considering the substantial differences in experimental conditions between the studies. A review of more recent data, comparing in vitro−in vivo human skin data under controlled conditions is urgently required to reevaluate this correlation.

B. In Vivo Model Systems

A wide variety of in vivo techniques have been developed to study kinetic, dynamic, and therapeutic (clinical) parameters of topical drug application. In the following section some techniques to sample drug(s) in the biological matrix and to measure drug effects in the skin will be reviewed and discussed. Owing to the large amount of data available, only a selection of information is presented. Although studies in animal models are useful in the development of topical formulations, the variation in percutaneous absorption between animals and humans requires that bioequivalence is confirmed in the latter.

1. Kinetic Models

a. Skin-Stripping. Tape-stripping is a technique that is useful in dermatological research for selectively and, at times, exhaustively removing the skin's outermost layer, the stratum corneum. Typically an adhesive tape is pressed onto the test site of the skin and is subsequently abruptly removed. The application and removal procedure may be repeated 10 to more than 100 times (82,83). Such stripped skin has been used as standardized injury in wound-healing research. The technique has been adapted for studying epidermal growth kinetics (84–88), and it may also be useful as a diagnostic tool in occupational dermatology to assess the quality of the horny layer (89,90).

The observation that the skin may serve as a reservoir for chemicals was originally noted in 1955 (91). The localization of this reservoir within the stratum corneum was demonstrated for corticosteroids (92) and confirmed by others (93–96). The introduction of the tape-stripping method to further investigate the reservoir and barrier functions of the skin was a significant expansion of this experimental tool in skin research (97–99). Some of the following questions are approached.

Differences in the permeability of intact and fully stripped skin have provided information on the diffusional resistance of the skin (100). It has been recognized that complete removal of the horny layer after 30–40 strippings was not possible (101), and after local application of hydrocortisone to stripped skin, a significant barrier function still exists (102,103).

The presumption that factors which improve percutaneous absorption also result in an increase in stratum corneum reservoir (92,96) made the tape-stripping method a possible approach for selecting or comparing vehicles for topical drugs. Data from tape-stripping experiments were, therefore, related to (a) chemical penetration into skin, (b) chemical permeation through skin, (c) chemical elimination from the skin, (d) pharmacodynamic parameters, and (e) clinical parameters.

In in vitro and in vivo investigations with the tape-stripping technique, drug penetration into the stratum corneum, as determined by quantification of radiolabeled chemical on the tape after being removed from the skin, was clearly vehicle-dependent (97,104–106). The tape-stripping technique has been validated and standardized (107–110). This stripping method determines the concentration of a chemical in the stratum corneum at the end of a short application period (30 min). A linear relation was found between the stratum corneum reservoir content and in vivo percutaneous absorption (total amount of drug permeated in 4 days) using the standard urinary excretion method (81,111,112). These investigations also showed that, for a variety of simple pharmaceutical vehicles, the percutaneous absorption of benzoic acid is vehicle-dependent and can be predicted from the amount of the drug within the stratum corneum 30 min after application. Once validated, the major advantages of the tape-stripping protocol are the elimination of urinary and fecal excretion to determine absorption, and the applicability to nonradiolabeled determination of percutaneous absorption because the skin strippings contain adequate chemical concentrations for nonlabeled assay methodology. Despite that the assay provides reliable prediction of total absorption for a group of selected compounds, mechanistic interpretations are still rare. From the data of Rougier and co-workers, Auton (113) recently presented the first mathematical approaches that may help explain some of the foregoing observations.

The tape-stripping technique has also been used to analyze biological activity, thus taking into account binding, decomposition, and metabolism of a given drug (82,114–119). Pershing and co-workers (120–123) simultaneously compared a skin-blanching bioassay with drug content in human stratum corneum following topical application of commercial 0.05% betamethasone dipropionate formulations (pharmaceutical *complex* vehicles). The rank order of the betamethasone dipropionate formulation potency is similar between the visual skin-blanching assay, the tape-stripping, and the "a scale" on the chromameter. The rank correlation between tape-stripping method and the skin-blanching response was moderate to good.

Owing to a series of still unanswered questions and the potential of the tape-stripping technique a variety of parameters, such as the properties of adhesive tapes, vehicle effects on the harvesting of stratum corneum, and tape application and removal procedures, are currently under systematic investigation (120,124–126). Recently a Draft Guidance for Industry Document has been published by the FDA (126a). This guidance, available on the World Wide Web (FDA, CDER) has caused a heated debate (126b).

b. Skin Biopsy. The most invasive—but still practicable—method to access skin compartments is the excision of skin tissue. In contrast to the other methods described, the punch and the shave biopsies allow a direct ingression into the compartment of interest. After removal (optional) of the stratum corneum from skin with an appropriate technique (tape-stripping, glue) the punch biopsy will represent parts of subcutaneous tissues, dermis, and epidermis, and the shave biopsy will represent mainly epidermis and some dermis. Parts of the stratum corneum may remain on the epidermis, depending on the method used for stratum corneum removal. Subcutaneous tissue can mechanically be divided from the dermis, and dermis can be separated from the epidermis by heat (127). Human skin samples larger than 100 mg are difficult to obtain, and the usual amount that can be harvested is less than 50 mg.

Surber and Laugier (128,129) compared the acitretin concentration (synthetic retinoid) in human skin after oral and topical dosing using three different skin-sampling techniques: punch biopsy, shave biopsy, and suction blister. All three skin-sampling techniques have been used by various investigators to quantitate drugs and xenobiotics in the skin. Each technique gives access to distinctive skin compartments (Fig. 4).

Despite that therapeutic acitretin concentrations were found after topical and oral treatment, no beneficial effects in psoriasis, or other disorders of keratinization, are observed following topical administration of this drug. Similar observations have been made with methotrexate and cyclosporine (130,131). One may postulate that drug concentration at a particular site within the skin following both routes of administration could be different owing to the direction of the drug concentration gradient. This hypothesis has also been postulated and illustrated by Parry et al. (118), comparing the clinical efficacy of topical and oral acyclovir. Model predictions and in vivo data agree that the topical administration of acyclovir results in a much greater total epidermal acyclovir concentration than after oral administration. However, mathematical modeling of the acyclovir concentration gradient through the

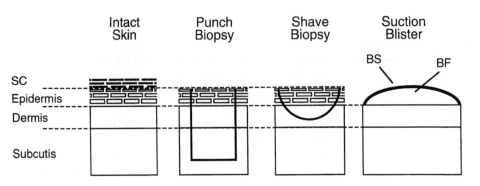

Figure 4   Schematic diagram of intact skin (left) and the skin specimens taken using three sampling techniques: BS, blister skin; BF, blister fluid.

epidermis revealed that the drug concentration at the target site of the herpes simplex infection, the basal epidermis, was two to three times less after topical administration than after oral administration. Furthermore, one may postulate that drug metabolism could be different, depending on the route of administration. Data supporting this hypothesis are still incomplete.

Despite skilled experimenters, sophisticated sampling techniques, and instrumentation, the information gained from tissue sampling is probably only an estimate of the chemical distribution within the skin. Because of possible interlaminate drug contamination, accurate and specific information on drug localization within a particular skin compartment following both routes of administration is not obtained by these methods. The following two techniques have been used to obtain skin sections, parallel to the outer skin surface to a depth of about 300–600 μm.

Semiautomated skin-sectioning technique.   With this approach the skin tissue from a punch biopsy is placed on a cryomicrotome table and 10- to 40-μm sections are cut parallel to the skin surface. This technique is excellently summarized in the form of a standard operating procedure (SOP) (132,133). When the cylinder-shaped punch biopsy is placed, dermis-side down, on a microtome chuck maintained at $-17°C$ it is essential that the stratum corneum surface is parallel to the cutting plane. This can be accomplished when several metal rings of a total thickness slightly greater than that of the skin sample are placed around the tissue, and the rings are filled with embedding medium. Subsequently, a flat, precooled surface (glass slide) is pressed against the stratum corneum, resulting in a flat outer surface. The whole assembly is then frozen in approximately 10 s and can be stored. The technique, however, does not take into account the dermal papillae, and a clear separation of histologically distinct compartments is not possible. The method has predominately been used to characterize the pharmacokinetic and metabolic behavior of topically applied drugs (15,104).

With excised human skin and tissue grafted to athymic mice, the in vitro and in vivo concentration profiles of salicylic acid and salicylate esters were obtained for the outer several hundred microns of the skin (15). The results show significant differences in the extent of enzymatic cleavage and distribution of metabolites between in vitro and in vivo studies. The data also suggest that in vitro results may overestimate metabolism because of increased enzymatic activity or decreased capillary removal (Fig. 5).

Manual skin-sectioning technique. With the development of the skin sandwich flap model (134), Pershing and co-workers (117) also proposed a new manual skin-sectioning technique. With this technique, the profile of a test compound's disposition in skin after topical administration is examined. The technique requires the use of radiolabeled compound and fresh skin, without freezing. The sectioning technique uses cyanoacrylic cement, Scotch tape, glass slides, 115-μm–thick glass coverslips, a thin razor blade, and a fresh 2-mm–skin punch biopsy. Briefly (Fig. 6), a 2-mm–skin punch biopsy is sectioned in 115-μm–thick sections starting from the stratum corneum (see Fig. 6a) completely through the biopsy, by dipping the stratum corneum end of the biopsy into cyanoacrylate adhesive (step 1) and placing the biopsy on a microscope slide between two 115-μm–thick microscope coverslips that are held in place on the microscope slide with tape (step 2). A single-edged razor blade is used to shave the stratum corneum side of the punch from the remainder of the biopsy. The remaining biopsy (see Fig. 6b–d) is then dipped again in the cyanoacrylate adhesive in the original orientation (stratum corneum face down) and placed in a new location on the microscope slide (see Fig. 6b). The remainder of the skin biopsy is sectioned similarly, producing multiple 115-μm sections on an individual microscope slide (step 3). After complete sectioning of the skin biopsy, the various skin wafers on the microscope slide are individually digested with tissue solublizer and submitted to liquid scintillation counting for drug quantification and, subsequently, a concentration gradient and disposition profile of the test compound in the skin is reconstructed. The authors also report that freezing the 2-mm biopsy results in redistribution of the drug from the stratum corneum to the middle sections (400- to 600-μm–depth) of the skin biopsy, thereby portraying an inaccurate distribution profile in the skin. Unfortunately, this statement is not further documented. Even though the redistribution phenomenon was described only for the 2-mm biopsy, it may also be relevant for the 4- to 6-mm biopsy. The statement makes a careful reevaluation of the skin preparation procedure necessary.

  c. Drug in Blood and in Excreta. In vivo percutaneous absorption may also be determined by indirect method of measuring radioactivity in excreta after topical application of a labeled compound. The compound, usually labeled with ^{14}C or ^{3}H, is applied and the total amount of radioactivity excreted in urine or urine plus feces is determined. The amount of radioactivity retained in the body or excreted by some other route not assayed (e.g., CO_2 or sweat) is corrected for by determining the amount of radioactivity excreted after parenteral administration. This final amount of radioactivity is then expressed as the percentage of applied dose that was absorbed (81,111,112). The limitation of determining percutaneous absorption from urinary or fecal radioactivity, or both, is that the method does not account for metabolism by the skin or other organs. The radioactivity in excreta is a mixture of parent compound and metabolites. This type of information is useful in defining the total disposition of the applied topical dose. Plasma radioactivity can also be measured and percutaneous absorption determined by the ratio of the AUC following topical and parenteral administration (135). The same limitations discussed for excreta also apply here.

  A comparative example of three methods (136) was performed using [^{14}C]-nitroglycerin in the rhesus monkey. Topical bioavailability estimated from urinary

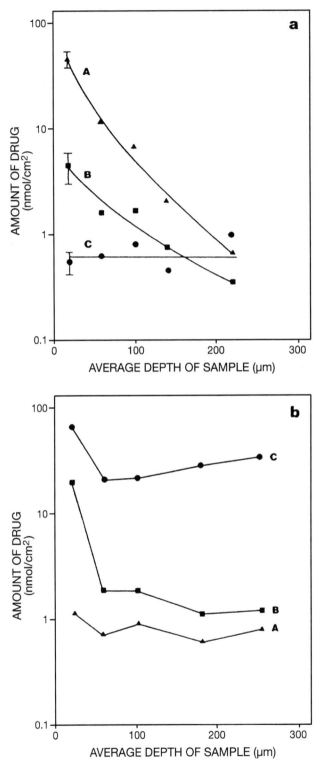

Figure 5 The concentration profile of (A) the diester, (B) monoester, and (C) salicylate, as a function of the depth within human skin either (a) grafted to athymic mice, or (b) in vitro. Results were obtained after 24 h of topical drug application. (From Ref. 15.)

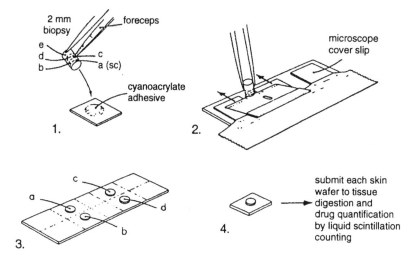

Figure 6 Manual sectioning technique of skin biopsies (see text for details). (From Ref. 117.)

excretion was $72.7 \pm 5.8\%$. This was similar to the $77.2 \pm 6.7\%$ estimated from the plasma total radioactivity AUC. The estimates from urinary ^{14}C and plasma ^{14}C were in agreement. The absolute bioavailablity estimated from the plasma's unchanged nitroglycerin AUCs was $56.6 \pm 2.5\%$. The difference in estimate between that of the absolute bioavailability (56.6%) and that of ^{14}C (72.7–77.2%) is the percentage of compound metabolized in the skin during absorption. For nitroglycerin this is approximately 20% (137).

Nitroglycerin is an example of a drug with a high absolute bioavailability. However, most topically applied drugs have low absolute bioavailability. These data demonstrate absolute bioavailability of topical steroids and retinoids ranging from 1 to 4% (91,103,111,138–142). Although not reviewed here, other series of compounds, including azole antifungals, antivirals, and vitamin D_3 derivatives also have absolute topical bioavailabilities in this same range. Where bioavailability is low there is potential for increased absorption owing to biological or pharmaceutical reasons (see Sec. IV).

2. Dynamic Models

a. Quantification of Skin Color Changes. Skin color is predominantly determined by pigments such as hemoglobin, melanin, bilirubin, and carotene. Those skin–color-determining components can be altered significantly by radiation (UV or infrared light) and several substances (drugs or irritants). The quantification of experimentally induced color changes is a widely used method in dermatological and cosmetic research, for the magnitude of the color response can be used as an indicator of skin properties (integrity of the skin barrier or sensitivity), drug properties (concentration and bioavailability), vehicle properties (formulation and enhancer), protection properties (sun screen, and such) (143,144).

Until recently, assessment of skin color was performed by visual grading, based on semiquantitative scales. Visual grading, however, needs a panel of trained experimenters. Thus, only a limited number of research centers were able to generate reproducible data. Despite that the human eye is highly sensitive in the discrimination between colors close to each other, it has also been postulated that the human brain is unable to memorize colors. As a consequence, the human eye is inadequate to make comparisons at different times or at distant sites. To improve color readings, standardized methods have been developed to express color in a more scientific way. In tristimulus reflectance colorimetry, the reflected spectrum is analyzed at three selected wavelengths, simulating the human color perception system. The three measured values can be converted to different color-indexing systems (Yxy system, Munsell system, and the L* a* b* system) (145). Skin color measurements based on the L* a* b* system proposed by the Commission Internationale de l'Eclairage (CIE) has become an accepted color-indexing system (144–146). Color is expressed in a three-dimensional coordinate system: L* represents brightness (varying between white and black), a* represents hue and color on the green–red axis, while b* represents hue and color along the blue–yellow axis. Today the Minolta Chromameter (Minolta Camera Co., Ltd., Osaka, Japan) is the most widely used instrument.

Erythema and tanning of the skin after photoprovocation with UVB were assessed using the L* and a* parameters (146). Seitz et al. (147) used the a* parameter to quantify the erythema, whereas the b* parameter was used for quantification of the tanning. A positive relation between the a* color parameter and the perfusion of the skin microcirculation in the quantification of the methyl nicotinate-induced erythema was also detected (148). Wilhelm and Surber (149) challenged the skin with different sodium lauryl sulfate concentrations. The irritation was evaluated with reflectance colorimetry, transepidermal water loss measurement, laser Doppler flow measurement, and visual scores. All evaluation methods detected a significant sodium lauryl sulfate dose dependency (dose–response). Moreover, a strong positive correlation was detected between the different evaluation techniques. A similar positive correlation was detected between transepidermal water loss measurements and the a* parameter in the evaluation of surfactant-induced stratum corneum irritation (150,151). Quantifing the erythema by means of skin color measurements (a* parameter) seems to be the most realiable method (152–154). Other approaches (e.g., laser Doppler or transepidermal water loss) produce results with greater variability than does the chromameter. The simultaneous use of these approaches is justified because information from a different depth of the skin is collected.

Use of tristimulus color analysis for the quantification of skin color changes induced by topical cortiosteroid prepartions has stimulated new research in bioscreening for measuring topical corticosteroids potency (154a–c). For further information see *Guidance; Topical Dermatological Corticosteroids: In Vivo Bioequivalence* (21).

    b.   Skin Blood Flow (Laser Doppler Velocimetry).   Laser Doppler velocimetry (LDV) is a noninvasive, optical procedure allowing continuous, real-time monitoring of skin blood flow. The measurement principle of the laser Doppler instrument is based on the Doppler shift when reflected from moving particles. A low-energy monochromatic laser beam (633 or 780 nm) is guided to the skin. The light illuminates a hemisphere with a diameter of 1 mm. The light is reflected by static structures (not Doppler-shifted) and by moving red blood cells (Doppler-shifted). The magnitude and frequency distribution of the backscattered light is related to the number and the velocity of the red blood cells moving in the illuminated volume (flux). The flux is measured in relative and dimensionless arbitrary units. The small size of the illuminated volume causes a moderate-to-low reproducibility when the probe has to be repositioned because of inhomogeneities in the microvascular architecture. To decrease the spatial variability, the probe can be positioned above the skin surface to increase the evaluated area and volume. Recently, the laser Doppler technique has been considerably improved. Instead of a single-point measurement (a major drawback of conventional laser Doppler flowmetry), readings are made with either integrating probes, which use the signals from various adjacent measuring volumes, or they are carried out over a scanned skin surface (up to 5 × 7 cm) to build up a laser Doppler image (155–157). The new scanning laser Doppler technique is very useful in patch testing and psoriasis studies, because both the degree and the area of increased blood flow can be determined.

The laser Doppler technique is particularly suitable for the assessment of local pharmacodynamic changes induced by topical administration of vasoactive chemicals. The technique has been extensively used to examine the vasoresponse in humans to topically applied esters of nicotinic acid (which are potent, local vasodilators). Furthermore, these experiments have been analyzed to provide (a) insights to percutaneous absorption process itself and (b) an understanding of the the effect of formulation on the topical bioavailability of an applied drug (Fig. 7) (158–161).

c. Rhino Mouse Model. Adult rhino mice are hairless mutants that carry the rhino gene, a recessive allele of the hairless gene (hr^{rh}/hr^{rh}). At birth, rhino mice are indistinguishable from their nonmutant littermates. Shortly afterward, and before the end of their first hair growth cycle, a defect in catagen results in irreversible hair loss. The pilary canals widen, accumulate keratin, and undergo a transformation into

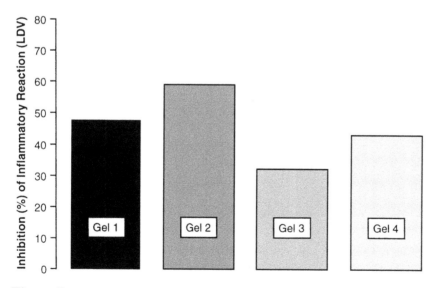

Figure 7 Percentage inhibition of the inflammatory reaction (LDV) induced by methyl nicotinate in indomethacin gel-pretreated skin: no significant difference was observed between formulations 1, 2, and 4, although urea (gel 2) did appear to have a greater enhancing effect than ethanol (gel 3; $p < 0.05$). (From Ref. 159.)

horn-filled utriculi that resemble the human noninflamed microcomedone. Comedo-lytic effects of various antiacne agents in the rhino mouse skin were first reported (162,163) with use of qualitative methods. Later, a standardized histological and image analysis method was introduced (164) to characterize and quantify the cuta-neous morphological effects of antiacne agents. The following microscopic param-eters can be assessed from skin slices obtained from skin biopsies before and after treatment (165) (Fig. 8): on each open epidermal comedo, the largest diameter of the utricle d, and the diameter taken at half depth D, can be used to calculate a comedo profile ($r = d/D$). This ratio gives a measure of the morphological aspects

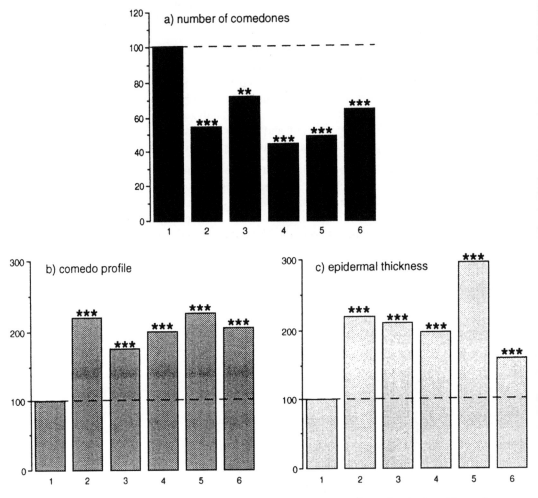

Figure 8   Effect of treatment with topical commercial preparations of retinoic acid: number of (a) epidermal comedones; (b) comedo profile; and (c) epidermal thickness were measured by image analysis methods (165). Animals were treated with (1) placebo, (2) Aberel gel, 0.025%; (3) Aberel cream, 0.05%; (4) Aberel lotion, 0.03%; (5) Retin A cream, 0.05%; and (6) Retacnyl cream 0.025%. 100% control values were 69 ± 1 epidermal comedones per centimeter length of stratum corneum, comedo profile 0.6 ± 0.02; epidermal thickness 21.0 ± 0.6 μm.

of the comedo. The surface of epidermis from an intercomedo area, the length of the corresponding basal layer, and the number of epidermal comedones are further parameters. These factors can be used for statistical analysis.

Commercially available topical preparations of retinoic acid, used clinically in the treatment of acne, presented similar comedolytic activities (number of epidermal comedones and comedo profile). The only significant differences observed among the Aberel formulations, Retin A, and Retacnyl were related to the degree of epidermal thickening. From the foregoing data, it would appear that comedolytic and epidermal thickening activities of retinoic acid can be dissociated by formulation changes. In consequence, differences in retinoic acid-induced skin irritation in formulations such as Retin A and Retacnyl may be anticipated.

Possible systemic effects of the treatment were not discussed. Repeated application ($3 \times 5 \times 50$ μL) of formulation to the body surface of the rhino mouse of 30–40% may produce systemic drug levels. Inadequate housing and animal behavior may result in ingestion of formulations.

d. UVA-Induced Neutrophil Infiltration. A variety of human and animal models have been developed for assessing the anti-inflammatory activity of topical corticosteroids. Vasoconstriction is a rapid and popular screening technique, although skin sites can vary in sensitivity to the assay (166). Human studies are troublesome for both technical and ethical reasons: individual responses are highly variable, and the tests are burdensome.

Based on an observation by Gilchrest et al. (167) that 50 J/cm^2 of long-UV– radiation (UVA) to human skin resulted in a modest infiltrate of neutrophils Woodbury et al. (168) and Kligman (169) irradiated hairless albino mouse skin with a single large dose (195 J/cm^2) of UVA. This produced a massive influx of neutrophils into the dermis. This dramatic and quantifiable event formed the basis of a new model for assaying topical corticosteroids.

Fifteen name brand corticosteroids (100 μL each) were evenly applied to dorsal trunk skin immediately postirradiation and again 12 h later. Biopsies were taken at 24 h after irradiation when the density of neutrophils was at peak. The number of neutrophils in the dermis was counted in ten fields per specimen (blinded) (Fig. 9) (168). Further refinement and validation of the test (169) confirmed the initial study.

The data from this model are very promising and further investigation on sensitivity and the influence of dosing regimens are necessary to prove the validity of the assay.

C. Therapeutic Models and Clinical Trials

It is assumed that therapeutic models and clinical trials are the least accurate, least sensitive, and least reproducible of the general approaches for measuring bioavailability or bioequivalence. In general, they are not acceptable for formulations intended for systemic availability if analytical methods are available to measure systemic drug concentrations. However, clinical trials may be the only means to determine the bioavailability or bioequivalence of dosage forms intended to deliver the active moiety locally. Examples include topical preparations for the skin, eye, and mucous membranes; oral dosage forms not intended to be absorbed (e.g., antacid, radiopaque medium); and bronchodilators administered by inhalation if onset and duration of pharmacological activity are defined (*Federal Register* 54:28941, 1989).

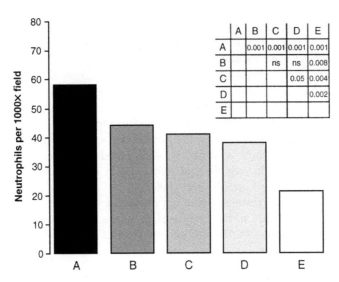

Figure 9 Comparison of name-brand and generic steroids: average number of neutrophils per 1000× field: (A) UVA only; (B) generic desoximethasone, 0.05%; (C) generic desoximethasone, 0.25%; (D) name-brand desoximethasone, 0.05%; (E) name-brand desoximethasone, 0.25%. Statistical analysis is shown in the inset. (From Ref. 169.)

In the following section some background into approaches that have been suggested to evaluate bioavailability or bioequivalence for topical products for the skin are discussed.

1. Grading of Disease States

Diseases that can be assessed and graded by objective means are rare in medicine. Thus, the definition of numerical severity indices to grade a disease is helpful and permits one to follow the course of a disease and to assess a particular treatment or treatment modality of the disease. Particularly in dermatology, scores and severity scales are widely used to grade a disease and to follow its course. A crucial part of any assessment is the choice of what is to be measured to assess each patient's response.

The most primitive form of measurement is the grouping of individual observations into qualitative classes. Such a nominal scale is used in all studies in which the patient's response is classified as "success" or "failure."

If the measurement classes can be arranged in order of magnitude (e.g., the overall grade of acne severity) (170), then one has an ordinal scale. In such a scale each observation can be placed only on one class and the classes can be arranged in order. Although ordinal measurements are often expressed as numbers, arithmetical manipulation may not truly summarize the response measured. That is to say, the mean score may be a poor way of summarizing responses measured on an improvement-rating scale; the median, a nonparametric measure of location, is usually better.

Quantitative measurements of lesion area, lesion thickness, or other properties, as determined by bioengineering techniques, duration to clearing, and such, add greatly to the discriminatory power of a trial, and an attempt should always be made to use them.

Frequently, the rating scales in dermatology are used only by their author, who assumes that the scale measurement correlates with the responses under investigation. In some areas of dermatological research important steps have been taken to develop a consensus on the most meaningful rating scales (e.g., for atopic dermatitis: *Consensus Report of the European Task Force on Atopic Dermatitis*; 171).

Lesion counts are a well-established method for assessing response in acne. A potential bioequivalence study design may require a double-blind, 12-week, multicenter, parallel study, including brand, generic, and vehicle in patients with mild-to-moderate acne (grade 2–3). The primary endpoint will be lesion counts (inflammatory and noninflammatory) and a global evaluation by the investigator. Patients are evaluated at baseline and at 1, 2, 4, 8, and 12 weeks. Irritation and sensitization rates will also be determined (*F-D-C Reports*. T&G-11, April 2, 1990; Topical antifungal and acne drug product bioequivalence workshop, Nov. 16, 1989, FDA, Rockville, MD). In several other studies the area, diameter, or thickness of lesion or the amount of skin involved, expressed as an "intensity score" or as a percentage of body area, have been used (172–174). Complicating factors include the need to instruct the patients thoroughly on the use of the drug; assessment of patient compliance (e.g., by periodically weighing the residues in the container); the influence of menstrual cycle in the course of acne; and the high placebo rate associated with acne treatment.

Published studies using photographic assessment are few, given the obvious dependence of dermatology on visual skills for assessment. Despite that the accuracy of photographic reproduction has been found unsatisfactory, presumably because the many variables are not easily standardized, attempts have been made to find a consensus (175,176).

In 1978 Fredriksson and Pettersson defined a "*Psoriasis Area and Severity Index*" (PASI) and successfully assessed clinical efficacy of three different dosing schedules of etretinate (177). The PASI score is based on severity and area of psoriatic lesions and is calculated by grading clinical symptoms (Eq. 1).

$$PASI = 0.1\overset{\text{head}}{(E_h + I_h + D_h)A_h} + 0.1\overset{\text{trunk}}{(E_t + I_t + D_t)A_t} +$$

$$0.1\overset{\text{upper extremities}}{(E_u + I_u + D_u)A_u} + 0.1\overset{\text{lower extremities}}{(E_l + I_l + D_l)A_l} \qquad (1)$$

For calculation of this, the four main body areas were assessed: the head (h), the trunk (t), the upper extremities (u), and the lower extremities (l), corresponding to 10, 20, 30, and 40% of the total body area, respectively. The area of psoriatic involvement of these four main areas (A_h, A_t, A_u, A_l) was given a numerical value: 0 = no involvement; 1 = <10%; 2 = 10–30%; 3 = 30–50%; 4 = 50–70%; 5 = 70–90%; and 6 = 90–100%. To evaluate the severity of the psoriatic lesions three target symptoms; namely, erythema (E), infiltration (I), and desquamation (D) were assessed according to a scale 0–4, where 0 means a complete lack of cutaneous involvement, and 4 represents the severest possible involvement.

This index was further modified by adding the grading of the clinical symptom "pustulosis" and by reducing the grading scale to 0–3 (178,179). These values can then be used for calculating a remission index (RI) giving the percentage improvement in the psoriasis at the time T_X, compared with the starting time T_O (Eq. 2) (180).

$$RI(\%) = \frac{\text{PASI } T_O - \text{PASI } T_X}{\text{PASI } T_O} \times 100 \qquad (2)$$

Despite that the PASI is easy to calculate, data collection remains inconvenient. The main difficulty is the grading of the severity of the clinical symptom and questions of reproducibility may arise. For example, the grading of the "erythema" depends highly on the grading of a previous observation. A score 2 will become 3 in the case of an aggravation and will become 1 in case of an improvement. However, it is impossible to say that a decline from score 3 to 2 depicts the same improvement as a decline from score 2 to 1. The problem of grading becomes even more complex when a scoring system is used in a multicenter study. The main problem of scoring systems in dermatology is the mixture of clincal parameters of very different nature that are more or less related and that entities of a very different nature are added and multiplied, therefore, may be questioned. The localization and extent of a lesion is easily assessed; however, the handicapping nature of the lesion remains in the dark. Therefore, it is of utmost importance that the scoring system of clinical signs is supplemented by patient judgments (e.g., pruritus, sleep loss, pain, or other). Percentages of change for an individual or a group expressed as numeric values from clinical-scoring systems, therefore, should be supplemented by patient judgments and, whenever possible, by additional means (colorimetry, histology, photography, or other). An excellent example is the study design of Kimbrough–Green et al. (181) for which the topical vehicle-controlled tretinoin treatment of melasma in black patients was studied. The melasma area and severity index (MASI) was supplemented by colorimetric, photographic, and histological means.

Despite the problems inherent in scoring (scoring systems are usually developed for a particular disease), these systems, when standardized, are valuable in assessing clinical signs more accurately than simple global evaluation. In some cases the scoring systems have been used to study the influence of pharmaceutical strategies (vehicle effect) by evaluating the pain-producing or pain-releasing effect [pain questionnaire (182); pinprick test (183)].

2. Allergic Contact Reaction

The induction of allergic contact dermatitis lesions in hypersensitive subjects (humans and animals) are replicas of the natural pathology (184). Therefore, multiple simultaneous experimental and therapeutic comparisons are possible in a double-blind, randomized manner.

In a recent study, the immunosuppressive effect of tacrolimus (FK 506), cyclosporine, dexamethasone, and clobetasol on dinitrofluorobenzene-induced contact dermatitis was tested in the domestic pig (Table 1) (185). Test sites were treated 0.5 and 6 h after challenge with test solutions of drug vehicle. Efficacy of each preparation was determined clinically for intensity, extent of erythema, and consistency. In addition, skin changes were biophysically characterized by measuring microvascular perfusion and skin color changes.

The data clearly show that the topical administration of the macrolide antibiotic tacrolimus (FK 506) and the corticosteroids have an immunosuppressive effect. Even though a weak dose–response relation is seen with FK 506 and clobetasole, the differences are not statistically significant over a dose range of 10.

Table 1 Influence of Topical Treatment with Drug and Placebo Formulations on the Allergic Contact Dermatitis Test Sites in Domestic Pigs

	Efficacy of drug formulation (% inhibition vs. placebo)		
Drug	Clinical evaluation	Skin color	Microvascular perfusion
FK 506, 0.4%	72	77	64
FK 506, 0.13%	66	68	51
FK 506, 0.04%	75	51	40
Cyclosporine, 10%	9	8	4
Dexamethasone, 1.2%	31	38	25
Clobetasole, 1.2%	53	87	44
Clobetasole, 0.13%	44	76	37

Source: Ref. 185.

One may conclude that numerous intraindividual, anti-inflammatory activity comparisons performed on a natural, controlled inflammatory disease using bio-physical techniques is suitable both to determine drug delivery to the target sites in the skin and to assess conventional therapeutic schedules. However, it seems unlikely that minor changes in formulation will be detectable by these means.

In purely clinical situations, the influence of the vehicle and its ingredients on the allergic contact dermatitis reaction has been observed (186). Several reports describe contact sensitization to topical acyclovir products (187–189). Extensive patch testing revealed that the influence of the vehicle and its ingredients may be responsible for the skin reaction. Baes et al. (190) describe two patients who had positive patch test reactions to Zovirax (acyclovir) cream, but were negative to all its constituents as well as to Zovirax ointment (paraffin as base). This phenomenon, which is defined as *compound allergy* and refers to contact allergy to a preparation with negative patch tests to all its constituents, is still rarely observed. However, the observations made so far illustrate the necessity of patch-testing patients with suspected contact allergy to their own topical products, as well as to their constituents, to avoid missing compound allergy.

3. Infectious Conditions

a. Viral Infections. The topical delivery of antiviral compounds formulated in various preparations has been studied in several cutaneous herpes models.

Eight different vehicles that were appropriate to deliver drug to the herpes simplex virus-2 (HSV-2)-infected mucocutaneous surface of the vagina were studied in mice and guinea pigs that were infected intravaginally with HSV-2. Efficacy of the formulations were determined using survivor increase, mean survival time increase, and reduction in mean vaginal lesion (Fig. 10).

To halt the viral replication and lateral spread in the basal layer during the prodromal stage, various formulations were used to deliver interferon to cutaneous herpes simplex 1 (HSV-1) lesions (192). The virus was inoculated with a spring-loaded vaccination instrument. To quantitate the effects of formulation factors on therapeutic effect a scoring system was used on a daily basis for up to 10 days. From

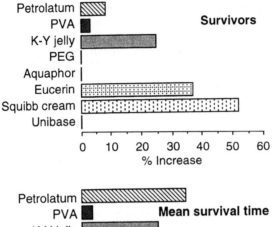

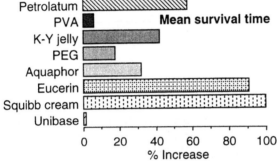

Figure 10   Comparison of effects of vehicles on 1% 2-acetylpyridine-semicarbazone against HSV-2 infections of mice. (From Ref. 191.)

the vehicles used (solution, emulsions, multilamellar and large unilamellar liposomes) only liposomes prepared by the hydration–rehydration method were effective. The authors hypothesize that the dehydration and subsequent rehydration of the liposomes facilitate partitioning of interferon into liposomal bilayers, where the drug is positioned for transfer into the lipid compartment of the stratum corneum. They further note that liposomes do not appear to function as permeation enhancers, but seem to provide the physicochemical environment needed for transfer of interferon into the skin. This report describes a therapeutic model in which the influence of a particular preparation technique (dehydration–rehydration) of a topical semisolid formulation was observed.

A novel method based on the measurement of free drug concentration (C^*), or thermodynamic activity of drug at the skin target site, has been developed (193,194) to assess bioavailability and to predict efficacy for antivirals and other dermatological formulations. To measure C^* following the administration of a topical formulation entailed, first, establishing a correlation between the steady-state dermal drug flux and an elicited efficacy. This was accomplished using a novel animal model in which hairless mice were infected in a small spot at a lumbar skin area with cutaneous HSV-1 (195,196). This induced, 3-days postinoculation, a narrow band of skin lesion development along the peripheral neural path toward the spinal cord. Taking advantage of this unique pattern of lesion development, an antiviral agent, such as acyclovir, was applied to a laurocapram (Azone)-pretreated skin area, dorsal to the virus inoculation site and in the predicted path of lesion development, to limit the lesion development. Five days after virus inoculation, the lesion development was scored for each mouse and two different antiviral efficacy parameters were assessed separately: (a) "topical" (local) efficacy, measured the antiviral activity of acyclovir delivery topically to the local skin area directly under the drug application site; and (b) "systemic" efficacy, measured the antiviral activity of acyclovir delivery through systemic circulation to the target site, presumably the epidermal basal layer (197). To quantify drug flux, a transdermal delivery system was developed in this animal model and the amount of acyclovir delivered to each infected animal could be controlled during the time period of drug therapy through a rate-controlling membrane. The actual (experimental) flux was determined at the end of each in vivo experiment by carrying out an extraction of the residual acyclovir in the transdermal delivery system. This extraction assay served to validate the expected (theoretical) flux or, alternatively, provided the bounds of uncertainty to the drug flux in the particular experiment. The results clearly showed a quantitative relation between the antiviral efficacy and the experimental flux of acyclovir obtained from in vivo experiments. Topical efficacy increased with increasing acyclovir flux in the range of $10–100$ $\mu g/$ $cm^2 day^{-1}$. Given the relatively high precision of topical efficacy results, it is believed that the quantitative nature of this animal model should be valuable in screening of new antiviral agents for topical treatment of cutaneous herpesvirus infections and the optimization of topical formulations. Two factors may limit the applicability of this elegant approach to other classes of dermatological formulations. First, the drug was delivered from a transdermal delivery system at a constant rate over several days (with semisolid formulations, formulation application and formulation changes are difficult to control) and, second, in the foregoing approach, the target for acyclovir (cutaneous HSV-1) and the localization of the target (epidermal basal layer) is known. Unfortunately, the target sites for many dermatological agents are still unknown.

b. Fungal Infections. In March 1990, the Division of Anti-infective Drug Products and the Division of Bioequivalence published a draft guidance for the performance of a bioequivalence study for topical antifungal products (198). During the course of developing a new antifungal drug, a minimum of two well-controlled clinical trials are required to show that the product is superior to the vehicle control. For a generic product, a single, randomized double-blind, three-way parallel group study has been proposed, comparing the test drug with the innovator, as well as with the vehicle used for the generic product. The vehicle control helps eliminate investigator bias and permits a better measure of the effectiveness of the generic product. If the reference product is approved for tinea pedis, the patients should have the disease, as confirmed by clinical examination and mycological testing. If the clinical efficacy is confirmed, then the generic product may be used for all of the conditions of use specified for the innovator product. The evaluation is to be conducted 4 weeks after starting treatment, at the time treatment is discontinued, and 2 weeks later. This last evaluation is the most important in determining bioequivalence. Assuming the test and reference products have identical success rates of 50%, it is estimated that at least 108 evaluable patients will be needed for the test and reference products, and 60 patients will suffice for the vehicle dose, as it should be easier to show a difference between active product and vehicle than between active and active product. If the success rate of the test drug is different from the reference drug (but still within the equivalence region), more patients will be needed.

If the reference drug is approved for tinea pedis, the patients enrolled should be infected with tinea pedis, proved by clinical examination and positive mycological testing. Patients may be entered into the study and randomized on the basis of clinical presentation and a KOH examination. Cultures should be taken, and only those patients with clinical signs of tinea pedis, a positive KOH, and a recovered dermatophyte will be evaluated. Clinical and mycological examination (KOH and culture) should be performed at the end of a 4-week treatment period and at 2-weeks posttreatment.

The evaluation of therapeutic equivalence will be based on the percentage of patients in each group who have a clinical cure and a mycological cure. A clinical cure means that the signs and symptoms of the disease have cleared. Both the KOH examination and culture must convert to negative to qualify as a mycological cure. Each of the clinical and mycological outcomes with the reference drug must be statistically comparable with those of the test drug. In addition, the cure rates for the test and for the reference drugs must each be statistically superior to that of the vehicle control for each measure of outcome for the study to be a valid evaluation of the test product. Although these comparisons should be evaluated at the end of treatment and at the 2-week follow-up visits, primary weight will be given to the 2-week follow-up evaluation in determining whether bioequivalence has been established.

A well-documented example of the foregoing types of studies that are appropriate in evaluating new formulations (199) involves bioequivalence studies of a lotion and a cream formulation of the antifungal drug cyclopirox. During the development of the products, in vitro studies were carried out using both human cadaver and domestic pig skin treated with two formulations (cream vs. lotion). In vivo studies included an animal model, during which guinea pigs were inoculated dorsally with *Trichophyton* (*tinea*[*sic*])*mentagrophytes* and after 3 weeks, the resulting lesions were treated with the lotion, cream, and the two vehicles. This test was followed by a human model, where four sites on the forearms of human volunteers were inoculated with *T. mentagrophytes*. Each site was treated with the lotion, cream, or the two vehicles. After 15 days of treatment the sites were examined clinically and mycologically. Finally, a multicenter, double-blind, parallel-group clinical trial was conducted comparing the lotion with the vehicle in patients with plantar, interdigital, or vesicular tinea pedis. Assessment was based on criteria that basically furnished the base of the FDA guidance (see foregoing).

The efficacy of cyclopirox (ciclopirox olamine) lotion 1% was clearly manifest in patients with interdigital tinea pedis. The plantar form, however, is difficult to manage and usually requires a longer duration of therapy than was provided in the trial.

c. Bacterial Infections. Topical and systemic antibiotic therapy is common in dermatology. Apart from considerations on the development of antibiotic resistance (200) or the development of allergic contact dermatitis (201), it is difficult to find a rationale for a particular drug administration route in some diseases. In several comparative studies on antibiotic therapy of impetigo (*Staphylococcus aureus* or *Streptococcus pyogenes*), the superiority of topical or systemic treatment regimens could not be found (202,203). In a rare trial, in which topical or oral tetracycline was allotted to acne patients at random, Schwanitz et al. (204) found no significant differences between the two routes of therapy. Since 1982, when Eady and associates (205) in an exhaustive review, felt that the question of topical versus oral acne therapy remains open, no new information can be added to prove the superiority of topical or systemic antibiotic treatment regimens.

4. Local Anesthesia

Topically induced anesthesia is of great value in clinical practice, particularly in pediatric patients. The need for the development of an effective topical anesthetic preparation prompted researchers to try various approaches.

The pharmaceutically unique eutectic mixture of the local anesthetics lidocaine and prilocaine provides dermal anesthesia–analgesia following topical application. The term *eutectic* refers to the phenomenon whereby the melting point of the two anesthetics is lower after mixing than that of either anesthetic alone. When lidocaine and prilocaine are mixed at 25°C, an oily liquid is formed that can be emulsified in water (1:1). Thus, eutectic lidocaine–prilocaine cream is an emulsion, with the mixture of local anesthetics as the dispersed phase and water as the continuous phase (Broberg/Evers, 1981; EU patent no. 0,002,425). Whereas the cationic form of a local anesthetic is responsible for blocking neurotransmission, it is the uncharged (base) form that diffuses into the tissues after topical administration. Because skin penetration is facilitated by the presence of an aqueous solvent, absorption of the anesthetic mixture is enhanced by the high water content. Unlike other topical local anesthetics, which are generally limited to use on mucosal membranes, eutectic lidocaine–prilocaine cream achieves effective dermal anesthesia or analgesia following application to intact skin under an occlusive dressing. The principal indications in which eutectic lidocaine–prilocaine cream have been successfully studied are the management of pain associated with venipuncture, intravenous cannulation, curettage of molluscum contagiosum lesions, and split-skin graft harvesting (206).

The standardized pinprick test (183,207) has usually been used to determine effectiveness of topical anesthetic preparations. Kushala and Zatz (208) successfully applied a new method to assess the anesthetic activity of various topical lidocaine formulations. A commercially available electrometric device to determine tooth pulp vitality can also be used to stimulate the nerves within the skin. The device delivers low-current electric stimulation in the form of bursts of pulses. Contact of the probe with the skin initiates a series of low-voltage bursts; successive bursts are of incrementally higher voltage (≥ 300 V). The voltage is transposed into arbitrary units from 0 to 80. The digital instrument reading is proportional to voltage. To apply this technique to humans, the subjects have to be instructed in the testing technique. Briefly, under supervision, the subject places the device against the skin of the forearm. The subject removes the probe from contact with the skin when a sensation, usually described as a mild buzz, is first felt. When the subject is comfortable with handling the device, the consistency of their response is evaluated. Subjects exhibiting inconsistent or nonreproducible results are rejected. This technique may be considered to be a promising alternative to the conventional pinprick-test (Fig. 11).

The comparison of the mean effect values, determined electrometrically, yielded one group of inactive formulations (two placebos) and two overlapping groups of active lidocaine formulations. Statistical analyses indicate that all the li-

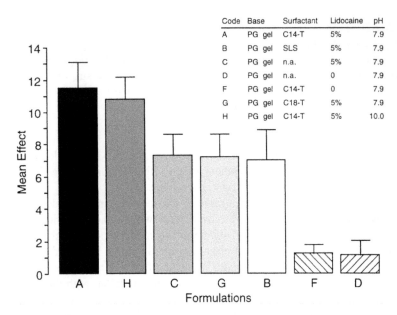

Code	Base	Surfactant	Lidocaine	pH
A	PG gel	C14-T	5%	7.9
B	PG gel	SLS	5%	7.9
C	PG gel	n.a.	5%	7.9
D	PG gel	n.a.	0	7.9
F	PG gel	C14-T	0	7.9
G	PG gel	C18-T	5%	7.9
H	PG gel	C14-T	5%	10.0

Figure 11   Mean effect for one lidocaine cream, five lidocaine suspension formulations, and two placebo formulations: C14-T, tetradecyltrimethylammonium bromide; SLS, sodium lauryl sulfate; C18-T, octadecyltrimethylammonium chloride. Bars indicate SEM. (From Ref. 209.)

docaine-containing suspensions performed at least as well as the plain formulation, which contained no surfactant. Changing the pH also had no significant effect on the in vivo performance. The pharmacodynamic data were compared with in vitro permeation data using human cadaver skin. The rank order for the suspension formulations was the same as that for steady-state permeation in the in vitro experiments. Application of the cream formulation produced greater effect in vivo than was anticipated from the in vitro flux values (209).

Pain associated with herpes zoster and postherpatic neuralgia has been difficult to manage. None of the many suggested medical, anesthestic, physical, or surgical methods have been widely accepted. Topical applications of analgesic and neuroleptic agents formulated in a variety of different vehicles, have recently been reported with greater frequency. Their efficacy, usually determined with a pain questionnaire, has, however, varied widely (210).

IV. FACTORS INFLUENCING THE RATE AND EXTENT OF DRUG ABSORPTION FROM TOPICAL PRODUCTS

Factors from the following three areas may potentially influence the rate and extent of drug absorption from topical products: (a) the skin conditions, (b) the composition of the topical vehicle, and (c) the application procedures or application conditions, because these affect the interaction of the topical vehicle with the skin. These have been reviewed to identify the primary factors responsible for variation in bioavailability, bioinequivalence, and therapeutic inequivalence of topical products.

A. Effect of Skin Conditions

1. Skin Microflora

The skin surface supports a microbial population that has the potential to carry out biotransformations of topically applied therapeutic agents. At present, there appears little in vivo evidence to suggest that the microbial transformation of compounds applied topically for percutaneous absorption have any greater significance than the metabolic action of the skin itself (211–215).

2. Skin pH

The pH of topical vehicles affects the extent of dissociation of ionizable drug molecules and, thus, their thermodynamic activity, partitioning, and skin penetration (216–218). Normal human skin has a surface pH of 4–6. A pH gradient exists within the skin (83). The influence of the pH of the skin surface or in the skin on penetration of an ionizable xenobiotic has not been directly studied. The use of pH-sensitive (or temperature-sensitive) delivery systems (e.g., liposomes) has been suggested and may also offer new therapeutic approaches in dermatology (219).

3. Skin Surface Lipids

The skin possesses sebaceous glands that secrete a mixture of lipids that form an irregular 0.4- to 4-μm–thick film on the skin surface (220–223). Recently, Cheng et al. (224) investigated the role of skin surface lipids on lidocaine permeation from pressure-sensitive adhesive (PSA) tapes and found that the skin lipids can dissolve solid drug in the PSA tape to decrease its thermodynamic activity and thus drug permeation.

4. Temperature

Temperature changes on or in the skin are always accompanied by other physiological reactions, such as increased blood flow, or increased moisture content of the horny layer. These factors themselves can contribute to higher percutaneous absorption. Furthermore, increase in temperature increases drug solubility in both vehicle and stratum corneum and increases diffusivity, both of which will lead to a further increase in percutaneous absorption (225,226). In clinical practice, temperature may be of minor importance, but in occupational dermatology, temperature may be an important stimulus (227,228).

5. Blood Flow

The transepidermal resorption process feeding into the cutaneous microcirculation brings compounds into the underlying tissues or the systemic circulation. Early studies with thyroxine and steroids (229,230) showed that the cutaneous vasculature apparently does not function as an infinite sink that transports all topically applied drugs to the systemic circulation. Part of the drug accumulates in the superficial dermis and diffuses into deeper parts of the skin (231–233). Under these conditions, cutaneous blood flow can modify the concentration levels and the accumulation of substances in the dermis or deeper parts of the skin.

Vasoactive drugs (topical and systemic), or blood flow decrease obtained by ligation, can modulate the transdermal delivery of drugs (234–240).

6. Diseased and Damaged Skin

Despite the limited data available, the belief that damaged skin enables chemicals to enter the body unhindered is certainly false. The data suggest that differences in absorption will exist for different drugs and for different disease conditions (16,241–250). In patients with erythroderma—a most severe form of cutaneous inflammation—percutaneous absorption of hydrocortisone ranged from 4 to 19%, compared with approximately 1% in normal skin (24). In psoriasis, definite formulation effects were observed (251). With treatment, a skin condition usually changes and thus affects drug delivery into the skin (252). Damage to the skin by tape-stripping of the stratum corneum leads to an approximately twofold increase in the permeation of hydrocortisone (103).

7. Influence of Anatomical Site

The influence of skin site on the extent of absorption of compounds has been reported (139,253–257). For example, the classic study of Feldmann (139) shows anatomical variation in the percutaneous absorption of hydrocortisone of over two orders of magnitude (foot arch 0.17%, forearm 1.04%, jaw angle 12.25%, scrotum 36.2% absorption). These studies are based mainly on the determination of a drug concentration in an accessible body fluid, usually urine. Pharmacodynamic studies, however, reveal an unequivocal picture. Hansen et al. (258) examined the response of volunteers to the application of nitroglycerin ointment at three body sites. Of the three sites studied (midforehead, left lower anterior chest, and medial left ankle), the forehead site uniformly produced the greatest response in terms of magnitude and time of onset of changes in systolic blood pressure and the incidence of subjective adverse effects. Moe and Amstrong (259) studied the influence of skin site on bioavailability of nitroglycerin using pharmacokinetic and pharmacodynamic parameters. Nitroglycerin ointment was administered to patients, who had severe congestive heart failure, in a randomized fashion to three skin sites: arm, chest, or thigh. Hemodynamic parameters and the arterial nitroglycerin concentrations were measured. No significant differences among the sites were observed in mean arterial pressure, left ventricular-filling pressure, right arterial pressure, or nitroglycerin concentration. This information confirms earlier pharmacokinetic data from Noonan and Wester (260), who studied percutaneous absorption of $[^{14}C]+$ nitroglycerin through a uniform area of the chest, arm, inner thigh, and postauricular region of the rhesus monkey. No difference in percentage absorption was observed among these skin sites. However, compared with many topical agents, nitroglycerin is well absorbed, and this will tend to reduce differences in extent of absorption between skin sites. A recent report on the percutaneous absorption of ketoprofen in healthy volunteers showed that absorption was similar when applied to either the back or the arm, but was lower when applied to the knee (261).

In the management of psoriasis with topical glucocorticosteroids, any given steroid formulation commonly results in a distinct grading of response of the psoriasis lesions, based on region. The dorsa of the feet and hands, elbows, and knees respond poorly, but better than the palms and soles, which rarely respond. Lesions on the upper thighs respond better than lesions on the lower legs; lesions on the chest better than those on the upper arms; and those on the face best of all (262).

Thus, some of the quantitative measurements of permeation through various regions correlate well with the clinical observations of regional response to topical gluco-corticosteroids. Nevertheless, other explanations may be possible. The number of "receptors" or sensitivity of receptors at the target site may be the main determinant of the pharmacodynamic response.

Overall, variation between and within anatomical sites is likely to be the most important "skin condition" factor influencing extent of absorption. Bioequivalence studies are almost always conducted on a single skin site, often the forearm. If we assume bioequivalence is established, the large variation in extent of absorption between the skin site used in the bioequivalence model and the clinical setting gives rise to potential for therapeutic inequivalance, unless the change in the extent of absorption with the skin site is the same for all products.

8. Influence of Appendages

Evidence to support or refute the hypothesis that appendageal pathways are significant can be identified. The arguments against—which are based on the surface area —are persuasively illustrated by historic examination and by various numerical calculations. On the other hand, studies using (a) autoradiography, (b) an animal model that totally lacks appendageal shunts, and (c) an alternative data analytical approach, all point to the significant participation of the appendages in the transport (particularly, the non–steady-state portion) of molecules across the skin. The importance of these observations and the clinical relevance (e.g., multiple application) remains vague. To truly ascertain the significance of follicular delivery, appropriate models and quantitative methods must still be developed (220,263–290). Others have made the case for direct drug targeting of the pilosebaceous glands and have developed special formulations to optimize this delivery route (269,291).

9. Effect of Skin Metabolism

Far from being a passive membrane for the passage of drugs, the skin has significant metabolic activity. As pointed out in several reviews, metabolic activity spans a broad range of oxidative, reductive, hydrolytic, and conjugative reactions (292,293), making the skin a source of extrahepatic metabolism of many xenobiotics and topically applied drugs. The transport and metabolism of drugs in the skin places two kinetic events in competition. Results of experimental (294) and theoretical (295–298) investigations have suggested that diffusional and metabolic processes in the skin are intimately related, with one often having a profound effect on the other (15,299).

As the physicochemical properties of drugs themselves are not always optimal for their delivery into or through skin, topical administration is not always possible. An elegant approach for overcoming these problems is to make a transient derivative —a prodrug of the drug—which imparts to the drug the desired transient change in its physicochemical properties (300). Once the prodrug has overcome the skin barrier, it is metabolized to the parent compound of established safety and efficacy.

Although the prodrug concept involves the chemical modification of a known pharmacologically active compound into an inactive or less active bioreversible form that becomes metabolically active within the skin layers, the softdrug concept involves metabolic deactivation of an active compound within the viable layers or in the systemic circulation (301).

10. Effect of Age

Despite much research into the mechanism of cutaneous aging and the identification of significant age-associated biological and biophysical changes within the skin, the question, "how does aging affect percutaneous absorption in vivo?" remains partly unanswered. Some data suggest that the diminished surface lipid content of old "skin" implies a diminished dissolution medium for compounds administered topically. It is reasonable to speculate that this physiological change most severely affects those permeants for which lipid solubility is low (302). Biological effect is generally decreased in the aged individual (303); therefore, pharmacodynamic parameters, suggesting reduced effect or penetration, have to be used with care. Skin permeability is greater in premature (or newborn) infants (304). Although there is certainly a potential for age-related effects to result in bioinequivalence in the premature or newborn for two formulations that appear equivalent in healthy adults, there is no direct evidence of any such occurrences (305).

11. Effect of Race

Racial differences in physicochemical properties of the skin, in in vivo percutaneous absorption and to various chemical stimuli have been reported (306–309); however, the data is not unequivocal. Except for one study (310), that indicates racial differences in nitroglycerin absorption after transdermal application, one may state within the limitations of the experiments that, if any differences exist, they are below the limit of detection of the currently established methods. Currently, no information is available on the responsiveness of a specific disease to topical therapy in different races.

12. Difference Between Subjects

Large differences exist in percutaneous absorption between subjects (140) as a result of drug–vehicle–skin interaction. This can result in subject–formulation interactions, such that the bioavailability of a test product relative to that of a reference product may be consistently different between individuals (311). The resulting issues of switchability and prescribability have been discussed elsewhere (29,312).

B. The Vehicle Effect

A vehicle can be defined either by its structural matrix (e.g., ointment, cream, gel, liposome, or other) or by its excipients (e.g., petrolatum, propylene glycol, isopropylmyristate, water, and the like). When discussing vehicle effects it is important to clearly distinguish between these two terms; for example, the data from Table 2 compares the potency of various 0.5% betamethasone dipropionate products (313). It is uncertain which contribution of matrix' or excipient's effects are responsible for the differences between formulations, although the excipient's effects will usually override.

1. Structural Matrix

The question is often asked whether a cream, a gel, an ointment, or a liposome is better to deliver a drug to the skin, to promote the drug's penetration and therapeutic effect. Certainly, cosmetic aspects of the delivery system, a result of the structural matrix, may have an influence on compliance and, they therefore, are of clinical

Table 2 Comparative Potency of Various 0.5%
Betamethasone Dipropionate Products

Drug	Vehicle type	Potency group (U.S.)
Diprolene	Ointment	I
Diprolene AF	Optimized cream	I
Diprosone	Ointment	II
Diprosone	Cream	III
Diprosone	Lotion	V

Source: Ref. 313.

relevance. Moreover, ointments or heavy creams may occlude the skin and increase drug penetration. However drug delivery to the skin is controlled by the vehicle excipients, as these effect partitioning into and diffusion through the stratum corneum, as described later. It remains speculative whether one can assign particular structural features of a vehicle to a specific effect. Furthermore, it is difficult to formulate a vehicle with the same ingredients and different structural features (314–316).

2. Formulation Excipients

The effects of formulation excipients on the rate and extent of drug absorption are greater with topical drug delivery than with any other route of drug administration. For example, comparing alternative formulations of the same drug, differences in extent of penetration of 10- to 50-fold and higher have been reported (317–320). To put this into perspective, 50–100% (up to onefold) differences in extent of absorption by the oral route are rare (7–9).

    The potential for large differences in the extent of absorption between topical formulations is due to the complex interactions between the drug, the vehicle, and the skin, which control partitioning into and diffusion through the stratum corneum barrier. However, it is the low extent of drug absorption under "normal" conditions, often in the range of 1–5% of dose applied, that allows the potential of these interactions to be realized as large differences in extent of absorption. Thus, one may consider the low extent of absorption—low absolute bioavailability—as being a significant factor leading to the potential for topical bioinequivalence and therapeutic inequivalence. The case for the use of rational dosing—reduced drug concentration in topical formulations—is developed in this chapter.

    Literature describing the effect of formulation on the rate and extent of percutaneous absorption is enormous. Here, we will only try to explain, on the basis of a simple physical model, how formulation may influence the various parameters that control the percutaneous absorption process and, thereby, affect bioavailability, bioequivalence, therapeutic equivalence, and potentially result in bioinequivalence and therapeutic inequivalence. For this latter point, which will be reviewed in more detail in Sec. IV, we will need to consider how formulations that have been declared as bioequivalent (i.e., in a volunteer study), may interact differently in the true clinical setting, for which variation in skin conditions, particularly anatomical site, and in dosage requirements of individual patients may exist.

In the following discussion, only transepidermal transport (i.e., through the stratum corneum) as distinct from transappendageal transport, will be considered. Relatively little is known about the effects of the vehicle on absorption by the appendages (see Sec. IV.A.8), although some of the factors discussed are also likely to be relevant to this route (321).

a. The Higuchi Physical Model. Figure 12 describes percutaneous absorption as a passive diffusional process through parallel membranes (321). By *passive*, it is meant that the absorption process is not energetically coupled to any biological process. Higuchi (321) considered two main examples of this general model: one, in which the rate-controlling step is in transport across the skin; the second, in which the rate-controlling step is in release from the topically applied formulation, as discussed in Sec. III.A.1. Figure 12 (right side) describes the concentration profile of the drug across the skin for the common case when the rate-controlling step is in transport across the skin and the barrier is in the stratum corneum.

The various cases of the Higuchi model, and particularly the subcases of case 1, are shown schematically in Fig. 13. Although it is useful to understand the logic of the entire Higuchi model, only the most common case (1.1.), when the barrier to absorption is in the stratum corneum, will be considered in detail. Case 2, in which the rate-controlling step is in the formulation, is relevant to only topical therapy when the skin barrier function is severely damaged; for example, severe burns. Case 1.2., for which the rate-limiting step is within the viable epidermis–dermis, applies only when the permeant is extremely hydrophobic (322,323).

b. Drug Structure–Permeation Relation. The stratum corneum barrier has been modeled as a bricks-and-mortar structure (324) of corneocyte bricks held to-

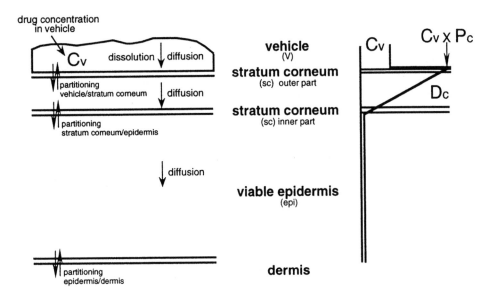

Figure 12 Schematic of percutaneous absorption as a diffusional process across parallel membranes. The right-hand side shows changes in drug concentration when the stratum corneum is the rate-limiting barrier. The driving force for diffusion is the drug concentration in the outer layer of the stratum corneum ($C_v \times P_c$).

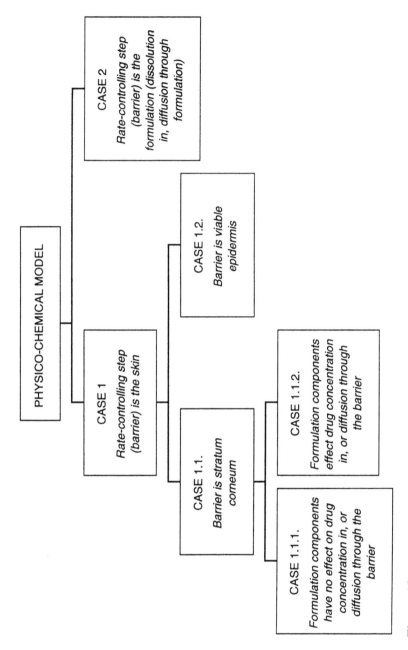

Figure 13 The hierarchical structure of the Higuchi physical model of percutaneous absorption. (From Ref. 321.)

gether with an extracellular lipid mortar. Most agents are believed to permeate the stratum corneum by the extracellular lipid route. The intrinsic permeation of a drug through the stratum corneum lipids is dependent on several physicochemical parameters (325), and it is useful to classify these into those factors that affect solubility in the stratum corneum lipids, such as partition coefficient and drug-melting point, and those that affect diffusivity through the barrier, such as molecular volume and hydrogen-bonding potential (326). Small, low-melting–point, hydrophobic compounds, such as nicotinates, salicylates, and nitroglycerin are, thus, relatively well absorbed; accordingly, they are less likely to suffer from problems of bioavailability and bioequivalence. Large, crystalline compounds, such as corticosteroids and retinoic acid derivatives, even though such structures are generally hydrophobic, are relatively poorly absorbed to the extent of 1–5% of dose applied; consequently, problems of bioavailabilty and bioequivalence may, and do, occur (327). Large, crystalline polar compounds (e.g., peptides) are predicted to be poorly absorbed and prone to problems of bioavailability.

 c. *Interactions Between Drug, Vehicle, and Skin.* Equation (3) describes those factors that control percutaneous absorption, based on case 1.1. (see Fig. 13) under conditions of nondepletion of donor compartment and sink conditions in the receptor compartment.

$$\frac{\delta_Q}{\delta_t} = \frac{C_v \times P_c \times D}{L} \tag{3}$$

where, δ_Q/δ_t is the steady-state rate of percutaneous absorption per unit area of skin; C_v, the drug concentration in solution in the vehicle; P_c, the partition coefficient of the drug between the vehicle and the stratum corneum; D, the diffusion coefficient of the drug in the stratum corneum; and L, the apparent thickness of the stratum corneum. For convenience, these interactions can be considered to fall into two areas, drug–vehicle (thermodynamic) effects and vehicle–stratum corneum (penetration enhancer) effects. In Eq. (3) the term C_v is an approximation for $[C_v(t = 0) - C_v(t = t)]$ that is used under conditions of nondepletion–infinite dose where $C_v(t_0 - t)$ approaches zero.

 Thermodynamic effect. The driving force for diffusion is the thermodynamic activity of the drug in the vehicle which, in turn, is related to C_v and to the activity coefficient of the drug in the vehicle. Thermodynamics being an illusive discipline, it is best to consider C_v and P_c (which is proportional to the reciprocal of drug solubility in the vehicle) as being the factors that control drug–vehicle interactions. The higher the value of the product of $C_v \times P_c$, the higher the thermodynamic activity and the higher the drug permeation.

 Effect of drug concentration in solution. From Eq. (3), in a single vehicle, thus, under conditions of constant P_c and D, change in C_v is predicted to produce a corresponding linear change in permeation. For example, in Fig. 14, Davis and Hadgraft (67) showed a linear increase in hydrocortisone acetate flux across a Silastic membrane as drug concentration in a single vehicle of propylene–water (56:44) was increased up to saturation at 0.02%. Many similar results have been reported, including the early classic studies of Poulsen et al. (328,329) and more recently (65,318,330).

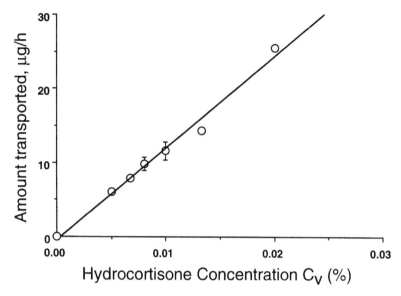

Figure 14 Transport of hydrocortisone acetate from a single propylene glycol–water (56: 44) vehicle across Silastic membrane. Demonstration of response to C_v. (From Ref. 67.)

Effect of vehicle partition coefficient. Relatively few studies have systematically investigated the effect of vehicle partition coefficient for a fixed C_v, although P_c is equally as important as C_v in determining the driving force for diffusion, the drug concentration in the outer layer of the stratum corneum. Saturated solubility in the vehicle, and thus P_c, may vary over several orders of magnitude (331). Davis and Hadgraft (67) showed a linear increase in hydrocortisone acetate flux across a Silastic membrane as a partition coefficient (expressed as reciprocal of saturated solubility) was increased for a fixed drug concentration of 0.02% (Fig. 15).

Flynn and Smith (65) found that, for a fixed concentration of *p*-aminoaceto-phenone, transport across a Silastic membrane increased linearly with increase in partition coefficient (reduction in propylene glycol fraction) up to the point at which saturation of the vehicle occurred. A further increase in partition coefficient (by a reduction in the propylene glycol fraction) resulted in formation of drug suspensions that showed no further increase in transport. It may be possible, for a fixed C_v, to reduce the partition coefficient, but to still maintain a drug in solution, such that supersaturated solutions are formed (67,317). Figure 16 shows that transport of 0.02% hydrocortisone acetate across a Silastic membrane, from saturated to eightfold supersaturated solutions, is linearly dependent on the degree of saturation, or at a fixed value of C_v, dependent on the partition coefficient of the vehicle. In terms of bioavailability and bioequivalence, for a fixed value of C_v, differences in thermodynamic activity (otherwise expressed as partition coefficient or solubility of the drug in the vehicle) may result in large differences in the extent of absorption and thus bioinequivalence.

Saturated solutions. C_{skin} is the driving force for the diffusional process of percutaneous absorption and

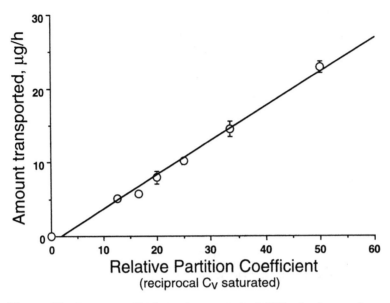

Figure 15   Transport of hydrocortisone acetate, 0.02% w/w, from various propylene glycol–water vehicles across Silastic membranes. Response to vehicle partition coefficient expressed as reciprocal saturated solubility. (From Ref. 67.)

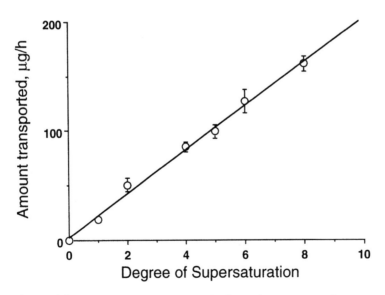

Figure 16   Transport of 0.02% w/w hydrocortisone acetate from supersaturated vehicles across Silastic membrane. Response to degree of supersaturation. (From Ref. 67.)

$$C_{\text{skin}} = C_v \times P_c \tag{4}$$

Under stable, equilibrium conditions, flux will be at a maximum when the outer layer of the skin is saturated and, by definition, this will occur when the vehicle is also saturated with solute. From Eq. (4), it is apparent that there is an inverse relation between the saturated solubility of the solute in the vehicle and the partition coefficient between that vehicle and the skin, such that for all saturated systems, their product is a constant: that is, the saturated solubility of the solute in the stratum corneum (328).

$$C_{v(\text{saturated})} \times P_c = \text{constant} \tag{5}$$

Equations (4) and (5) are the basis for the statement commonly found in the literature that all saturated vehicles of the same permeant will give the same flux, independently of concentration, provided that the vehicle components do not alter the barrier function of the skin and that significant depletion of the permeant from the vehicle does not occur.

Many authors have published experimental data to support this argument (65,67,328–330,332,333). Figure 17 shows that transport of methylparaben through Silastic membranes is the same for all saturated solutions despite a 95-fold difference in drug concentration in solution.

Figure 18 shows data (333) on the vasoconstrictor response of betamethasone benzoate from five vehicles of differing ratios of mineral oil:myglyol (85:15; 60:40; 40:60; 20:80; 10:90, respectively). Addition of myglyol (a triglyceride) increases saturated solubility of betamethasone benzoate and decreases P_c. In Fig. 18, open symbols represent subsaturated solutions and filled symbols the single, saturated solution and suspension systems for each of the five vehicles. This data allows an excellent summary of the importance of drug–vehicle interactions. First, in relation to the potential for bioinequivalence, for a fixed C_v, vasoconstriction is strongly

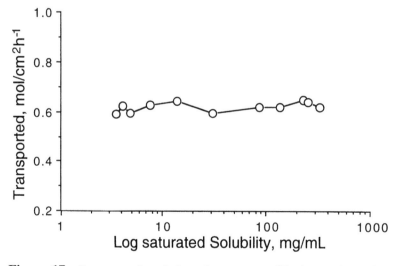

Figure 17 Transport of methyl paraben across a Silastic membrane from simple water–glycol-saturated solutions. Transport is dependent on thermodynamic activity, otherwise expressed as $C_v \times P_c$. (From Ref. 65.)

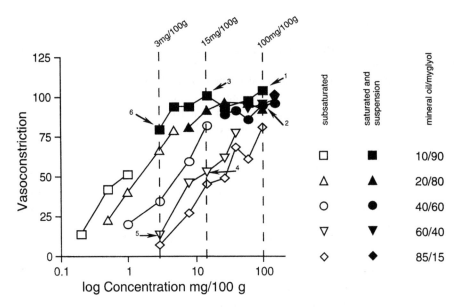

Figure 18 Relative vasoconstrictor response from subsaturated solutions (open symbols) and saturated solution and suspension (filled symbols) formulations of betamethasone benzoate in mineral oil–myglyol. As an example, at 100 mg/100 g, formulations 10:90 and 60:40 give vasconstriction scores of 103.4 ($\rightarrow$1) and 94.7 ($\rightarrow$2) units, respectively and would be considered bioequivalent. However, 15 mg/100 g and 3 mg/100 g vasoconstriction scores are different at, respectively, 100.5 ($\rightarrow$3) versus 52.1 ($\rightarrow$4) and 78.8 ($\rightarrow$5) versus 12.7 ($\rightarrow$6), and at these lower strengths the formulations would be considered bioinequivalent. (From Ref. 333.)

dependent on the vehicle (P_c), as discussed previously. For example, any vertical line (fixed C_v) drawn between the 1- and 10-unit doses shows a large variation in vasoconstrictor response depending on the vehicle composition.

The response to increase in C_v is complex, again depending on the vehicle. In any single vehicle, an increase in C_v results in an increase in vasoconstriction up to saturation (this strongly suggests that the limit of vasoconstrictor response is controlled by drug absorption and not saturation of the pharmacodynamic response in these systems). In suspension, a further increase in drug concentration beyond the saturated concentration does not lead to any further increase in penetration or vasoconstrictor response in a single vehicle. Across all the vehicles, saturated solutions (the first filled symbol in each series) give essentially the same response. Thus, a dose–response to increase in concentration will not always occur, as has been previously reported (327). Pharmacokinetic and pharmacodynamic control of dose–response is discussed further in Sec. V.

Bioequivalence is usually based on single-strength comparisons. Thus, assuming bioequivalance at a single strength, a further issue is one of the potential for bioequivalence—therapeutic inequivalence when topical products available in a range of concentrations have different dose–response relations because of formulation differences, as illustrated in Fig. 18.

Vehicle–stratum corneum (penetration enhancer) effects. In reality, many vehicle components do interact with the stratum corneum, and some—penetration enhancers—are included in formulations specifically for this purpose. Figure 19 shows data (334) on the in vitro permeation of water across epidermal membranes from a variety of vehicles. To rule out water–vehicle thermodynamic factors, these authors used half-saturated water solutions. The horizontal line shows the theoretical flux, assuming no interaction of the vehicle with the stratum corneum. Although flux from many vehicles is close to the line, others are above the line, indicating effects of the vehicle on the stratum corneum barrier because of the study's control of thermodynamic activity.

Depending on the mechanism of interaction of the specific enhancer, the parameters P_c (through change in drug solubility in the stratum corneum), or D_c (by fluidizing the stratum corneum lipid barrier), or both (335), are affected. The effect of enhancers is most marked with polar compounds, and this is likely owing to the effect of enhancers that reduce the hydrophobicity of stratum corneum lipids, thereby increasing solubility of these compounds. For example, Yamane et al. (320) studied the effects of pretreatment with terpenes and oleic acid–propylene glycol on the in vitro permeation of fluorouracil across the epidermal membrane. As shown in Fig. 20, up to approximately 100 times enhancement of penetration was observed.

In terms of bioavailability and bioequivalence, at a fixed thermodynamic activity, differences in the enhancer activity between vehicles may result in large dif-

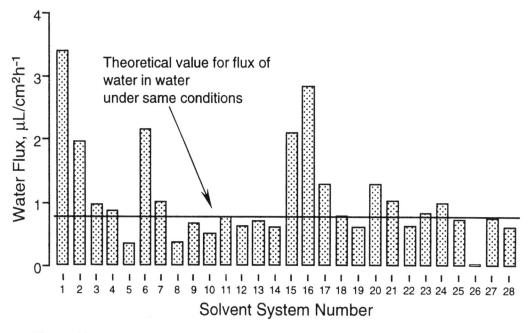

Figure 19 Transport of water across human stratum corneum from a series of miscible vehicles in which the water was at half saturation. The horizontal line shows the estimated transport of water from water at half saturation. Vehicles with water transport significantly above the line (> ×2) are considered to have interacted with the stratum corneum to reduce the barrier function. (From Ref. 334.)

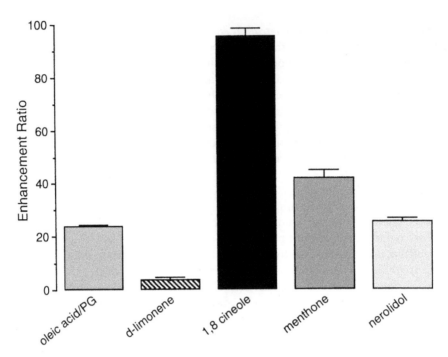

Figure 20 Effect of 12-h pretreatment on enhancement of transport of 5-fluorouracil across epidermal membranes in vitro. (From Ref. 320.)

ferences in permeation and, thus, bioinequivalence. When differences in both drug–vehicle interaction and enhancer activity exist between vehicles, large differences in bioavailability of up to 100-fold can occur (317).

Summary of interactions between drug, vehicle, and skin. For a given drug concentration, drug–vehicle thermodynamic effects and vehicle–skin enhancers effects, both separately and together, may cause large differences in bioavailability, up to about 10–50-fold and beyond, and result in gross bioinequivalence. Although these interactions between the drug, vehicle, and skin give rise to the potential for large differences in permeation and bioavailability, it is the low absolute bioavailability, put another way, the relatively high concentrations, in many current topical formulations that allow this potential to be realized.

C. Influence of Application Conditions

1. Rubbing

The data available are insufficient to confirm or deny the assumption that rubbing has an effect on drug uptake into skin (336,337).

2. Occlusion

Occlusion prevents loss of moisture from the skin and from aqueous delivery vehicles to the atmosphere. This trapped moisture, either endogenous or exogenous, extensively hydrates the stratum corneum, causing it to swell appreciably (338). The increased clinical efficacy of topical drugs caused by covering the site of application with water-impermeable material is a long-known fact (339–341). The enhanced pharmacodynamic effect of topical corticosteroids under occlusion was demonstrated by vasoconstriction studies (342,343). Occlusion has also been reported to increase the percutaneous absorption of various topically applied compounds. However, from the data available, one may suppose that occlusion does not necessarily increase percutaneous absorption of all substances, particularly the permeation of hydrophilic compounds. Recent studies also indicate that the extent of percutaneous absorption may depend on the occlusive system used (138,344,345). Occlusion, per se, can cause local skin irritation that, again, may lead to increased drug absorption. In addition, Vickers (92) suggested a definite correlation between skin occlusion and the extent of the drug reservoir formation in the stratum corneum after percutaneous absorption.

3. Loss of Vehicle

Despite its importance, the problem of loss of vehicle from an application site or the translocation of an applied dose from treated to untreated sites is rarely discussed.

A single report presents a physiologically based mathematical model that describes removal processes on the surface of the skin, including the effects of washing and desquamation, and rubbing off onto clothing. This model was applied to human in vitro and in vivo percutaneous absorption data of an aqueous formulation of the herbicide fluazifop-butyl (346).

Marples and Kligman (475) report that the spread of a drug from the site of application to other areas of skin may give rise to false interpretations in paired comparison tests of efficacy. For example, in infected dermatoses treated with topical neomycin, the placebo-treated site improved clinically and there was almost as much reduction in the *Staphylococcus aureus* population. These clinical observations were visualized using tetracycline hydrochloride containing fluorescent ointments, creams, lotions, and tinctures from various sites of applications (347). The topically applied medications did not remain confined to sites of initial application. The anatomical sites of initial application determined the pattern of transfer. Certain patterns of transfer were characteristic among groups of patients and for specific patients. The vehicle of a topical preparation greatly influenced the extent and sites of transfer. Ointments and creams were transferred considerably more readily than lotions and tinctures.

4. Metamorphosis (Change) of the Vehicle

In experimental and clinical situations, most dermatological vehicles (structural matrix and excipients) undergo considerable changes after they are removed from the primary container and applied to the skin. The initial structural matrix of the vehicle may change owing to rubbing (thixotropic effects) or as evaporation of volatile ingredients increases viscosity. These events can offer benefits that enhance the performance of vehicle(s) and drug(s). Rubbing certain vehicles onto the skin may have emulsifying effects and evaporation of certain ingredients may produce the desired cooling effect (348). Furthermore, evaporation from topical formulations is a required quality of repellents (349–351).

Evaporation of solvents may, depending on the polarities of the volatile solvent(s) and the drug, reduce or increase the solubility of the drug in the residual phase and, thereby, the drug thermodynamic activity and drug permeation (66).

Loss of volatile solvents, such as ethanol and isopropyl alcohol, because they are better solvents than the residual phase, often results in an increase in thermo-dynamic activity until saturation is achieved, which normally sets the maximum thermodynamic activity. At this time, the drug is anticipated to precipitate, where-upon its flux should remain constant as set by the driving force of the saturated solution. However, the drug may supersaturate and flux levels significantly greater those set by a saturated solution may be obtained (352,353). Clearly, in such for-mulations, there are large differences in drug penetraton between initial and residual states and thus also under occluded and open conditions. Especially for these types of vehicles (alcoholic gels and lotions), biopharmaceutical studies in volunteers should not be conducted under occlusion unless this is also used in clinical practice. Furthermore, uptake of water from the skin, brought about by occlusion of micro-emulsion vehicles, may result in supersaturation (354).

Cheng et al. (224) have shown that skin-surface lipids may mix with vehicles to change drug solubility at the vehicle–skin surface–lipid interface and thus influ-ence permeation.

Because of the complexity of theoretical models and experimental setups, only limited data is available on the metamorphosis of the vehicle (98,355–358).

5. Factors Influencing Dose

In most medical and toxicological investigations, the dose administered is precisely defined. Unfortunately, this is not always possible when a chemical is applied to the skin. Because the dose is a reference value and thus of utmost importance in bio-availability and bioequivalency considerations, the issue of dose is thoroughly dis-cussed. Dermal exposures (amount of drug absorbed) are defined in terms of dose (amount of drug applied), surface area, time of exposure, and frequency of application.

a. Dose Prescribed.   The dose prescribed in a clinical sense includes amount (e.g., number or quantity [weight]) of the dosage form, the strength, and an instruc-tion on the application modalities. The quantities prescribed are usually governed by the total surface areas to be treated, the application schedule, and the likely duration of treatment. The total dose prescribed includes a special consideration relative to possible toxicity (e.g., local or systemic effects of topical steroids).

Textbooks and guidelines in clinical dermatology give various recommenda-tions (359–365). Polano (359) recommends 10 g of cream or ointment per appli-cation per day as the minimum amount feasible for whole-body application ($\sim$0.6 mg/cm^2 day^{-1}). Schlagel et al. (360) found 12 g/day ($\sim$0.7 mg/cm^2 day^{-1}) the min-imum amount necessary, and "liberal" applications of emollients may entail using more than 100 g/day ($\sim$6 mg/cm^2 day^{-1}). Griffiths et al. (361) (Table 3) give the most practical information by presenting recommendations for a 7-day treatment period.

b. Dose Applied.

Dose applied in clinical situations.   The dose applied in clinical situations reflects the compliance of the patients with the dose prescribed or recommended. Lynfield et al. (366) determined the amount of dosage form that is applied in "clin-ical" practice by both "patient" and by health care professionals according to the following instruction: "Put this on thinly, covering. . . ." (Table 4). The data clearly

Table 3 Dosing Recommendations for a 7-Day Treatment

Site	To use sparingly	To use liberally	Lotions
Whole-body	100 g	250–500 g	500 mL
	($\sim$0.8 mg/cm^2, day)	($\sim$2–4 mg/cm^2, day)	($\sim$4 mg/cm^2, day)
Localized disease	15–30 g	50–100 g	25–100 mL

Source: Ref. 361.

reflect the variation that is present in clinical practice. This table's data confirm results of others (367).

The sun protection factor (SPF) is the ratio that estimates the protective efficacy of a sunscreen against sunburn. The generally accepted methods that are used to determine the SPF of a sunscreen require that the amount of formulation applied is 1.5 mg/cm^2 (DIN-Norm) or 2 mg/cm^2 or 2 μL/cm^2 (FDA), forming a homogeneous film of a defined thickness (20 μm) (368). The applied thickness of a sunscreen is important for the degree of photoprotection (Beer's law). In a field test, Bech–Thomsen and Wulff (369) found that the amount of the applied sunscreen was on average 0.5 mg/cm^2. A sunbather's application of sunscreen is, therefore, probably inadequate to obtain the sun protection factor assigned to the preparation. Previously, in 1985, Stenberg's data indicated that the sun protection factor under ad libitum conditions was only 50% of what would be achieved using 2 mg/cm^2 (476).

Another important aspect of applying a semisolid formulation on to the skin has been addressed (370). In this report recent observations were noted that although the frequency and amount of sunscreen may be adequate, the application technique may be inadequate. The investigators observed that the pattern of coverage was often incomplete and was dependant on the region treated (Fig. 21).

Dose applied in experimental situations. The dose applied in experimental in vitro or in vivo percutaneous absorption studies may be expressed as mass per unit area of the neat substance or vehicle. According to the recommendations of the Topical Therapeutic Products Workshop (March 26–28, 1990; Crystal City, VA) 1–3 mg or μL of formulation per square centimeter should be applied to the skin, usually with inunction, corresponding to films of 10–30 μm in thickness (film thick-

Table 4 Amount of Dosage Form Applied May Be Influenced by Formulation (Consistency), the Dispenser, and Person Applying the Dosage Form

Formulation	Dispenser	Weight (g)	Application	g/m^2	mg/cm^2
O/W cream	Tube	30	Self	7.1 $\pm$ 3.1	>0.7
O/W cream	Jar	450	Self	17.0 $\pm$ 17.4	>1.7
O/W cream	Jar	450	Nurse	9.1 $\pm$ 3.2	>0.9
Ointment	Tube	30	Self	13.6 $\pm$ 8.9	>1.4
Liquid	Squeeze bottle	210	Self	9.4 $\pm$ 5.5	>0.9
Lotion (liquid)	Squeeze bottle	210	Self	12.8 $\pm$ 12.7	>1.3

Source: Ref. 366.

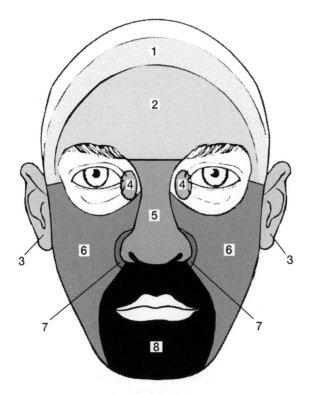

Figure 21 Sunscreen application pattern: anatomical regions: 1, hairline; 2, forehead; 3, ears; 4, periorbital; 5, nose; 6, cheeks; 7, nasolabial; and 8, perioral. Density of shading indicates coverage with sunscreen: dark (e.g., 8), good coverage; light gray (e.g., 1), incomplete coverage. (From Ref. 370.)

ness is an estimated value calculated from the dose volume and the estimated body surface). This guidance reflects usage in the clinical situation. The skin should be left open to the atmospheric conditions if this mimics the clinical use situation (371).

The amount of dosage form applied in experimental human in vivo studies to investigate percutaneous absorption are in general considerably higher than the amount recommended by Shah et al. (371) and Griffiths (361) or for that found by Lynfield and Schechter (366) (Table 5).

Table 5 Examples for Dose Applied in Human In Vivo Studies

Formulation	Dose (vehicle)/cm^2	Ref.
Solution	6.6 μL/cm^2	372
Solution	10 μL/cm^2	142
Ointment	5 mg/cm^2	373
Gel, cream, ointment	8.9 (2.2) mg/cm^2	374
Cream	5 mg/cm^2	375

c. Dosing Technique. Historically, the most frequently used in vitro method for studying absorption is the so-called infinite-dose technique. The skin is mounted as a barrier between two well-stirred, fluid-filled chambers. The compound under investigation is added to the solution on one side (donor) of the (skin) membrane, and absorption is assessed by serially sampling and assaying the concentration in the bathing solution (receptor) on the other side. Generally, the amount of solute that permeates during the course of an experiment is small relative to the total amount available, and there is no appreciable reduction in solute concentration in the donor solution. In effect, it appears to be infinite.

Several objections can be raised to the use of the *infinite*-dose technique as a predictive model for living skin. The most obvious, and certainly that of greatest significance, is the use of fully hydrated skin; both sides of the (skin) membrane are bathed by a donor and receptor solution. Furthermore, changes in composition due to loss of volatile components may not occur so readily under conditions of infinite dose. Equally blatant is that under clinical or "use" conditions, the amounts of material applied are of the order of a few milligrams (of vehicle) per square centimeter skin.

To avoid the shortcomings of the previously described approach, the so-called finite-dose technique was introduced (376–378). The (skin) membrane is mounted in specially constructed diffusion chambers. The dermis is bathed by thermostatically controlled receptor medium that can be replaced at various time intervals or continuously. A few milligrams of test vehicle are applied to the skin and can be throughly rubbed into it. The amount of applied material can be determined exactly. In contrast with the infinite-dose technique, as the absorption proceeds, the concentration of drug on the surface of the skin in a finite-dose experiment is depleted and flux falls during the experiment.

It is generally assumed that the finite-dose technique realistically mimics the in vivo situation. There are several publications addressing the finite-dose situation on topical therapy (64,296,379–384). The finite-dose situation differs from the infinite-dose situation (e.g., most transdermal drug-delivery devices) in that transient effects (non–steady-state ones) may be much more important and, in fact, may dominate the bioavailability issues.

Despite that in vitro percutaneous absorption data, determined by the finite-dose technique (377), suggest that the in vitro experiments readily mimic in vivo percutaneous absorption as determined by the Feldmann–Maibach protocol (81,111), there is an observation that may raise new questions.

As documented by many authors (103,111,138–142,245,385–391), the absolute bioavailability of topical corticosteroids and many other topical compounds, including retinoids, vitamin D derivatives, antifungals, antiacne, and antivirals, is generally in the range of a few percent of the dose applied. Provided that the experimental conditions permit, a mass balance following termination of the experiment [a quality control step that should be made with all experiments (138,392)], one usually observes that the dose recovered on the surface is in the range of 60–95% of the dose applied. Given this observation, one may argue that, under these experimental conditions, the delivery system (vehicle) is far from being exhausted (depleted) and even under clinical, hence, less-controlled conditions, the dose (amount of drug) that can be recovered after an adequate time interval is high. Therefore, one may conclude that both in experimental and clinical situations one is actually closer to an infinite-dose situation than to that of a finite-dose. This observation asks for a revision of older statements or a refined definition of finite or infinite-dose situations. To summarize, there are three dosing technique situations (Table 6), with the clinical situation for most drug being in category 2. For the few, well-absorbed molecules, such as nitroglycerin, phenols, and salicylate and nicotinate esters, the clinical situation is in category 3.

The infinite-dose of most compounds in clinical usage is another way of stating the problem of low absolute bioavailability, or put another way, the use of excessively high drug concentrations in most topical formulations. This brings us back to the proposal for rational dosing—reduced drug concentration—in topical formulations (see Sec. IV.B.2.c).

The use of infinite dose leads to the potential for bioinequivalence as discussed. Furthermore, use of infinite doses complicates the efforts to meet regulatory requirements of showing dose–response relations (21,22,123,393). For further discussion see Sec. V.

d. Variation of (a Single) Dose. A single dose can be varied using four general approaches (123):

1. *Concentration method*: varying the drug concentration in the topical product for a constant time on a constant surface area.
2. *Film-thickness method*: varying the surface area of application of the same volume or weight and the same concentration test formulation for the same duration of time.
3. *Surface method*: varying the surface area and volume of formulation applied with the same concentration formulation for the same duration of time.
4. *Duration method*: varying the duration of application of the same concentration formulation to the same surface area of the skin.

Table 6 Dosing Situations in Topical Drug Application

Category	Dosing technique	Loading	Comment	Bioavailability (estimate)
1	Infinite dose	High-loading/cm^2 (>10 mg to several 100 mg)	Two-chambered cells in in vitro experiments	Less than 5% (typically 1–2%)
2	Infinite dose	In-use loading/cm^2 (0.5–5 mg)	Poorly absorbed molecules in clinical situations, (e.g., corticosteroids)	Less than 5% (typically 1–2%)
3	Finite dose	In-use loading/cm^2 (0.5–5 mg)	Well-absorbed molecules in clinical situations (e.g., nitroglycerin)	More than 10% (typically 25–50%)

Sources: Categories 2 and 3, Refs. 361, 366.

In the following paragraph the four methods are presented and referenced. Whenever possible some recent examples are given of a (pharmaco)kinetic (K), (pharmaco)dynamic (D), or clinical (C) investigation.

Concentration method

(K). For many compounds, in vitro and in vivo, the relation between dose (concentration) and the dose absorbed is roughly linear over a broad range (75,112,394–396, and others) provided that the compound is in solution (354).

(K). Application of various concentrations of betamethasone dipropionate (0.020, 0.040, 0.050, and 0.063%) for 6 h was associated with a statistically significant ($p < 0.05$) linear increase in drug uptake into treated human stratum corneum up to the 0.050% concentration. Increasing the drug concentration applied beyond 0.050 up to 0.063%, however, did not further increase drug uptake ($p > 0.05$) (123).

(D). The increasing drug concentration (betamethasone dipropionate: 0.020, 0.040, 0.050, and 0.063%) was not associated with a corresponding increase in pharmacodynamic response over the 24-h observation period. Plotting the maximal visual skin-blanching response, as a function of the drug concentration applied, demonstrated that the maximal pharmacodynamic response to this corticosteroid occurs at strengths substantially lower than the strength marketed for clinical use (123).

(D). A similar disparity between drug concentration applied and pharmacodynamic response was previously observed with the use of topical corticosteroid formulations (374,397–399). Several authors used the dilution method to create a dose–response reaction. This, however, has the inherent danger of altering the physicochemical parameters of the drug in the vehicle, particularly the thermodynamic activity of the drug, which will influence drug partitioning and permeation (400).

(C). Double-blind studies directly comparing clinical effect of the different concentrations of the same steroid in the same vehicle are rare. One clinical study comparing 2.5% with 1.0% hydrocortisone cream in eczema revealed no difference between the two concentrations (313).

Film-thickness method

(K). In carefully controlled in vitro percutaneous absorption experiments with halcinonide, Walker et al. (401) observed, at clinically low levels of application (i.e., less than 5 mg/cm^2 of a cream), a dose-dependent low rate of permeation. With applications of 5 mg/cm^2 and higher (increasing vehicle thickness) the rate of permeation appeared constant, although the total amount permeating the membrane was dependent on the dose.

(D). Jackson et al. (402) studied five different marketed betamethasone valerate 0.1% creams and six different marketed triamcinolone acetonide 0.1% creams, in two groups of 12 subjects each. The subjects received five 10-μL portions of each cream spread over different skin surface areas to yield doses of 20, 10, 2, 1, 0.6 μL/cm^2. The area of application was encompassed by a Plexiglas ring open to the air. The creams remained in place for 6 h. Statistically significant differences were found among creams containing both drugs, suggesting a lack of equivalence. The authors report a diminution of vasoconstriction with increased area of application. In addition, particularly for the triamcinolone acetonide creams, the largest differences among the six creams were at the lowest "dilution" (largest application area). This suggests that testing of topicals should not be confined to a single surface area or dilution.

(D).   Pershing et al. (123) performed a similar experiment. Under unoccluded conditions a constant volume of a 0.05% betamethasone diproprionate cream formulation (10 mg) was applied to an increasing surface area of the skin (0.5, 2.0, 3.8, and 5.1 cm^2) yielding doses of 20, 5, 2.63, and 1.96 mg of cream per square centimeter, or 100, 25, 13, and 10 μg betamethasone diproprionate per square centimeter forming a film thickness of 200, 50, 26, and 20 μm, respectively. Visual skin blanching responses and objective measurements using the Minolta-Chromameter did not reveal significant differences (p > 0.05) between the various film thicknesses.

(D).   The independence of film thickness applied on the resulting maximal pharmacodynamic response in the Pershing study (123) differs from those in pharmacodynamic observations by Stoughton (22) with the same drug in a different cream formulation. Stoughton found that there are differences in vasoconstriction between doses of 1–4 mg/cm^2 but not those higher, to as much as 50 mg/cm^2. This was true for three different potent or midpotency topical glucocorticosteroids (Lidex cream, 0.05%; Kenalog cream, 0.1%; Diprosone cream, 0.05%). Differences between the Stoughton and the Pershing studies with the same drug likely reflect the influence of one or more of the following parameters: (a) vehicle composition (with propylene glycol in the Pershing "cream" versus no propylene glycol in Stoughton "cream"); (b) the efficiency with which the doses were spread over the skin (370); (c) the duration of drug application (6 vs. 16 h, respectively); (d) the time at which the pharmacodynamic response was measured (8 vs. 18 h, respectively); and (e) the method used to calculate the composite or average skin-blanching response in the population studied.

(D).   Barry and Woodford (403) showed that there was no difference in vasoconstriction between 6 and 16 mg/cm^2 when betamethasone valerate formulation was used, whereas Magnus et al. (404) reported a proportionate increase in vasoconstriction when 3.2–9.8 mg/cm^2 (1.6–4.8 mg/49 mm^2) formulation (betamethasone-17-valerate 0.1%) was applied, but no greater activity at doses higher than 9.8 mg/cm^2 (4.8 mg/49 mm^2).

(C).   There are no controlled studies available on the effect of film thickness applied on the resulting clinical outcome. However, certain dosage forms, such as pastes, cataplasms, and the like, are often thickly applied to the skin.

Surface area method

(K).   Unoccluded application of 10, 40, 80, or 100 μL of a 0.05% betamethasone diproprionate cream formulation to 0.5, 2.0, 3.8, or 5.1 cm^2 on human ventral forearm skin produced a constant film thickness of 200

μm over successively larger surface areas. Increasing the skin surface area from 0.5 to 5.1 cm^2, to which a constant film thickness is applied, decreased the mean amount of drug uptake per 1-cm^2 area twofold, but was not significantly different ($p > 0.05$). These data suggest that stratum corneum uptake of betamethasone dipropionate is independent of the surface area of the skin treated with constant film thickness (123).

(D).   Increasing volumes over successively larger skin surface area maintaining a constant film thickness, did not result in any significant change in the corticosteroid pharmacodynamic response at various time points of more than a 24-h period (123).

(C).   The surface area of skin treated with a drug may require special consideration relative to possible toxicity (e.g., local or systemic effects of topical steroids). Frequency of application (see next section) and the treated surface-area/body-weight relation are important factors (405).

Duration method

(K).   Increasing the unoccluded duration of 10 mg of 0.05% betamethasone dipropionate cream over the same skin surface area from 0.5 to 16 h, in vivo in humans, resulted in a linear increase of drug uptake into the human stratum corneum up to 2 h. Increasing the duration beyond 2–6 and 16 h did not significantly alter the drug concentration in the tape-stripped skin sample ($p \geq 0.05$). The time required for maximal drug uptake into human stratum corneum averaged close to 6 h (123).

(K).   Increasing the unoccluded duration of 30 μL of an ethanolic solution of hydrocortisone (200 nmol/cm^2) over the same skin surface area, from 0.5 to 6 h, in vivo in rats, resulted in no increase of drug uptake into the stratum corneum (126).

(K).   Two oil-in-water creams and two gels, containing 10 or 5% ibuprofen, respectively, were applied for 0.5, 1, and 2 h on excised human skin. The application time had no influence on the epidermal drug concentration, whereas the two gel formulations produced concentrations approximately twice those obtained with the emulsions (406).

(K).   In a human in vivo study, 1 mL of a minoxidil solution was applied twice daily over 150 cm^2 of bald scalp to each subject for 6 days. Unabsorbed drug was washed off the scalp after 1, 2, 4, and 11.5 h of contact time in each of four treatments. The extent of minoxidil absorption, expressed as steady-state urinary excretion of unchanged minoxidil, minoxidil glucuronide, or the sum of these, increased in a disproportionate manner with increase in contact time of the drug on the scalp. Relative to the amount absorbed after a contact time of 11.5 h, absorption was approximately 50% complete by 1 h and <75% complete by 4 h. This suggests that minoxidil absorption from the vehicle into skin occurs rapidly relative to diffusion through skin (372).

(D).   Increasing the duration of drug application from 0.5 to 2, and to 6 h, using the same concentration of betamethasone dipropionate, produced similar visual skin-blanching response profiles over time, suggesting that even more brief durations of application may be necessary with potent corticosteroids to achieve a reliable dose–response relation (123). These observations confirm data from Stoughton (393) and others (407).

(D). Stoughton and Wullich (393) (Table 7) assessed bioavailability over different time periods of exposure when 0.05% clobetasol propionate cream (Temovate) (class I) was applied and left on for periods of 0.5, 1.0, 1.5, or 16.0 h and subsequently washed off. Maximal responses were achieved after 1.5 h of exposure, but there was no significant difference in intensity of vasoconstriction between 1.0, 1.5, and 16.0 h of exposure before washing the sites. Exposures to 0.05% clobetasol propionate cream for 0.5 h were not significantly different from 16 h to that of 0.05% fluocinonide cream (Lidex) (class II), but exposures to 0.05% clobetasol propionate cream for 1.0, 1.5, and 16.0 h, all resulted in significant increases in vasoconstriction responses compared with fluocinonide cream applied and left on for 16 h.

These observations (123,393), and results from other investigators (71,374,407), suggest that the basic methodology of the Stoughton–McKenzie vasoconstrictor assay needs additional validation.

(C). Dithranol is estabished as a highly effective treatment for psoriasis (e.g., Lassar's paste) (408,409). The disadvantages of its use are the staining of clothing, linen, and skin, irritation, and long application time. Schaefer and co-workers (410–413) were able to show through their laboratory and clinical work that short-duration therapy (e.g., 10 min; "minute"-therapy) with higher dithranol concentrations (1–3%) presented all the advantages of dithranol and avoided most of its previous drawbacks. This modification of treatment has made the use of dithranol at home generally possible. This type of therapy requires a high degree of compliance.

(C). Jaeger (414) showed, in a double-blind, placebo-controlled study in patients with histopathologically verified psoriasis, that the application of betamethasone dipropionate, 0.05% for 3–5 min, combined with occlusion for 20 min was as effective as long-term corticoid treatment (415).

(C). A trial of three treatment schedules, consisting of 1% γ-benzene hexachloride (GBH) lotion, applied head to toe, and left on the body for 2, 6, or 12–24 h was conducted on an island of more than 2000 persons, approximately 70% of whom had clinical evidence of scabies. Examination at 1 month after therapy showed that both the 6-h and 12- to 24-

Table 7 Vasoconstrictor Assay in 30 Subjects

Name	Clobetasole propionate (Temovate) U.S. class I				Fluocinonide (Lidex) U.S. class II	Placebo cream
Time	0.5 h	1.0 h	1.5 h	16 h	16 h	16 h
Total scores	43	58	68	67	30	4

Source: Ref. 393.

h–cure rate was high (96 and 98%). There was a significantly lower cure rate in the 2-h group, in which only 82% were cured (416).

e. Multiple Doses. The dose can also be defined in terms of frequency of application. Intermittent therapy can be one, two, or more exposures per day. Prolongation of the dosing interval up to several days is also possible and may help avoid adverse drug effects in cases for whom long-term application is necessary. Most percutaneous absorption studies have employed a single administration of the compound under investigation; however, the relevant clinical (and toxicological) situations usually involve multiple contacts of drug (xenobiotic) with the same skin site. Despite this obvious relevance of topical pharmacokinetics following multiple topical application, there has been only limited investigation of the subject.

(K). The in vivo percutaneous absorption of hydrocortisone through the skin of the ventral forearm of the rhesus monkey, quantified by measuring ^{14}C in aliquots of urine over 5 days, was no different when 13.3 $\mu g/cm^2$ was applied as a single dose or when 13.3 $\mu g/cm^2$ was applied three times, totalling 40 $\mu g/cm^2$. However, when 40 $\mu g/cm^2$ was applied as a single dose, absorption was substantially increased over that of 13.3 $\mu g/cm^2$ applied either once or three times. Additionally, when the skin was washed between applications to remove previously applied material in the three-application experiment, there was a statistically significant increase over not washing the skin (417). The study of Wester et al. (417) has recently been repeated in humans. When a single dose (13.3 $\mu g/cm^2$) was tripled (40 $\mu g/cm^2$) the amount delivered through the skin increased by nearly three times. Three serial doses, with and without soap-and-water washing between the doses, increased percutaneous absorption remarkably (142) (Fig. 22).

(K). With a similar experimental protocol (417), the short- and long-term (8 days) administration of hydrocortisone in acetone or in an emulsion ointment base was studied in the rhesus monkey (418). Absorption significantly increased during long-term administration, whether applied in an acetone or in an emulsion ointment base. A placebo study in which only an acetone vehicle was applied for a long period followed by [^{14}C]hydrocortisone application, showed no enhanced absorption. It was suggested that long-term application of hydrocortisone alters the skin barrier, resulting in enhanced absorption. After short- and long-term (8 days) administration of a methylprednisolone aceponate ointment (Advantan), percutaneous absorption, assessed by cumulative urinary excretion of ^{14}C-labeled substances (<1% of dose applied), were not statistically different, suggesting that repeated application of the vehicle does not change the barrier and reservoir functions of human skin (373,419). Azone cream was topically dosed on the ventral forearm of humans for 21 consecutive days. On days 1, 8, and 15, the Azone cream contained ^{14}C-labeled Azone (375). The skin application site was washed with soap and water after each 24-h dosing. Percutaneous absorption, determined by urinary radioactivity excretion after repeated application (8 days), nearly doubled (p < 0.002), but stayed the same after continued repeated application (15 days). It is concluded, that repeated application of Azone

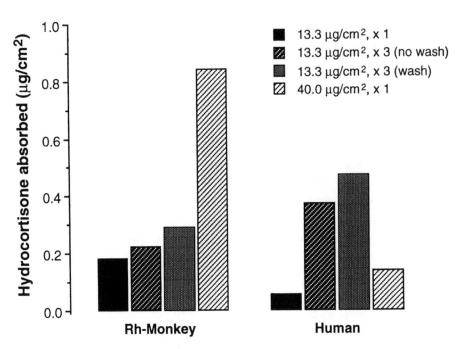

Figure 22 Percutaneous absorption of [^{14}C]hydrocortisone through the ventral forearm of rhesus monkeys and humans. (From Refs. 142,417.)

results in an initial self-absorption enhancement, and steady-state percutaneous absorption is established after this change.

(K). Percutaneous absorption of diflucortolone-21-valerate in an ointment base, with and without salicylic acid, was determined following a large-area skin treatment (20 g ointment twice a day for 8 days) in two groups of healthy volunteers. No differences were found either in percutaneous absorption of diflucortolone-21-valerate, or in effects on the pituitary–adrenal axis between the two treatment groups, suggesting that the addition of salicylic acid to the vehicle did not significantly alter the skin barrier function during the 8-day treatment (420).

Despite some in-depth investigations a clear answer pertaining to the effect of repeated application on percutaneous absorption is still unavailable. A variety of differences of experimental conditions make comparison of the data difficult. Because of the therapeutic and toxicological relevance of the repeated-dose situation, resolution of the problem is urgently needed. From a large amount of clinical and experimental evidence, one may conclude that the vehicle influences the skin barrier and skin reservoir function. It is obvious that pretreatment of the skin with vehicle alone or reapplication of vehicle to a drug-treated area may also influence drug absorption (421,422).

(K/C). The concentration of iododeoxyuridine (antiviral) found in the stratum corneum of guinea pig skin by tape-stripping at different time points after single and multiple topical doses of the drug, was studied. With each dosing frequency, the cumulative amount of drug in the stratum corneum correlated with the strength of the test formulation and with efficacy in the animal. For each of the three formulations, increasing the number of daily doses from one up to three led to progressive increases in cumulative stratum corneum iododeoxyuridine levels and clinical efficacy. An increase in the number of daily applications to four had little effect on drug efficacy and was associated with a plateau in stratum corneum iododeoxyuridine levels (82).

(C). In a double-blind multicenter, placebo-controlled study in patients with psoriasis and in patients with atopic dermatitis, Fredriksson et al. (423) showed that the application of halcinonide 0.1% once daily was equally as effective as the cream applied three times daily. However, the onset of action was more rapid when the cream was applied three times daily. With a similar study design Sudilovsky et al. (424), under the same clinical conditions, showed that a once-daily regimen can be an effective treatment; however, the three-times daily regimen was superior overall, and the authors recommend this regimen as treatment of choice in severe psoriasis.

f. Intermittent Dosing

(C). Actinic keratoses often occur on the sun-exposed skin of adults. A standard treatment for multiple actinic keratoses is topical 5-fluorouracil applied twice daily for 2–4 weeks until maximal inflammation is achieved. Although effective, extreme local irritation makes it unacceptable to many patients. Pearlman (425) showed that weekly "pulse" dosing of topical 5-fluorouracil on mutiple facial actinic keratoses produced the same benefit, with much less local irritation, than a conventional daily-dosing schedule. In view of its efficacy, relative comfort, lower cost, and simplicity of use, weekly pulse dosing may be preferable.

(C). Adverse drug effects have also been used to study the influence of different application schedules. Several authors found a correlation between the therapeutic efficacy and the skin-thinning effect of topical corticosteroids (426–430). Potent corticosteroids produce marked skin-thinning, whereas weak corticosteroids induce only a slight reduction of skin thickness. To avoid steroid-induced dermal thinning and other adverse effects, various strategies of therapy have been developed (431). Various studies show that a topical treatment with glucocorticosteroids two or three times daily has an efficacy similar to once-daily treatment (423,424,432). Similarly, the skin-thinning effect of a clobetasol-17-propionate treatment (41 days) is identical when the drug is administered once daily or at an interval of 72 h (433). This was also shown for fluprednidene acetate, fluocinolone acetonide, and bctamethasone-17,21,-dipropionate (434,435). Prolongation of the dosing interval up to 8 days diminished the skin-thinning effect, but the effect could not be totally avoided (433,436). Results from various studies in psoriatic patients suggest that intermittent pulse dosing with potent corticosteroids can offer an efficacious method for extended maintenance therapy with fewer adverse effects (437).

D. Conclusion

From this review of effects of skin conditions, vehicle effects and their interactions, and consideration of dose, we can identify certain primary factors that influence bioavailability, bioequivalence, and therapeutic equivalence of topical products.

1. *Bioavailability*: Under most clinical and experimental conditions, the use of infinite (and hypertherapeutic) doses leads to low absolute bioavailability, often in the range 1–5% of dose applied.

2. *Bioequivalence and bioinequivalence*: The low, absolute bioavailability of topical products significantly increases the likelihood of bioinequivalence when formulation differences lead to differences in drug partitioning into, or diffusion through, the stratum corneum. From this, a rational approach is to reduce the dose, which in the clinical situation, requires reduction of drug concentration. Low-concentration topical products with absolute bioavailabilities in the target range of 50–100%, yet bioequivalent to current products, are suggested as a future standard. This is discussed in the final section of this chapter.

3. *Bioequivalence and therapeutic equivalence*: Even when bioequivalence has been established in a volunteer model (e.g., vasoconstrictor assay), there is the potential for therapeutic inequivalence (11–13). One concern is that the allowable limits for bioequivalence (e.g., ±20%) may lead to therapeutic difference in compounds with a steep dose–response curve (438). Also, there is the possibility of therapeutic inequivalence owing to differences in absorption between the volunteer model and clinical situation (13); for example, because of differences in absorption between anatomical sites and on diseased skin, as discussed in Sec. III. Finally, there is the complex issue of dose–response, when topical products are available in a range of strengths. We have seen that bioequivalence at one strength is no guarantee of bioequivalence or therapeutic equivalence at another. These points are discussed further in Sec. V.

V. BIOAVAILABILITY, BIOEQUIVALENCE, AND THERAPEUTIC EQUIVALENCE

This final section will briefly review standards of bioequivalence in currently marketed topical dermatological products using glucocorticosteroids as the example. Moreover, as a result of initiatives to introduce generic drug products, therapeutic equivalence has been defined on the basis of bioequivalence plus pharmaceutical equivalence (439). Thus, this section will also examine the proposition that bioequivalence predicts therapeutic equivalence.

To accomplish this, one needs to consider the dose–response relation and the sources of variation that lead to inter- and intraindividual differences.

A. "Biological" Variation

Interindividual differences ("biological" variation) exist in virtually all steps of drugs interacting with the human body. Application of pharmacokinetic (absorption–distribution–metabolism–elimination; ADME) or pharmacodynamic (concentration–

effect) principles (440,441) to pharmaceutical drug development or to applied therapeutics for patient care, requires an understanding of these processes and their variation.

In general, pharmacokinetic principles relate specifically to changes of drug concentration in the body relative to time. Specifically, there are two pharmacokinetic "dose" relations:

1. Between the dose applied and the dose absorbed: this relation is particularly important in topical therapy, but is rarely considered. The two are related by the extent of the dose absorbed: the absolute bioavailability.
2. Between the dose absorbed and the drug concentration at the target site: the two are related by the pharmacokinetic factors that govern distribution: metabolism and elimination.

The pharmacodynamic relationship is between the drug concentration at the target site and some cellular response, the classic biological concentration–effect relation.

An understanding of the concept and an appreciation of the merits of these approaches can be formed by focusing on the basic tenet of clinical pharmacology: a dose (concentration)–response relation. Irrespective of the shape (relation) of these curves, there are four parameters: slope, potency, maximal effect, and variation, that define such a curve (Fig. 23).

Although the slope, maximal effect, and potency parameters are commonly considered an inherent property of a drug molecule, it is well recognized that all three parameters (slope, maximal effect, and potency) are greatly influenced by the processes governing the absorption, distribution, metabolism, and excretion of the drug. Variation also exists in the pharmacodynamic process owing to differences in biology. Levy (443,444) has provided strong evidence that variation in the pharmacodynamic process is at least as important as that in pharmacokinetics. It is accepted that these pharmacokinetic and pharmacodynamic processes contribute significantly toward the observed variablility or so-called biological variation commonly seen in dose (concentration)–effect curves.

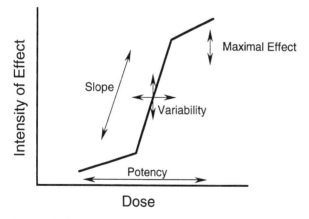

Figure 23 The log dose–effect relation with its four characterized variables. (From Ref. 442.)

Inter- and intraindividual variations in response occur because of both controllable and uncontrollable factors. Inappropriate compliance (445) and placebo effects (446–449) are primary contributers to the latter. There are two controllable factors:

1.   Control of variation in extent of absorption from the vehicle. This relates to bioavailability and bioequivalence in both the volunteer and clinical setting.
2.   Individualization of dosing to adjust for variation in distribution, metabolism, excretion, and concentration–effect relationships (dose titration).

Sampling site or drug quantitation procedures may also be a source of variability. Furthermore, pharmacodynamic measurements are often difficult and prone to errors, and the extent of variability may also vary from drug to drug.

B.   Variation and Dose (Concentration)–Effect Relationship

Various mathematical models have been developed to describe the concentration–effect relation, the most widely used being the sigmoid E_{max} model as shown in Eq. (6).

$$E = \frac{E_{max} \times C^n}{EC_{50}^n + C^n} \qquad (6)$$

where
E is the observed effect
E_{max} is the maximum effect
EC_{50} is the concentration (dose) that elicits 50% of the maximal effect
C is drug concentration (dose)
n is the Hill exponent that governs the slope effect

Figure 24 shows a schematic of the sigmoid E_{max} response for a drug having both wanted and unwanted effects over the dose range of interest.

Levy (438) has described how the same change in dose, for example, a 50% difference in bioavailability between two products, has a relatively greater effect on response in the lower range of the response curve than at the top. For example, in Fig. 24 the same change in dose D results in differences in wanted responses R_1 and R_2. Change in dose D can also result in relatively greater changes in unwanted effects than in wanted effects: R_3 versus R_2. Olson et al. (450) have illustrated how the steepness (sigmoidicity) of the concentration–response curve is important in determining the effect of a given variation on the extent of absorption on drug response. Selection of optimum dose and limits on criteria for bioequivalence close to this, ideally, should be based on dose–effect relations. Such data now exists for relatively few systemic drugs, but this should increase through a focus on development of surrogate markers for clinical response (451).

C.   Variation in Extent of Absorption

In Sec. IV, factors affecting the rate and extent of percutaneous absorption were reviewed. Figure 25 shows examples of variation in the extent of absorption owing to variation among subjects, skin site, skin condition, and formulation. The low

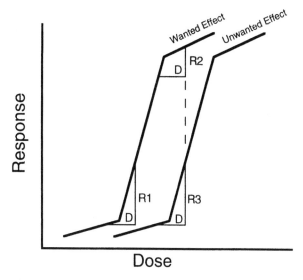

Figure 24 Schematic of sigmoidal E_{max} relation: effect of change in dose D on response R depending on position on concentration–effect curve. (From Ref. 438.)

extent of absorption of many topical products, including corticosteroids and retinoids, in the range of 1–5% of the dose applied, is a major contributory factor that allows the potential for variation caused by these factors to be translated into bioinequivalence and therapeutic inequivalence. Intuitively, if the extent of absorption is low, there is a potential for variation and, thus, for bioinequivalence and therapeutic inequivalence. In the following Secs. C.1 and C.2 the consequences of variation in the extent of absorption on bioequivalence and therapeutic equivalence are reviewed.

1. Bioavailability and Bioequivalence of Topical Glucocorticosteroids

After reviewing the theoretical potential for bioinequivalence in Sec. II, what are the actual standards of bioequivalence in currently marketed topical dermatological products? A brief review of glucocorticosteroids is of interest, as the bioequivalence of these compounds has been relatively widely studied. In these studies the major source of variation in absorption is between formulations.

Early studies gave the first indications for the potential for bioinequivalence of corticosteroids formulated into standard pharmacopeial cream and ointment bases (452–454). Following the development of topical formulations of hydrocortisone, triamcinolone acetonide, fluocinolone acetonide, and other synthetic corticosteroids, Barry and Woodford conducted a series of studies throughout the mid-1970s to compare the bioavailability of different formulations of the same corticosteroids (403, 455–458). In summary, these studies show many examples of 50–100% differences in bioavailability of the same corticosteroid at the same strength in the same vehicle type. In 1987, Stoughton published results on a series of proprietary topical glucocorticoids (459) showing significant differences in bioavailablility between commercial cream and ointment formulations containing the same active substance at the same concentration. In the worst case, triamcinolone acetonide 0.1%, vasoconstrictor response differed 2.5-fold within the creams and almost 4-fold (see Fig. 25D) within

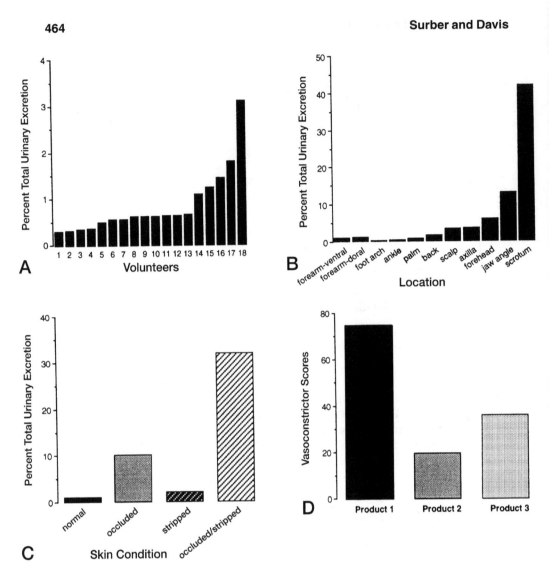

Figure 25 (A) Variation in percentage of total urinary excretion of hydrocortisone from the forearm site in 18 individuals; (B) variation in percentage of total urinary excretion of hydrocortisone applied to different skin sites in humans; (C) effect of forearm skin condition on urinary excretion of hydrocortisone; (D) bioequivalence of triamcinolone acetonide, 0.1%, in various ointment formulations.

the ointments. Shah et al. (460), Jackson et al. (402), and Olsen (461) have since confirmed bioinequivalence within commercial topical steroid formulations containing the same active substance at the same concentration.

All of the foregoing studies used the vasoconstrictor method to measure bioequivalence. Relatively few data are currently available for corticosteroids determining bioequivalence using drug level measurements in vivo (120,121). Recently, Agrawal et al. (462) compared the bioavailability of hydrocortisone from seven commercial products using a standard in vitro technique. Total permeation over 48 h ranged from 3.0 μg (1% cream) to 0.13 μg (1% ointment), with large differences between 1% cream formulations, depending on manufacturer. This group, under similar conditions, compared the bioavailability of hydroquinone from topical over-the-counter (OTC) and prescription products (463a). The extent of absorption varied tenfold, with some of the 2% OTC products delivering more drug than the 4% prescription products.

From this review, it appears that bioinequivalence is common within topical dermatological products. However, this is to be anticipated given the low extent of absorption and the potential for extremely large variation between formulations in drug partitioning into and diffusion through the skin.

2. Bioequivalence and Therapeutic Equivalence

Even if we assume that bioequivalence is established between two or more topical dermatological formulations in volunteers, there is still potential for therapeutic inequivalence. In this context, variation is in the extent of absorption between skin sites (e.g., forearm and face) and skin conditions (e.g., open and occluded, healthy or diseased) in the volunteer experimental versus the clinical setting.

a. Potential for Bioequivalence but Difference in Efficacy. Figure 26 illustrates the hypothetical case in which, even though two formulations are bioequivalent, variation in extent of absorption occurs between the bioavailability model and the clinical setting. The curves represent wanted and unwanted effects (see Fig. 24) and, for simplicity, similar shapes for the effect curves in both volunteers and the clinical setting have been assumed.

Data to exemplify the hypothetical cases in Fig. 26 are still rare (313). However, as outlined previously (see Sec. III), it is likely that there are changes in extent of absorption in volunteers and patients owing to change in skin site (forearm in volunteer studies versus any skin site in patients) and change in skin condition (healthy vs. diseased skin, and occluded conditions in many experimental situations vs. the open or occluded conditions in clinical situations). The following examples, comparing the vasoconstrictor activity of betamethasone valerate in cream and ointment formulations under occluded or open test situations illustrates the hypothetical cases (Table 8).

Case A. If Betnovate and Betnovate N,C were tested in volunteers under *open* conditions, they would be approximately bioequivalent, whereas under *occluded* conditions in patients Betnovate is supraavailable.

Case B. If Betnovate and Benovate A,C,N were tested in volunteers under *occluded* conditions, they would be approximately bioequivalent, whereas under *open* conditions in patients Benovate-A would be inferior.

Thus, two or more formulations that are bioequivalent in volunteers may give different clinical beneficial responses.

b. Effect of Finite and Infinite Dose: Control of Overdosing. Occlusion may give rise to modest increases in percutaneous absorption and thus changes in efficacy. Also of clinical importance is that gross increase in the extent of absorption between the bioavailability model and clinical setting may occur and will result in overdosing and, possibly, severe adverse effects. For example, Aalto–Korte (24) has described the absorption of systemically active doses of hydrocortisone following topical ap-

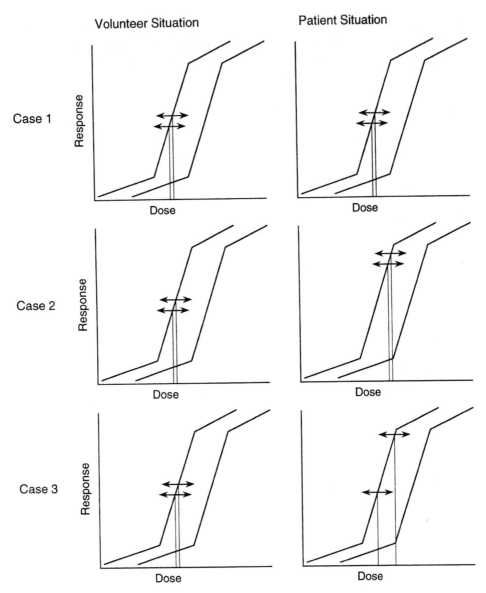

Figure 26 Pictogrammed hypothetical cases in which two formulations are bioequivalent in volunteers, but variation in extent of absorption in a therapeutic setting may result in the potential for therapeutic inequivalence in patients. In case 1 no difference in extent of absorption in volunteers and patients for both products is observed, resulting in therapeutic equivalence (comparable with Synalar formulations; 455). In case 2 significant change in extent of absorption in patients is observed for both products, also resulting in therapeutic equivalence (comparable with Propaderm formulations; 403). In case 3 significant change in the extent of absorption in patients is observed for only one of the formulations resulting in therapeutic inequivalence (comparable with Betnovate cream and ointment formulations, 403,455).

Table 8 Cumulative Vasoconstrictor Values for Various Betamethasone Valerate Formulations (Betamethasone) Under Occluded and Open Conditions

Case A				Case B			
Formulation (creams)	Occluded conditions	Open conditions	Ratio (%)	Formulation (ointments)	Occluded conditions	Open conditions	Ratio (%)
—	—	—	—	Betnovate-A	2240	1230	54.9
Betnovate-C	1210	1220	100.0	Betnovate-C	2170		74.7
Betnovate-N	1300	1130	86.9	Betnovate-N	2140	1630	76.2
Betnovate	1670	1060	63.6	Betnovate	2080	1870	89.90

Sources: Case A, 403; case B, 455.

plication to patients with erythroderma. As outlined in detail in Sec. IV.C.5.c, under experimental conditions, the delivery system (vehicle) is generally far from being exhausted (depleted) and even under clinical, hence, less controlled, conditions the dose (amount of drug) that can be recovered after an adequate time interval is high. Therefore, one may conclude that in both experimental and clinical situations, one is actually closer to an *infinite* dose situation than to a *finite* dose situation.

Figure 27 shows a schematic of the pharmacokinetic–pharmacodynamic dose (concentration)–response, and variation in this, from a high, infinite, dose and a low, finite, dose. On average, the two doses are bioequivalent, which requires the use of an enhanced delivery system to increase bioavailability from the finite dose as described later. With the infinite dose, variation in skin diffusivity causes the dose absorbed to vary into that range where significant—and unjustified—unwanted effects may occur. For the finite dose, control by the dose on the upper limit of

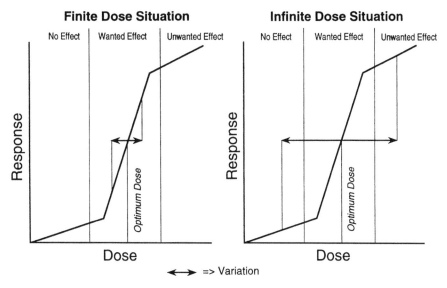

Figure 27 Schematic showing the effect of *finite* and *infinite* dose situations.

absorption limits the potential for unwanted effects. Use of finite–dose-enhanced delivery systems are described further in the following.

Among individuals, for a given skin site and skin condition, there is approximately a 10-fold variation. For any given individual, there is approximately a 50-fold variation in the dose absorbed between skin sites. Variation in skin condition leads to further variation (see Fig. 25A–C). Such differences in the extent of absorption between formulations, depending on skin site and condition, will also lead to disproportionate increases in unwanted effect. This situation is probably common in topical therapy, particularly for compounds such as corticosteroids that are applied to body sites of differing permeabilties.

c. Conventional High-Concentration and Enhanced Low-Concentration Formulations. The overemphasis of drug concentration in topical vehicle design has been mentioned, as has the fact that many current topical formulations are poorly bioavailable; that is, they contain high concentrations of active agent compared with the fractional amount absorbed and required. One may therefore define these formulations as *conventional high-concentration formulations.* The application of these formulations may create an infinite-dose situation (see Table 6) on the skin.

In Sec. IV.B.2.c, the potential of pharmaceutical techniques, such as thermodynamic activity and chemical enhancement, to increase percutaneous absorption are outlined. In the following section, arguments for increasing the bioavailability—extent of absorption—of topical products using such techniques, with concurrent reduction of drug concentration in these formulations will be developed. In contrast to the aforementioned conventional high-concentration formulations, a new type of formulation will be defined as *enhanced low-concentration formulations.* In Fig. 28 the concept of the enhanced low-concentration formulations to control variation in the extent of absorption is developed.

As discussed earlier and illustrated in Figs. 25–28, one usually has little or no control over the dose absorbed with the current conventional high-concentration formulations. This fact is also pictographed in Fig. 28B. With current conventional high-concentration formulations, the bioavailability of topical dermatological agents are generally low (1–5%), leading to a potential for 20- to 100-fold increase in absorption, depending on skin permeability. Increasing bioavailability by pharmaceutical

Figure 28 Schematic presentations of the effect of *enhanced low-concentration formulations*: (A) Chemical enhancers or supersaturation are pharmaceutical means to enhance drug absorption through the skin. Chemical enhancers act by fluidizing the stratum corneum lipid barrier (pictogrammed as opening the funnel), supersaturation acts by increasing the thermodynamic activity of the drug in the vehicle (pictogrammed as arrows). (B) Variation in skin site and skin condition, including disease and disease stage, as well as the individual, influence topical bioavailability. Bioavailability of topical dermatological agents are generally low, leading to a potential for dramatic increase in absorption depending on skin permeability. (C) Increasing bioavailability by pharmaceutical means (e.g., supersaturation) does not eliminate variation owing to skin site, skin condition, including disease and disease stage, or the individual. (D) Variation in bioavailability may be reduced by reducing drug concentration in the delivery system and, at the same time, increasing the delivery efficiency (e.g., supersaturation) from the delivery system.

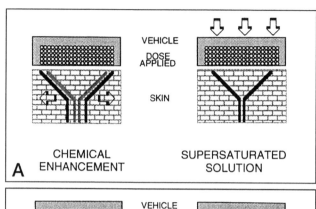

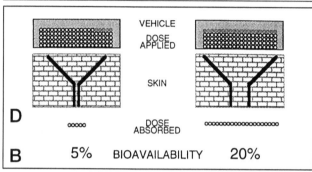

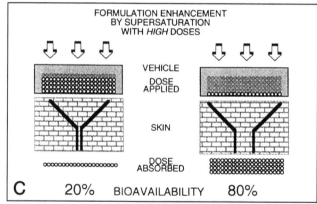

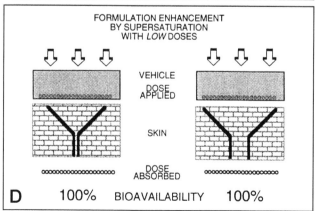

means (e.g., chemical enhancers or supersaturation) is possible (see Fig. 28A); however it does not account for the different permeability of the skin (see Fig. 28C) and, therefore, does not avoid differences in bioavailability (high variation). To avoid this source of variation, one may reduce drug concentration in the delivery system and at the same time increase delivery efficiency from the delivery system (see Fig. 28D). Reduction of the drug concentration in the delivery system reduces the potential for excessive delivery. This maneuver is important for individuals with highly permeable skin (newborns), for body sites with highly permeable skin (intertrigo), or for individuals with impaired barrier function. Increasing the delivery efficiency of the delivery system increases drug delivery in individuals with less permeable skin (adults), increases drug delivery in skin sites with low skin permeability (forearm) or with increased barrier function. The skillful combination of these two strategies—reduction of the drug concentration and increased delivery efficiency—therefore, will most likely ensure equivalent therapy and lead to a drug delivery system with highly increased bioavailability, preferably in the range of 50–100%. By definition, the variation in dose absorbed, depending on skin permeability, will decrease dramatically. Pharmaceutical technologies, such as supersaturation or chemical enhancers, are required to deliver low doses to the required bioavailability standards of 50–100%, and further work to apply these to enhancement of low-concentration formulations is required.

Marks *et al.* (464) have established that these enhanced low-concentration formulations are as efficacious as conventional high-concentration formulations (Fig. 29). In study AI/AII the vasoconstrictor responses of the 1% hydrocortisone acetate (HA) and 0.02% eightfold-saturated HA are not significantly different from each other, yet each of these is significantly different from the control treatments ($p < 0.05$, Wilcoxon-matched pairs-signed ranks test). Ideally, these comparisons should be made under nonoccluded conditions because occlusion will increase absolute bioavailability. Nonoccluded vasoconstrictor studies on HA are not yet feasible owing to the relatively weak response of HA in the model. However, these results demonstrate bioequivalence of the 1% HA and 0.02% eightfold-saturated HA formulations under the conditions of this study. In study BI/BII—the surfactant-induced erythema test—the significance level was at $p < 0.05$ using the Wilcoxon-matched pairs-signed ranks test.

From our recent in vitro membrane permeation experiments (463b), it was concluded that drug flux can be increased supraproportionally with increasing donor concentration (drug [super]saturation [proportional]) beyond what would be anticipated based on ideal donor concentrations and partition coefficient considerations alone. These findings could not be confirmed in an in vivo investigation, probably owing to additional vehicle effects (e.g., enhancement, irritation, or drug-binding) that must be expected and could have altered the integrity of the stratum corneum and, thereby, topical bioavailability.

D. Variation in the Distribution, Metabolism, and Excretion Process and Concentration–Effect Response: Individualization of Dosing

If there is variation in pharmacokinetic (distribution, metabolism, and excretion process) or pharmacodynamic response (concentration–effect response) among volun-

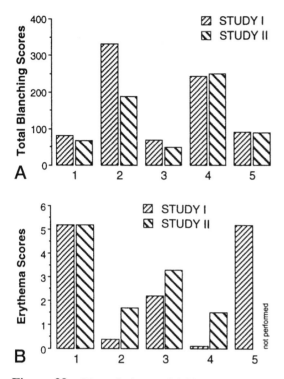

Figure 29 Bioequivalence of 0.02% w/w supersaturated hydrocortisone acetate (HA) with conventional 1% w/w HA cream: study AI/AII, blanching-test in volunteers; study BI/BII, surfactant-induced erythema in volunteers. Treatments: 1, untreated; 2, 0.02% w/w supersaturated HA in gel; 3, gel base; 4, 1% w/w HA in cream; 5, cream base. (From Ref. 52.)

teers or patients, then one may titrate to that specific dose of drug that best suits an individual patient. For example, in concentration-controlled trials (CCTs), the dose is varied between patients to obtain a specific tissue, usually plasma, concentration. CCTs address the issue of variation in pharmacokinetic factors. Levy *et al.* (444) have discussed the importance of variation in pharmacodynamic factors and have proposed the strategy of varying the dose to obtain a specific pharmacological end-point, the so-called *effect-controlled trials* (ECT). They provide convincing evidence from the literature that, contrary to current paradigm, variation in pharmacodynamic factors are the major source of variation in therapy. Thus, there is a growing belief that the current, fixed-dose, paradigm is outdated, and that one should be developing new dosing strategies for dose titration and improved health care (444,465,466). Whether this is possible with our current armamentarium of topical dermatological products will now be discussed.

1. Dose Titration with Conventional High-Concentration Formulations

Despite some single experiments in which dose–response is observed; for example, (a) pharmacokinetically with hydrocortisone by varying the concentration of the dose applied (142,417); (b) pharmacodynamically with corticosteroids by varying the dose absorbed by duration of dosing (123); and (c) clinically with 5-fluorouracil by varying the dose applied by frequency (intermitting dosing) (425). Generally, no such relations with topical dermatological products are observed. Again, using the corticosteroids as a well-studied example, there are many reports of lack of, or poor, dose–response (393,397,399,467–472).

For an increase in dose applied to result in an increase in response, both the dose applied–dose absorbed relation (pharmacokinetic) and the dose absorbed–response relation (pharmacodynamic) should be linear or approximately so. Thus, when the response from increasing the dose applied is flat, this can either be due to pharmacokinetic control (dose applied–dose absorbed is flat) or pharmacodynamic control (dose absorbed–response relation is flat), or both. Pharmacokinetic versus pharmacodynamic control of flat topical dose–response has been the subject of some discussion (333,460,470,473). Variation in pharmacodynamic dose–response is wide among subjects (466), and this will confound the use of pharmacodynamic parameters to predict pharmacokinetics, including bioavailability and bioequivalence. As part of a program to improve the sensitivity of the vasoconstrictor test, the recent *Guidance Topical Dermatologic Corticosteroids: In Vivo Bioequivalence* (21) proposes the screening of subjects into detectors and hypo- and hyperresponders. Detectors, subjects showing linear pharmacodynamic response to dose absorbed, are selected for bioavailability comparisons.

Barry et al. (470), in a study designed to show pharmacokinetic control, have shown no difference between the vasoconstrictor response and clinical antipsoriatic response to 0.05 and 0.1% desonide creams. Both concentrations of desonide were saturated in the vehicle and, thus, as described in Sec. IV, were predicted to give rise to the same extent of drug penetration. Because there is no increase in dose absorbed as dose applied is increased, there can be no difference in vasoconstriction or clinical activity. What is not fully appreciated is the importance of lack of depletion in the flat relation between dose applied and dose absorbed. Saturated vehicles saturate the skin and thus the initial rate of absorption is the same. Only because they are infinite doses do these vehicles remain essentially saturated and thus continue to give the same rate of drug absorption. To achieve a pseudo-dose–response (increase in dose absorbed with dose applied) with current high-concentration formulations it is necessary to increase the thermodynamic activity of the drug in the vehicle as concentration is increased. For example, if the highest drug concentration in a given vehicle is just saturated, then lower concentrations in this same vehicle will give a dose–response (333) (see Sec. IV, Fig. 17 and discussion). Although increase in thermodynamic activity can be used to generate a pseudo-dose–response with high-concentration formulations, it is not (theoretically) possible to provide a dose–response to increase in drug dose varied by the amount of product per skin area (52). Again this is because of lack of depletion. The key to provision of a robust dose–response between dose applied and dose absorbed is depletion or, put another way, the use of rational, low but therapeutic, doses with optimized delivery to give bioavailability in the range of 50–100%.

2. Dose Titration with Enhanced Low-Concentration Formulations

On the basis of simple pharmacokinetic modeling, low, yet rational, doses formulated in enhanced low-concentration topical delivery systems, are predicted to produce a strong relation between dose applied and dose absorbed (52). Although this is almost self-evident, there are, currently, few in vivo experimental studies to support this prediction. Figure 30 shows experimental data in volunteers for the irritant response (on the forehead) to increasing concentrations of a sixfold supersaturated topical retinoic acid system. Despite significant between-subject variation in response, including the presence of hyporesponders, there was a clear dose (concentration)–applied response indicative of a strong dose-applied–dose-absorbed relation from this type of enhanced low-concentration formulation (52,474).

VI. CONCLUSIONS

Standards in topical bioavailability, bioequivalence, and therapeutic equivalence, both in methods and protocols for testing, and also in biopharmaceutical parameters of current products, are some 10–20 years behind those standards that exist in other (e.g., oral) therapies.

Developments in methods and protocols for bioavailability and bioequivalence, including new statistical evaluations, growing knowledge in design of studies, and the extraordinary range of analytical assays, provoked the revision of many older regulatory guidances. Significant attempts have been made by the authorities of many countries to eliminate the vagueness in older guidelines concerning the investigation of bioavailability and bioequivalence. The various recommendations emerged from a series of international symposias (e.g., 2nd EUFEPS Nürnberg Conference: Symposium on Quality and Interchangeability of Topical Products for Local Action, December 8–9, 1995, Nuremberg, Germany; or FDA/AAPS, October, 1996, Washington DC) help to aim for a coherent body of terminology on, and understanding of, bioavailability and bioequivalence that can be used in international harmonization on this subject.

Nevertheless, numerous issues and areas of research still require further indepth discussion and experimentation. The vasoconstrictor test has been utilized as a tool to assess bioavailability and bioequivalence of topical corticosteroids for several decades. However, work is still needed to establish the mechanism of action to determine to what extent skin blanching is actually related to the therapeutic use of such drug products. The vasoconstrictor test has also provoked a heated debate on the possibilities and procedures to show dose–response relation with topical corticosteroids to meet some of the regulatory requirements. Topical corticosteroids represent just one class of dermatological therapeutics that can be judged by biopharmaceutical means; however, for most other dermatological remedies, less-developed or no bioassays exist.

Therefore, clinical trials appear to be the only means of assessing bioavailability–bioequivalence of many topical dosage forms in the foreseeable future. Establishment of bioavailability–bioequivalence criteria tailored to specific diseases, to specific groups of chemicals with the same indication, or to special patient populations are currently being discussed.

Topical bioavailability of most currently marketed formulations is low, in the region of 1–5% of dose applied. From this, given the profound effect that formulation can have on drug partitioning into and diffusion through the skin, bioinequivalence can be anticipated and, from the data available, seems widespread. Even when bioequivalence is established in volunteers, the known occurrence of large variations in absorption between skin sites and depending on skin condition, give little assurance of therapeutic equivalence in the clinical situation. Major improvements in the

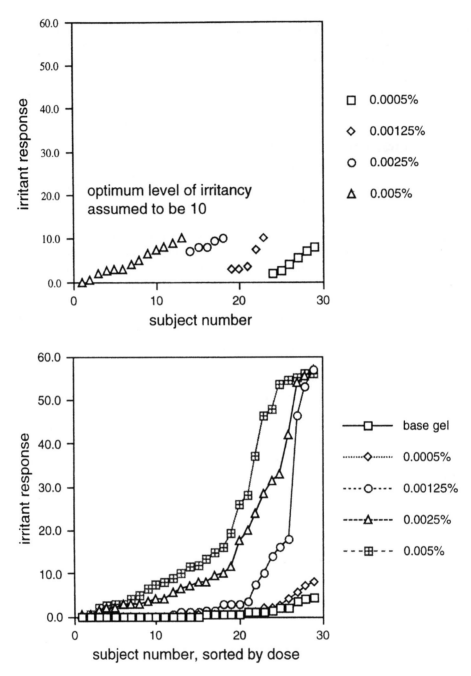

Figure 30   (Bottom) Dose (concentration)–response of forehead irritancy in 30 volunteers to increasing doses of enhanced low-concentration retinoic acid formulation. (From Ref. 52.)

quality of therapy are predicted as a result of individualization of dosing. However, a dose–response from topical dermatological products is poor, confounded by pharmacodynamic variation, but controlled by the pharmacokinetics of current, high-concentration, formulations. Low-concentration, pharmaceutically optimized formulations of 50–100% bioavailability seem, from a therapeutic perspective, to set future standards.

Many scientific and regulatory questions remain to be answered. The artful therapies and dosage forms of the past have definitely fallen into disuse and have been replaced by experimentally tested and proved therapeutics housed in more elegant delivery vehicles. Nevertheless, strong threads of ancient art in today's topical treatment remain, and intuition and feel, trial-and-error, still surrogate for science in topical product development and dermatological practice.

REFERENCES

1. Melnick D, Hochberg M, Oser BL. Physiological availability of vitamins. I. The human bioassay technique. J Nutr 30:67–79, 1945.
2. Morrison AB, Chapman DG, Campbell JA. Further studies on the relation between in vitro disintegration time of tablets and the urinary excretion rates of riboflavin. J Am Pharm Assoc Sci Ed 48:634–637, 1959.
3. Morrison AB, Campbell JA. The relationship between physiological availability of salicylates and riboflavin and in vitro disintegration time of enteric coated tablets. J Am Pharm Assoc Sci Ed 49:473–478, 1960.
4. Levy G. Comparison of dissolution and absorption rates of different commercial aspirin tablets. J Pharm Sci 50:388–392, 1961.
5. Middleton EJ, Davies JM, Morrison AB. Relationship between rate of dissolution, disintegration time, and physiological availability of riboflavin in sugar-coated tablets. J Pharm Sci 53:1378–1380, 1964.
6. Tyrer JH, Eadie MJ, Sutherland JM, Hooper WD. Outbreak of anticonvulsant toxicity in an Australian city. Br Med J 4:271–273, 1970.
7. Danon A, Horowitz J, Ben–Zvi Z. An outbreak of digoxin intoxication. Clin Pharmacol Ther 21:643–646, 1977.
8. Melikian AP, Straughn AB, Slywka GWA, Whyatt PL, Meyer MC. Bioavailability of eleven phenytoin products. J Pharmacokinet Biopharm 5:133–146, 1977.
9. Glazko AJ, Kinkel AW, Alegnani WC, Holmes EL. An evaluation of absorption characteristics of different chloramphenicol preparations in normal human subjects. Clin Pharmacol 9:472–483, 1968.
10. Flynn GL, Weiner ND. Topical and transdermal delivery—provinces of realism. In: Gurny R, Teubner A, eds. Dermal and Transdermal Drug Delivery. Stuttgart: Wissenschaftliche Verlagsgesellschaft, pp 33–65, 1993.
11. Somberg JC. Bioequivalence or therapeutic equivalence. J Clin Pharmacol 26:1, 1986.
12. Dettelbach HR. A time to speak out on bioequivalence and therapeutic equivalence. J Clin Pharmacol 26:307–308, 1986.
13. Lamy PP. Generic equivalents: issues and concerns. J Clin Pharmacol 26:309–316, 1986.
14. Dupuis D, Rougier A, Roguet R, Lotte C, Kalopissis G. In vivo relationship between horny layer reservoir effect and percutaneous absorption in human and rat. J Invest Dermatol 82:353–356, 1984.
15. Guzek DB, Kennedy AH, McNeill SC, Wakshull E, Potts RO. Transdermal drug transport and metabolism. I. Comparison of in vitro and in vivo results. Pharm Res 6:33–39, 1989.

16. Schalla W, Jamoulle J–C, Schaefer H. Localization of compounds in different skin layers and its use as an indicator of percutaneous absorption. In: Bronaugh RL, Maibach HI, eds. Percutaneous Absorption. 2nd ed. New York: Marcel Dekker, pp 283–312, 1989.

17. Surber C. Drug concentration in the skin. In: Maibach HI, ed. Dermatologic Research Techniques. Boca Raton: CRC Press, pp 151–178, 1996.

18. Metzler CM. Statistical criteria. In: Welling PG, Tse FLS, Dighe SV, eds. Pharmaceutical Bioequivalence. New York: Marcel Dekker, pp 35–66, 1991.

19. Hauck WW. Bioequivalence studies of topical preparations: statistical considerations. Int J Dermatol 31(suppl 1):29–33, 1992.

20. Rauws AG. Bioequivalence: a European community regulatory perspective. In: Welling PG, Tse FLS, Dighe SV, eds. Pharmaceutical Bioequivalence. New York: Marcel Dekker, pp 419–441, 1991.

21. FDA, Center for Drug Evaluation and Research. GUIDANCE: Topical Dermatologic Corticosteroids: In Vivo Bioequivalence. Rockville, MD: Division of Bioequivalence, 1995.

22. Stoughton RB. Vasoconstrictor assay: specific applications. In: Maibach HI, Surber C, eds. Topical Corticosteroids. Basel: S Karger, pp 42–53, 1992.

23. Smith EW, Meyer E, Haigh JM. Accuracy and reproducibility of the multiple-reading skin blanching assay. In: Maibach HI, Surber C, eds. Topical Corticosteroids. Basel: S Karger, pp 65–73, 1992.

24. Aalto–Korte K. Quantifying systemic absorption of topical hydrocortisone in erythroderma. Br J Dermatol 133:403–408, 1995.

25. Anderson S, Hauck WW. A new procedure for testing equivalence in comparative bioavailability and other clinical trials. Commun Stat Theor Method 12:2663–2692, 1983.

26. Hauck WW, Anderson S. A new statistical procedure for testing equivalence in two-group comparative bioavailability trials. J Pharmacokinet Biopharm 12:83–91, 1984.

27. Schuirmann DJ. A comparison of the two one-sided tests procedure and the power approach for assessing the equivalence of average bioavailability. J Pharmacokinet Biopharm 15:657–680, 1987.

28. Steinijans VW, Hauschke D, Jonkman JHG. Controversies in bioequivalence studies. Clin Pharmacokinet 22:247–253, 1992.

29. Hauck WW, Anderson S. Types of bioequivalence and related statistical considerations. Int J Clin Pharmacol Ther Toxicol 30:181–187, 1992.

30. Skelly JP, Shah VP, Maibach HI, Guy RH, Wester RC, Flynn GL, Yacobi A. FDA and AAPS report of the workshop on principles and practices of in vitro percutaneous penetration studies: relevance to bioavailability and bioequivalence. Pharm Res 4: 265–267, 1987.

31. Higuchi WI. Analysis of data on the medicament release from ointments. J Pharm Sci 51:802–804, 1962.

32. Shah VP, Elkins J, Hanus J, Noorizadeh C, Skelly JP. In vitro release of hydrocortisone from topical preparations and automated procedure. Pharm Res 8:55–59, 1991.

33. Neubert R, Wohlrab W. In vitro methods for biopharmaceutical evaluation of topical formulations. Acta Pharm Technol 36:197–206, 1990.

34. FDA, Center for Drug Evaluation and Research. Interim Guidance. Topical Corticosteroids: In Vivo Bioequivalence and In Vitro Release Methods. Rockville, MD: Division of Bioequivalence, 1992.

35. United States Pharmacopeia. Vol. XXII. Rockville, MD: United States Pharmacopeial Convention, 1990.

36. Shah VP, Tymes NW, Yamamoto LA, Skelly JP. In vitro dissolution profile of transdermal nitroglycerin patches using paddle method. Int J Pharm 32:243–250, 1986.

37. Shah VP, Tymes NW, Skelly JP. Comparative in vitro release profiles of marketed nitroglycerin patches by different dissolution methods. J Control Release 7:79–86, 1988.
38. Shah VP, Tymes NW, Skelly JP. In vitro release profiles of clonidine transdermal therapeutic systems and scopolamine transdermal patches. Pharm Res 6:346–351, 1989.
39. Reddish G, Wales H. Antiseptic action of *USP* and *NF* ointments. J Am Pharm Assoc 18:576–578, 1929.
40. Christoff K. Einfluss der oberflächenaktiven Substanzen OAS auf die strukturrheologischen Eigenschaften der Salbengrundlagen und die Abgabeschwindigkeit von Arzneimitteln. Pharmazie 22:251–253, 1967.
41. Clark G, Davies G. A new method for measuring diffusion of antiseptics from ointment bases. Br J Dermatol 1:521–525, 1949.
42. Unna P. Zur frage der salbengrundlagen. Dtsch Med Wochenschr 51:197–199, 1926.
43. Poulsen BJ, Young E, Coquilla V, Katz M. Effect of topical vehicle composition on the in vitro release of fluocinolone acetonide and its acetate ester. J Pharm Sci 57:928–933, 1968.
44. Loth H, Holla–Benninger A. Untersuchungen der arzneiliberation aus Salben. 1. Mitteilung: entwicklung eines in vitro Liberationsmodells. Pharm Ind 40:256–261, 1978.
45. De Vos A, Vervoort L, Kinget R. Release of indomethacin from transparent oil–waters gels. J Pharm Sci 83:641–643, 1994.
46. Bottari F, Di Colo G, Nannipieri E, Saettone MF, Serafini MF. Influence of drug concentration on in vitro relase of salicylic acid from ointment bases. J Pharm Sci 63:1779–1783, 1974.
47. Delonca H, Chanal JL, Maillols H, Ghebbi R. Méthode d'étude in vitro des préparations cutanées semi-solides. Comparisons in vitro–in vivo. Pharm Acta Helv 52:51–64, 1977.
48. Guy RH, Fleming R. The estimation of diffusion coefficients using the rotating diffusion cell. Int J Pharm 3:143–149, 1979.
49. Touitou E, Abed L. The permeation behavior of several membrane with potential use in the design of transdermal devices. Pharm Acta Helv 60:193–198, 1985.
50. Hadgraft J, Walters KA, Wotton PK. Facilitated percutaneous absorption: a comparison and evaluation of two in vitro models. Int J Pharm 32:257–263, 1986.
51. Velissaratou AS, Papaioannou G. In vitro release of chlorpheniramine maleate from ointment bases. Int J Pharm 52:83–86, 1989.
52. Davis A. Novel formulations for improved therapy. PhD dissertation, University of Wales, College of Cardiff, 1995.
53. Shah VP, Elkins J, Lam SY, Skelly JP. Determination of in vitro drug release from hydrocortisone creams. Int J Dermatol 53:53–59, 1989.
54. Shah VP, Elkins J, Skelly JP. Relationship between in vivo skin blanching and in vitro release rate for betamethasone valerate creams. J Pharm Sci 81:104–106, 1992.
55. Shah VP, Skelly JP. Practical considerations in developing a quality control (in vitro release) procedure for topical drug products. In: Shah VP, Maibach HI, eds. Topical Drug Bioavailability, Bioequivalence, and Penetration. New York: Plenum Press, pp 107–117, 1993.
56. Behme RJ, Kensler TT, Brooke D. A new technique for determining in vitro release rates of drugs from creams. J Pharm Sci 71:1303–1305, 1982.
57. Corbo M, Schultz TW, Wong GK, Van Buskirk GA. Development and validation of in vitro release testing methods for semisolid formulations. Pharm Technol (Sept):112–128, 1993.
58. Guy RH, Hadgraft J. On the determination of drug release rates from topical dosage forms. Int J Pharm 60:R1–R3, 1990.

59. Rolland A, Demichelis G, Jamoulle J–C, Shroot B. Influence of formulation, receptor fluid, and occlusion, on in vitro drug release from topical dosage forms, using an automated flow-through diffusion cell. Pharm Res 9:82–86, 1992.

60. Fares HM, Zatz JL. Measurement of drug release from topical gels using two types of apparatus. Pharm Technol (Jan):54–56, 1995.

61. Flynn GL. Comparisons between in vitro techniques. Acta Pharm Suec 20:54–55, 1983.

62. Poulsen BJ, Flynn GL. In vitro methods used to study dermal delivery and percutaneous absorption. In: Bronaugh RL, Maibach HI, eds. Percutaneous Absorption. New York: Marcel Dekker, pp 431–459, 1985.

63. Barry BW. Dermatological Formulations. New York: Marcel Dekker, 1983.

64. Addicks W, Weiner N, Flynn G, Curl R, Topp E. Topical drug delivery from thin applications: Theoretical predictions and experimental results. Pharm Res 7:1048–1054, 1990.

65. Flynn GL, Smith RW. Membrane diffusion III: Influence of solvent composition and permeant solubility on membrane transport. J Pharm Sci 61:61–66, 1972.

66. Tanaka S, Takashima Y, Murayama H, Tsuchiya S. Studies on release from ointments: V. Release of hydrocortisone butyrate propionate from topical dosage forms to silicone rubber. Int J Pharm 27:29–38, 1985.

67. Davis AF, Hadgraft J. Effect of supersaturation on membrane transport: 1. Hydrocortisone acetate. Int J Pharm 76:1–8, 1991.

68. Hadgraft J, Ridout G. Development of model membranes for percutaneous absorption measurements. II. Dipalmitoyl phosphatidylcholine, linoleic acid and tetradecane. Int J Pharm 42:97–104, 1988.

69. Buchmann S. Entwicklung eines in vitro absorptionmodelles auf der basis von zellkulturen. Basel: University of Basel, 1987.

70. Steinsträsser I. The organized HaCaT cell culture sheet: a model approach to study epidermal peptide drug metabolism. Zürich: ETH Zürich, 1994.

71. Williams RL. Bioequivalence of topical corticosteroids: a regulatory perspective. Semin Dermatol 31(suppl 1):2–5, 1992.

72. Beastall J, Guy RH, Hadgraft J, Wilding I. The influence of urea on percutaneous absorption. Pharm Res 3:294–297, 1986.

73. Lippold BC, Teubner A. Biopharmazeutische qualität von Arzneiformen, insbesondere für lokale Anwendung, abgeleitet aus Wirkungsmessungen. Pharm Ind 43:71–73, 1981.

74. Lippold BC, Teubner A. Einfluss verschiedener salbengrundlagen auf die Wirkung von Nicotinsäurebenzylester in lösungssalben. Pharm Ind 43:1123–1133, 1981.

75. Lippold BC, Reimann H. Wirkungsbeeinflussung bei lösungssalben durch Vehikel am beispiel von Methylnicotinat. Acta Pharm Technol 3:136–142, 1989.

76. Katz M, Poulsen BJ. Corticoid, vehicle, and skin interaction in percutaneous absorption. J Soc Cosmet Chem 23:565–590, 1972.

77. Caron D, Queille–Roussel C, Shah VP, Schaefer H. Correlation between the drug penetration and the blanching effect of topically applied hydrocortisone creams in human beings. J Am Acad Dermatol 23:458–462, 1990.

78. Kaiho F, Nomura H, Makabe E, Kato Y. Percutaneous absorption of indomethacin from mixtures of fatty alcohol and propylene glycol (FAPG base) through rat skin: effects of fatty acid added to FAPG base. Chem Pharm Bull 35:2928–2934, 1987.

79. Tojo K. Concentration profile in plasma after transdermal drug delivery. Int J Pharm 43:201–205, 1988.

80. Franz TJ. Percutaneous absorption. On the relevance of in vitro data. J Invest Dermatol 64:190–195, 1975.

81. Feldmann RJ, Maibach HI. Absorption of some organic compounds through the skin in man. J Invest Dermatol 54:399–404, 1970.

82. Sheth NV, McKeough MB, Spruance SL. Measurement of stratum corneum drug reservoir to predict the therapeutic efficacy of topical iododeoxyuridine for herpes simplex. J Invest Dermatol 89:598–602, 1987.

83. Öhman H, Vahlquist A. In vivo studies concerning a pH gradient in human stratum corneum and upper epidermis. Acta Derm Venereol (Stockh) 74:375–379, 1994.

84. Erikson G, Lamke L. Regeneration of human epidermal surface and water barrier function after stripping. Acta Derm Venereol (Stockh) 51:169–178, 1971.

85. Wilhelm D, Elsner P, Maibach HI. Standardized trauma (tape-stripping) in human vulvar and forearm skin. Acta Derm Venereol (Stockh) 71:123–126, 1991.

86. Downes AM, Matoltsy AG, Sweeney TM. Rate of turnover of the stratum corneum in hairless mice. J Invest Dermatol 49:400–405, 1967.

87. Pinkus H. Examination of the epidermis by the strip method of removing horny layers. I. Observations on thickness of the horny layer, and on mitotic activity after stripping. J Invest Dermatol 16:383–386, 1951.

88. Pinkus H. Examination of the epidermis by the strip method of removing horny layers. II. Biometric data on regeneration of human epidermis. J Invest Dermatol 16:431–447, 1951.

89. Piérard GE, Piérard–Franchimont C, Saint–Léger D, Kligman AM. Squamometry: the assessment of xerosis by colorimetry of D-Squame adhesive discs. J Soc Cosmet Chem 47:297–305, 1992.

90. Schatz H, Kligman AM, Manning S, Stoudemayer T. Quantification of dry (xerotic) skin by image analysis of scales removed by adhesive discs (D-Squame). J Soc Cosmet Chem 44:53–63, 1993.

91. Malkinson FD, Ferguson EH. Percutaneous absorption of hydrocortisone-4-^{14}C in two human subjects. J Invest Dermatol 25:281–285, 1955.

92. Vickers CFH. Existence of reservoir in the stratum corneum. Arch Dermatol 88:20–23, 1963.

93. Stoughton RB. Dimethylsulfoxide (DMSO) induction of a steroid reservoir in human skin. Arch Dermatol 91:657–660, 1965.

94. Carr RD, Wieland RG. Corticosteroid reservoir in the stratum corneum. Arch Dermatol 94:81–84, 1966.

95. Carr RD, Tarnowski WM. The corticosteroid reservoir. Arch Dermatol 94:639–642, 1966.

96. Munro DD. The relationship between percutaneous absorption and stratum corneum retention. Br J Dermatol 81(suppl 4):92–97, 1969.

97. Lücker P, Nowak H, Stüttgen G, Werner G. Penetrationskinetik eines Tritium-markierten 9Alpha-fluor-16 methylen-prednisolonesters nach epicutaner Applikation beim Menschen. Arzneimittelforschung 18:27–29, 1968.

98. Tsai J–C, Cappel MJ, Flynn GL, Weiner ND, Kreuter J, Ferry J. Drug and vehicle deposition from topical applications: use of in vitro mass balance technique with minoxidil solutions. J Pharm Sci 81:736–743, 1992.

99. Tojo K, Lee AC. A method for predicting steady-state rate of skin penetration in vivo. J Invest Dermatol 92:105–108, 1989.

100. Moon KC, Wester RC, Maibach HI. Diseased skin models in the hairless guinea pig: in vivo percutaneous absorption. Dermatologica 180:8–121, 1990.

101. Holoyo–Tomoka MT, Kligman AM. Does cellophane tape stripping remove the horny layer? Arch Dermatol 106:767–768, 1972.

102. Malkinson FD. Studies on the percutaneous absorption of ^{14}C-labelled steroids by use of glass-flow cell. J Invest Dermatol 31:19–28, 1958.

103. Feldmann RJ, Maibach HI. Penetration of ^{14}C hydrocortisone through normal skin: effect of stripping and occlusion. Arch Dermatol 91:661–666, 1965.

104. Zesch A, Schaefer H. Penetrationskinetik von radiomarkierten Hydrocortisone aus verschiedenartigen salbengrundlagen in die Menschliche haut. Arch Dermatol Forsch 252:245–256, 1975.

105. Zesch A, Schaefer H, Hoffmann W. Barriere und reservoirfunktion der einzelnen hornschichtlagen der Menschlichen haut für lokal aufgetragene Arzneimittel. Arch Dermatol Forsch 246:103–107, 1973.

106. Zesch A. Reservoirfunktion der hornschicht. In: Klaschka F, ed. Stratum corneum: Struktur und Funktion. Berlin: Grosse Verlag, pp 63–76, 1981.

107. Dupuis D, Rougier A, Roguet R, Lotte C. The measurement of the stratum corneum reservoir: a simple method to predict the influence of vehicles on in vivo percutaneous absorption. Br J Dermatol 115:233–238, 1986.

108. Rougier A, Lotte C, Maibach HI. The hairless rat: a relevant animal model to predict in vivo percutaneous absorption in humans? J Invest Dermatol 88:577–581, 1987.

109. Rougier A, Lotte C, Dupuis D. An original predictive method for in vivo percutaneous absorption studies. J Soc Cosmet Chem 38:397–417, 1987.

110. Rougier A, Rallis M, Krien P, Lotte C. In vivo percutaneous absorption: a key role for stratum corneum/vehicle partitioning. Arch Dermatol Res 282:498–505, 1990.

111. Feldmann RJ, Maibach HI. Percutaneous penetration of steroids in man. J Invest Dermatol 52:89–94, 1969.

112. Feldmann RJ, Maibach HI. Percutaneous penetration of some pesticides and herbicides in man. Toxicol Appl Pharmacol 28:126–132, 1974.

113. Auton TR. Skin stripping and science: a mechanistic interpretation using mathematical modelling of skin deposition as a predictor of total absorption. In: Scott RC, Guy RH, Hadgraft J, Boddé HE, eds. Prediction of Percutaneous Penetration. Vol. 2. London: IBC Technical Services, pp 558–576, 1990.

114. Knight AG. The activity of various topical griseofulvin preparations and the appearance of oral griseofulvin in the stratum corneum. Br J Dermatol 91:49–55, 1974.

115. Dittmar W, Jovic N. Laboratory technique alternative to in vivo experiments for studying the liberation, penetration and fungicidal action of topical antimycotic agents in the skin, including ciclopirox olamine. Mykosen 30:326–342, 1986.

116. Pershing LK, Corlett J, Jorgensen C. In vivo pharmacokinetics and pharmacodynamics of topical ketoconazole and miconazole in human stratum corneum. Antimicrob Agents Chemother 38:90–95, 1994.

117. Pershing LK, Krueger GG. Human skin sandwich flap model for percutaneous absorption. In: Bronaugh RL, Maibach HI, eds. Percutaneous Absorption. New York: Marcel Dekker, pp 397–414, 1989.

118. Parry GE, Dunn P, Shah VP, Pershing LK. Acyclovir bioavailability in human skin. J Invest Dermatol 98:856–863, 1992.

119. Michel C, Purmann T, Mentrup E, Seiller E, Kreuter J. Effect of liposomes on percutaneous penetration of lipophilic materials. Int J Pharm 84:93–105, 1992.

120. Pershing LK, Silver BS, Krueger GG, Shah VP, Skelly JP. Feasibility of measuring the bioavailability of topical betamethasone dipropionate in commercial formulations using drug in skin and a skin blanching bioassay. Pharm Res 9:45–51, 1992.

121. Pershing LK, Lambert LD, Shah VP, Lam SY. Variability and correlation of chromameter and tape-stripping methods with the visual skin blanching assay in quantitative assessment of topical 0.05% betamethasone dipropionate bioavailability in humans. Int J Pharm 86:201–210, 1992.

122. Pershing LK, Corlett JL, Lambert LD, Poncelet CE. Circadian activity of topical 0.05% betamethasone dipropionate in human skin in vivo. J Invest Dermatol 102:734–739, 1994.

123. Pershing LK, Lambert L, Wright ED, Shah VP, Williams RL. Topical 0.05% beta-methasone dipropionate: pharmacokinetic and pharmacodynamic dose–response studies in humans. Arch Dermatol 130:740–747, 1994.

124. Tsai JC, Cappel MJ, Weiner ND, Flynn GL, Ferry J. Solvent effects on the harvesting of stratum corneum from hairless mouse skin through adhesive tape stripping in vitro. Int J Pharm 68:127–133, 1991.

125. Tsai JC, Weiner ND, Flynn GL, Ferry J. Properties of adhesive tapes used for stratum corneum stripping. Int J Pharm 72:227–231, 1991.

126. Henn U, Surber C, Schweitzer A, Bieli E. D-Squame adhesive tapes for standardized stratum corneum stripping. In: Brain KR, James VJ, Walters KA, eds. Prediction of Percutaneous Penetration. Vol. 3b. Cardiff: STS Publishing, pp 477–481, 1993.

126a. FDA, Center for Drug Evaluation and Research. Draft-Guidance for Industry: Topical Dermatological Drug Product NDAs and ANDAs—In Vivo Bioavailability, Bioequivalence, In Vitro Release, and Associated Studies. Rockville, MD: Division of Bioequivalence, 1998.

126b. Surber C, Schwarb FP, Smith EW. Tape-stripping technique. In: Bronaugh RL, Maibach HI, eds. Percutaneous Absorption: Drugs—Cosmetics—Mechanisms—Methodology, 3rd ed. New York: Marcel Dekker, pp 395–409, 1999.

127. Surber C, Wilhelm KP, Hori M, Maibach HI, Guy RH. Optimization of topical therapy: partitioning of drugs into stratum corneum. Pharm Res 7:1320–1324, 1990.

128. Surber C, Wilhelm KP, Berman D, Maibach HI. In vivo skin penetration of acitretin in volunteers using three different sampling techniques. Pharm Res 10:1291–1294, 1993.

129. Laugier J–P, Surber C, Bun H, Geiger J–M, Wilhelm K–P, Durand A, Maibach HI. Determination of acitretin in the skin, in the suction blister, and in plasma of human volunteers after multiple oral dosing. J Pharm Sci 83:623–628, 1994.

130. Fry L, McMinn RMH. Topical methotrexate in psoriasis. Arch Dermatol 96:483–488, 1967.

131. Surber C, Itin P, Büchner S. Clinical controversy on the effect of topical ciclosporin [*sic*] A: what is the target organ. Dermatology 185:242–245, 1992.

132. Schaefer H, Lamaud E. Standardization of experimental models. In: Shroot B, Schaefer H, eds. Skin Pharmacokinetics. Basel: Karger, pp 77–80, 1987.

133. Schaefer H, Stüttgen G, Zesch A, Schalla W, Gazith J. Quantitative determination of percutaneous absorption of radiolabeled drugs in vitro and in vivo by human skin. Curr Probl Dermatol 7:80–94, 1978.

134. Wojciechowski Z, Pershing LK, Huether S, Leonard L, Burton SA, Higuchi WI, Krueger GG. An experimental skin sandwich flap on an independent vascular supply for the study of percutaneous absorption. J Invest Dermatol 88:439–446, 1987.

135. Wester RC, Noonan PK. Topical bioavailability of a potential anti-acne agent (SC-23110) as determined by cumulative excretion and areas under plasma concentration time curves. J Invest Dermatol 70:92–94, 1978.

136. Wester RC, Noonan PK, Smeach S, Kosobud L. Pharmacokinetics and bioavailability of intravenous and topical nitroglycerin in the rhesus monkey: estimate of percutaneous first-pass metabolism. J Pharm Sci 72:745–748, 1983.

137. Wester RC, Maibach HI. Cutaneous pharmacokinetics: 10 steps to percutaneous absorption. Drug Metab Rev 14:169–205, 1983.

138. Bucks DAW, McMaster JMA, Maibach HI, Guy RH. Bioavailability of topically administered steroids: a "mass balance" technique. J Invest Dermatol 91:29–33, 1988.

139. Feldmann RJ, Maibach HI. Regional variation in percutaneous penetration of ^{14}C cortisol in man. J Invest Dermatol 48:181–183, 1967.

140. Maibach HI. In vivo percutaneous penetration of corticoids in man and unresolved problems in their efficacy. Dermatologica 152(suppl 1):11–25, 1976.

141. Wester RC, Maibach HI. Relationship of topical dose and percutaneous absorption in rhesus monkey and man. J Invest Dermatol 67:518–520, 1976.

142. Melendres JL, Bucks DAW, Camel E, Wester RC, Maibach HI. In vivo percutaneous absorption of hydrocortisone: multiple-application dosing in man. Pharmacol Res 9: 1164–1167, 1992.

143. Wheeler DA, Moyler DA, Thirkettle JT. Instrumental colour assessment: some practical experience. J Soc Cosmet Chem 27:15–43, 1976.

144. Agache P, Girardot I, Bernengo JC. Optical properties of the skin. In: Leveque JL, ed. Cutaneous Investigation in Health and Disease: Noninvasive Methods and Instrumentation. New York: Marcel Dekker, pp 241–274, 1989.

145. Weatherall IL, Coombs BD. Skin color measurements in terms of CIELAB color space values. J Invest Dermatol 99:468–473, 1992.

146. el-Gammal S, Hoffmann K, Steiert P, Grassmüller J, Dirschka T, Altmeyer P. Objective assessment of intra- and inter-individual skin colour variability: an analysis of human skin reaction to sun and UVB. In: Marks R, Plewig G, eds. The Environmental Threat to the Skin. London: Martin Dunitz, pp 99–116, 1992.

147. Seitz JC, Whitemore CG. Measurement of erythema and tanning responses in human skin using a tri-stimulus colorimeter. Dermatologica 177:70–75, 1988.

148. Clarys P, Buchet E, Barel AO. Evaluation of different topical vasodilatory products with non-invasive techniques. In: Scott RC, Guy RH, Hadgraft J, Boddé HE, eds. Prediction of Percutaneous Penetration. Vol. 2. London: IBC Technical Services, pp 46–59, 1991.

149. Wilhelm K–P, Surber C, Maibach HI. Quantification of sodium lauryl sulfate irritant dermatitis in man: comparison of four techniques: skin color reflectance, transepidermal water loss, laser Doppler flow measurement and visual scores. Arch Dermatol Res 281:293–295, 1989.

150. Queille–Roussel C, Duteil L, Padilla J–M, Poncet M, Czernielewski J. Objective assessemnt of topical anti-inflammatory drug activity on experimentally induced nickel contact dermatitis: comparison between visual scoring, colorimetry, laser Doppler velocimetry and transepidermal water loss. Skin Pharmacol 3:248–255, 1990.

151. Zhou J, Mark R, Stoudemayer T, Skar A, Lichten JL, Gabriel KL. The value of multiple instrumental and clinical methods, repeated patch applications, and daily evaluations for assessing stratum corneum changes induced by surfactants. J Soc Cosmet Chem 42:105–128, 1991.

152. Lathi A, Kopola H, Harila A, Mullylä A, Hannuksela M. Assessment of skin erythema by eye, laser Doppler flowmeter, spectroradiometer, two-channel erythema meter and Minolta chromameter. Arch Dermatol Res 285:278–282, 1993.

153. Treffel P, Gabard B, Bieli E. Relationship between the in vitro diclofenac epidermal level and the in vivo anti-inflammatory efficacy in the methylnicotinate test. In: Brain KR, James VJ, Walters KA, eds. Prediction of Percutaneous Penetration. Vol. 3b. Cardiff: STS Publishing, pp 520–529, 1993.

154. Clarys P, Barel AO, Gabard B. Objective evaluation of the skin blanching induced by corticosteroids. Influence of corticosteroid concentration and body region. In: Brain KR, James VJ, Walters KA, eds. Prediction of Percutaneous Penetration. Vol. 3b. Cardiff: STS Publishing, pp 502–509, 1993.

154a. Schwarb FP, Smith EW, Haigh JM, Surber C. Analysis of chromameter results obtained from corticosteroid-induced skin blanching assay: comparison of visual and chromameter data. Eur J Pharmacol Biopharmacol 47:261–267, 1999.

154b. Demana PH, Smith EW, Walker RB, Haigh JM, Kanfer I. Evaluation of the proposed FDA pilot dose–response methodology for topical corticosteroid bioequivalence testing. Pharm Res 14:303–309, 1997.

154c. Smith EW, Walker RB, Haigh JM. Analysis of chromameter results obtained from corticosteroid-induced blanching. I. Manipulation of data. Pharm Res 15:280–285, 1998.

155. Troilius A, Wardell K, Bornmyr S, Nilson GE, Ljunggren B. Evaluation of port wine stain perfusion by laser Doppler imaging and thermography before and after argon laser treatment. Acta Derm Venereol (Stockh) 72:6–10, 1992.

156. Quinn AG, McLelland J, Essex T, Farr PM. Quantification of contact allergic inflammation: a comparison of existing methods with a scanning laser Doppler velocimeter. Acta Derm Venereol (Stockh) 73:21–25, 1993.

157. Speight EL, Essex TJH, Farr PM. The study of plaques of psoriasis using a scanning laser Doppler velocimeter. Br J Dermatol 128:519–524, 1993.

158. Guy RH, Tur E, Bugatto B, Gaebel C, Sheiner LB, Maibach HI. Pharmacodynamic measurements of methyl nicotinate percutaneous absorption. Pharm Res 1:76–81, 1984.

159. Poelman MC, Piot B, Guyon M, Leveque JL. Assessment of topical non-steroidal anti-inflammatory drugs. J Pharm Pharmacol 41:720–722, 1989.

160. Hori M, Ohtsuka S, Sunami M, Guy RH, Maibach HI. Cutaneous pharmacodynamics of transdermally delivered isosorbide dinitrate. Pharm Res 7:1298–1301, 1990.

161. Treffel P, Gabard B. Feasibility of measuring the bioavailability of topical ibuprofen in commercial formulations using drug content in epidermis and methyl nicotinate skin inflammation assay. Skin Pharmacol 129:286–275, 1993.

162. Van Scott EJ. Experimental animal integumental models for screening potential dermatologic drugs. In: Montagna W, Van Scott EJ, Stoughton RB, eds. Pharmacology and the Skin. Advances in Biology of the Skin. New York: Appleton Century Croft, pp 523–533, 1972.

163. Kligman LH, Kligman AM. The effect on rhino mouse skin of agents which influence keratinization and exfoliation. J Invest Dermatol 73:354–358, 1979.

164. Mezick JA, Bhatia MC, Shea LM. Anti-acne activity of retinoids in the rhino mouse. In: Maibach HI, Lowe NJ, eds. Models in Dermatology. Basel: S Karger, pp 59–63, 1985.

165. Bouclier M, Chatelus A, Ferracin J, Delain C, Shroot B, Hensby CN. Quantification of epidermal histological changes induced by topical retinoids and CD271 in the rhino mouse model using a standarized image analysis technique. Skin Pharmacol 4:65–73, 1991.

166. Meyer E, Smith EW, Haigh JM. Sensitivity of different areas of the flexor aspect of the human forearm to corticosteroid-induced skin blanching. Br J Dermatol 127:379–381, 1992.

167. Gilchrest BA, Soter NA, Hawks JLM. Histologic changes associated with ultraviolet-A induced erythema in normal skin. J Am Acad Dermatol 9:213–219, 1983.

168. Woodbury RA, Kligman LH, Woodbury MJ, Kligman AM. Rapid assay of anti-inflammatory activity of topical corticosteroids by inhibition of a UVA-induced neutrophil infiltration in hairless mouse skin. I. The assay and its sensitivity. Acta Derm Venereol (Stockh) 74:15–17, 1994.

169. Kligman LH. Rapid assay of anti-inflammatory activity of topical corticosteroids by inhibition of a UVA-induced neutrophil infiltration in hairless mouse skin. II. Assessment of name brand versus generic potency. Acta Derm Venereol (Stockh) 74:18–19, 1994.

170. Michaelsson G, Juhlin L, Vahlquist A. Effects of oral zinc and vitamin A in acne. Arch Dermatol 133:31–136, 1977.

171. Consensus Report of the European Task Force on Atopic Dermatits. Severity scoring of atopic dermatitis: the SCORAD Index. Dermatology 186:23–31, 1993.

172. Marks R, Barton SP, Shuttleworth D, Finlay AY. Assessment of disease progress in psoriasis. Arch Dermatol 125:235–240, 1989.

173. Berardesca E, Vignoli GP, Farinelli N, Vignini M, Distante F, Rabbiosi G. Noninvasive evaluation of topical calicipotriol versus clobetasol in the treatment of psoriasis. Acta Derm Venereol (Stockh) 74:302–304, 1994.

174. Ramsay B, Lawrence AM. Measurement of involved surface area in patients with poriasis. Br J Dermatol 124:565–570, 1991.

175. Larnier C, Ortonne J–P, Venot A, Faivre B, Béani J–C, Thomas P, Brown TC, Sendagorta E. Evaluation of cutaneous photodamage using a photographic scale. Br J Dermatol 130:167–173, 1995.

176. Seitz JC, Spencer TS. Clinical research photography. In: Maibach HI, Lowe NJ, eds. Models in Dermatology. Basel: S Karger, pp 63–70, 1989.

177. Fredriksson T, Pettersson U. Severe psoriasis—oral therapy with a new retinoid. Dermatologica 157:238–244, 1978.

178. Wolska H, Jablonska S, Bounameaux Y. Etretinate in severe psoriasis. Results of a double-blind study and maintenance therapy in pustilar psoriasis. J Am Acad Dermatol 9:883–889, 1983.

179. Geiger J–M, Czarnetzki BM. Acitretin (Ro 10-1670): overall evaluation of clinical studies. Dermatologica 176:182–190, 1988.

180. Altmeyer PJ, Matthes U, Pawlak F, Hoffmann K, Frosch PJ, Ruppert P, Wassilew SW, Horn T, Kreysel HW, Lutz G, Barth J, Rietzschel I, Joshi RK. Antipsoriatic effect of fumaric acid derivatives. J Am Acad Dermatol 30:977–981, 1994.

181. Kimbrough–Green CK, Griffith CEM, Finkel LJ, Hamilton TA, Bulengo–Ransby SM, Ellis CN, Voorhees JJ. Topical retinoic acid (tretinoin) for melasma in black patients. Arch Dermatol 130:727–733, 1994.

182. Surber C, Lüdin E, Flückiger A, Dubach UC, Ziegler WH. Pain assessment after intramuscular injection. Arzneimittelforschung 44:1389–1394, 1994.

183. Santos DJ, Juneja M, Bridenbaugh PO. A device for uniform testing of sensory neural blockade during regional anesthesia. Anesth Analg 66:581–582, 1987.

184. Funk JO. Horizons in pharmacologic intervention in allergic contact dermatitis. J Am Acad Dermatol 31:999–1014, 1994.

185. Meingasser JG, Stütz A. Immunosuppressive macrolides of the type FK 506: a novel class of topical agents for treatment of skin diseases? J Invest Dermatol 98:851–855, 1992.

186. Hannuksela M. Propylene glycol promotes allergic patch test reactions. Boll Dermatol Allergol Prof 2:40–44, 1987.

187. Camarasa JG, Serra–Baldrich E. Allergic contact dermatitis from acyclovir. Contact Dermatitis 19:235–236, 1988.

188. Goh CL. Compound allergy to Spectraban 15 lotion and Zovirax cream. Contact Dermatitis 22:61–62, 1990.

189. Gola M, Francalanci S, Brusi C, Lombardi P, Sertoli A. Contact sensitization to acyclovir. Contact Dermatitis 20:394–395, 1990.

190. Baes H, Van Hecke E. Contact dermatits from Zovirax cream. Contact Dermatits 23: 200–201, 1990.

191. Sidwell RW, Huffman JH, Schafer TW, Shipman C. Influence of vehicle on topical efficacy of 2-acetylpyridine thiosemicarbazone and related derivates on in vivo type 2 herpes simplex virus infections. Chemotherapy 36:58–69, 1990.

192. Weiner N, Williams N, Birch G, Ramachandran C, Shipman C, Flynn G. Topical delivery of liposomally encapsulated interferon evaluated in a cutaneous herpes guinea pig model. Antimicrob Agents Chemother 33:1217–1221, 1989.

193. Imanidis G, Song W, Lee PH, Su MH, Kern ER, Higuchi WI. Estimation of skin target site acyclovir concentrations following controlled (trans)dermal drug delivery

in topical and systemic treatment of cutaneous HSV-1 infections in hairless mice. Pharm Res 11:1035–1041, 1994.

194. Lee PH, Su MH, Kern ER, Higuchi WI. Novel animal model for evaluating topical efficacy of antiviral agents: flux versus efficacy correlations in acyclovir treatment of cutaneous herpes simplex virus type 1 (HSV-1) infections in hairless mice. Pharm Res 9:979–989, 1992.

195. Gonsho A, Imanidis G, Vogt P, Kern ER, Tsuge H, Su MH, Choi SH, Higuchi WI. Controlled (trans)dermal delivery of an antiviral agent (acyclovir). I: An in vivo animal model for efficacy evaluation in cutaneous HSV-1 infections. Int J Pharm 65: 183–194, 1990.

196. Lee PH, Su MH, Ghanem AH, Inamori T, Kern ER, Higuchi WI. An application of the C* concept in predicting the topical efficacy of finite dose acyclovir in the treatment of cutaneous HSV-1 infections in hairless mice. Int J Pharm 93:139–152, 1993.

197. Price RW. Neurobiology of human herpes virus infections. Crit Rev Clin Neurobiol 2:61–123, 1986.

198. FDA, Center of Drug Evaluation and Research. Draft guidance for the performance of a bioequivalence study for topical antifungal products. Rockville, MD: Division of Bioequivalence.

199. Aly R, Maibach HI. Ciclopirox olamine lotion 1%: bioequivalence to ciclopirox olamine cream 1% and clinical efficacy in tinea pedis. Clin Ther 11:290–303, 1989.

200. Eady EA, Cove JH. Topical antibiotic therapy; current status and future prospects. Drugs Exp Clin Res 16:423–433, 1990.

201. Held JL, Kalb RE, Ruszkowski AM, DeLeo V. Allergic contact dermatitis from bacitracin. J Am Acad Dermatol 17:592–594, 1987.

202. Villiger JW, Robertson WD, Kanji K. A comparison of the new topical antibiotic mupirocin (Bactroban) with oral antibiotics in the treatment of skin infections in general practice. Curr Med Res Opin 10:339–345, 1986.

203. Mertz PM, Marshall DA, Eaglestein WH. Topical mupirocin treatment of impetigo is equal to oral erythromycin therapy. Arch Dermatol 125:1069–1073, 1989.

204. Schwanitz HJ, Macher E. Internal versus topical tetracycline therapy in acne. Z Hautkrankh 59:1515–1521, 1984.

205. Eady EA, Holland KT, Cunliffe WJ. Should topical antibiotics be used for the treatment of acne vulagris? Br J Dermatol 107:235–246, 1982.

206. Buckley MM, Benfiled P. Eutectic lidocaine/prilocaine cream. Drugs 46:126–151, 1993.

207. Foldvari M. In vitro cutaneous and percutaneous delivery and in vivo efficacy of tetracaine from liposomal and conventional vehicles. Pharm Res 11:1593–1598, 1994.

208. Kushla GP, Zatz JL. Evaluation of noninvasive method for monitoring percutaneous absorption of lidocaine in vivo. Pharm Res 7:1033–1037, 1990.

209. Kushla GP, Zatz JL, Mills OH, Berger RS. Noninvasive assessment of anesthetic activity of topical lidocaine formulations. J Pharm Sci 82:1118–1122, 1993.

210. King RB. Topical aspirin in chloroform and the relief of pain due to herpes zoster and postherapeutic neuralgia. Arch Neurol 50:1046–1053, 1993.

211. Brookes FL, Hugo WB, Denyer SP. Transformation of betamethasone-17-valerate by skin microflora. J Pharm Pharmacol 34:61P, 1982.

212. Abu Shamat MS, Beckett AH. Glyceryl trinitrate: metabolism by intestinal flora. J Pharm Pharmacol 35:71P, 1983.

213. Denyer SP, Hugo WB, O'Brien M. Metabolism of glyeryl trinitrate by skin staphylococci. J Pharm Pharmacol 36:61P, 1984.

214. Denyer SP, Guy RH, Hadgraft J, Hugo WB. The microbial degradation of topically applied drugs. Int J Pharm 26:89–97, 1985.

215. Martin RJ, Denyer SP, Hadgraft J. Skin metabolism of topically applied compounds. Int J Pharm 39:23–32, 1987.

216. Shaw JE, Chandrasekaran SK. Controlled topical delivery of drugs for systemic action. Drug Metab Rev 8:223–233, 1978.

217. Samuelov Y, Donbrow M, Friedman M. Effect of pH on salicylic acid permeation through ethyl cellulose–PEG 4000 films. J Pharm Pharmacol 31:120–121, 1979.

218. Jack L, Cameron BD, Scott RC, Hadgraft J. In vitro percutaneous absorption of salicylic acid; effect of vehicle pH. In: Scott RC, Guy RH, Hadgraft J, Boddé HE, eds. Prediction of Percutaneous Penetration. Vol. 2. London: IBC Technical Services, pp 515–518, 1991.

219. Özer AY, Farivar M, Hincal AA. Temperature- and pH-sensitive liposomes. Eur J Pharm Biopharm 39:97–101, 1993.

220. Tregear RT. Physical Functions of Skin. London: Academic Press, 1966.

221. Wheatley VR. Proc Sci Soc Toilet Goods Assoc 39:25, 1963.

222. Wheatley VR, Flech P, Esoda EGJ, Coon WM, Mandol R. Studies of the chemical composition of the horny layer lipids. J Invest Dermatol 43:395–405, 1964.

223. Wilkenson DI. Variability in composition of surface lipids. J Invest Dermatol 54:339–343, 1969.

224. Cheng Y–H, Hosoya O, Sugibayashi K, Morimoto Y. Effect of skin surface lipid on skin permeation of lidocaine from pressure sensitive adhesives. Biol Pharm Bull 17:1640–1644, 1994.

225. Cummings EG. Temperature and concentration effects on penetration of N-octyamine through human skin in situ. J Invest Dermatol 53:64–70, 1969.

226. Katz M, Poulsen BJ. Absorption of drugs through skin. In: Brodie BB, Gillette JR, eds. Handbook of Experimental Pharmacology, 28. New York: Springer, pp 103–174, 1971.

227. Rothenborg HW, Menné T, Sjølin KE. Temperature dependent primary irritant dermatitis from lemon perfume. Contact Dermatitis 3:37–48, 1977.

228. Emilson A, Lindberg M, Forslind B. The temperature effect on in vitro penetration of sodium lauryl sulfate and nickel chloride through human skin. Acta Derm Venereol (Stockh) 73:203–207, 1993.

229. James M, Marty J–P, Wepierre J. Diffusion localisée des substances absorbées par voie percutanée. CR Acad Sci (Paris) 278(Ser. D):2063–2066, 1974.

230. James M, Wepierre J. Absorption percutanée de la thyroxine ^{125}I et de la triiodothyronine ^{125}I chez le rat in vivo. Ann Pharm Fr 32:633–640, 1974.

231. Guy RH, Maibach HI. Drug delivery to local subcutaneous structures following topical administration. J Pharm Sci 72:1375–1380, 1983.

232. Marty J–P, Guy RH, Maibach HI. Percutaneous penetration as a method of delivery to muscle and other tissue. In: Bronaugh RL, Maibach HI, eds. Percutaneous Absorption: Mechanism—Methodology—Drug Delivery. New York: Marcel Dekker, pp 511–529, 1989.

233. Singh P, Roberts M. Deep tissue penetration of bases and steroids after dermal application in rat. J Pharm Pharmacol 46:956–964, 1994.

234. Benowitz NL, Jacob P, Olsson P, Johansson CJ. Intravenous nicotine retards transdermal absorption of nicotine: evidence of blood flow-limited percutaneous absorption. Clin Pharmacol Ther 52:223–230, 1992.

235. Benowitz NL, Jacob P, Jones RT, Rosenberg J. Interindividual variability in metabolism and cardiovascular effects of nicotine in man. J Pharmacol Exp Ther 221:368–372, 1982.

236. Monteiro–Riviere NA. Altered epidermal morphology secondary to lidocaine iontophoresis: in vivo and in vitro studies in porcine skin. Fundam Appl Toxicol 15:174–185, 1990.

237. Riviere JE, Sage BH, Monteiro–Riviere NA. Transdermal lidocaine ionophoresis in isolated perfused porcine skin. J Toxicol Cutan Ocul Toxicol 8:493–504, 1990.

238. Auclair F, Besnard M, Dupont C, Wepierre J. Importance of blood flow to the local distribution of drugs after percutaneous absorption in the bipediculated dorsal flap of the hairless rat. Skin Pharmacol 4:1–8, 1991.

239. Nakashima E, Noonan PK, Benet LZ. Transdermal bioavailability and first-pass skin matabolism: a preliminary evaluation with nitroglycerin. J Pharmacokinet Biopharm 15:423–437, 1987.

240. Upton RN. Regional pharmacokinetics. I. Physiological and physicochemical basis. Biopharm Drug Dispos 11:647–662, 1990.

241. Turpeinen M, Mashkilleyson N, Bjorksten F, Salo OP. Percutaneous absorption of hydrocortisone during exacerbation and remission of atopic dermatitis in adults. Acta Derm Venereol (Stockh) 68:331–335, 1988.

242. Turpeinen M. Influence of age and severity of dermatitis on percutaneous absorption of hydrocortisone in children. Br J Dermatol 118:517–522, 1988.

243. Turpeinen M. Absorption of hydrocortisone from the skin reservoir in atopic dermatitis. Br J Dermatol 124:358–360, 1991.

244. Turpeinen M, Lehtokoski–Lehtiniemi E, Leisti S, Salo OP. Percutaneous absorption during and after the acute phase of dermatitis in children. Pediatr Dermatol 5:276–279, 1988.

245. Wester RC, Bucks DAW, Maibach HI. Percutaneous absorption of hydrocortisone in psoriatic patients and normal volunteers. J Am Acad Dermatol 8:645–647, 1983.

246. Wester RC, Mobayen M, Ryatt K, Bucks D, Maibach HI. In vivo percutaneous absorption of dithranol on psoriatic and normal volunteers. In: Farber E, ed. Psoriasis. Amsterdam: Elsevier, pp 429, 1987.

247. Schaefer H, Zesch A, Stüttgen G. Skin Permeability. Berlin: Springer, 1982.

248. Erlanger M, Martz G, Ott F, Storck H, Rider J, Kessler S. Cutaneous absorption and urinary excretion of 6-^{14}C-5-fluorouracil following application in an ointment to healthy and diseased human skin. Dermatologica (suppl 1):7–14, 1970.

249. Foreman M, Clanchan I, Kelly I. The diffusion of nandrolone through occluded and non-occluded human skin. J Pharm Pharmacol 30:152–157, 1978.

250. Shani J. Skin penetration of minerals in psoriasis and guinea pig bathing in hypertonic salt solutions. Pharmacol Res Commun 17:501–511, 1985.

251. Wang JCT, Patel BG, Ehmann CW, Lowe N. The release and percutaneous permeation of anthralin products, using clinically involved and uninvolved psoriatic skin. J Am Acad Dermatol 16:812–821, 1987.

252. Juhlin L, Hägglund G, Evers H. Absorption of lidocaine and prilocaine after application of eutectic mixture of local anesthetics (EMLA) on normal and diseased skin. Acta Derm Venereol (Stockh) 69:18–22, 1989.

253. Maibach HI, Feldmann RJ, Milby TH, Serat WF. Regional variation in percutaneous penetration in man. Arch Environ Health 23:208–211, 1971.

254. Elias PM, Cooper ER, Korc A, Brown BE. Percutaneous transport in relation to stratum corneum structure and lipid composition. J Invest Dermatol 76:297–301, 1981.

255. Scott RC, Corrigan MA, Smith F, Mason H. The influence of skin structure on permeability: an intersite and interspecies comparison with hydrophilic penetrants. J Invest Dermatol 96:921–925, 1991.

256. Ebihara A, Fujimura A, Ohashi K, Shiga T, Kumagai Y, Nakashima H, Kotegawa T. Influence of application site of a new transdermal clonidine, M-5041T, on its pharmakokinetics and pharmacodynamics in healthy subjects. J Clin Pharmacol 33:1188–1191, 1993.

257. Hopkins K, Aarons L, Rowland M. Absorption of clonidine from transdermal therapeutic system when applied to different body sites. In: Weber MA, Mathias CJ, eds. Mild Hypertension. Darmstadt: Steinkopf. pp 143–147, 1984.

258. Hansen MS, Woods SL, Willis RE. Relative effectiveness of nitroglycerin ointment according to site of application. Heart Lung 8:716–720, 1979.

259. Moe G, Armstrong PW. Influence of skin site on bioavailability of nitroglycerin ointment in congestive heart failure. Am J Med 81:765–770, 1986.

260. Noonan PK, Wester RC. Percutaneous absorption of nitroglycerin. J Pharm Sci 69: 385–386, 1980.

261. Shah AK, Wei G, Lamman RC, Bhargava VO, Weir SJ. Percutaneous absorption of ketoprofen from different anatomical sites in man. Pharm Res 13:168–172, 1996.

262. Stoughton RB. Percutaneous absorption of drugs. Annu Rev Pharmacol Toxicol 29: 55–69, 1989.

263. Scheuplein RJ. Mechanism of percutaneous absorption. II. Transient diffusion and the relative importance of various routes of skin penetration. J Invest Dermatol 48:79–88, 1967.

264. Albery WJ, Hadgraft J. Percutaneous absorption: in vivo experiments. J Pharm Pharmacol 31:140–147, 1979.

265. Pratzel HG. Iontophorese. Berlin: Springer, 1987.

266. Mackee GM, Sulzberger MB, Herrmann F, Bare RL. Histologic studies on percutaneous penetration with special reference to the effects of vehicles. J Invest Dermatol 6:43–61, 1945.

267. Tregear RT. Relative penetrability of hair follicles and epidermis. J Physiol 156:307–313, 1961.

268. Grasso P. Some aspects of the role of skin appendages in percutaneous absorption. J Soc Cosmet Chem 22:523–534, 1971.

269. Schaefer H, Watts F, Brod J, Illel B. Follicular penetration. In: Scott RC, Guy RH, Hadgraft J, eds. Prediction of Percutaneous Penetration. Vol. 1. London: IBC Technical Services, pp 163–173, 1990.

270. Lauer AC, Lieb LM, Ramachandran C, Flynn GL, Weiner ND. Transfollicular drug delivery. Pharm Res 12:179–186, 1995.

271. Scheuplein RJ. Mechanism of percutaneous absorption. J Invest Dermatol 45:334–346, 1965.

272. Scheuplein RJ, Blank IH, Brauner GJ, MacFarlane DJ. Percutaneous absorption of steroids. J Invest Dermatol 52:63–70, 1969.

273. Shelly WB, Melton FM. Studies on absorption through normal human skin. FASEB 6:199–200, 1947.

274. Behl CR, Bellantone NH, Flynn GL. Influence of age on percutaneous absorption of drug substances. In: Bronaugh RL, Maibach HI, eds. Percutaneous Absorption: Mechanism—Methodology—Drug Delivery. New York: Marcel Dekker, pp 183–212, 1985.

275. Suzuki M, Asaba K, Komatsu H, Mochizuka M. Autoradiographic study on percutaneous absorption of several oils useful for cosmetics. J Soc Cosmet Chem 29:265–282, 1978.

276. Nicolau G, Baughman RA, Tonelli A, McWilliams W, Schiltz J, Yacobi A. Deposition of viprostol (a synthetic PGE_2 vasodilator) in the skin following topical administration to laboratory animals. Xenobiotica 17:1113–1120, 1987.

277. Borelli, Metzger M. Fluorescenzmikroskopische untersuchungen über die perkutane penetration fluorescierender stoffe. Hautarzt 8:261–266, 1957.

278. Kao J, Hall J, Helman G. In vitro percutaneous absorption in mouse skin: influence of skin appendages. Toxicol Appl Pharmacol 94:93–103, 1988.

279. Wallace SM, Barnett G. Pharmacokinetic analysis of percutaneous absorption: evidence of parallel penetration pathways for methotrexate. J Pharmacokinet Biopharm 6:315–325, 1978.

280. Keister JC, Kasting GB. The use of transient diffusion to investigate transport pathways through skin. J Control Release 4:111–117, 1986.

281. Corroller M, Didry JR, Siou G, Wepierre J. Sebaceous accumulation of linoleic acid following topical application in the hairless rat its mathematical treatment. In: Shroot B, Schaefer H, eds. Skin Pharmacokinetics. Basel: Karger, pp 111–120, 1987.

282. Illel B, Schaefer H, Wepierre J, Doucet O. Follicles play an important role in percutaneous absorption. J Pharm Sci 80:424–427, 1991.

283. Illel B, Schaefer H. Transfollicular percutaneous absorption. Acta Derm Venereol (Stockh) 68:427–457, 1988.

284. Behl CR, Wittkowsky A, Barett M, Pierson CL, Flynn GL. Technique for preparing appendage-free skin (scar) on hairless mouse. J Pharm Sci 70:835–837, 1981.

285. Hueber F, Schaefer H, Wepierre J. Role of transepidermal and transfollicular routes in percutaneous absorption of steroids: in vitro studies on human skin. Skin Pharmacol 7:237–244, 1994.

286. Hueber F, Besnard M, Schaefer H, Wepierre J. Percutaneous absorption of estradiol and progesterone in normal and appendage-free skin of the hairless rat: lack of importance of nutritional blood flow. Skin Pharmacol 7:245–256, 1994.

287. Hueber F, Wepierre J, Schaefer H. Role of transepidermal and transfollicular routes in percutaneous absorption of hydrocortisone and testosterone: in vivo study on hairless rat. Skin Pharmacol 5:99–107, 1992.

288. Mukherji E, Millenbaugh NJ, Au J. Percutaneous absorption of 2',3'-dideoxyinosine in rats. Pharm Res 11:809–815, 1994.

289. Bamba FL, Wepierre J. Role of appendageal pathway in the percutaneous absorption of pyridostigmine bromide in various vehicles. Eur J Drug Metab Pharmacokinet 18: 339–348, 1993.

290. Tur E, Maibach HI, Guy RI. Percutaneous penetration of methyl nicotinate at three anatomic sites: evidence for appendageal contribution to transport? Skin Pharmacol 4:230–234, 1991.

291. Rolland A, Wagner N, Chatelus A, Shroot B, Schaefer H. Site-specific drug delivery to pilosebaceous structures using polymeric microspheres. Pharm Res 10:1738–1744, 1993.

292. Chan SY, Po ALW. Prodrugs for dermal delivery. Int J Pharm 55:1–16, 1989.

293. Kao J, Carver MP. Cutaneous metabolism of xenobiotics. Drug Metab Rev 22:363–410, 1990.

294. Lodén M. The in vitro hydrolysis of di-isopropyl fluorophosphate during penetration through human full-thickness skin and isolated epidermis. J Invest Dermatol 85:335–339, 1985.

295. Potts RO, Guy RH. A predictive algorithm for skin permeability: the effects of molecular size and hydrogen bond activity. Pharm Res 12:1628–1633, 1995.

296. Hadgraft J. Theoretical aspects of metabolism in the epidermis. Int J Pharm 4:229–239, 1980.

297. Guy RH, Hadgraft J, Maibach HI. A pharmacokinetic model for percutaneous absorption. Int J Pharm 11:119–129, 1982.

298. Tojo K, Valia KH, Chotani G, Chien YW. Long-term permeation kinetics of estradiol: (IV) a theroretical approach to the simultaneous skin permeation and bioconversion of estradiol. Drug Dev Ind Pharm 11:1175–1193, 1985.

299. Potts RO, McNeill SC, Desbonnet CR, Wakshull E. Transdermal drug transport and metabolism: II. The role of competing kinetic events. Pharm Res 6:119–124, 1989.

300. Sloan KB. Prodrugs: Topical and Ocular Drug Delivery. New York: Marcel Dekker, 1992.

301. Bodor N, Kaminski JJ, Selk S. Soft drugs. I. Labile quaternary ammonium salts as soft antibacterials. J Med Chem 23:469–474, 1980.

302. Roskos KV, Maibach HI, Guy RH. The effect of aging on percutaneous absorption in man. J Pharmacokinet Biopharm 17:617–630, 1989.

303. Harvell JD, Maibach HI. Percutaneous absorption and inflammation in aged skin: a review. J Am Acad Dermatol 31:1015–1021, 1994.

304. Micalis S, Bhatt RH, Distefano S, Caltabiano L, Cook B, Fischer J, Solomon LM, West DP. Evaluation of transdermal theophylline pharmacokinetics in neonates. Pharmacotherapy 13:386–390, 1993.

305. Barrett DA, Rutter N. Transdermal delivery and the premature neonate. Crit Rev Ther Drug Carrier Syst 11:1–30, 1994.

306. Weigand DA, Gaylor JR. Irritant reaction in Negro and Caucasian skin. South Med J 67:548–551, 1974.

307. Johnson LC, Corah NL. Racial differences in skin resistance. Science 139:766–767, 1960.

308. Reinertson RP, Wheatley VR. Studies on the chemical composition of human epidermal lipids. J Invest Dermatol 32:49–59, 1959.

309. Berardesca E, Maibach HI. Racial differences in sodium lauryl sulphate induced cutaneous irritation: black and white. Contact Dermatitis 17:12–17, 1987.

310. Williams RL, Thakker KM, John V, Lin ET, Gee WL, Benet LZ. Nitroglycerin absorption from transdermal systems: formulation effects and metabolic concentrations. Pharm Res 8:744–749, 1991.

311. Levy G. The clay feet of bioequivalence testing. J Pharm Pharmacol 47:975–977, 1995.

312. Hauck WW, Anderson S. Measuring switchability and prescribability: when is average bioequivalence sufficient? J Pharmacokinet Biopharm 22:551–564, 1994.

313. Cornell RC, Stoughton RB. Correlation of the vasoconstriction assay and clinical activity in psoriasis. Arch Dermatol 121:63–67, 1985.

314. Ferreira LAM, Seiller M, Grossiord JL, Marty JP, Wepierre J. Vehicle influence on in vitro release and percutaneous absorption of glucose: role of W/O/W multiple emulsion. Proc Int Symp Control Release Bioact Mater 21:453–454, 1994.

315. Ferreira LAM, Doucet J, Seiller M, Grossiord JL, Marty JP, Wepierre J. In vitro percutaneous absorption of metronidazole and glucose: comparison of O/W, W/O/W and W/O systems. Int J Pharm 121:169–179, 1995.

316. Friedman DI, Schwarz JS, Weisspapir M. Sub-micron emulsion vehicle for enhanced transdermal delivery of steroidal antiinflammatory drugs. J Pharm Sci 84:324–329, 1995.

317. Kondo S, Yamanaka C, Sugimoto I. Enhancement of transdermal delivery by superfluous thermodynamic potential. III. Percutaneous absorption of nifedipine in rats. J Pharmacobiodyn 10:743–749, 1987.

318. Megrab NA, Williams AC, Barry BW. Oestradiol permeation through human skin and Silastic membrane: effects of propylene glycol and supersaturation. J Control Release 36:277–294, 1995.

319. Sato K, Sugibayashi K, Morimoto Y. Effect of mode of action of aliphatic esters on the in-vitro skin penetration of nicorandil. Int J Pharm 43:31–40, 1988.

320. Yamane MA, Williams AC, Barry BW. Effects of terpenes and oleic acid as skin penetration enhancers towards 5-fluorouracil as assessed with time; permeation, partitioning and differential scanning calorimetry. Int J Pharm 116:237–251, 1995.

321. Higuchi T. Physical chemical analysis of percutaneous absorption process from creams and ointments. J Soc Cosmet Chem 11:85–97, 1960.

322. Reifenrath WG, Hawkins GS, Kurtz MS. Percutaneous penetration and skin retention of topically applied compounds: an in vitro–in vivo study. J Pharm Sci 80:526–532, 1991.

323. Surber C, Wilhelm K–P, Maibach HI. In vitro skin pharmacokinetics of acitretin: percutaneous absorption studies in intact and modified skin of three different species. J Pharm Pharmacol 43:836–840, 1991.

324. Elias PM. Epidermal lipids, membranes, and keratinization. Int J Dermatol 2:1–19, 1981.

325. Potts RO, Guy RH. Predicting skin permeability. Pharm Res 9:663–669, 1992.

326. Pugh WJ, Roberts MS, Hadgraft J, Degim T. Relationship between H-bonding of penetrants to stratum corneum lipids and diffusion within the stratum corneum. In: Brain KR, James VJ, Walters KA, eds. Prediction of Percutaneous Penetration. Vol. 4b. Cardiff: STS Publishing, pp 48–51, 1996.

327. Stoughton RB. Are generic formulations equivalent to trade name topical glucocorticoids? Arch Dermatol 123:1312–1314, 1987.

328. Poulsen BJ, Young E, Coquilla V, Katz M. Effect of topical vehicle composition on the in vitro release of fluocinolone acetonide and its acetate ester. J Pharm Sci 57:928–933, 1968.

329. Ostrenga J, Steinmetz C, Poulsen B. Significance of vehicle composition I: relationship between topical vehicle composition, skin penetrability, and clinical efficacy. J Pharm Sci 60:1175–1179, 1971.

330. Twist JN, Zatz JL. Influence of solvents on paraben permeation through idealized skin model membranes. J Soc Cosmet Chem 37:429–444, 1986.

331. Yalkowsky SH, Valvani SC, Amidon GL. Solubility of nonelectrolytes in polar solvents IV: nonpolar drugs in mixed solvents. J Pharm Sci 65:1488–1494, 1976.

332. Hadgraft J, Hadgraft JW, Sarkany I. The effect of glycerol on percutaneous absorption of methyl nicotinate. Br J Dermatol 87:30–36, 1972.

333. Lippold BC, Schneemann H. The influence of vehicles on the local bioavailability of betamethasone-17-benzoate from solution and suspension-type ointments. Int J Pharm 22:31–43, 1984.

334. Dugard PH, Scott RC. A method of predicting percutaneous absorption rates from vehicle to vehicle: an experimental assessment. Int J Pharm 28:219–227, 1986.

335. Goodman M, Barry BW. Lipid–protein-partitioning (LPP) theory of skin enhancer activity: finite dose technique. Int J Pharm 57:29–40, 1989.

336. Lamaud E, Schalla W, Schäfer H. Effect of repeated rubbing on the in vivo penetration of hydrocortisone. In: Schäfer H, Schalla W, Shroot B, eds. Advances in Skin Pharmacology and Skin Pharmacokinetics; Nice: Centre International de Recherche Dermatologiques. pp 23–24, 1986.

337. McMaster J, Maibach HI, Wester RC, Bucks DAW. Does rubbing enhance in vivo dermal absorption? In: Bronaugh RL, Maibach HI, eds. Percutaneous Absorption: Mechanism—Methodoloy—Drug Delivery. New York: Marcel Dekker, pp 359–361, 1985.

338. Idson B. Percutaneous absorption. J Pharm Sci 64:901–924, 1975.

339. Garb J. Nevus verrucosus unilateris cured with podophyllin ointment. Arch Dermatol 81:606–609, 1960.

340. Scholtz JR. Topical therapy of psoriasis with fluocinolone acetonide. Arch Dermatol 84:1029–1030, 1961.

341. Sulzberger MB, Witten VH. Thin pliable plastic films in topical dermatologic therapy. Arch Dermatol 84:1027–1028, 1961.

342. McKenzie AW. Percutaneous absorption of steroids. Arch Dermatol 86:91–94, 1962.

343. McKenzie AW, Stoughton RB. Method for comparing percutaneous absorption of steroids. Arch Dermatol 86:608–610, 1962.

344. Guy RH, Bucks DAW, McMaster JR, Villaflor DA, Roskos KV, Hinz RS, Maibach HI. Kinetics of drug absorption across human skin in vivo: developments in methodology. Skin Pharmacol 1:70–76, 1988.

345. Kim HO, Wester RC, McMaster JA, Bucks DAW, Maibach HI. Skin absorption from patch test systems. Contact Dermatitis 17:178–180, 1987.

346. Auton TR, Westhead DR, Woollen BH, Scott RC, Wilks MF. A physiologically based mathematical model of dermal absorption in man. Hum Exp Toxicol 13:51–60, 1994.

347. Johnson R, Nusbaum BP, Horwitz SN, Frost P. Transfer of topically applied tetracycline in various vehicles. Arch Dermatol 119:660–663, 1983.

348. Haupenthal H, Heilmann P, Moll F. Kühlsalben. Krankenhauspharm 15:671–674, 1994.

349. Reifenrath WG, Robinson PB. In vitro skin evaporation and penetration characteristics of mosquito repellents. J Pharm Sci 71:1014–1018, 1982.

350. Hawkins GS, Reifenrath WG. Influence of skin source, penetration cell fluid, and partition coefficient on in vitro skin penetration. J Pharm Sci 75:378–381, 1986.

351. Reifenrath WG, Spencer TS. Evaporation and penetration from skin. In: Bronaugh RL, Maibach HI, eds. Percutaneous Absorption: Mechanism—Methodolgy—Drug Delivery. New York: Marcel Dekker, pp 313–334, 1989.

352. Coldman MF, Poulsen BJ, Higuchi T. Enhancement of percutaneous absorption by the use of volatile:nonvolatile systems as vehicles. J Pharm Sci 58:1098–1102, 1969.

353. Chiang C–M, Flynn GL, Weiner ND, Szpunar GJ. Bioavailability assessment of topical delivery systems: effect of vehicle evaporation upon in vitro delivery of minoxidil from solution formulations. Int J Pharm 55:229–236, 1989.

354. Kemken J, Ziegler A, Müller BW. Influence of supersaturation on the pharmacodynamic effect of bupranolol after dermal administration using microemulsions as vehicle. Pharm Res 9:554–558, 1992.

355. de Carvalho M, Falson–Rieg F, Eynard I, Rojas J, Lafforgue C, Hadgraft J. Changes in vehicle composition during skin permeation studies. In: Brain KR, James VJ, Walters KA, eds. Prediction of Percutaneous Penetration. Vol. 3b. Cardiff: STS Publishing, pp 251–254, 1993.

356. Guy RH, Hadgraft J. Percutaneous absorption kinetics of topically applied agents liable to surface loss. J Soc Cosmet Chem 35:103–113, 1984.

357. Flynn GL. Topical drug absorption and topical pharmaceutical systems. In: Banker GS, Rhodes CT, eds. Modern Pharmaceutics. New York: Marcel Dekker, pp 263–325, 1990.

358. Williams PL, Brooks JD, Inman AO, Monteiro–Riviere NA, Riviere JE. Determination of physicochemical properties of phenol, p-nitrophenol, acetone and ethanol relevant to quantitating their percutaneous absorption in porcine skin. Res Commun Chem Pathol Pharmacol 83:61–75, 1994.

359. Polano MK. Topical Skin Therapeutics. Edinburgh: Churchill Livingstone, 1984.

360. Schlagel CA, Sanborn EC. The weights of topical preparations required for total and partial body inunction. J Invest Dermatol 42:253–256, 1964.

361. Griffiths WAD, Wilkinson JD. Topical therapy. In: Rook A, Wilkinson DS, Ebling FJG, Champion RH, Burton JL, eds. Textbook of Dermatology. London: Blackwell Scientific, pp 3037–3084, 1992.

362. Habif TP. Clinical Dermatology: A Color Guide to Diagnosis and Therapy. St. Louis: CV Mosby, 1990.

363. Long CC. The rule of hand: 4 hand areas = 2 FTU 1 g. Arch Dermatol 128:1129–1130, 1992.

364. Lutz UL, Weirich E. Untersuchungen zur hautflächen-Dosis von Dermatopharmakotherapeutika. Z Hautkr 50:753–769, 1975.

365. Willis I, Kligman AM. Evaluation of sunscreen by human assay. J Soc Cosmet Chem 20:639–651, 1969.

366. Lynfield YL, Schechter S. Choosing and using a vehicle. J Am Acad Dermatol 10: 56–59, 1984.

367. Kaidberg KH, Kligman AM. Laboratory methods for appraising the efficacy of sunscreens. J Soc Cosmet Chem 29:525–536, 1978.

368. FDA. Sunscreen drug products for over-the counter human use. Fed Reg 43:38206–38269, 1978.

369. Bech–Thomsen N, Wulff HC. Sunbather's application of sunscreen is probably inadequate to obtain the sun protection factor assigned to the preparation. Photodermatol Photoimmunol Photomed 9:242–244, 1993.

370. Loesch H, Kaplan DL. Pitfalls in sunscreen application. Arch Dermatol 130:665–666, 1994.

371. Shah VP, Behl CR, Flynn GL, Higuchi WI, Schaefer H. Principles and criteria in the development and optimization of topical therapeutic products. J Pharm Sci 81:1051–1054, 1992.

372. Ferry JJ, Shepard JH, Szpunar GJ. Relationship between contact time of applied dose and percutaneous absorption of minoxidil from topical solution. J Pharm Sci 79:483–486, 1990.

373. Täuber U, Matthes H. Percutaneous absorption of methylprednisolone aceponate after single and multiple dermal application as ointment in male volunteers. Arzneimittelforschung 42:1122–1124, 1992.

374. Gao HY, Li Wan Po A. Topical formulations of fluocinolone acetonide. Are creams, gels and ointments bioequivalent and does dilution affect activity? Eur J Clin Pharmacol 46:71–75, 1994.

375. Wester RC, Melendres J, Sedik L, Maibach HI. Percutaneous absorption of Azone following single and multiple doses to human volunteers. J Pharm Sci 83:124–125, 1994.

376. Carslaw HS, Jaeger JC. Conduction of Heat in Solids. London: Oxford University Press, 1959.

377. Franz TJ. Percutaneous absorption. On the relevance of in vitro data. J Invest Dermatol 64:190–195, 1975.

378. Franz TJ. The finite dose techniques as a valid in vitro model for the study of percutaneous absorption in man. Curr Probl Dermatol 7:58–68, 1978.

379. Fox JL, Yu C–D, Higuchi WI, Ho NFH. General physical model for simultaneous diffusion and metabolism in biological membranes. The computational approach for the steady-state case. Int J Pharmaceut 2:41–57, 1979.

380. Okamoto H, Yamashita F, Saito K, Hashida M. Analysis of drug penetration through the skin by two-layer skin model. Pharm Res 6:931–937, 1989.

381. Addicks WJ, Flynn GL, Weiner N, Curl RA. A mathematical model to describe drug release from thin topical application. Int J Pharm 56:243–248, 1989.

382. Kubota K, Yamada T. Finite dose percutaneous absorption: theory and its application to in vitro timolol permeation. J Pharm Sci 79:1015–1019, 1990.

383. Kubota K. Finite dose percutaneous drug absorption: a BASIC program for the solution of the diffusion equation. Comput Biomed Res 24:196–207, 1991.

384. Seta Y, Ghanem AH, Higuchi WI, Borsadia S, Behl CR, Malick AW. Physical model approach to understanding finite dose transport and uptake of hydrocortisone in hairless guinea pig skin. Int J Pharm 81:89–99, 1992.

385. Malkinson FD, Ferguson EH. Percutaneous absorption of hydrocortisone-4-C^{14} in two subjects. J Invest Dermatol 25:281–283, 1956.

386. Feldmann RJ, Maibach HI. Percutaneous penetration of hydrocortisone with urea. Arch Dermatol 109:58–59, 1974.

387. Britz MB, Maibach HI, Anjo DM. Human percutaneous absorption of hydrocortisone: the vulva. Arch Dermatol Res 267:313–316, 1980.

388. Bucks DAW, Maibach HI, Guy RH. Percutaneous absortion of steroids: effect of repeated application. J Pharm Sci 74:1337–1339, 1985.

389. Maibach HI, Feldmann RJ. Percutaneous penetration in man: comparison of some steroids. Clin Res 14:270, 1966.

390. Franz TJ, Lehman PA. Percutaneous absorption of fluocinolone in man: assessment of relative bioequivalence of 0.05% Lidex ointment and solution. Pharm Res 7(suppl 194):PDD 7331, 1990.

391. Wickrema SAJ, Shaw SR, Weber DJ. Percutaneous absorption and excretion of tritium-labeled diflorasone diacetate, a new topical corticosteroid in the rat, monkey and man. J Invest Dermatol 71:372–377, 1978.

392. Bucks DAW, Maibach HI, Guy RH. Mass balance and dose accountability in percutaneous absorption studies: development of a nonocclusive application system. Pharm Res 5:313–314, 1988.

393. Stoughton RB, Wullich K. Relation of application time to bioactivity of a potent topical glucocorticosteroid formulation. J Am Acad Dermatol 22:1038–1041, 1990.

394. Maibach HI, Feldmann R. Effect of applied concentration on percutaneous absorption in man. J Invest Dermatol 52:382, 1969.

395. Roberts MS, Horlock E. Effect of repeated skin application on percutaneous absorption of salicylic acid. J Pharm Sci 67:1685–1687, 1978.

396. Lippold BC, Reimann H. Wirkungsbeeinflussung bei lösungssalben durch Vehikel am beispiel von Methylnicotinat. Acta Pharm Technol 35:128–135, 1989.

397. Stoughton RB, Wullich K. The same glucocorticoid in brand-name products. Arch Dermatol 125:1509–1511, 1989.

398. Guin JD, Wallis MS, Wall MS, Lehmann PA, Franz TJ. Quantitative vasoconstrictor assay for topical corticosteroids: the puzzling case of fluocinolone acetonide. J Am Acad Dermatol 29:197–202, 1993.

399. Gibson JR, Hough JE, Marks PM. Effect of concentration on the clinical pharmacology of corticosteroid ointment formulations. In: Maibach HI, Surber C, eds. Topical Corticosteroids. Basel: S Karger, pp 74–92, 1992.

400. Gibson JR, Kirsch J, Darley CR, Burke CA. An attempt to evaluate the relative clinical potencies of various diluted and undiluted proprietary corticosteroid preparations. Clin Exp Dermatol 8:489–493, 1983.

401. Walker M, Chambers LA, Hollingsbee DA, Hadgraft J. Significance of vehicle thickness to skin penetration of halcinonide. Int J Pharm 70:167–172, 1991.

402. Jackson DB, Thompson C, McCormack JR, Guin JD. Bioequivalence (bioavailability) of generic topical corticosteroids. J Am Acad Dermatol 20:791–796, 1989.

403. Barry BW, Woodford R. Comparative bioavailability of proprietary topical corticosteroid preparations; vasoconstrictor assays on thirty creams and gels. Br J Dermatol 91:323–338, 1974.

404. Magnus AD, Haigh JM, Kanfer I. Assessment of some variables affecting the blanching activity of betamethasone-17-valerate cream. Dermatologica 160:321–327, 1980.

405. Zesch A. Zur Bioäquivalenz und Bioverfügbarkeit topisch applizierter Arzneimittel. Pharm Ind 50:746–749, 1988.

406. Treffel P, Gabard B. Ibuprofen epidermal levels after topical application in vitro: effect of formulation, application time, dose variation and occlusion. Br J Dermatol 127:286–291, 1993.

407. Claris P. Non-invasive Biophysical Measurements on the Skin: The Use of Some Bioengineering Methods for the Assessment of Percutaneous Penetration. Brussels: Vrije Universiteit Brussel; 1994.

408. Unna PG. Cigolin als heilmittel der Psoriasis. Dermatol Wochenschr 6:116–137; 150–163; 175–183, 1916.

409. Farber FM, Harris DR. Hospital treatment of psoriasis. A modified anthralin therapy. Arch Dermatol 101:381–389, 1970.

410. Schaefer H, Farber EM, Goldberg L, Schalla W. Limited application period for dithranol in psoriasis. Br J Dermatol 102:571–573, 1980.

411. Schalla W, Bauer E, Wesendahl C, Goldberg L, Farber EM, Schaefer H. Arch Dermatol Res 267:203, 1980.

412. Runne U, Kunze J. Short-duration ("minutes") therapy with dithranol for psoriasis: a new out-patient regimen. Br J Dermatol 106:135–139, 1982.

413. Schaefer H, Schalla W, Shroot B. Anthralin—facts, trends and unresolved problems. In: Hornstein OP, Schnyder UW, Schönfeld J, eds. Neue Entwicklungen in der Dermatologie. Berlin: Springer, pp 82–91, 1984.

414. Jaeger L. Psoriasis treatment with betamethasone dipropionate using short-term application and short-term occlusion. Acta Derm Venereol (Stockh) 66:84–87, 1986.

415. Leibsohn E. Comparison of diflorasone diacetate and betamethasone dipropionate ointment in the treatment of psoriasis. J Int Med Res 10:22–27, 1982.

416. Taplin D, Rivera A, Walker JG, Roth WI, Reno D, Meinking T. A comparative trial of three treatment schedules for the eradication of scabies. J Am Acad Dermatol 9: 550–554, 1983.

417. Wester RC, Noonan PK, Maibach HI. Frequency of application on percutaneous absorption of hydrocortisone. Arch Dermatol 113:620–622, 1977.

418. Wester RC, Noonan PK, Maibach HI. Percutaneous absorption of hydrocortisone increases with long-term administration. Arch Dermatol 116:186–188, 1980.

419. Täuber U. Skin pharmacokinetics of the new topical glucocorticosteroid MPA. In: Scott RC, Guy RH, Hadgraft J, eds. Prediction of Percutaneous Penetration. Vol. 1. London: IBC Technical Services, pp 37–48, 1990.

420. Täuber U, Weiss C, Matthes H. Does salicylic acid increase the percutaneous absorption of diflucortolone-21-valerate. Skin Pharmacol 6:276–281, 1993.

421. Tsai JC, Flynn GL, Weiner N, Ferry JJ. Influence of application time and formulation reapplication on the delivery of minoxidil through hairless mouse skin as measured in Franz diffusion cells. Skin Pharmacol 7:270–277, 1994.

422. Sherertz EF, Sloan KB, McTiernan RG. Effect of skin pretreatment with vehicle alone or drug in vehicle on flux of a subsequently applied drug: results of hairless mouse skin and diffusion cell studies. J Invest Dermatol 89:249–252, 1987.

423. Fredriksson T, Lassus A, Bleeker J. Treatment of psoriasis and atopic dermatitis with halcinonide cream applied once and three times daily. Br J Dermatol 102:575–577, 1980.

424. Sudilovsky A, Muir JG, Bococo FC. A comparison of single and multiple applications of halcinonide cream. Int J Dermatol 20:609–613, 1981.

425. Pearlman DL. Weekly pulse dosing: effective and comfortable topical 5-fluorouracil treatment of multiple facial actinic keratoses. J Am Acad Dermatol 25:665–667, 1991.

426. Frosch PJ, Behrenbeck EM, Frosch K, Macher E. The Duhring chamber assay for corticosteroid atrophy. Br J Dermatol 104:57–65, 1981.

427. Lehmann P, Zheng P, Lavker RM, Kligman A. Corticosteroid atrophy in human skin. A study by light, scanning, and transmission electron microscopy. J Invest Dermatol 81:169–176, 1983.

428. Lavker RM, Schechter NM, Lazarus GS. Effects of topical corticosteroids on human dermis. Br J Dermatol 115(suppl 31):101–107, 1986.

429. Woodbury R, Kligman AM. The hairless mouse model for assaying the atrophogenicity of topical corticosteroids. Acta Derm Venereol (Stockh) 72:403–406, 1989.

430. Lubach D, Kietzmann M. Anti-inflammatory effect and thinning of the skin: models for testing topically applied corticosteroids on human skin. In: Maibach HI, Surber C, eds. Topical Corticosteroids. Basel: S Karger, pp 26–41, 1992.

431. Pflugshaupt C. Diskontinuierliche topische Kortikoidtherapie. Zbl Haut Geschlechtskrankh 148:1229–1367, 1983.

432. English JS, Bunker CB, Ruthven K, Dowd PM, Greaves MW. A double-blind comparison of efficacy of betamethasone dipropionate cream twice daily versus once daily in the treatment of steroid responsive dermatoses. Clin Exp Dermatol 14:32–34, 1989.

433. Lubach D, Rath J, Kietzmann M. Steroid-induced dermal thinning: discontinuous application of clobetasol-17-propionate ointment. Dermatology 185:44–48, 1992.

434. Lubach D, Bensmann A, Bornemann U. Steroid-induced dermal atrophy. Dermatologica 179:67–72, 1989.

435. Wozel G, Barth J. Fluocinolonacetonid und Psoriasis—kontinuierliche und alternierende Therapie. Dermatol Monatsschr 172:620–623, 1986.

436. Lubach D, Rath J, Kietzmann M. Alteration of skin thickness induced by initial continuous followed by intermittent topical application of clobetasol. Dermatology 190: 51–55, 1995.

437. Katz HI, Hien NT, Prawer SE, Scott JC, Grivna EM. Betamethasone dipropionate in optimized vehicle: intermittent pulse dosing for extended maintenance treatment of psoriasis. Arch Dermatol 123:1308–1311, 1987.

438. Levy G. Bioavailability, clinical effectiveness, and public interest. Pharmacology 8: 33–43, 1972.

439. Shah VP, Van Zwieten–Boot B. Characterisation of topical products for local action —quality, efficacy and safety. Symposium on Quality and Interchangeability of Topical Products for Local Action, Dec 8; Nürnberg: European Federation for Pharmaceutical Sciences, 1995.

440. Yacobi A, Skelly JP, Shah V, Benet LZ. Integration of Pharmacokinetics, Pharmacodynamics, and Toxicokinetics in Rational Drug Development. New York: Plenum Press, 1993.

441. Peck CC, Collins JM. First time in man studies; a regulatory perspective—art and science of phase I trials. J Clin Pharmacol 30:218–222, 1990.

442. Narang PK, Bianchine JR. Phase I studies: initial evaluation of drug toxicity, dose response, and efficacy. In: Max M, Portenoy R, Laska E, eds. Advances in Pain Research and Therapy: 18, New York: Raven Press, pp 9–31, 1991.

443. Levy G. Concentration-controlled versus concentration-defined clinical trials. Clin Pharmacol Ther 53:387–388, 1993.

444. Levy G, Ebling WF, Forrest A. Concentration- or effect-controlled clinical trials with sparse data. Clin Pharmacol Ther 56:1–8, 1994.

445. Urquart J. How much compliance is enough? Pharm Res 13:10–11, 1996.

446. Meyer UA, Kindli R. Plazebos und nozebos. Ther Umsch 46:544–554, 1989.

447. Pearce JMS. The placebo enigma. Q J Med 88:215–220, 1995.

448. Ernst E, Resch KL. Concept of true and perceived placebo effects. Br Med J 311: 551–553, 1995.

449. Golomb BA. Paradox of placebo effect. Nature 375:530, 1995.

450. Olson SC, Eldon MA, Toothaker RD, Ferry JJ, Colburn WA. Controversy. II: bioequivalence as an indicator of therapeutic equivalence: modeling the theoretic influence of bioinequivalence on single-dose drug effect. J Clin Pharmacol 27:342–345, 1987.

451. Lee JW, Hulse JD, Colburn WA. Surrogate biochemical markers: precise measurement for strategic drug and biologics development. J Clin Pharmacol 35:464–470, 1995.

452. Barrett CW, Hadgraft JW. The effect of particle size and the vehicle on the percutaneous absorption of fluocinolone acetonide. Br J Dermatol 77:576–578, 1965.

453. Sarkany I, Hadgraft JW, Caron GA, Barrett CW. The role of vehicles in the percutaneous absorption of corticosteroids. Br J Dermatol 77:569–575, 1965.

454. Christie GA, Moore–Robinson M. Vehicle assessment—methodology and results. Br J Dermatol 82(suppl. 6):93–98, 1970.

455. Barry BW. Comparative bio-availability and activity of proprietary topical corticosteroid preparations: vasoconstrictor assays on thirty-one ointments. Br J Dermatol 93: 563–567, 1975.

456. Barry BW. Proprietary hydrocortisone creams, vasoconstrictor activities and bio-availabilities of six preparations. Br J Dermatol 95:423–425, 1976.

457. Barry BW, Woodford R. Vasoconstrictor activities and bioavailabilities of seven proprietary corticosteroid creams assessed using a non-occluded multiple dosage regimen; clinical implications. Br J Dermatol 97:555–560, 1977.

458. Woodford R, Barry BW. Activity and bioavailability of amcinolone and triamcinolone acetonide in experimental and proprietary topical preparations. Curr Ther Res 21: 877–886, 1977.

459. Stoughton RB, Cornell RC. Review of super-potent topical corticosteroids. Semin Dermatol 6:72–76, 1987.

460. Shah VP, Peck CC, Skelly JP. "Vasoconstriction"—skin blanching-assay for glucocorticosteroids—a critique. Arch Dermatol 125:1558–1561, 1989.

461. Olsen E. A double-blind controlled comparison of generic and trade-name topical steroids using the vasoconstriction assay. Arch Dermatol 127:197–201, 1991.

462. Agrawal N, Lehman PA, Franz TJ. Bioavailability of hydrocortisone from seven topical formulations using the in vitro finite dose method. Pharm Res 12:S-277, 1995.

463a. Lehman PA, Franz TJ. Bioavailability of hydroquinone from six topical formulations using the in vitro finite dose method. Pharm Res 9:S-233, 1992.

463b. Schwarb FP, Imanidis G, Smith EW, Haigh JM, Surber C. Effect of concentration and degree of saturation of topical fluocinonide formulations on in vitro membrane transport and in vivo availability on human skin. Pharm Res 16:917–923, 1999

464. Marks R, Dykes PJ, Gordon J, Halan G, Davis A. Percutaneous penetration from a low dose supersaturated hydrocortisone acetate formulation. Presented at the British Society of Investigative Dermatology Annual Meeting, Sheffield, 1992.

465. Sanathanan LP, Peck CC, Temple RT, Lieberman R, Pledger G. Randomization, PK-controlled dosing, and titration: an integrated approach for designing clinical trials. Drug Inform J 25:425–431, 1991.

466. Levy G. Patient-oriented pharmaceutical research: focus on the individual. Pharm Res 12:943–944, 1995.

467. Stoughton RB. Penetration of drugs through skin. Dermatologica 152(suppl 1):27–36, 1976.

468. Blackwelder WC. Proving the null hypothesis in clinical trials. Controlled Clin Trials 3:345–353, 1982.

469. Gibson JR, Kirsch JM, Darley CR, Harvey SG, Burke CA, Hanson ME. An assessment of the relationship between vasoconstrictor assay findings, clinical efficacy and skin thinning effects of a variety of undiluted and diluted corticosteroid preparations. Br J Dermatol 111(suppl 27):204–212, 1984.

470. Barry BW, Fyrand O, Woodford R, Ulshagen K, Hogstad G. Control of bioavailability of a topical steroid; comparison of desonide creams 0.05% and 0.1% by vasoconstrictor studies and clinical trials. Clin Exp Dermatol 12:406–409, 1987.

471. Martin GP. The effect of dose under a hydrocolloid patch on the human bioavailability of a topical corticosteroid. J Pharm Pharmacol 41:129P, 1989.

472. Kecskés A, Jahn P, Wendt H, Lange L, Kleine Kuhlmann R. Dose–response relationship of topically applied methylprednisolone aceponate (MPA) in healthy volunteers. Eur J Clin Pharmacol 43:157–159, 1992.

473. Gibson JR. Effect of concentration on the clinical potency of corticosteroid ointment formulations. Skin Pharmacol 1:214–222, 1987.
474. Surber C, Davis A, Rufli T. Factors influencing topical bioavailability: New concept of topical dosing. Dermatology 191:173, 1995.
475. Marples and Kligman. 1973. Rec No 26240
476. Stenberg C, Larko O. Sunscreen application and its importance for the sun protection factor. Arch Dermatol 121:1400–1402, 1985.

9

Scale-up of Dermatological Dosage Forms: A Case for Multivariate Optimization and Product Homogeneity

OREST OLEJNIK and BRUCE A. FIRESTONE

Allergan, Inc., Irvine, California

I. INTRODUCTION

The objective of any drug-delivery system is to deliver in vivo the active drug moiety to the target tissue or receptor site. Whether the delivery system is simple or sophisticated, it is the responsibility of the formulator, and other product development support functions, to design a drug-delivery system capable of consistently achieving the desired pharmacokinetic–pharmacodynamic profile as an outcome of formulation and manufacturing process development and optimization. To achieve a robust formulation and reproducible product manufacturing in which product quality specifications are consistently met is not always an easy and straightforward exercise. Driven by formulation obstacles, processing complexities, an ever-increasing regulatory environment, health care cost-containment measures, and other pharmacoeconomic challenges can certainly influence the approach to and outcome of both product and process development. To achieve rapid and successful development of a drug-delivery system, there is an increasing focus on experimental design techniques, product scale-up planning, and process technology transfer activities. Additionally, and often considered the bane of many projects, is cost. This financial aspect is mentioned because there is a growing requirement within the pharmaceutical industry in specifying target values for manufacturing costs as early as possible within the life cycle of a project that could constrain the final production process. A knowledge of the effect of cost, estimated or otherwise, ensures not only that the criteria for efficacy, safety, and quality of a product are achieved, but also that the economic and environmental implications in product manufacturing are understood and deemed acceptable. It further augments the necessity in the implementation of well-planned, systematic methods, and the definitive need for understanding the performance characteristics and the critical aspects of manufacture of a product, particularly in its scale-up and transition to the production environment.

To address all these factors leading up to a successfully developed and manufacturable drug-delivery product that meets predefined criteria or target values would be a major undertaking. Although not dealt with here, their influence in any drug development project should not be viewed as insignificant. One aspect is important to stress, which centers on multivariate optimization. This technique, when used in experimental design, offers valuable information in understanding and improving formulation and process development. To the reader the application of design optimization techniques may be well understood and considered commonplace; however, it may be surprising to learn that these enabling techniques are infrequently applied within the pharmaceutical industry. Although available information from a survey on 68 pharmaceutical companies was somewhat dated, 14% consistently utilized optimization methods, with 35% reporting occasional use (1). This translated into 51% of those companies who, it must be surmised, have adopted some other iterative or time-consuming trial-by-error system in developing their drug-delivery systems. Product quality and cost under these circumstances must surely be negatively affected. Either way, these approaches are inefficient and can cause problems. Beyond phase I, attention to scale-up and process technology transfer becomes critical to the extent that reliance and implementation of multivariate optimization techniques is considered a necessity, as well as the practice of good science. It is, therefore, considered meaningful in discussing multivariate optimization, that is, experimental design methods with the intent not only to maintain an awareness of, but also to promote their utilization, especially when the intent is to conduct process scale-up studies.

In the broad context, the practices in the processing of pharmaceutical dosage forms have been reviewed elsewhere (2). Additionally, a workshop summary report on the scale-up of liquid and semisolid disperse systems provides useful guidance (3). The focus, in the second part of this chapter, is on a particular aspect of scale-up—namely, product homogeneity which, if not adequately understood and controlled, can have a catastrophic effect on product quality and performance. Homogeneity of topical products is a subject that has received scant discussion in the pharmaceutical literature compared with solid oral dosage forms. Because of the complex nature of many topical products as being multicomponent semisolid formulations, there is rising awareness of the importance and challenges in obtaining acceptable product homogeneity. This is particularly true for topical products that are difficult to mix (viscous systems), involve dispersed solids or emulsified liquids, or contain low concentrations of potent active substances. Achieving acceptable product homogeneity places prime importance on choice of manufacturing equipment, manufacturing process design, validation, and appropriate product stability monitoring to assure consistent product quality and performance. From this perspective the mixing operations, mixing equipment, and mixing parameters are highlighted. A final section deals with the verification of bulk homogeneity and content uniformity of topical products.

II. STUDY OPTIMIZATION BY EXPERIMENTAL DESIGN

Application of multivariate optimization techniques allows a structured and efficient way in finding the set of conditions that either cause or prevent a product or process from performing in an optimal manner. From the numerous methods (e.g., evolutionary operation [EVOP], or Taguchi methods) available to the formulation and processing scientist, experimental design and simplex optimization are the most frequently used (4–7). Regardless of the selected method, the process of optimization is based on a mathematical approach that searches for the conditions from several variables, n of a function $[f(x_1, x_2, x_3, \cdots x_n)]$ that optimize $f(x)$. The simplest approach, a univariate search, is taking a single variable or factor at a time to determine an optimal solution to the problem being analyzed. An example would be in the study of the effect of varying temperature on a formulation at a constant pH. Unfortunately, this procedure is inherently flawed, for it does not consider any of the interdependencies that may exist among the variables. For pharmaceutical formulations interdependencies between variables occur and are generally the rule, rather than the exception; accordingly; a multivariate optimization approach is preferred (8).

A. Factorial Design

When the number of variables are limited, generally to five or fewer, a two-level factorial design can provide a significant amount of information from only a few experiments. Such a design allows one to measure the influence of several variables simultaneously. This particular factorial design consists of all possible combinations of two levels of each of the variables. If two variables are involved, the full factorial design involves 2^2 experiments (i.e., a total of 4). The expression for the full factorial design is given as

$$N = L^k$$

where N is the number of experiments, L is the variable level, and k is the number of variables.

A two-level, three-variable design requires 2^3 or 8 experiments. These eight experiments can be represented in four different ways, in terms of:

Plus $(+)$ and minus $(-)$ notations
High and low levels of the variables
Treatment combinations
Vertices of a cube

To better illustrate this, designate the three variables as X, Y, and Z; and use the notation approach, in which $(-)$ signifies the low level of a variable and $(+)$ signifies the high level of a variable, the eight experimental trials can then be represented as

Trial no.	Variable X	Variable Y	Variable Z
1	−	−	−
2	+	−	−
3	−	+	−
4	+	+	−
5	−	−	+
6	+	−	+
7	−	+	+
8	+	+	+

The second approach, using "high" and "low" descriptors for each of the variables, can be represented as follows:

Trial no.	Level of X	Level of Y	Level of Z
1	Low	Low	Low
2	High	Low	Low
3	Low	High	Low
4	High	High	Low
5	Low	Low	High
6	High	Low	High
7	Low	High	High
8	High	High	High

Interpreting these high and low or (+) and (−) descriptors obviously requires definition of the conditions under which the specific variable is being evaluated. For example, in studying the homogeneity effect of a product under two different mixing times (60 and 30 min), 60 min is then the high or (+) level of the variable of interest (i.e., mixing time) and 30 min is the low or (−) level of mixing time.

The eight experimental trials, as presented, cover all the combinations of a two-level, three-variable design by which these combinations can be shown in the form of a factorial block:

		1		**z**	
1	1	Trial 1	(1)	Trial 5	z
	x	Trial 2	x	Trial 6	xz
y	1	Trial 3	y	Trial 7	yz
	x	Trial 4	xy	Trial 8	xyz

In the absence of a letter, or if the number 1 is present, this refers to the low level of a variable. If a letter is designated, then this represents the high level of the variable (e.g., x represents the high level of variable X). When interpreting this factorial block, consider the treatment combination xyz. This xyz block is in the x-row, the y-row, and the z-column. Consequently, the test sample is required to be prepared at the high level of X, the high level of Y, the high level of Z, which is trial number 8.

The fourth, is a geometric approach in representing the 2^3 factorial design. A three-dimensional schematic is used, in which the vertices of the cubical form represent the experimental trials as follows:

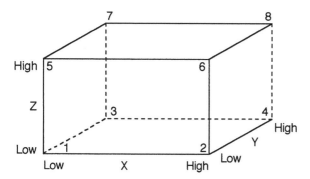

Each of the four ways described in the foregoing offers its own merits, but selection is frequently based on preference. For example, in the scale-up of a formulation the mixing of components to achieve product homogeneity may involve the following variables: temperature, mixing time, and excipient supplier. It is also assumed that the study is to determine the effects on some response of interest: here, product homogeneity, of two levels of temperature (40° and 20°C), two levels of mixing time (30 and 60 min), and two different suppliers (supplier S_1 and supplier S_2). Accepting the designations for temperature, mixing, and supplier as variables X, Y, and Z, respectively, then the designations of the three variables can be described as follows:

Variable X (temperature)
1. 20°C is designated as the low, or $(-)$, or (1) level of variable X.
2. 40°C is designated as the high, or $(+)$, or (x) level of variable X.

Variable Y (time)
1. 30 min is designated as the low, or $(-)$, or (1) level of variable Y.
2. 60 min is designated as the high, or $(+)$, or (y) level of variable Y.

Variable Z (supplier)
1. Supplier S_1 is designated as the low, or $(-)$, or (1) level of variable Z.
2. Supplier S_2 is designated as the high, or $(+)$, or (z) level of variable Z.

In taking the previous example of block xyz, experimental trial 8 can be described by any of the four following designations:

	Variable	
X	Y	Z
High	High	High
+	+	+
x	y	z
40°C	60 min	Supplier S_2

Similarly, all of the eight factorial blocks (treatment combinations), previously described, can be defined in this manner. By taking an individual sample at each of the eight treatment combinations any observed difference between result 2 and result 1 would be assigned to the high level of X versus the low level of X, assuming the absence of any other experimental variables. On this basis, the same would be true for samples 4 vs. 3, 6 vs. 5, and 8 vs. 7. Here the effect of variable X would have been measured a total of four times. By theoretical computation, this number can be shown to be a number necessary in making a valid decision. Additional and more detailed information are provided in other texts that deal with complete factorial designs, to which the reader is referred (9,10).

B.   Fractional Factorial Design

From the factorial equation given earlier in this text, it is obvious that as the number of variables increases, the number of experiments increases exponentially. Fractional factorial designs, a subset of full factorials, promotes efficiencies in conducting the relevant experimental studies through exploitation of any inherent redundancies that may exist. On a priori grounds some factor effects may be insignificant, with only the estimated main effects and some lower-order interactions evaluated. When high-order interactions occur, their effects are negligible and can be ignored because they are unlikely to bias the results. The mathematical expression for fractional factorial design is given by:

$$N = L_r^{k-f}$$

where r is the resolution code of a fractional factorial design; f is the fraction of the full factorial; N, L, and k are the terms presented previously. For example, a half-fractional ($f = 1$), two-level, five-factor design would be expressed as 2^{5-1} resulting in a total of 16 experimental trials. If the fraction is one-quarter ($f = 2$) then 2^{5-2} trials are required (i.e., eight experiments, or all of the experiments of the full design are conducted. The limitation to this type of design for testing purposes is that there is a loss in the degrees of freedom. A calculation of three-way and higher-order interaction effects would not be possible owing to insufficient degrees of freedom.

A half-factorial design for five factors at two levels was used by Schwartz et al. (11) in the optimization of an existing solid dosage form. In this study, three additional levels were selected, resulting in a total of 27 experimental trials. Data from the study were subjected to computed regression analysis to determine the fit to a second-order model. Despite limitations in restricting the model to second-order polynomials, the resulting equations were good predictors in subsequent computerized experimentation. Associated with computer-generated graphs, such as those of a given response variable as a function of an independent variable, allows an opportunity to rapidly simulate situations from which the formulator can gain an insight into the outcome of the process being investigated or optimized. Several steps are listed and defined, as necessary, to achieve a successful optimization program.

1. System selection
2. Variable selection
    a. Independent
    b. Dependent
3. Experimental trial
4. Statistical and regression analysis of data
5. Set specifications before feasibility search
6. Select constraints for grid search
7. Evaluate search results
8. Generate and evaluate
    a. Derivative plots
    b. Contour plots
9. Implement and monitor optimization decision(s)

This approach was effectively used in troubleshooting a formulation at the production scale (12). Moreover, it emphasized the value in utilizing preproduction optimization data in defining and facilitating the improvements necessary to achieve a product that consistently meets the quality specifications.

Experimental design studies can evolve into significantly larger exercises, as would be evident with a four-variable, five-level, full-factorial design resulting in a need to conduct 625 experiments. However, the goals before optimization must be clear and understood to avoid unnecessary experimentation. In the preceding design, the goal would not be to find the optimal settings for the variables but, rather, to gain an understanding of the system (formulation or process). For different phases of a project, experimental design would provide the wherewithal in guiding the project to an increasing chance of success and optimization.

Although our proceeding discussion on homogeneity centers on those parameters and factors that affect the uniform dispersion of excipient and the activity of a topical product, it is understood that any studies that are conducted must be based on a foundation of good experimental design. Only then can one achieve a solid understanding and a capability for optimization in the formulation and process control of a pharmaceutical product.

III. HOMOGENEITY

A. Product Physicochemical Properties

Important to the assurance of product homogencity is the design of manufacturing processes that are capable of producing a well-mixed, homogeneous product at the bulk stage, and that this design is maintained throughout product transfer, filling, transport, and shelf-life. This requires a thorough understanding of those physicochemical properties of the product that drive the choice of equipment and the design of process unit operations that compose the overall manufacturing process. For topical products the following properties are often considered:

Viscosity (rheology)
Shear sensitivity
Settling and Sedimentation rates
Solubility and dissolution of the active components
Stability of components (actives, antioxidants, etc.)
Particle, droplet, or globule size
Heating or cooling requirements
pH
Other physicochemical properties (e.g., crystal morphology, solvation, polymorphism of active component)

B. Mixing Operations

The design of bulk–product-mixing operations should consider the type of mixing that should be applied to the product to obtain the desired outcome. A diverse range of mixing operations are commonly employed during manufacture of topical products including the following:

Blending of multicomponent systems
Dispersing of solids into liquids
Emulsifying one liquid into another
Promotion of heat transfer between the product and vessel wall

Each of these operations requires suitably designed mixing equipment and the application of the appropriate type of mixing under a specified set of mixing parameters (rates of component addition, mixing speeds, temperatures, mixing times, etc.). Mixing can be thought of in terms of the applicable mixing scale required for a particular process. The distinction between macroscale mixing versus microscale mixing has been made for the purposes of discussion as now defined (13):

1. Macroscale mixing: Mixing that accomplishes adequate product flow in all regions of the mixing vessel both side-to-side and top-to-bottom to prevent stratification and ensure macroscopic homogeneity.
2. Microscale mixing: Mixing that accomplishes separation or dispersion of individual components (particles, droplets, etc.) to obtain the desired particle size distribution.

Macroscale and microscale mixing can be considered pumping and shearing processes, respectively. It follows from this discussion that virtually all multicomponent products will require adequate macroscale mixing (pumping) to be rendered homogeneous. However, macroscale mixing, while necessary, may not be sufficient for some products. Dispersed systems may, in addition, require microscale mixing (shearing) to render the desired particle size distribution.

C. Mixing Equipment

A diverse array of mixing equipment is available to use for liquids and semisolids in topical product manufacture. Generally, a cylindrical tank is employed that is equipped with one impeller agitator or more to accomplish the desired mixing operation. The following types of impellers are commonly employed.:

Paddles: Two-bladed and four-bladed paddles mounted on a vertical shaft and operated at low speeds are sometimes used for simpler mixing or heat transfer operations. The blades may be pitched to provide axial-flow patterns with low-applied shear. Scraper-blades that sweep the vessel surface are used for viscous products to prevent surface build-up and to facilitate heat transfer.

Propellers: Three- or four-bladed propellers are used for high speed, axial mixing of low-viscosity liquids. Generally achieves good pumping action.

Turbines: Used for high-speed mixing of liquids over a range of viscosities with blades that can be straight or curved, pitched or vertical. Impeller diameter is generally smaller than for paddles. Combines both pumping and shear mixing.

Dispersers: High-speed mixers using a sawblade or "Cowles" disk impeller for the dispersion of solids into liquids. Accomplishes good wetting and deagglomeration of solid particles through the applied shear.

Rotorstators: Mixer in which a high-speed impeller (rotor) is oriented in close proximity to a stationary housing (stator) leaving a narrow "gap" in-between. Designed for high-shear mixing applications including emulsification of liquids and dispersion of solids.

A vessel equipped with one or more of the mixers just listed may be required to provide the needed balance between macroscale mixing (pumping) and microscale mixing (shear).

D. Mixing Parameters

Once the mixing equipment has been chosen to provide the appropriate type of mixing for a particular product, the determination of optimal mixing parameters (rates of component addition, mixing speeds or tip speeds, temperatures, mixing times, and such) needs to be made to ensure that a homogeneous product is consistently obtained. Because undermixing or overmixing can have undesirable effects on the product, such as stratification, clumping of solids, incorrect particle size, or viscosity loss through excessive shearing, process development studies should be conducted to establish the optimal mixing conditions, including definition of all parameter ranges. This activity may also be required for in-process subparts that may be produced, depending on their physicochemical properties and the criticality of the mixing operation. The mixing parameters, often under worst-case conditions, are then formally demonstrated during process validation.

From a well-known mixer power consumption law, the following equation relates the power P consumed as a function of the mixer speed N, and the impeller diameter D where K is a constant (14):

$$P = KN^3D^5$$

This equation can be further partitioned into terms describing pumping and shearing as follows:

$$P = (K_1N^2D^2)(K_2ND^3)$$
$$\underbrace{}_{\text{shear}}\;\underbrace{}_{\text{pumping}}$$

The foregoing equation shows that a change in mixer speed N has a strong influence on the shear component of mixing, whereas a change in impeller diameter D affects the pumping component most strongly.

The considerations of product physicochemical properties, type of mixing required (pumping or shearing), equipment selection, and specific mixing parameters (speed, time, temperature, etc.) should guide the development of optimal mixing unit operations. This will ultimately lead to a manufacturing process that consistently produces homogeneous product that can be easily scaled-up and validated.

E. Verification of Bulk Homogeneity

As part of process development and validation, the homogeneity of the bulk product must be established after final mix (to establish final mixing conditions) by suitable sampling of the bulk and subsequent analytical testing.

Several sampling methods are available to establish the bulk product homogeneity after final mix. These techniques include

1. Sampling product from various locations within the tank with a suitable sampling device (i.e., sampling thief)
2. Sampling product from one or more valves (e.g., discharge valve) as a function of mixing time
3. Sampling from a transfer line at the beginning, middle, and end of batch transfer to a holding tank

Careful consideration must be given to the sampling method(s) chosen to establish bulk product homogeneity. The methods must produce samples that are representative of actual product at that stage in its manufacture and should not introduce any artifacts or perturbations that could be falsely interpreted. For example, it may not be appropriate to sample from various locations within the tank if to do so means opening a sealed tank and exposing the product to undesirable conditions (e.g., light or oxygen). Use of a sampling thief should be done only if the device is totally inert to the product (no drug binding, pH alteration, or particle adsorption). In fact more than one sampling technique may be applied especially for difficult-to-mix products (high viscosity) when stratification (settling) may occur over time, or for products containing small quantities of a finely divided drug.

F. Content Uniformity Testing of Topical Products

Testing of topical product homogeneity should also be performed over the shelf-life in the final container–closure system, as is often done for solid oral dosage forms. This information is important in establishing that the product remains homogeneous from beginning to end of the filling operation, after prolonged storage times under various storage conditions, and that phase separation leading to variations in dose uniformity does not occur.

Inherent in this discussion is deciding what is the relevant "scale of scrutiny" of the product to be tested for dose or content uniformity and how individual product containers should be sampled. A most reasonable scale of scrutiny for dosage forms is based on the size of a single dose, which is easily applied for dosage forms that are discrete (e.g., tablets, capsules, and suppositories). The same definition may be applied to topical products, as discussed by Train (15,16), but becomes harder to quantitatively define because of the bulk nature of the product in the final container–closure system and the imprecise nature of product application in the hands of the patient. Clearly, what is needed for content uniformity testing of topical products is a standardization of dose uniformity.

Hersey and Cook discussed the standardization of content uniformity for topical products (17), although formal standards have not yet been adopted by the pharmacopeia for topicals as they have for solids. Orr et al. defined a scale of scrutiny for ointments based on a "minimum ointment thickness" necessary for a therapeutic effect and a "minimum discrete area" of topical treatment that can be considered as a separate entity from surrounding areas of skin (18). Although they recognized that a minimum thickness of ointment film cannot be exactly measured, they reasoned that it would fall in to the range of $10–100$ μm. They also suggested that the minimum discrete area for an ointment layer 50-μm thick should not be greater than 1 mm^2.

In actuality, the minimum discrete area should be defined on the basis of the product's clinical indication, as greatly differing areas may be involved whether one using a product to treat acne, skin rash, or psoriasis. Estimates of total affected areas and the minimum film thickness for a given product will permit the definition of the appropriate scale of scrutiny. Provided that this dose level is practical from an analytical perspective (weighing or assay sensitivity, etc.), the single-dose quantity is suitably defined for the purposes of content uniformity testing. This sampling quantity can then be used to establish content uniformity for product from multiple locations within a single tube (closure-end, middle, and seal-end of expelled product or center-versus-side wall product after cutting tubes open) as a function of product age and storage conditions.

If we take into account that all the foregoing elements and requirements can lead to a well-defined and controlled product, achieving and maintaining the homogeneous state of a topical, semisolid formulation through scale-up and beyond is without question a desirable goal.

IV. CONCLUSIONS

In the real world, demands for the rapid and successful development of pharmaceutical formulations and product-manufacturing processes require the application of enabling tools, such as those of good experimental designs. Unfortunately, the tools and techniques available to the industrial pharmaceutical scientist are not always effectively applied or, indeed, considered. From a regulatory perspective, discussions on the guidance for chemistry, manufacturing, and control changes have emphasized the need for sound scientific and engineering practices during any scale-up or process change, which certainly underscores that effective and meaningful scientific practices are followed (19). Through an awareness of experimental design, a solid foundation on which the actual scale-up and process manufacture of products can be accomplished. A further emphasis placed on product homogeneity adds and expands the ability of the pharmaceutical scientist in achieving a successful outcome in the control and consistent manufacture of the final product.

REFERENCES

1. Shangraw RF, Demarest DA. A survey of current industrial practices in the formulation and manufacture of tablets and capsules. Pharm Technol 17:32, 1993.
2. Lachman L, Lieberman HA, Kanig JL. The Theory and Practice of Industrial Pharmacy. 3rd ed. Philadelphia: Lea & Febiger, 1986.
3. Van Buskirk GA, Shah VP, Adair D, et al. Workshop III report: scale-up of liquid and semisolid disperse systems. Pharm Res 11:1216–1220, 1994.
4. Fonner DE Jr, Buck JR, Banker GS. Mathematical optimization techniques in drug product design and process analysis. J Pharm Sci 59:1587, 1970.
5. Schwartz JB. Optimization techniques in product formulation. J Soc Cosmet Chem 32: 287, 1981.
6. Wehrle P, Korner D, Benita S. Sequential statistical optimization of a positively-charged submicron emulsion of miconazole. Pharm Dev Technol 1:97, 1996.
7. Taguchi G. Robust technology design. Mech Eng 115:60, 1993.
8. Bohidar NR, Restaino FA, Schwartz JB. Selecting key parameters in pharmaceutical formulations by principal component analysis. J Pharm Sci 64:966, 1975.
9. Anderson VL, McLean RA. Design of Experiments: a Realistic Approach. New York: Marcel Dekker, 1974.
10. Anderson RL. Practical Statistics for Analytical Chemists. New York: Van Nostrand Reinhold, 1987.
11. Schwartz JB, Flamholz JR, Press RH. Computer optimization of pharmaceutical formulations I: general procedure. J Pharm Sci 62:1165, 1973.
12. Schwartz JB, Flamholz JR, Press RH. Computer optimization of pharmaceutical formulations II: application in trouble shooting. J Pharm Sci 62:1518, 1973.
13. Scott RR. A practical guide to equipment selection and operating techniques. In: Lieberman HA, Rieger MM, Banker GS, eds. Pharmaceutical Dosage Forms: Disperse Systems, Vol. 2. New York: Marcel Dekker, pp 1–71, 1989.
14. McCabe WL, Smith JC, Harriott P. Unit of Operations of Chemical Engineering, 5th ed. New York: McGraw–Hill, 1993.
15. Train D. Pharm J 185:129, 1960.
16. Hersey JA. Drug Dev Commun 1:119–132, 1974–1975.
17. Hersey JA, Cook PC. J Pharm Pharmacol 26:126–133, 1974.
18. Orr NA, Smith EA, Smith JF. Int J Pharm Technol Prod Manuf 1:4–10, 1980.
19. Lucisano LJ, Franz RM. Pharm Technol 19:30, 1995.

10

Safety Considerations for Dermal and Transdermal Formulations

PETER J. DYKES

Cutest (Skin Toxicity Testing Company), Cardiff, Wales

ANTHONY D. PEARSE

University of Wales College of Medicine and Cutest (Skin Toxicity Testing Company), Cardiff, Wales

I. INTRODUCTION

Unlike our ancestors, we now live in a complex environment full of a variety of synthetic molecules. On a day-to-day basis contact is made with a wide range of compounds and almost everything that comes into contact with the skin has the potential for inducing adverse cutaneous reactions. A brief review of the dermatological journals dealing with contact dermatitis will reveal that someone, somewhere has reacted to something that they have been in contact with for short or long periods of time. Contact dermatitis reactions may be immediate or delayed, chronic or acute, irritant or allergic. An additional cofactor in the development of cutaneous reactions may be ultraviolet radiation (UVR). UVR has the capacity to energize molecules, and in this photoactive state, these molecules can produce phototoxic (photoirritant) and photosensitive (photoallergic) reactions. With topical formulations the cause of adverse reactions may be the vehicle and not the drug itself. Similarly with transdermal delivery systems the adhesives used to produce intimate contact with the skin may be a source of cutaneous reactivity. An additional problem with transdermal delivery systems is that the physical process of removal from the skin can induce mechanical trauma to the skin surface leading to erythema and edema. A further complication may be that the mechanical trauma leads to altered barrier function of the skin which, in turn, could enhance the percutaneous penetration of materials at or near the skin surface. Whatever the mechanism, adverse cutaneous reactions are a frequent problem with materials that come into contact with the skin. The resulting contact dermatitis can be a source of considerable discomfort and inconvenience and limit the usefulness of a topical formulation or transdermal delivery system.

II. TYPES OF CONTACT DERMATITIS

Some materials that come into contact with the skin penetrate the stratum corneum and cause cell death or injury. This effect is visualized as a local inflammatory skin reaction that is characterized primarily by erythema and edema; an irritant dermatitis. These types of reactions may be further classified into acute toxic contact dermatitis and irritant contact dermatitis. Acute toxic contact dermatitis may be provoked by a single or, sometimes, repeated application of strongly toxic materials. The association with the material and development of a reaction is usually quite obvious. These types of reaction occur after contact with materials such as acids, alkalis, solvents, and cleansers. This type of reaction is only rarely associated with topical application of a drug or cosmetic product.

In contrast with acute toxic contact dermatitis, irritant contact dermatitis occurs after repeated exposure. It is a localized, superficial, nonimmunologically based reaction. Many substances, including topical pharmaceutical agents, cause irritant contact dermatitis after repeated application. Often there is little indication of a reaction after the initial contact. However, after repeated exposure, the skin becomes erythematous, with drying and cracking of the skin surface. In more severe reactions edema may occur with papule and vesicle formation. Reactions are usually localized to the site of contact and, unlike allergic reactions, usually diminish fairly quickly after the stimulus has been removed. Irritant contact dermatitis will develop in all individuals given sufficient time and intensity of exposure. Reactions develop more quickly in situations where the material is held in intimate contact with the skin and in conditions where the barrier function of the skin is altered (e.g., in flexural creases or under occlusion with polyethylene).

Some materials penetrate the skin and stimulate an immune response. Any future contact results in an inflammatory immune reaction that can be intense, with erythema, edema, and vesiculation, possibly spreading beyond the original point of contact: an allergic contact dermatitis. Two main types of allergic contact dermatitis are involved in adverse reactions to topically applied materials: immediate-type and delayed-type hypersensitivity. Immediate type hypersensitivity (type I) is the result of antibody–allergen interaction occurring in the skin, the reaction that develops being known as allergic contact urticaria. Delayed-type hypersensitivity (type IV) is the result of cell-mediated immunity and is the most frequently reported side effect of topical drugs.

Some materials require ultraviolet radiation (UVR) to activate them before they are able to elicit adverse reactions. These photosensitive reactions can be phototoxic (photoirritant; i.e., nonimmunological) or photosensitization (photoallergic) reactions. In phototoxic reactions the chemical or drug is altered by UVR such that it becomes an irritant and causes inflammatory changes in the skin. In photosensitization reactions a reactive species is formed that is an allergen, or one that is capable of reacting with another moiety present in the skin to form an allergen. This can then stimulate the immune system and produce an allergic response such as a type I immediate-, or type IV delayed-hypersensitivity, reaction.

To minimize the risks of an excessive number of cutaneous reactions in the marketplace, pharmaceutical companies and cosmetic and toiletry manufacturers have developed a series of safety tests to which products are subjected before launch. Some of these tests are used for historical or legislative reasons and are not necessarily the best scientifically. Many involve experimental animals. However, percutaneous penetration is an all-important determinant of a product's potential for contact dermatitis and, as human skin is unique because of its epidermal thickness and barrier function, poor correlation between animal data and human experience is often seen. As the trend is toward using fewer animal-based test procedures, for both scientific and moral reasons, and because we believe that the most valid data is gained from the species for which the treatment is designed, only human safety tests will be discussed in this chapter.

III. PRETEST INFORMATION

Before any human testing is undertaken, every possible risk to the subjects must be assessed. This evaluation must be based on all known information, be it animal or *in vitro* based, that is present in the literature. To do otherwise would be considered unethical. The major requirements before any substances can be topically applied to humans are that they are not acute poisons or corrosive agents, and that there is no evidence that they are potent sensitizers. Much of this information can be obtained without the use of animal experiments. Various *in vitro* systems exist that will indicate whether a new chemical is an acute poison or is corrosive. For example, cell culture systems are widely used for screening (1). A system based on an epidermal slice technique has also been used successfully as a method for identifying strong irritants and corrosive materials (2). A more complex cell culture system, the skin equivalent, has also been proposed for use as a screening procedure for topically applied materials (3). In this system a three-dimensional reconstruction of human skin is prepared, using human epidermal cells layered on top of a collagen lattice containing fibroblasts. A "stratum corneum-like" structure is formed in this system and materials, including those not normally suitable for conventional cell culture procedures, can be applied directly to this surface. The release of inflammatory mediators, such as prostaglandins, from the dermal surface, thought to be central to the development of cutaneous irritancy *in vivo*, can subsequently be monitored.

Once satisfactory preliminary safety data has been obtained using these or similar *in vitro* systems, then it is possible to progress directly to human volunteer studies to determine the potential for contact dermatitis. The sequence of testing usually follows the route of irritancy tests followed by sensitization tests followed by photosensitivity tests, if deemed necessary. This is a logical approach because any product that is an irritant will probably not be developed further and the later tests are more complex and costly to perform. Any tendency toward irritancy would also cause problems in interpreting the results of other tests.

IV. IRRITANCY TESTS

Ethically, any human test procedure must be approached with the safety of the subjects as a priority. To this end, procedures have been developed that minimize the risk of serious cutaneous reactions (4,5). These procedures use a stepwise approach of increasing times of exposure and bioengineering methods to minimize the danger and discomfort to the subjects involved and to minimize the potential risk of irreversible skin damage. In most situations, however, the product being tested is not novel and is related to previously studied formulations. In this situation, the irritant potential can be determined by simple patch tests, such as the 48-h patch test or the 2- to 3-week cumulative irritancy test.

A. 48-Hour Irritancy Tests

The 48-h irritancy test consists of two sequential 24-h applications under occlusion to the same site. The break in the application is to premit assessment and possible termination of the application, should unexpected severe reactions occur. Assessments are made at the end of the 48-h period, and again at 72 h. The latter assessment time serves to check that any reactions are subsiding satisfactorily and to catch any late reactions that may develop only several hours after the removal of the patch. It may be appropriate to include both negative and positive controls in this type of test, particularly if experience of human testing with the class of product is minimal. Negative controls usually consist of the empty application chamber or the adhesive base without active agent. Positive controls may be known irritants, such as 0.5% sodium lauryl sulfate. It may also be desirable to compare results with existing or marketed products at this stage, so that if some reactions occur the relation to what is currently used is clear. An important variable, which is not possible to exclude from this type of test, is intraindividual variability in terms of cutaneous reactivity. This may be due to an individual's skin type, age, or hormonal status (6). To compensate for this, a reasonable number of individuals, usually 25–30, must be used in this type of study.

The 48-h irritancy test is sometimes referred to as a primary irritancy test. This is based on the definition that a *primary* irritant produces a reaction on first contact, and a *secondary* irritant produces a reaction only after repeated exposure (7). This is confusing because, depending on the concentration, a product can be both a primary and secondary irritant. Irritant potential is always dependent on conditions of exposure and is never absolute. Prolonged application of something apparently as innocuous as water can produce a dermatitis, given the right circumstances.

The interpretation of the results of the 48-h irritancy test is important. A negative result does not necessarily mean that the product is safe to use. Many irritants reveal themselves only after repeated application. As many products are, by virtue of their function, intended to be repeatedly applied to the skin, this short-term type of test is patently inadequate as a safety test. The 48-h test does however serve as a simple screening device and this is its primary function. It will indicate those products with a high potential for causing cutaneous irritancy. Several products can be tested at the same time, and information can be obtained quickly during product development.

B. Cumulative Irritancy Tests

Cumulative irritancy assays permit a more-sophisticated ranking of products and enable the comparison and classification of the weaker types of irritants, into which category most topical formulations and transdermal systems fall. There are several variations on the cumulative irritancy assay (8–10). The main variations are the type of application patch, the number of subjects, and the duration of the study. The central feature of these assays is that the product is repeatedly applied to the same site such that the exposure is exaggerated. In addition, the site is occluded, such that the barrier function of the skin is compromised, allowing greater penetration of potential irritants. By virtue of this exaggerated exposure, it is hoped that the information gained from a relatively small panel of individuals (usually 25–30 subjects) will predict the likely behavior under conditions of everyday consumer use.

The cumulative irritancy assay normally consists of either a 14- or 21-day application period. There is still some debate over whether the shorter time period is sufficient to detect very weak irritants, and the decision between the two time periods is usually made according to the product being tested and its proposed usage. Thus, for products that are in prolonged daily contact with the skin (e.g., dermal and transdermal therapeutic systems), the longer test procedure may be desirable to give the necessary reassurance of product safety.

Another variable in the cumulative assay is the number of applications over the test period. Some protocols call for daily applications and assessments (Monday to Friday, but not weekends), others for application and assessment every 2 to 3 days (Monday, Wednesday, and Friday). In both protocols the subjects continue to wear the patches over the weekend. There is very little to choose between the two schedules because both give continuous contact over the test period. The advantage of the daily regimen is that if severe reactions develop quickly, then the application can be stopped and, therefore the subject is exposed to minimal risk. However, this is not likely with the types of products generally tested, especially if they have been subjected to a 48-h test previously. The more frequent patch application and removal schedule has the distinct disadvantage that trauma to the skin surface by repeatedly applying and removing adhesive dressings may lead to severe "plaster" reactions that mean that the subject is no longer able to continue with the study. This is obviously more likely to happen in the 21-day protocol.

The nature of the application patch is also a variable in this assay. Many of the original assays used gauze pads (Webril pads) covered with impermeable tape such as Blenderm. We prefer an alternative system, 12-mm aluminum Finn chambers on Scanpore tape. The advantages of this system are as follows: Scanpore tape is not occlusive and minimizes the "plaster" reactions seen with occlusive tapes, such as Blenderm; the chambers have a fixed volume and when used with filter papers can be used for liquids; an impression of the raised rim is clearly seen on removal of the patch, giving the reassurance that occlusion was continuous and that the subject had complied with the conditions of the study. The chambers are available commercially as strips of chambers at fixed intervals that are perforated to allow subdivision. A possible disadvantage of the Finn chamber system is that the area of application (12-mm diameter = 113 mm^2) is less than the gauze pad. However, although the area of application is important in terms of the intensity of the response (11), above a certain level (approximately 100 mm^2) this effect is minimal.

With transdermal systems the patch itself is occlusive, and by its nature will produce cutaneous reactions such as folliculitis. An additional problem when testing transdermal systems is the pharmacological load on the subject. Topical formulations for treating skin diseases are designed to deliver relatively small amounts of drug locally. In contrast, transdermal systems are designed to deliver large doses systematically through a relatively small area of skin. The amount absorbed from a topical agent in a 12-mm–Finn chamber is usually negligible as far as the wellbeing of the subject is concerned. In contrast, a transdermal system will deliver a large dose over a short time period, and this must be taken into account when designing the study. Thus, a transdermal estrogen product is ideally tested for irritancy in postmenopausal women. In this situation the study almost becomes a phase I clinical study and pre- and poststudy medical screening is advisable. An alternative approach is to reduce the size of the transdermal patch, such that the pharmacological dose carries negligible risk, but the area is such that the irritant properties can be assessed.

Evaluation and interpretation of any irritant response also varies according to the test procedure. Grading scales for erythema vary from the simplest 0–4 scale (none, mild, moderate, severe), to more complex scales with half-point divisions. The scale we used (Table 1) has evolved over several years experience in assessing these types of reaction. The irritancy potential of a product is generally assessed by the number and severity of the reactions occurring during the study, in particular the number of grade 2 or higher reactions. A cumulative index of irritancy can be derived, and this can be related to previous assays and assays on marketed or competitor products. An alternative approach is to determine the time at which half of the subjects have reacted to a product in a procedure analogous to the LD_{50} determination (8,12). A range of concentrations needs to be tested and the proportion of subjects reacting to each concentration is plotted against the time of reaction and, by a curve-fitting procedure, the time when half the subjects have shown a reaction is recorded. A concentration at which half of the subjects react in this type of test can be determined and absolute comparisons made.

C. Facial-Stinging Test

Products may pass tests such as the cumulative irritancy test, but still cause problems for the consumer. Disagreeable sensations may occur, particularly when products are applied to sensitive areas such as the face. A not infrequent complaint is that of stinging or burning or itching after application. Signs of irritation, such as erythema or scaliness, may not necessarily occur in this situation. Part of the reason for such sensations is that facial skin is very permeable and has a rich nerve supply. It is also exposed to weathering and a constant bombardment of cosmetics and cleansers. These may all contribute to increased sensitivity of this area. A method to assess the stinging capacity of topical materials has been described (13). Subjects who are sensitive to the stinging sensation of lactic acid are selected as panelists for testing

Table 1   Human Patch Test-Grading Scale

0	No reaction
0.5	Slight, patchy erythema
1	Slight uniform erythema
2	Moderate, uniform erythema
3	Strong erythema
4	Strong erythema, spreading outside patch
5	Strong erythema, spreading outside patch with either swelling or vesiculation
6	Severe reaction with erosion

new products. Their subjective responses are noted, and a cumulative score is derived that is considered indicative of the potential for stinging in the general population.

V. SENSITIZATION TESTS

Allergic reactions to topically applied materials are much less common than irritant reactions. Many sensitizers will produce allergy only in a small percentage of the population, possibly less than 1%. As such, it is difficult to devise a test that will accurately reflect the incidence of sensitization when used by the population at large. The need for predictive tests was recognized more than 30 yeas ago, and the essential features of such tests were established (14). More recent examples of sensitization protocols indicate only small variations on the original procedures (15–17).

Essentially all protocols are based on the guinea pig assays developed to predict allergens and the original human studies by Draize et al. (18). Thus, there is an induction period, during which repeated exposure to the product occurs. This is followed 1–2 weeks later by a challenge to a previously unexposed site to determine sensitization. The test site for the induction phase is usually the back, although the upper arm is favored in some laboratories. The challenge should take place on previously unexposed skin, either at a site on the back, well removed from the original test site, or preferably on the upper arm. A control application of either vehicle without the active ingredient or a transdermal patch without drug is essential in interpreting any reactions that occur.

A. Induction Phase

The induction phase varies according to the protocol, but essentially consists of multiple applications, under occlusion, that are either continuous, as in the cumulative irritancy protocol, or intermittent (24-h on–24-h off). Typical examples are as follows:

1. An induction period of 42 days consisting of 21 × 48-h applications, each to a fresh site, as used for a transdermal nicotine patch (16).
2. An induction period of 21 days consisting of 9 × 24-h applications, 24-h on–24-h off, applications to the same site (17).
3. An induction period of 15 days consisting of 6 × 48–72-h applications to the same site (i.e., continuous application with inspection and reapplication).

We favor the latter variation because it gives continuous exposure to the same site and is probably equivalent, in terms of the immunological challenge, to an intermittent exposure over a longer period or to the longer exposure to fresh sites. Continuous exposure to the same site under occlusion maximizes the percutaneous penetration of any potential allergen. Intermittent exposure allows for barrier recovery and, therefore, may not be as severe a provocation.

A major variation to this induction procedure is the use of chemical agents to damage the skin before patch testing. The most common departure is the use of sodium lauryl sulfate in the maximization procedure (19). Briefly, an area of skin is treated with a 5% sodium lauryl sulfate solution for 24 h to induce a moderate inflammatory response. This is followed by a 48-h application of the test material to the same site. The sequence of 24-h irritant and 48-h test material application is repeated five times to the same site. This procedure is designed to maximize penetration of potential allergens. However, severe cutaneous reactions can develop with this approach, and this may be unacceptable to volunteer subjects. In addition the possibility of interaction between any sodium lauryl sulfate that has penetrated the skin and the test material cannot be ruled out. Such an interaction may alter the immunogenic potential of the test material.

B. Challenge Phase

In the challenge phase there is less variation in protocols. The main differences are the use of a single 48-h challenge or two consecutive 48-h challenges (back-to-back). The latter is claimed to be more sensitive in terms of detecting weak allergens (20). What is perhaps more critical is the use of appropriate controls and the time of assessment. Controls consisting of vehicle or transdermal system without active drug must be tested alongside the product of interest to interpret any reactions. At least two observations should be made, one assessment within 15–30 min after patch removal, and a second assessment 48 h later to determine late reactions. Sensitization challenges are usually carried out under occlusion and must be at sites remote from the initial induction site (e.g., the upper arm if the back has been used in the induction phase). The only other variation is the maximization procedure, which again uses pretreatment of the challenge site with sodium lauryl sulfate solutions.

C. Population Size

The number of subjects used in sensitization procedures varies from 25, in the maximization test, to 200 in the Draize test. The sample size must be large enough to give confidence, in terms of predicting what will happen in general use, but not so large that adoption of the procedure becomes expensive and time-consuming. The mathematics of extrapolating from a small test population to many consumers has been discussed (21). Briefly, if there are no reactions in a panel of 200 subjects, then as many as 15 out of every 1000 of the general population could react (95% confidence limits). For 100 and 50 subjects the figures are 30 and 58 out of every 1000 of the population. The possibility that the sensitization test may not predict a potential sensitizer may result from a variety of factors, such as skin site, variation in the frequency of application during the induction phase, poor release from the vehicle, and the dose applied. Clearly, to reduce the level of potential reactors in the general population to fewer than 1:1000 would require an enormous number of volunteers. The compromise is to use 100 or 200 subjects, which gives a reasonable level of confidence in the test. The choice of 100 or 200 subjects rests, to a certain extent, on the type of product. A cosmetic product that may be a reformulation or that does not contain any new chemical entities may require only 100 subjects. It may be more prudent to test a new pharmaceutical agent in a panel of 200 subjects. Any possible regulatory requirement of countries where the product is to be sold must also be taken into account.

D. Dosage

Undoubtedly the induction and elicitation of an contact sensitization reaction is dose-dependent. The higher the concentration, the more likely it is that individuals who are predisposed will become sensitized. With final formulations the test concentration is predetermined. When a range of therapeutic doses is being considered, the highest dose likely to be used clinically should be tested. With raw ingredients, the highest nonirritating concentration may be tested, depending on the likely exposure. As with irritancy, the area over which an allergen is applied is important. With a potent sensitizer, such as 2,4-dinitrochlorobenzene (DNCB), higher dose–response curves were obtained for a constant dose per unit area when the area of application was increased from 8 to 80 mm^2 (22). However, in a separate study with DNCB, at a similar dose per unit area, no difference was found between areas of 180 and 710 mm^2, neither in terms of the number of subjects sensitized nor in the challenge dose–response curves (23). Thus, with the potent sensitizer DNCB, the critical surface threshold area appears to be between 80 and 180 mm^2.

VI. PHOTOSENSITIVITY

It is possible that topically applied or systemically administered chemicals or drugs that have little or no potential to promote an irritant or allergic reaction in the skin, may do so in the presence of sunlight. Although photosensitive reactions may be rare, relative to irritant or sensitization reactions, the development of such reactions can lead to withdrawal from the market (e.g., the antirheumatic drug, benoxaprofen) (24).

A. Sunlight

Electromagnetic radiation is classified according to wavelength ranges. Electromagnetic radiation emitted by the sun ranges from very short wavelength cosmic rays to very long wavelength radio waves. Short-wavelength, high-energy forms of radiation such as gamma and x-rays, are termed ionizing radiation, whereas longer lower-energy radiation (more than 800 nm) is nonionizing and increases molecular motion, giving a thermal effect. Wavelengths between 200 and 800 nm are capable of causing chemical changes if absorbed by molecules in the skin. Ultraviolet radiation falls into this category and is between 100 and 400 nm. Fortunately for us, no wavelengths shorter than about 288 nm reach the earth's surface, because these are filtered out by the earth's atmospheric ozone layer. Damage to this important protective layer by man-made chemicals such as chlorofluorocarbons (CFCs) is currently under review and the subject of much discussion. UVR is subdivided into UVC (100–280 nm), UVB (280–320 nm), and UVA (320–400 nm). UVA can be further subdivided into UVA II (320–340 nm) and UVA I (340–400 nm) (25) (Table 2).

Table 2 Nonionizing Radiation

UVC	100–280 nm[a]
UVB	280–320 nm
UVA	320–400 nm
UVA II	320–340 nm
UVA I	340–400 nm
Visible	400–800 nm

[a]Does not reach the earth's surface.

Although sunlight that reaches the earth is essential to most living organisms and has many beneficial effects for humans, it is UVB and UVA that are responsible for most of the cutaneous photobiological events (26). These include DNA and RNA damage, inhibition of protein synthesis, damage to liposomal and cellular membranes, mutagenic and carcinogenic effects and erythematous responses. Chronic UV exposure is also responsible for dermal connective tissue injury (elastosis) leading to the changes known as photoaging (27).

B. UV-Induced Erythema

It is important to remember that sunlight is a continuous spectrum and that it is in its entirety that it causes erythema or sunburn in exposed skin. The paler the skin the greater the erythematous response for any given dose of UVR. Thus, skin type-I burns far more readily than skin type-VI (Table 3).

UVB is primarily responsible for most of the erythema seen in human skin following excessive sun exposure. This response is apparent approximately 6 h following exposure and is maximal at 20–24 h. In contrast to UVB, UVA can induce an immediate erythema that usually diminishes within 2 h of exposure (28). It can also give a delayed erythematous response that reaches a peak at 6 h. UVA contributes to about 10–20% of sunburn and passes through window glass, in contrast to UVB, which is essentially blocked by window glass. UVA can also induce an immediate pigment darkening response (IPD) (29). This reaction appears within seconds of exposure and fades within a few minutes. The pigment darkening is due to a photochemical reaction involving the oxidation of a low molecular weight form of melanin in melanosomes. The response is clearest in skin type-III, -IV, or -V individuals who have higher levels of melanin. Prolonged exposure to UVA can lead to persistent pigment darkening (PPD) or tanning of the skin (30).

The effects described so far are the results of UVR itself; however, photosensitivity reactions can occur when an exogenous or endogenous chemical (chromophore) absorbs UVR or visible light. These reactions may be either phototoxic (photoirritant) or photosensitive (photoallergic).

C. Phototoxicity Reactions

Phototoxicity refers to skin irritation that is produced through the interaction of chemical substances and radiant energy in the ultraviolet and visible ranges. Phototoxic or photoirritant effects are immediate and nonimmunological. When testing for the phototoxic potential of topically applied chemicals, the output from the required radiation source is UVA only. This is best accomplished by using a suitable

Table 3   Photo–Skin Types

Skin type I	Always burns and never tans
Skin type II	Always burns and tans with difficulty
Skin type III	Often burns and tans moderately
Skin type IV	Burns minimally and tans easily
Skin type V	Rarely burns and tans profusely
Skin type VI	Insensitive, never burns, deeply pigmented

filtered solar simulator. The clinical identification of phototoxic reactions in humans relies on both morphology and the clinical evidence or suspicion of the presence of possible phototoxic chemicals. Phototoxicity usually causes erythema or even bullae, increased skin temperature and pruritus and is followed by hyperpigmentation that, in some instances, may be long-lasting.

A range of phototoxic agents have been reported in the literature, many of which are naturally occurring plant substances, such as the furocoumarins, including the psoralens (31). Psoralens occur in a variety of plants, such as parsley, celery, and citrus fruits. A plant-induced phototoxic reaction is known as phytophotodermatitis. Perfumes that contain bergapton, which is a component of bergamot oil, a well known photoirritant, can be phototoxic. Other phototoxic agents include coal-tar derivatives, pyrene, anthracene, and fluoranthene (32). The cardiac antiarrhythmic agent amiodarone has produced phototoxic effects (33), as have quinoline antimalarial drugs (34). Tetracyclines, including demethylchlortetracycline, doxycycline, and chlortetracycline may also prove phototoxic when taken orally (35). The thiazide diuretics have also exhibited a phototoxic potential in experiments involving cardiovascular patients with hypertension and heart failure (36). Some nonsteroidal antiinflammatory drugs (NSAIDs) have a phototoxic potential, both when administered orally or topically (37). This is particularly true of propionic acid-related NSAIDs, that produce a unique wheal-and-flare response (38,39).

Most of the information gained on the underlying mechanisms of phototoxic reactions has been derived from animal or in vitro models. These have included *Candida albicans* (40,41), photohemolysis of red blood cells, and isolated normal human fibroblasts (42). Such in vitro models are, in our opinion, unreliable, as they lack the complexity of the whole animal. Recently, a human skin equivalent model has been used in an attempt to predict the phototoxic potential of topically applied personal care products (43). Known irritants and photoirritants were applied to the epidermal side of the "skin" for up to 24 h and then exposed to 2.9 J/cm^2 of UVA. The MTT reduction assay was used to assess cytotoxicity. The authors claim an 83% prediction with known phototoxic agents using this method. However, false-negative results were seen with well-known phototoxic agents such as amiodarone hydrochloride, 6-methycoumarin, bithionol, and piroxicam. Methods such as these may become more accurate predictors if they rely on more than one marker of cell damage for the prediction of chemical irritancy or, in this case, photoirritancy. It is our belief that the most reliable results are obtained when using the whole animal as a model. If the goal is to predict phototoxicity in humans, then the preferable test method should also use, when possible, human subjects.

D. Phototoxicity Assays

The development of procedures for assessing both the phototoxic (and photoallergic) potential of chemicals has, in general, been carried out in laboratory animals. A wide range of animal species have been used, including rabbits, mice, guinea pigs, squirrel monkeys, opossums, and swine. All methods require the skin to be irradiated with UVA following application of the test substance. However the measurable end result is not always similar to that obtained from human experiments. Increases in guinea pig (44) or mouse ear thickness (45,46) has been used to quantify phototoxic responses. Dermatitis and the increase in weight of the mouse tail has also been used

(47). However, hairless mice and albino guinea pigs have been used where simple erythema was the toxicological endpoint for assessing a phototoxic response (48). The results obtained from such experiments may be false-negative or false-positive when extrapolated to humans.

Kaidbey and Kligman (49) suggested a method for identifying potential topical phototoxic agents in humans, although the title of this publication misguidedly refers to photosensitizing, rather than phototoxic agents. Human testing is ethical as only small areas of skin are irradiated and clinical experience of phototoxic reactions indicate that when the stimulus is removed the dermatitis subsides. As with all clinical trials the informed consent of the subject and ethical committee approval must be obtained. The method we recommended is based on that of Kaidbey and Kligman and is suitable for assessing the phototoxic potential of topically applied drugs, chemicals, transdermal, and skin care products.

In the protocol for phototoxicity testing, a test panel consisting of a minimum of 12 healthy white adults with untanned back skin is required. As the dermatitis of a phototoxic agent can be produced in almost every subject, given sufficient exposure, it is not necessary to employ a large test panel. Two sets of test products and appropriate controls (usually vehicle or base alone) are applied occlusively to the midback area (one set on the left and one set on the right). In our experience 12-mm aluminum Finn chambers on Scanpore tape are suitable for product application. They are occlusive and ensure optimal product contact with the skin. An empty chamber should also be fixed to each side of the back. Six hours after administration of the materials and chambers, these should be removed and one set of applications only irradiated immediately with UVA. The chemical-to-skin contact time is relatively short compared with, for example, the 24 h needed in a vasoconstriction assay. If longer times are used (e.g., 24 h), light exposure will often produce negative results. The test sites should be irradiated with 20 J/cm^2 of UVA. The radiation source should ideally be a xenon arc lamp solar simulator.

Such a system could typically consist of a 1000-W ozone-free xenon arc lamp, the output from which is filtered with a Schott WG 345 filter of 2-mm thickness. This filter will block all erythemogenic UVC and UVB wavelengths below 320 nm. In addition unwanted longer wavelength visible and infrared radiation can be removed using a combination of a suitably coated dichroic mirror, water filter, and UG11 filter. If this is not accomplished then the subjects may feel heat and or pain from the irradiations. It is often convenient to deliver the UVA radiation to the skin surface using an 8-mm−diameter liquid light guide. A broad-spectrum thermopile should be used to measure the output of energy from the solar simulator, typically expressed in milliwatt per square centimeter (mW/cm^2). The thermopile should be calibrated against a known standard from, for example, The National Physics Laboratory in the United Kingdom. To calculate the time of irradiation necessary to administer a dose of 20 J/cm^2 to the skin, the formula in Table 4 should be used.

Skin assessments should be made immediately following irradiation and at 24 and 48 h after photoexposure. The grading system in Table 5 is suitable for recording any cutaneous reactions. The irradiated sites should be compared in each subject with the nonirradiated sites. If the response in any one subject at the irradiated site is greater than that seen at the nonirradiated site, then that product or chemical is deemed phototoxic. Phototoxicity is relatively easy to detect and therefore prevent;

Table 4   Calculation of Irradiation Times

$$t(\text{s}) = \frac{\text{mJ/cm}^2}{\text{mW/cm}^2}$$

Therefore, if a reading of 200 mW/cm^2 is obtained from the thermopile and a dose of 20 J/cm^2 (20,000 mJ/cm^2) is required then:

$$t(\text{s}) = \frac{20,000}{200}$$

and

$$t(\text{s}) = 100 \text{ s}$$

however, phototoxic reactions may sometimes mask or contribute to photoallergic reactions.

E. Photoallergic Reactions

Photoallergy (or photosensitization) refers to a dermatitis that is produced through the interaction of chemical substances and radiant energy, in the UVR and visible wavelengths, to produce an allergen. The chemical substance may be orally ingested or topically applied (photocontact allergy). Unlike phototoxic reactions, photoallergic reactions are often delayed, immunological, and less dose-dependent. Photoallergic and particularly photocontact photoallergic reactions are relatively uncommon when compared with phototoxic reactions. Clinically, photocontact allergic reactions produce a dermatitis that resembles allergic contact dermatitis, appearing as an acute dermatitis affecting primarily, but not exclusively, light-exposed skin. A characteristic histological feature of photocontact allergy is a dense perivascular round cell infiltrate in the dermis, which helps distinguish this dermatitis from a phototoxic reaction (50). A second and rare type of photoallergic reaction is solar urticaria (51). This occurs after only brief exposure to light and is characterized by an immediate urticarial wheal-and-flare reaction within minutes of exposure. The reaction usually subsides within 1–2 h, is associated with degranulation of mast cells at the site of exposure, and the release of neutrophil chemotactic factors and histamine into venous blood near the reaction sites. Photoallergic reactions, when they occur, may apparently be

Table 5   Phototoxicity Grading Scale

Grade	Cutaneous rection
0	No reaction
1	Mild erythema, possibly with scaling
2	Moderate or strong erythema
3	Moderate or strong erythema with a papular response
4	As grade 3, but with definite edema
5	Vesicular or bullous eruption

triggered by irradiation alone (in the absence of known sensitizers) or may be due to exogenous chemicals and UVR. Photoallergy, whether photocontact allergic dermatitis (delayed hypersensitivity: type IV reaction) or solar urticaria (immediate hypersensitivity: type I reaction), is an acquired reactivity dependent on cell-mediated hypersensitivity or antigen–antibody interaction (52).

Test procedures designed to identify potential photosensitizing chemicals were developed in response to the outbreak of reactions caused by the use of antibacterial halogenated salicylanilides in the early 1960s (53). A minority of affected individuals developed a persistent photodermatitis that lasted several years despite the avoidance of contact with photosensitizing phenolic compounds (54). It therefore became clear that there was a requirement for a laboratory test to detect potential photosensitizing agents and avoid such situations.

Several photocontact sensitizers have been identified, including coumarins and coumarin derivatives, musk ambrette, fentichlor, bromochlorosalicylanilide, chloro-2-phenylphenol, and benzocaine (55–57). Certain sunscreens have also been reported to produce photocontact allergic dermatitis, most notably *para*-aminobenzoic acid and derivatives, benzophenone 3, mexenone, and cinnamates (58–60). Several essential oils have also produced photoallergic reactions (e.g., sandalwood oil; 61). Also, 8-methoxypsoralen, which is used in PUVA (psoralen + UVA), can also be photoallergic as well as phototoxic (62).

F. Photoallergenicity Assays

Test procedures to identify potential photosensitizers have received less attention than methods designed to detect either ordinary contact sensitizers or phototoxic chemicals. Landsteiner and Chase (63) demonstrated that low molecular weight haptens can produce contact dermatitis in guinea pigs. They also observed that allergic contact dermatitis could be conferred on immunologically naive guinea pigs by the passive transfer of mononuclear cells from nonsensitized animals. Furthermore guinea pigs develop edema and erythema after contact with topically applied sensitizers and, to some extent, develop a response similar to the clinical response in humans. These observations became the cornerstone of photocontact allergic dermatitis research over many years and led to the guinea pig becoming the most commonly used animal in photoallergy studies (64,65). On the other hand, mouse ear swelling is claimed to be a more sensitive model (66), but is even further removed from the human response than that of the guinea pig. To induce contact photosensitivity in any animal it has to be repeatedly exposed to the test molecule in the presence of UVR. For this induction phase a broad-spectrum source is necessary, which should include UVB as well as UVA. The period of induction should be similar to that for testing contact sensitizers, followed by a rest period and then a challenge on a previously untested site with the test chemical and UVA alone.

The photomaximization test for the prediction of photosensitizers is conducted on humans and is similar in design to a sensitization test, but with the addition of exaggerated UVR exposure of both the chemical and the skin to which it is applied. In an ideal world, this type of test would be carried out on a large number of subjects (>100) to more accurately predict the incidence of photosensitization reactions in the population at large. From a practical point of view this is not possible because of the demanding nature of the protocol; therefore, a test panel of 26 is normally recommended. The method described here is based on that of Kaidbey and Kligman (67) and is suitable for identifying topical photocontact sensitizers.

The photomaximization test is a 6-week study and is divided into an induction phase and an elicitation phase. A solar simulator, as described previously, is an ideal source of UVR. During the induction phase a 1-mm–WG320 filter and a 2-mm–UG11 filter should be used, allowing both UVA and UVB (290–400 nm) to reach the subject's skin. For the elicitation phase, a 2-mm–WG345 and a 1-mm–UG11 filter should be used to allow only UVA (320–400 nm) to reach the skin. The test chemical together with the vehicle control is applied occlusively to the midback of each subject using 12-mm aluminum Finn chambers on Scanpore tape. Twenty-four hours later, the patches are removed and the test sites wiped clean and allowed to air-dry for approximately 30 min. Each site is then exposed to twice the subjects minimal erythema dose (MED) of solar-simulated UVR. The sites are then left uncovered and exposed to the air for approximately 48 h. This procedure of application and irradiation is repeated such that each subject has six applications and irradiations over a 6-wk period. The sites are evaluated 24 h after each irradiation and, if the reaction becomes severe such that further application and irradiation is undesirable, then application of the material and subsequent irradiation is carried out at an adjacent site.

Following completion of the induction phase, there is a 2-wk rest period, with no applications or irradiations. Approximately 14 days after the end of the induction phase, two sets of test materials are applied to previously untreated sites on the midback, again using 12-mm–Finn chambers. Twenty-four hours later the patches are removed, the skin wiped dry with a gauze swab, and one set of applications irradiated with only 4 J/cm^2 of UVA. The sites are then evaluated at 24, 48, and 72 h after the elicitation irradiation. The grading system for all irradiations (both induction and elicitation) can be seen in Table 6. If one or more subjects develops a

Table 6   Grading System for Photoallergy Test: Induction and Elicitation Phase

0	No reaction.
1	Reaction readily visible, but mild unless letter grade appended (see E and F grades). Mild reactions include weak, but definite erythema, and weak superficial skin responses such as glazing, cracking, or peeling.
2	Definite papular response (appended E, F, or S if appropriate).
3	Definite edema (appended E, F, or S if appropriate).
4	Definite edema and papules (appended E, F, or S if appropriate).
5	Vesicular–bullous eruption (appended E, F, or S if appropriate).
E	Strong erythema at patch site.
F	Strong effects on superficial layers of the skin including fissures, a film of dried serous exudate, small petechial erosions, or scabs.
S	Reaction spreading beyond test site.
I	Itching.
B/S	Burning or stinging.

(Applications must be either terminated or moved to an adjacent nonirradiated site if a reaction score of 2 or higher occurs).
Descriptive letter designations may be added to the numerical score if experienced at the test site.
(Any other signs or symptoms (e.g., wheal-and-flare responses) may be described separately).

reaction at an irradiated site during the elicitation phase that is greater than the corresponding unirradiated site, then that chemical is considered to be a photosensitizer. In practical terms there are usually many or no reactors in a test panel making the decision as to whether a product is a photoallergen relatively easy.

REFERENCES

1. Duffy PA, Flint OP. In vitro dermal irritancy tests. In: Atterwill C, Steele C, eds. In Vitro Methods in Toxicology. London: Cambridge University Press, p 279, 1987.
2. Oliver GJA, Pemberton MA. An in vitro epidermal slice technique for identifying chemicals with potential for severe cutaneous effects. Food Chem Toxicol 23:229, 1985.
3. Dykes PJ, Edwards MJ, O'Donovan MR, Merrett V, Morgan HE, Marks R. In vitro reconstruction of human skin: the use of skin equivalents as potential indicators of cutaneous toxicity. Toxicol In Vitro 5:1, 1991.
4. Basketter DA, Whittle E, Griffiths HA, York M. The identification and classification of skin irritation hazard by a human patch test. Food Chem Toxicol 32:769, 1994.
5. Dykes PJ, Black DR, York M, Dickens AD, Marks R. A stepwise procedure for evaluating irritant materials in normal volunteer subjects. Toxicol In Vitro 14:204, 1995.
6. Patil S, Maibach H. Effect of age and sex on the elicitation of irritant contact dermatitis. Contact Dermatitis 30:257, 1994.
7. Shelanski HA. Experiences with and considerations of the human patch test method. J Soc Cosmet Chem 2:324–331, 1951.
8. Kligman AM, Wooding WM. A method for the measurement and evaluation of irritants in human skin. J Invest Dermatol 49:78–94, 1967.
9. Phillips L, Steinberg M, Maibach HI, Akers WA. A comparison of the rabbit and human skin response to certain irritants. Toxicol Appl Pharmacol 21:369, 1972.
10. Carabello FB. The design and interpretation of human skin irritation studies. J Toxicol Cutan Ocul Toxicol 4:61–71, 1985.
11. Dykes PJ, Hill S, Marks R. The effect of area of application on the intensity of response to a cutaneous irritant. Br J Dermatol 125:330, 1991.
12. Bahmer FA. In vivo assessments of irritants made simple and subject friendly. Contact Dermatitis 33:210, 1995.
13. Frosch PJ, Kligman AM. A method for appraising the stinging capacity of topically applied substances. J Soc Cosmet Chem 28:197, 1977.
14. Kligman AM. The identification of contact allergens by human assay. J Invest Dermatol 47:369–374, 1966.
15. Robinson MK, Stotts J, Danneman PJ, Nusair TL. A risk assessment procedure for allergic contact sensitization. Food Chem Toxicol 27:479, 1989.
16. Jordan WP. Clinical evaluation of the contact sensitization potential of a transdermal nicotine system (Nicoderm). J Fam Pract 34:709, 1992.
17. Gerberick GF, Robinson MK, Stotts J. An approach to allergic contact sensitization risk assessment of new chemicals and product ingredients. Am J Contact Dermatitis 4:205, 1993.
18. Draize JH, Woodard G, Calvery HD. Methods for the study of the irritation and toxicity of substances applied topically to the skin and mucous membranes. J Pharmacol Exp Ther 83:377–390, 1944.
19. Kligman AM. The identification of contact allergens by human assay. III The maximisation test: a procedure for screening and rating contact sensitizers. J Invest Dermatol 47:393–409, 1966.
20. Jordan WP. 24-, 48-, and 48/48-hour patch tests. Contact Dermatitis 6:151, 1980.

21. Henderson CR, Riley EC. Certain statistical considerations in patch testing. J Invest Dermatol 6:227–232, 1945.

22. Rees JL, Friedmann PS, Matthews JNS. The influence of area of application on sensitization by dinitrochlorobenzene. Br J Dermatol 122:29, 1990.

23. White SI, Friedmann PS, Moss C, Simpson JM. The effect of altering area of application and dose per unit area on sensitization by DNCB. Br J Dermatol 115:663–668, 1986.

24. Allen B. Benoxaprofen and the skin. Br J Dermatol 109:361–364, 1983.

25. Fitzpatrick TB. Ultraviolet induced pigmentary changes: benefits and hazards. In: Honigsmann H, Stingl G, eds. Therapeutic Photomedicine. Basel: Karger, pp 25–38, 1986.

26. Epstein JH. Photomedicine. In: Smith KC, ed. The Science of Photobiology, 2nd ed. Boca Raton: CRC Press, 1988.

27. Pathak MA. Sunscreens: topical and systemic approaches for protection of human skin against harmful effects of solar radiation. J Am Acad Dermatol 1:285, 1982.

28. Whitman GB, Leach EE, Deleo VA, Harber LC. Comparative response to UVA radiation in humans and guinea pigs. In: Urbach F, Gange RW, eds. The Biological Effects of UVA Radiation. New York: Praeger, pp 79–86, 1986.

29. Kaidbey KH, Barnes A. Determination of UVA protection factors by means of immediate pigment darkening in normal skin. J Am Acad Dermatol 25:262, 1991.

30. Pathak MA. Activation of the melanocyte system by ultraviolet radiation and cell transformation. Ann NY Acad Sci 453:328, 1985.

31. Pathak MA. Phytophotodermatitis. In: Pathak MA, Harber L, Seiji M, Kukita A, eds. Sunlight and Man: Normal and Abnormal Photobiologic Responses. Tokyo: University of Tokyo Press, pp 495–513, 1974.

32. Kochevar I, Armstrong RB, Einbinder J, Walther RR, Harber L. Coal tar phototoxicity: active compounds and action spectra. Photochem Photobiol 36:65–69, 1982.

33. Rappersberger K, Honigsmann H, Ortel B, Tanew A, Konrad K, Wolff K. Photosensitivity and hyperpigmentation in amiodarone treated patients: incidence, time course and recovery. J Invest Dermatol 93:201–209, 1989.

34. Ljunggren B, Winestrand L. Phototoxic properties of quinine and quinidine: two quinoline methanol isomers. Photodermatology 5:133–138, 1988.

35. Bjellerup M, Ljunggren B. Photohaemolytic potency of tetracyclines. J Invest Dermatol 84:262–264, 1985.

36. Diffey BL, Langtry J. Phototoxic potential of thiazide diuretics in normal subjects. Arch Dermatol 125:1355–1358, 1989.

37. Stern RS. Phototoxic reactions to piroxicam and other nonsteroidal anti-inflammatory agents. N Engl J Med 309:186–187, 1983.

38. Diffey BL, Daymond JJ, Fairgreaves H. Phototoxic reactions to piroxicam, naproxen, and tiaprofenic acid. Br J Rhematol 22:239–242, 1983.

39. Kaidbey K, Mitchell F. Photosensitizing potential of certain non-steroidal antiinflammatory agents. Arch Dermatol 125:783–786, 1989.

40. Daniels F. A simple microbiological method for demonstrating phototoxic compounds. J Invest Dermatol 44:259–263, 1965.

41. Mitchell JC. Psoralen type phototoxicity of tetramethylthiur-ammonosulphide for *Candida albicans*: not for man or mouse. J Invest Dermatol 56:340, 1971.

42. Lock SO, Friend JV. Interaction of ultraviolet light, chemicals and cultured mammalian cells: photobiological reactions of halogenated antiseptics, drugs and dyes. Int J Cosmet Sci 5:39–49, 1983.

43. Edwards SM, Donelly TA, Sayre RM, Rheims LA. Quantitative in vitro assessment of phototoxicity using a human skin model, Skin^2. Photodermatol Photoimmunol Photomed 10:111, 1994.

44. Stott CW, Stasse J, Bonomo R, Campbell AH. Evaluation of the phototoxic potential of topically applied agents using longwave ultraviolet light. J Invest Dermatol 55:335–338, 1970.

45. Cole C, Sambuco C, Forbes P, Davies R. Response to ultraviolet radiation: ear swelling in hairless mice. Photodermatology 1:114–118, 1984.

46. Gerberick G, Ryan C. A predictive mouse ear-swelling model for investigating topical phototoxicity. Food Chem Toxicol 27:813–819, 1989.

47. Ljunggren B, Muller H. Phototoxic reaction to chlorpromazine as studied with the quantitative mouse tail technique. Acta Derm Venereol 56:373, 1976.

48. Forbes PD, Urbach F, Davies RE. Phototoxicity testing of fragrance materials. Food Cosmet Toxicol 15:55–60, 1977.

49. Kaidbey KH, Kligman AM. Identification of topical photosensitizing agents in humans. J Invest Dermatol 70:149, 1978.

50. Epstein JH. Photoallergy and photoimmunology. In: Stone J, ed. Dermatologic Immunology and Allergy. St Louis: CV Mosby, 1985.

51. Sams WM. Solar urticaria: studies of the active serum factor. J Allergy Clin Immunol 45:295, 1970.

52. Epstein JH. Phototoxicity and photoallergy in man. J Am Acad Dermatol 8:141, 1983.

53. Wilkinson DS. Photodermatitis due to tetrachlorosalicylanilide. Br J Dermatol 73:213, 1961.

54. Smith SZ, Epstein JH, Photocontact dermatitis to halogenated salicylanilides and related compounds. Arch Dermatol 113:1372, 1977.

55. Kaidbey KH, Kligman AM. Photosensitization by coumarin derivatives: structure activity relationships. Arch Dermatol 117:258, 1981.

56. Raugi GJ, Storrs FJ, Larsen WG. Photoallergic contact dermatitis to mens perfume. Contact Dermatitis 5:251, 1979.

57. Epstein JH. Photoallergy: a review. Arch Dermatol 106:741, 1972.

58. Thune P. Contact and photocontact allergy to sunscreens. Photodermatology 1:5, 1984.

59. Szczurko C, Dompmartin A, Michel M, Moreau A, Leroy D. Photocontact allergy to oxybenzone: ten years of experience. Photodermatol Photomed 10:144, 1994.

60. Parry EJ, Bilsland D, Morley WN. Photocontact allergy to 4-*tert*-butyl-4'-methoxy-dibenzoylmethane (Parsol 1789). Contact Dermatitis 32:251, 1995.

61. Starke JC. Photoallergy to sandlewood oil. Arch Dermatol 96:62, 1967.

62. Fulton JE, Willis I. Photoallergy to methoxsalen. Arch Dermatol 98:455, 1968.

63. Landsteiner K, Chase MW. Experiments on transfer of cutaneous sensitivity to simple compounds. Proc Soc Exp Biol Med 49:688, 1942.

64. Harber LC, DeLeo VA. Guinea pigs and mice as effective models for the prediction of immunologically mediated contact photosensitivity. In: Maibach HI, Lowe NJ, eds. Models in Dermatology, Dermatopharmacology and Toxicology. New York: Karger, 1985.

65. Harber LC, Targovnik SE, Baer RW. Contact photosensitivity patterns to halogenated salicylanilides in man and guinea pigs. Arch Dermatol 96:646, 1967.

66. Maguire HC, Kaidbey KH. Experimental photoallergic contact dermatitis: a mouse model. J Invest Dermatol 73:147–152, 1982.

67. Kaidbey KH, Kligman AM. Photomaximization test for identifying photoallergic contact sensitizers. Contact Dermatitis 6:161, 1980.

11

Transdermal Delivery and Cutaneous Reactions

JAGDISH SINGH

North Dakota State University, Fargo, North Dakota

HOWARD I. MAIBACH

University of California School of Medicine, San Francisco, California

I. INTRODUCTION

Drugs and excipients have different sensitization capacities for inducing contact allergy. The risk of skin reactions produced by chemicals depends on their inherent allergenicity and ability to penetrate into the normal skin or damaged skin. As fully described in earlier chapters, the penetration of chemicals into the skin depends on skin condition, anatomical site, chemical characteristics of the substance, lipid solubility and concentration of the chemical. Penetration is also influenced by external factors, especially solvents, surface-active agents, alkalies, moisture, temperature, extreme dehydration, and mechanical effects. The length of time that a substance contacts the skin is of great importance. Skin irritation influences the cells of the skin and results in an increased sensitization risk. Such cell damage can be produced by variety of chemicals or by mechanical means.

Irritation is the nonimmunological evocation of normal or exaggerated reaction in a tissue by application of a stimulus. Irritation may be subjective or objective. *Subjective* irritation refers to transient pruritus, stinging, burning, or related sensations without subsequent visible inflammation (e.g., alcohol on an open wound). A chemical substance that evokes inflammation on initial exposure is called an *acute (primary) irritant* but, on repeated exposure to an identical site, will cause *cumulative irritation*. In the past, soaps, cosmetic materials, and pest control chemicals have

529

been recognized as potential sources of cutaneous irritation. More recently it has been recognized that a multitude of occupational and environmental factors, such as organic dyes and solvents or industrial waste material contribute to this topical skin disorder.

The most common reaction consists of a local inflammatory response characterized by erythema or edema, or a corrosive reaction characterized by local tissue destruction or necrosis. Other reactions, sometimes referred to as irritation, do not display visible signs of inflammation. Subtle increases in epidermal thickness, without visible or histological inflammation, may be produced by a variety of substances usually thought to be nonirritating (1).

The occlusive nature of many transdermal delivery systems provides an ideal model for inducing sensitization. Potential allergens include the adhesive, the diffusion membrane, the solvent, the enhancer, and the drug. Allergic contact dermatitis with redness, swelling, and sometimes vesiculation, is the most overt presentation of skin sensitivity from transdermal delivery systems. The reaction is usually localized to the site of application of the current patch, but may also occur at the sites of previous applications, the flare-up reaction (2). Spread of the eczematous reaction to sites not associated with the application of patches may occur (3). Urticaria and angioedema are rare allergic reactions to transdermal therapeutic systems (4).

This chapter deals with the skin reactions caused by topical drug delivery systems.

II. PREDICTIVE TESTING

A. Irritation

There is not yet an adequately validated in vitro model available to predict skin irritation of topical chemicals. Details of the current state of development of these potentially useful assays were summarized (5). The standard method to forecast skin irritation is by so-called predictive tests on humans or animals. The most widely used test for predicting potential skin irritants to humans, using animal models, was published by Draize et al. (6), and has been refined by many groups (7). The test was initially designed to classify chemicals that cause primary (acute) irritation. However, in designing a test that would eliminate false-negative reactions (type 2 errors), Draize permitted the introduction of a significant number of false-positive reactions (type 1 errors). The rabbit Draize test, properly performed and interpreted by experienced scientists, still remains valuable.

Transepidermal water loss (TEWL) is a well-accepted method for quantifying alterations in stratum corneum function (8), and it provides a robust method for assessing stratum corneum damage. Irritation tends to reduce the efficiency of the stratum corneum barrier function and may result in an increase in TEWL. This is sometimes associated with a decrease in skin water content (9). Hence, measurement of skin capacitance or skin hydration (10) may also be used to assess irritation (11,12). Measurements of carbon dioxide emission from human skin can also be used to determine the degree of irritation (13). Rates of carbon dioxide emission from irritated skin increase roughly in proportion to the degree of irritation (14).

Four techniques (skin color reflectance, TEWL, laser Doppler flow (LDF) measurement, and visual scores) have been compared for their ability to quantify sodium lauryl sulfate irritant dermatitis in humans (9). The study concluded that, although TEWL measurements may be an accurate and sensitive method in evaluating skin irritation when stratum corneum damage is present, color reflectance measurements may be a useful complimentary tool in the evaluation of skin damage. Detailed documentation on these bioengineering tools can be found in recent text books (15–17).

B. Allergic Contact Dermatitis

Allergic contact dermatitis testing is widely performed, with both human subjects and laboratory animals, to determine the irritant potential of various chemicals. The oldest of these assays is the Draize guinea pig test. The Draize test with animal models requires careful planning and performance. Buehler (18) and Magnusson and Kligman (19) used five chemicals (benzocaine, formalin, monobenzyl ether of hydroquinone, potassium chromate, and tetrachlorosalicylanilide) to compare sensitization rate by the Draize test, closed patch test, and guinea pig maximization test (GPM test). The percentage of sensitized animals was about 5% with the Draize test, 38% with the closed patch test, and 61% with the GPM test. This provides a rough estimate of the relative capacity of the three techniques to identify contact sensitizers.

Marzulli et al. (20) tested eight compounds using various modifications of the Draize guinea pig and human sensitization techniques (21). Skin sensitization was observed both in humans and guinea pigs with p-phenylenediamine and dinitrochlorobenzene, and in humans, but not in guinea pigs, with neomycin, benzocaine, hexachlorophene, furacin, and a mixture of methyl and propyl parabens. The authors stated that a negative result with guinea pigs provide an insufficient basis for concluding that a human is not likely to be sensitized by a substance.

The GPM test, the human maximization test, and the Draize repeat insult patch test (22) have been used for extensive comparisons of contact sensitizers in guinea pigs and humans (19). Contact sensitizers were rated in the five grades such that weak (I), mild (II), moderate (III), strong (IV), and extensive (V) corresponded to 0–8%, 9–28%, 29–64%, 65–80%, and 81–100% sensitized. The results are given in Table 1. Neither technique produced sensitization with very weak allergens, such as hexachlorophene and lanolin. Substances that did not sensitize humans, such as aluminum chloride, sodium lauryl sulfate, and polysorbate 80, did not sensitize the guinea pigs either. Quantitative structure activity relations (QSAR) analysis provides a powerful tool for predicting not only sensitization potential, but also how to define appropriate testing parameters (23–26).

Guinea pig testing constitutes the first step in evaluating the allergenicity of new compounds or products. There is a reasonable degree of correspondence between the results obtained with the GPM test and the human maximization test. These two tests rate the allergenicity of common human sensitizers in a similar fashion. Substances that sensitize in the human test also do so in the animal test (27). Detailed discussions of animal and human sensitization assays and interpretation can be found (28,29). Skin reactions from topical drug delivery systems, including chemicals, metals, and textiles, have been extensively investigated and may be found in series of publications (30–73).

Table 1 Comparative Sensitization in Humans and Guinea Pigs by the Guinea Pig Maximization Test and Human Maximization Test from the Following Chemicals

Chemicals	Guinea pig maximization test		Human maximization test	
	Positive (%)	Grade	Positive (%)	Grade
Aluminum chloride	0	I	0	I
Apresoline	80	IV	100	IV
Atabrine	90	V	78	IV
Benzocaine	28	II	22	II
Formalin	80	IV	72	IV
Hexachlorophene	0	I	0	I
Lanolin	0	I	0	I
Malathion	54	III	100	V
Mercaptobenzothiazole	40	III	38	III
Mercuric chloride	32	III	92	V
Monobenzyl ether of hydroquinone	50	III	92	V
Neomycin	72	IV	28	II
Nickel sulfate	55	III	48	III
Penicillin G	100	V	67	IV
Polysorbate 80	0	I	0	I
Potassium dichromate	75	IV	100	V
Sodium lauryl sulfate	0	I	0	I
Sulfathiazole	36	III	4	I
Streptomycin	72	IV	80	IV
Tetrachlorosalicylanilide	72	IV	88	IV
Turpentine	64	III	72	IV
Vioform	20	II	0	I

Source: Ref. 8.

III. REACTIONS TO DRUG DELIVERY SYSTEMS

A. Transdermal Therapeutic Systems

Transdermal drug delivery systems for systemic effect are feasible for small, potent, and lipophilic drug molecules (74–76). Transdermal drug delivery systems are presently marketed in the United States for seven drugs (estradiol, clonidine, nitroglycerin, scopolamine, nicotine, fentanyl, and testosterone), and others are under development. As the drug is the most frequently identified allergen, human subjects can be patch-tested with the drug at an appropriate concentration in a suitable vehicle (Table 2). Skin irritation at the application site is the most common adverse effect accompanying the use of transdermal therapeutic systems, occurring in as many as 5–24% of women (77–89). Although generally mild and transient, it appears to be the most common reason for discontinuation of treatment during published efficacy and tolerability studies (90). The following are adverse reactions related to the use of commercially available transdermal drug delivery systems.

Table 2 Concentration of Drug in Appropriate Vehicle Used in Patch Tests

Drug	Vehicle	Concentration (%)	Ref.
Clonidine	Petrolatum	1	76
	Petrolatum	9	77
Estradiol	Petrolatum	5	75
Fentanyl[a]			
Nicotine	Water	10	84
Nitroglycerin	Petrolatum	2	79,80
	Water	0.02	81–83
Scopolamine	Petrolatum	1.8	4
	Water	0.25	78
Testosterone	Petrolatum	5	

[a]To be determined.

1. Estradiol

Estradiol is available as transdermal therapeutic systems, licensed for hormone replacement in postmenopausal women. After assessing data from those trials involving more than 100 patients (and up to 15,194 patients), the reported incidence of skin reactions to the transdermal therapeutic system was between 5 and 35% (91–96). Most reactions consisted of mild erythema or pruritus at the application site, which generally resolved after system removal. However, a small percentage of cases have been of sufficient severity to cause patients to discontinue treatment. Erkkola et al. (92) noted that 8.8% of patients withdrew from transdermal estradiol therapy because of skin irritation, although the number of patients withdrawing from treatment for this reason in other studies has been less than 5% (91,93,95–97). The most common adverse effect observed using transdermal estradiol was local irritation at the application site (98,99).

Similar results have been found in long-term (1-year) studies. Nachtigall (100) reported skin irritation in 14% of 138 patients receiving transdermal estradiol therapy; 3% of patients discontinued treatment for this reason. Randall (101) reported on 29 patients, 10% withdrawing because of skin irritation.

Unpublished tolerability data involving 11,562 patients using estradiol transdermal therapeutic systems have shown a comparable incidence of dermatological adverse experiences. Treatment was either cyclic or continuous and, in some cases, included concomitant oral administration of progestogen. Duration of treatment varied from 8 to 52 weeks. Pooled results showed that, on average, the incidence of local skin reactions was 14.2%. Skin reactions were the most commonly reported adverse experience, accounting for 47% of all reported adverse experiences. These reactions caused 6.3% of the patients, on average, to withdraw from treatment (data on file, Ciba Geigy).

Several studies have specifically investigated the effects of the estradiol transdermal therapeutic system on skin. In many cases, the cutaneous adverse effects reported have been overcome by changing application site. Allergic contact dermatitis has been induced by the components of the patch, as well as from the estrogen. The components of the patch, such as adhesive (102), hydroxypropyl cellulose (103), enhancer, such as alcohol, present in the reservoir (102,104); as well as the estrogen (102,105,106) have been shown to cause contact dermatitis, but this is uncommon (107,108).

2. Clonidine

Clonidine is a centrally acting α-agonist used primarily as an antihypertensive agent. A common adverse effect associated with transdermally administered clonidine is the development of local skin reactions to the clonidine preparation. Reports of such dermatological reactions range in incidence between 5 and 42% (109–112). These reactions vary in severity from mild erythema and pruritus to vesiculation and inflammatory infiltration of the skin beneath the transdermal patch. Rarely, development of a generalized maculopapular rash has also been reported to occur following transdermal clonidine therapy. The majority of the skin reactions requiring discontinuation of therapy are mediated by a delayed-type IV hypersensitivity reaction (allergic contact dermatitis), which can be confirmed with patch testing using components of the clonidine transdermal device. In most of these patients the allergic reaction is due to clonidine specifically, whereas in other patients, a specific component of the transdermal system (polyisobutylene) functions as the allergen (79,109).

There is an effect of race and gender on the irritation rates from clonidine patch systems. For example, occlusive transfermal clonidine patch systems show sensitization rates of 34% in white women, 18% in white men, 14% in black women, and 8% in black men (113). Itchiness under the patch and contact dermatitis were reported from clonidine transdermal patches (114). The long-term safety and efficacy of transdermal clonidine was evaluated in 102 patients for over 5 years. Transient local side effects occurred, mainly between weeks 4–26; thereafter, the incidence clearly diminished and adverse events did not cause any withdrawal related to skin reactions from 1 year up to 5 years. Overall the long-term transdermal clonidine treatment was highly accepted and was well tolerated by the patients (115). It is important to point out that predictive patch testing for allergic contact dermatitis potential requires special techniques, not only for clonidine systems but for transdermal systems in general (79,116).

3. Scopolamine

Scopolamine, a belladonna alkaloid indicated for nausea and vomiting associated with motion, radiotherapy, anesthesia, and surgery, was the first drug approved for use as a transdermal patch-type delivery system. There are three reports of allergic contact dermatitis to scopolamine. In a group of 164 sailors, 10% developed allergic contact dermatitis after 1.5–15 months of use (117). Patch testing with 1.8% scopolamine in petrolatum (2) or 0.25% in water (80) has been used to confirm allergic contact dermatitis. Patch testing with structurally related alkaloids has failed to demonstrate cross-sensitivity, indicating the specific nature of the antigenic site for scopolamine (2,118).

4. Nitroglycerin

Nitroglycerin is an organic nitrate used for the prevention and treatment of angina pectoris caused by coronary artery disease. Erythema under the nitroglycerin transdermal patch is frequent and represents the capacity of nitroglycerin to cause vasodilation. Rubefaction at the margins of covered skin, noticed with a similar frequency in placebo and active nitrate patches, is indicative of mild irritation (119). Irritant reddening disappeared spontaneously within a few hours. A more severe reaction, localized to the site of nitroglycerin and subsequently placebo patches, has been described (120).

Allergic contact dermatitis to nitroglycerin, both in ointments and patch-type transdermal drug delivery systems, has been reported (81,83,85,121,122). Some delivery systems have employed acrylate adhesives, which have been implicated as the allergen in several of these cases (123).

5. Nicotine

The pharmacological side effects of transdermally absorbed nicotine have assumed greater significance following recent research in alkaloid delivery through the skin. Percutaneous administration of nicotine may reduce the craving experienced during abstinence from cigarette smoking and, thereby, serve as useful supplementation regimen during the behavioral modification process (124). Percutaneous nicotine administration induces predominant sudorific and rubiform responses in the skin that may be accompanied by subtle pyloerection, hyperalgesia, and pruritus (125).

The most common adverse effects of nicotine patches are cutaneous, characterized by itching (16–29% of patients), erythema (7–25%), and edema (2–7%). Poor cutaneous tolerability is a significant problem, resulting in withdrawal of the treatment in 2–5% of patients (126,127). Bircher et al. (86) investigated 14 volunteers with a history of adverse skin reactions to nicotine transdermal therapeutic systems. Five of 14 demonstrated contact sensitivity to 10% aqueous nicotine solution. Irritant reactions in 9 individuals were due to occlusion. The safety, tolerability, and efficacy of transdermal nicotine patch was studied in 80 patients who smoked. Side effects, such as itch and local erythema, were reported in 4 patients (128).

6. Testosterone

Testosterone transdermal therapeutic systems are used in the treatment of hypogonadism in men. One system is designed to be applied to the scrotum and requires changing daily. Three male subjects of nine reported transient pruritus with the placebo patch; however, none reported this with the use of testosterone transdermal therapeutic systems (129). An alternative system for application to glabrous skin was recently been commercialized in the United States. The package insert lists blister development in 11.5% of the phase I–III clinical study population (130).

7. Fentanyl

Fentanyl is a narcotic analgesic used for medication before surgical procedures. Adverse effects on skin (erythema) have been reported (131,132). The physicochemical properties and adverse effects of transdermally administered fentanyl have been described. Dermatological reactions to the fentanyl patch are generally transient and mild (133).

B. Iontophoresis

Iontophoresis increases the penetration of ionized substances into or through a tissue by application of an electric field (76,134–137). Iontophoresis has the potential to overcome many limitations associated with conventional transdermal systems and could be feasible for ionic, hydrophilic, and higher molecular weight drugs. Skin irritation, however, has been reported following iontophoresis, but extensive toxicological studies are still required (138). Such studies are underway in our laboratory (139).

1. Barrier Properties and Skin Reactions

Skin irritation and stratum corneum integrity following 1- and 4-h saline iontophoresis in human subjects were evaluated using several response measurements:

1. A visual scoring system
2. Transepidermal water loss
3. Skin capacitance
4. Skin color
5. Skin temperature

Saline iontophoresis for 1 or 4 h did not produce significant changes in skin water loss and skin water content, suggesting that skin function was unaffected by transcutaneous electrical stimulation. However, the occurrence of transient changes in skin structure (papules) was observed (unpublished data).

Papules are observed following iontophoresis, indicating that the electrical current occasionally induced short-term, transient changes in skin. Direct effects of electrical stimulation on vascular permeability were reported. For example, macromolecular capillary leakage was demonstrated following stimulation of the hamster cheek pouch and the rabbit tibia with direct current of 5–50 μA for 30–160 min (140). Several types of sweat retention (miliaria) have been described. Iontophoresis produces miliaria rubra with distilled water after 10 min of current delivery at 0.5 mA/cm^2 (141). The same study showed that vesicles had a different aspect than those we have observed, in that they were uniformly scattered and their walls were fragile enough to be rubbed off with a towel. Also, physiological saline did not produce vesicles under the same conditions.

The effect of 4-h saline iontophoresis at the current density of 0.2 mA/cm^2 was investigated on skin barrier function and irritation in four ethnic groups (whites, Hispanics, blacks, and Asians) (142). The results suggested that iontophoresis was well tolerated in all four groups, and that skin barrier function, as determined by TEWL and skin capacitance measurements, was not irreversibly affected by iontophoresis in any group. There was no significant difference ($p > 0.05$) in skin temperature, compared with baseline at all observation points in the ethnic groups. No edema was observed in any group. However erythema was higher than the baseline owing to iontophoresis in all the four groups (Table 3). The subjects also demonstrated papules. The highest number of subjects exhibiting papules were in the Asian group followed by Hispanics, whites, and blacks (142). The results of skin reactions to iontophoresis in four ethnic groups are given in Table 4. Details of differences in ethnic skin can be found in Berardesca and Maibach (143).

Solvents remove intercellular lipids resulting in cutaneous barrier disruption. In a study on the effect of alcohol, acetone, and electrode gel swabbing and iontophoresis on skin irritation, the skin integrity was not affected. However, erythema and papules were observed, but these were virtually resolved 24 h after patch removal (144).

Table 3 Draize Erythema Scores[a] from Iontophoresis in Ethnic Groups

Ethnic group	Active site	Control site
White	8E1 (3E1); 2E2	2E1
Black	9E1 (3E1); 1E2	2E1
Hispanic	3E1 (4E1); 7E2	2E1
Asian	4E1; 6E2	1E1

[a]Entries are frequencies of subjects experiencing the level of the erythema (E). Entries are listed such that for example: 8E1 (3E1) indicates eight subjects developed erythema score of "1" which was not resolved in three subjects 1440 min after patch removal. Erythema score was not significantly different ($p > 0.05$) among ethnic groups.
Source: Ref. 142.

2. Sensation and Itching

The range of sensations evoked by transcutaneous electrical stimulation have varied from tactile (touch, vibration, or other) to pricking pain and itch. Thermal sensations, however, have rarely been reported (145–152). A high-voltage low-current transcutaneous electrical stimulating device was tested for prickle sensation in 162 subjects. The initial sensation experienced by subjects was prickle (153).

Itching is felt in certain subjects during and after iontophoresis. The way that itch is signaled to the central nervous system (CNS) remains incompletely under-

Table 4 Skin Reactions[a] to 4-h Iontophoresis: Influence of Ethnic Group

Ethnic group	Observation time		
	Patch removal	60 min	1440 min
White	PA 1p		
	AA 3p	AA 2p	AA 1p
Black	AA 3p	AA 3p	AA 1c
Hispanic	AA 8p	AA 7p	AA 3c
	AC 1p		
Asian	AA 8p	AA 10p	AA 6c
	AC 5p	AC 5p	

[a]Entries are frequencies of subjects ($n = 10$) experiencing papules (p) and papules in dry state [i.e., crust (c)]. Entries are listed as active anode, AA; active cathode, AC; passive anode, PA; and passive cathode, PC: for example, PA 1p at patch removal immediately after iontophoresis indicates one subject developed papules at the passive anode site at observation time immediately after patch removal and AA 1c at 1440 min (24 h) indicates one subject still had papules in dry state (crust) 1440 min after iontophoresis at the active anode site.
Source: Ref. 142.

stood. A general theory proposes that the whole spectrum of cutaneous sensations is signaled by differences in the patterns of activity; hence, any particular neuron can signal a variety of sensory modalities. However, the finding that high-frequency, electrical stimulation of large myelinated axons in the peripheral nerves of conscious humans consistently evokes painless sensations argues against such models (154). An alternative view proposes that an individual neuron transmits a specific type of sensory information. Itch sensation evoked by percutaneous microiontophoresis of histamine on hairy human skin was studied (155). Iontophoresis of histamine evoked sensations of itch in human subjects; therefore, itch sensation may be implicated owing to release of histamine during iontophoresis.

Under proper conditions, touch sensations, such as thumping, vibration, and pulsing, can be elicited by electrical stimulation of hairy skin (156–158). In contrast with hairy skin, the threshold sensations on glabrous skin are touch instead of pruritus (159). With electrode paste used on six subjects as a conductive medium, half reported painful sensation, similar to acid burn, but with saline they felt itch (160). Changes in electrode size may alter the quality of sensation from itch to pain (161). Recent advances in defining C-fiber function is described (162).

3. Burns

Burns occur if the patient uses excessive stimulation with small-area electrodes or if the interface between the skin and electrode is dry (163). Shealy and Maurer (164) demonstrated that the electrode surface area must be more than 4 cm^2 for a 500-μs 85-mA–pulse, 185-pps stimulus. The heat produced must be less than 250 mcal/cm^2 s^{-1} to avoid localized burns. Burton (165) described another type of injury, micropunctate burns. The explanation was that current flow is not distributed over a wide surface area, but is concentrated in small punctate areas (usually hair follicles). Because of the concentration of the large volume of current in small areas, current density is high, resulting in skin burns. These micropunctate skin burns represent true thermal damage to the skin, but Burton (165) feels that they are of little clinical significance, in themselves, as with allergic reactions, simple discontinuation of use of the electrodes permit recovery.

4. Virus Activation

In humans, one isolated case of an outbreak of molluscum contagiosum, a DNA virus of the pox group, at the site of hydrocortisone iontophoresis has been described (166). It appears that the phenomenon is more related to the drug being delivered than to iontophoretic mode of delivery.

C. Electroporation

Electroporation involves alteration of lipid bilayers when transient and pulsed electric fields lead to the reversible formation of nonlamellar lipid phases: a pore. Iontophoresis utilizes existing pathways, such as hair follicles or sweat glands. These sweat glands and hair follicles comprise only about 0.1% of the total skin surface area. Thus, a high-charge density occurs around sweat glands and hair follicles, which may potentially lead to localized skin irritation. In electroporation, the other 99.9% of the skin's surface area is reversibly altered using a brief pulse of electricity. As a consequence, the current density is distributed more uniformly across the surface; thus, potential for irritation may be reduced (167).

The effect of current and voltage on pig skin was evaluated under conditions of iontophoresis and electroporation (167). Pigs were treated with either an iontophoresis or an electroporation protocol. The study evaluated irritation, not drug delivery. Current densities used were in the range from 0 to 10 mA/cm^2, and the applied voltage ranged from 0 to 1000 V. The potential was a single pulse followed by 30 min of iontophoresis. Irritation was measured at 0 and 4 h after treatment. Skin biopsies were taken for histological examination. Irritation was measured by the visual scoring system of Draize et al. (6). Use of conventional iontophoresis, when there was no applied voltage pulse and current density was near 0.2 mA/cm^2, resulted in no significant difference from the no-pulse values in either of these measures. The skin response was measured in terms of erythema at the anode and the cathode. Again iontophoresis produced a value not significantly different from that of a pulse plus iontophoresis at both the cathode and the anode. These results showed that a pulse voltage of up to 1000 V had no effect on erythema or edema. Erythema and edema are equivalent for iontophoresis and electroporation. Thus, one can conclude that electroporation under these test conditions produced no measurable damage to skin or tissue.

IV. CONCLUSIONS

Adequate evaluation of irritation potential of chemical substances depends on a thorough understanding of the variables influencing the irritant response. Guinea pig testing and the local lymph node assays constitute a first step in evaluating the allergenicity of new compounds or products. With the traditional Draize test, potent irritants can be detected. Substances that irritate in humans also do so in some animals. More sensitive animal tests will identify weak irritants. The comparative sensitivity of these various tests is still under examination. There is a reasonable degree of correlation between the GPM test and the human maximization test.

Transdermal therapeutic systems have proved to be a useful adjunct for administration of systemic medications. Their potential for future applications seems excellent. However, the systems carry a risk of either irritant or allergic skin sensitivity. Avoidance of reapplication of patches directly over the previous site, should help minimize the incidence and severity of such irritation. Keep in view that with the delayed onset of some allergic reactions, safety data based on short-term experience should be considered with caution.

There is increasing interest in the use of iontophoresis. Such therapy may require long-term delivery and the extended wearing of delivery systems. Irritations in such patients may be greater than those found with the more brief applications for which iontophoresis is most widely used today. There is no doubt that iontophoresis can be a safe and effective method of drug delivery by the innovative application of modern electronics and material science; however, extensive skin toxicological studies are warranted. Alternatively, electroporation followed by iontophoresis can be used to lower the skin irritation. With electroporation, new pathways are created. As a consequence, there is more even distribution of charge; hence, there may be a lower potential for irritation.

REFERENCES

1. Sarkany I, Gaylarde PM. Thickening of guinea pig epidermis due to application of commonly used ointment bases. Trans St John's Hosp Dermatol Soc 59:241, 1973.
2. Trozak DJ. Delayed hypersensitivity to scopolamine delivered by a transdermal device. J Am Acad Dermatol 13:247–251, 1985.
3. Weickel R, Frosch PJ. Kontaktallergie auf Glycerotrinitrat (Nitroderm TTS). Hautarzt 37:511, 1986.
4. Chandraratna PAN, O'Dell RE. Allergic reactions to nitroglycerin ointment: a report of five cases. Curr Ther Res 25:481, 1979.
5. Rougier A, Goldberg A, Maibach HI, eds. In Vitro Skin Toxicology: Irritation, Phototoxicity and Sensitization. New York: Mary Ann Liebert, 1994.
6. Draize JH, Woodard G, Calvery HO. Methods for the study of irritation and toxicity of substances applied topically to the skin and mucous membranes. J Pharmacol Exp Ther 83:377–390, 1944.
7. Patrick E, Maibach HI. Predictive assays: animal and man, and in vitro and in vivo. In: Textbook of Contact Dermatitis. Berlin: Springer-Verlag, p 105, 1995.
8. Idson DR. In vivo measurement of transepidermal water loss. J Soc Cosmet Chem 29:573, 1978.
9. Wilhelm KP, Surber C, Maibach HI. Quantification of sodium lauryl sulfate irritant dermatitis in man: comparison of four techniques: skin color reflectance, transepidermal water loss, laser Doppler flow measurement and visual scores. Arch Dermatol Res 281:1–3, 1989.
10. Tagami H, Ohi H, Iwatsuki K, Kanamaru Y, Yamada M, Ichijo B. Evaluation of the skin surface hydration in vivo by electrical measurement. J Invest Dermatol 75:500–507, 1980.
11. Thiele FAJ, Malten KE. Evaluation of skin damage. Br J Dermatol 89:373, 1973.
12. Serban GP, Henry SM, Cotty VF, Marcus AD. In vivo evaluation of skin lotions by electrical capacitance 1. The effect of several lotions on the progression of damage and healing after repeated insult with sodium lauryl sulfate. J Soc Cosmet Chem 32:407–419, 1981.
13. Malten KE, Thiele FAJ. Evaluation of skin damage. II. Water loss and carbon dioxide release measurements related to skin resistance measurements. Br J Dermatol 89:565, 1973.
14. Thiele FAJ. Measurements on the Surface of the Skin. Nijmegen, Netherlands: Drekkeij van Mammeren, p 81, 1974.
15. Elsner P, Berardesca E, Maibach HI, eds. Bioengineering of the Skin: Water and the Stratum Corneum. Boca Raton: CRC Press, 1994.
16. Berardesca E, Elsner P, Maibach HI, eds. Bioengineering of the Skin: Cutaneous Blood Flow and Erythema. Boca Raton: CRC Press, 1994.
17. Berardesca E, Elsner P, Wilhelm K, Maibach HI, eds. Bioengineering of the Skin: Methods and Instrumentation. Boca Raton: CRC Press, 1995.
18. Buehler EV. Delayed contact hypersensitivity in the guinea pig. Arch Dermatol 91:171–177, 1965.
19. Magnusson B, Kligman AM. The identification of contact allergens by animal assay. The guinea pig maximization test. J Invest Dermatol 52:268–276, 1969.
20. Marzulli FN, Carson TR, Maibach HI. Delayed contact hypersensitivity studies in man and animals. Proc Joint Conf Cosmet Sci, p 107, 1968.
21. Draize JH. Dermal toxicity. Appraisal of the safety of chemicals in foods, drugs and cosmetics. The Association of Food and Drug Officials of the United States. Texas State Department of Health, p 46, 1959.

22. Kligman AM. The identification of contact allergens by human assay. III. The maximization test: a procedure for screening and rating contact sensitizers. J Invest Dermatol 47:393–409, 1966.

23. Lepoittevin J–P, Benezra C, Sigman CC, Bagheri D, Fraginals R, Maibach HI. Molecular aspects of allergic contact dermatitis. In: Textbook of Contact Dermatitis. Berlin: Springer-Verlag, p 105, 1995.

24. Menne T, Flyvholm MA, Maibach HI. Prevention of allergic contact sensitization. In: Vos JG, Younes M, Smith E, eds. Allergic Hypersensitivities Induced by Chemicals. Boca Raton: CRC Press, p 287, 1996.

25. Marzulli FN, Maibach HI. Allergic contact dermatitis. In: Marzulli FN, Maibach HI, eds. Dermatotoxicology. 5th ed. Washington: Hemisphere, p 143–146, 1996.

26. Marzulli FN, Maibach HI. Test methods for allergic contact dermatitis in humans. In: Marzulli FN, Maibach HI, eds. Dermatotoxicology. 5th ed. Washington: Hemisphere, pp 477–483, 1996.

27. Magnusson B, Kligman AM. Usefulness of guinea pig tests for detection of contact sensitizers. In: Marzulli FN, Maibach HI, eds. Advances in Modern Toxicology. Vol 4. Dermatotoxicology and Pharmacology. New York: John Wiley & Sons, p 551, 1977.

28. Marzulli FN, Maibach HI, eds. Dermatotoxicology. 5th ed. Washington: Hemisphere, 1996.

29. Anderson R, Maibach HI, eds. Guinea Pig Sensitization Assays. New York: Karger, 1987.

30. Welthriend S, Kwangsukstith C, Maibach HI. Contact urticaria from cucumber pickle and strawberry. Contact Dermatitis 32:173–174, 1995.

31. Hardy MP, Maibach HI. Contact urticaria syndrome from sorbitan sesquioleate in a corticosteroid ointment. Contact Dermatitis 32:114, 1995.

32. Maibach HI. Contact urticaria syndrome from mold on salami casing. Contact Dermatitis 32:120–121, 1995.

33. West I, Maibach HI. Contact urticaria syndrome from multiple cosmetic components. Contact Dermatitis 32:121, 1995.

34. Elsner P, Maibach HI, eds. Irritant Dermatitis, New Clinical and Experimental Aspects. Current Problems in Dermatology. Vol. 23. Basel: Karger, 1995.

35. Holness DL, Nethercott JR, Adams RM, Belsito D, Deleo V, Emmett EA, Fowler J, Fisher AA, Larsen WG, Maibach HI, Marks J, Reitschel RL, Rosenthal LE, Schorr WF, Storrs FJ, Taylor JS. Concomitant positive patch test results with standard screening tray in North America 1985–1989. Contact Dermatitis 32:289–292, 1995.

36. Maibach HI. Dermatologic vehicles: science and art. Cutis 55(4S):4, 1995.

37. Funk JO, Dromgoole SH, Maibach HI. Sunscreen intolerance: contact sensitization, photocontact sensitization, and irritancy of sunscreen agents. Dermatol Clin 13:473–481, 1995.

38. Hatch KL, Maibach HI. Textile dermatitis: an update (1) resins, additives and fibers. Contact Dermatitis 32:319–326, 1995.

39. Hardy M, Maibach HI. Contact urticaria syndrome from sorbitan sesquioleate in a corticosteroid ointment. Contact Dermatitis 32:114, 1995.

40. Lee CH, Maibach HI. The sodium lauryl sulfate model: an overview. Contact Dermatitis 33:1–7, 1995.

41. Hostynek JJ, Lauerma AI, Magee PS, Maibach HI. A local lymph-node assay validation study of a structure–activity relationship model for contact allergens. Arch Dermatol Res 287:567–571, 1995.

42. Hui X, Wester RC, Magee PS, Maibach HI. Partitioning of chemicals from water into powdered human stratum corneum (callus): a model study. In Vitro Toxicol 8:159–167, 1995.

43. Tur E, Aviram G, Zeltser D, Brenner S, Maibach HI. Regional variations of human skin blood flow response to histamine. In: Exogenous Dermatology, Advances in Skin-Related Allergology, Bioengineering, Pharmacology and Toxicology. Basel: Karger, p 59, 1995.

44. Ale SI, Maibach HI. Clinical relevance in allergic contact dermatitis, an algorithmic approach. Occup Environ Dermatoses 43:119–121, 1995.

45. Liden C, Maibach HI, Wahlberg JE. Skin. In: Metal Toxicology. New York: Academic, p 447, 1995.

46. Kwangsukstith C, Maibach HI. Contact urticaria from polyurethane membrane hypoallergenic gloves. Contact Dermatitis 33:200–201, 1995.

47. Patil S, Harvell J, Maibach HI. In vitro skin irritation assays: relevance to human skin. In: Gali CL, Goldberg AM, Marinovich M, eds. Modulation of Cellular Responses in Toxicity. Berlin: Springer-Verlag, p 283, 1995.

48. Effendy I, Maibach HI. Surfactants and experimental irritant contact dermatitis. Contact Dermatitis 33:217–225, 1995.

49. Lauerma AI, Aioi A, Maibach HI. Topical *cis*-urocanic acid suppresses both induction and elicitation of contact hypersensitivity in BALB/C mice. Acta Derm Venereol 75: 272–275, 1995.

50. Andersen PH, Maibach HI. Skin irritation in man: a comparative bioengineering study using improved reflectance spectroscopy. Contact Dermatitis 33:315–322, 1995.

51. Tur E, Eshkol Z, Brenner S, Maibach HI. Cumulative effect of subthreshold concentrations of irritants in humans. J Contact Dermatitis 6:216–220, 1995.

52. van der Valk PGM, Maibach HI, eds. The Irritant Contact Dermatitis Syndrome. Boca Raton: CRC Press, 1995.

53. Ophaswongse S, Maibach HI. Allergic contact cheilitis. Contact Dermatitis 33:365–370, 1995.

54. Effendy I, Loeffler H, Maibach HI. Baseline transepidermal water loss in patients with acute and healed irritant contact dermatitis. Contact Dermatitis 33:371–374, 1995.

55. Kwangsukstith C, Maibach HI. Effect of age and sex on the induction and elicitation of allergic contact dermatitis. Contact Dermatitis 33:289–298, 1995.

56. Frosch PJ, Pilz B, Andersen KE, Burrows D. Carnarasa JG, Dooms–Goossens A, Ducombs G, Fuchs T, Hannuksela M, Lachapelle JM, Lahti A, Maibach HI, Menne T, Rycroft RJG, Shaw S, Wahlberg JE, White IR, Wilkinson JD. Patch testing with fragrances: results of multicenter study of the European Environmental and Contact Dermatitis Research group with 48 frequently used constituents of perfumes. Contact Dermatitis 33:333–342, 1995.

57. Maibach HI. Possible cosmetic dermatitis due to mercaptobenzothiazole. Contact Dermatitis 34:72, 1996.

58. Patil S, Singh P, Maibach HI. Radial spread of sodium lauryl sulfate after topical application. Pharm Res 12:2018–2023, 1995.

59. Amin S, Maibach HI. Cosmetic intolerance syndrome: pathophysiology and management. Cosmet Dermatol 9:34, 1996.

60. Smith JD, Odom RB, Maibach HI. Contact urticaria from cobalt chloride. Arch Dermatol 111:1610, 1995.

61. Effendy I, Kwangsukstith C, Lee JY, Maibach HI. Functional changes in human stratum corneum induced by topical glycolic acid: comparisons with all-*trans*-retinoic acid. Acta Derm Venereol 75:455, 1995.

62. Nangia A, Andersen PH, Berner B, Maibach HI. High dissociation constants (pKa) of basic permeants are associated with in vivo skin irritation in man. Contact Dermatitis 34:237–242, 1996.

63. Maibach HI. Acne necrotica varioliformis (*P. acnes* folliculitis); resolution with isotretinoin. Eur J Dermatol 6:153, 1996.

64. Mitchell JC, Maibach HI. Difficulties to investigate dermatitis from plants, a practical approach to office-based diagnosis. Dermatosan 44:29, 1996.

65. Effendy I, Weltfriend S, Patil S, Maibach HI. Differential irritant skin responses to topical retinoic acid and sodium lauryl sulfate: alone and crossover design. Br J Dermatol 134:424, 1996.

66. Flyvholm MA, Menne T, Maibach HI. Skin allergy: exposure and dose–response relationships. In: Vos JG, Younes M, Smith E, eds. Allergic Hypersensitivities Induced by Chemicals. Boca Raton: CRC Press, p 261, 1996.

67. Weltfriend S, Bason M, Lammintausta K, Maibach HI. Irritant dermatitis (irritation). In: Marzulli FN, Maibach HI, eds. Dermatotoxicology. 5th ed. Washington: Hemisphere, pp 87–118, 1996.

68. Menné T, Veien N, Maibach HI. Systemic contact-type dermatitis. In: Marzulli FN, Maibach HI, eds. Dermatotoxicology. 5th ed. Washington: Hemisphere, pp 161–175, 1996.

69. Patil SM, Hogan DJ, Maibach HI. Transdermal drug delivery systems: adverse dermatologic reactions. In: Marzulli FN, Maibach HI, eds. Dermatotoxicology. 5th ed. Washington: Hemisphere, pp 389–396, 1996.

70. Amin S, Lahti A, Maibach HI. Contact urticaria and the contact urticaria syndrome (immediate contact reactions). In: Marzulli FN, Maibach HI, eds. Dermatotoxicology. 5th ed. Washington: Hemisphere, pp 485–503, 1996.

71. Amin S, Lauerma AI, Maibach HI. Diagnostic tests in dermatology: patch and photopatch testing and contact urticaria. In: Marzulli FN, Maibach HI, eds. Dermatotoxicology. 5th ed. Washington: Hemisphere, pp 505–513, 1996.

72. Marzulli FN, Maibach HI. Photoirritation (phototoxicity) testing in humans. In: Marzulli FN, Maibach HI, eds. Dermatotoxicology. 5th ed. Washington: Hemisphere, pp 531–533, 1996.

73. Toro JR, Engasser PG, Maibach HI. Cosmetic reactions. In: Marzulli FN, Maibach HI, eds. Dermatotoxicology. 5th ed. Washington: Hemisphere, pp 607–642, 1996.

74. Guy RH, Hadgraft J, Bucks DAW. Transdermal drug delivery and cutaneous metabolism. Xenobiotica 17:325–343, 1987.

75. Roberts MS, Singh J, Yoshida N, Currie KI. Iontophoretic transport of selected solutes through human epidermis. In: Scott RC, Hadgraft J, Guy RH, eds. Prediction of Percutaneous Penetration. Vol. 1. London: IBC Technical Services, pp 231–241, 1990.

76. Singh S, Singh J. Transdermal drug delivery by passive diffusion and iontophoresis: a review. Med Res Rev 13:569, 1993.

77. Nater JP, de Groot AC, eds. Unwanted Effects of Cosmetics and Drugs Used in Dermatology. Amsterdam: Elsevier, 1985.

78. Grattan CEH, Kennedy CTC. Allergic contact dermatitis to transdermal clonidine. Contact Dermatitis 12:225–226, 1985.

79. Maibach HI. Oral substitution in patients sensitized by transdermal clonidine treatment. Contact Dermatitis 16:1–8, 1987.

80. van der Willigen AH, Oranje AP, Stolze E, van Joost T. Delayed hypersensitivity to scopolamine in transdermal therapeutic systems. J Am Acad Dematol 18:146–147, 1988.

81. Harari Z, Sommer I, Knobel B. Multifocal contact dermatitis to Nitroderm TTS 5 with extensive postinflammatory hypermelanosis. Dermatologica 174:249–252, 1987.

82. Zugerman C, Zheutlin T, Giacobetti R. Allergic contact dermatitis secondary to nitroglycerin in Nitro-Bid ointment. Contact Dermatitis 5:27, 1979.

83. Rosenfeld AS, White WB. Allergic contact dermatitis secondary to transdermal nitroglycerin. Am Heart J 108:1061, 1984.

84. Sausker WF, Frederick FD. Allergic contact dermatitis secondary to topical nitroglycerin. JAMA 239:1743–1744, 1978.

85. Topaz O, Abraham D. Severe allergic contact dermatitis secondary to nitroglycerin in a transdermal therapeutic system. Ann Allergy 59:365–366, 1987.

86. Bircher AJ, Havald H, Rufli T. Adverse reaction to nicotine in a transdermal therapeutic system. Contact Dermatitis 25:230, 1991.

87. Sitruk–Ware R. Estrogen therapy during menopause: practical treatment recommendations. Drugs 39:203–217, 1990.

88. Utian WH. Transdermal oestradiol: a recent advance in oestrogen therapy. Drugs 36: 383–386, 1988.

89. Youngkin EQ. Estrogen replacement therapy and the Estraderm transdermal system. Nurse Pract 15:19, 1990.

90. Balfour JA, Heel RC. Transdermal estradiol: a review of its pharmacodynamic and pharmacokinetic properties, and therapeutic efficacy in the treatment of the menopausal complaints. Drugs 40:561–582, 1990.

91. Buvat J, Buvat–Herbaut M, Desmons F, Cuvelier C, Vasseur C. Les patchs à l'estradiol. Un progrès dans le traitement de la ménopause. Gynécologie 40:53–59, 1989.

92. Erkkola R, Holma P, Jarvi T, Nummi S, Punnonen R. Transdermal oestrogen replacement therapy in a Finnish population. Maturitas 13:275–281, 1991.

93. Grall JY. Estrogenothérapie substitutive par voie transdermique: sur quels critères choisir la posologie initale? Tempo Méd 410(suppl):1–7, 1990.

94. Janaud A. Estraderm TTS dans le traitement des troubles liés à la ménopause: résultats d'une étude long terme chez 324 femmes. Gaz Méd 97:1–7, 1990.

95. Kerzel C, Keller PJ. Ein transdermales therapeutisches System zur hormonellen Therapie klimakterischer ausfallserscheinungen. Multizentrische Studie. Geb Frauenheilkd 47:565–568, 1987.

96. Mück AO. Transdermales therapeutisches System zur physiologischen Östrogensubstitution. Therapie 40:41–51, 1990.

97. Rabe T, Mück AO, Runnebaum B. Östrogensubstitution in der postmenopause mit Östradiol durch ein transdermales therapeutisches System In: Rietbrock, ed. Die Haut als Transportorgan für Arzneistoffe. Darmstadt: Steinkopff Verlag, pp 81–92, 1990.

98. Balfour JA, McTavish D. Transdermal estradiol: a review of its pharmacological profile, and therapeutic potential in the prevention of postmenopausal osteoporosis. Drug Aging 2:487–507, 1992.

99. Gordon SF, Thompson KA, Ruoff GE, Imig JR, Lane PJ, Schwenker CE. Efficacy and safety of a seven-day transdermal drug delivery system: comparison with conjugated estrogens and placebo. The Transdermal Estradiol Patch Study Group. Int J Fertil Stud 40:126, 1995.

100. Nachtigall LE. Longer-term study of Estraderm TTS. In: Birdwood GF, ed. Transdermal Estrogen Replacement for Menopausal Women. Toronto: Hans Huber, pp 25–28, 1998.

101. Randall S. Clinical experience with transdermal hormone replacement therapy. In: Birdwood, ed. Transdermal Estrogen Replacement for Menopausal Women. Toronto: Hans Huber, pp 42–44, 1988.

102. Fisher AA. Contact dermatitis: highlights from the 1987 meeting of the American Academy of Dermatology, San Antonio, Texas. Cutis 41:87, 1988.

103. Schwartz BK, Clendenning WE. Allergic contact dermatitis from hydroxypropyl cellulose in a transdermal estradiol patch. Contact Dermatitis 18:106, 1988.

104. Ducros B, Bonnin JP, Navaranne A, Colomb D. Eczema due to contact with ethanol in oestradiol transdermal patch (Estraderm TTS 50). Nouv Dermatol 8:21, 1989.

105. Carmichael AJ, Foulds IS. Allergic contact dermatitis to oestradiol in oestrogen patches. Contact Dermatitis 26:194, 1992.

106. McBurney EI, Noel SB, Collins JH. Contact dermatitis to transdermal estradiol system. J Am Acad Dermatol 20:503–510, 1989.

107. Davis GF, Winter L. Cumulative irritation study of placebo estrogen patches. Curr Ther Res 42:712–719, 1987.
108. Hogan DJ, Maibach HI. Adverse dermatologic reactions to transdermal drug delivery systems. J Am Acad Dermatol 22:811–814, 1990.
109. Groth H, Vetter H, Kneusel J. Transdermal clonidine in essential hypertension: problems during long term treatment. In: Weber MA, Drayer JIM, Kolloch R, eds. Low Dose Oral and Transdermal Clonidine Therapy of Hypertension. Darmstadt: Steinkopff, pp 60–65, 1985.
110. Weber MA, Drayer JIM. Clinical experience with rate controlled delivery of antihypertensive therapy by a transdermal system. Am Heart J 108:231–236, 1984.
111. Popli S, Stroka G, Ing TS. Transdermal clonidine for hypertensive patients. Clin Ther 5:624–628, 1983.
112. Boekhorst JC. Allergic contact dermatitis with transdermal clonidine. Lancet 2:1031–1032, 1983.
113. Boehringer Ingelheim. Catapress-TTS (clonidine) package insert. Ridgefield, CT, 1987.
114. Corazza M, Montovani L, Virgili A, Strumia R. Allergic contact dermatitis from a clonidine transdermal delivery system. Contact Dermatitis 32:246, 1995.
115. Breidthardt J, Schumacher H, Mehlburger L. Long-term (5 year) experience with transdermal clonidine in the treatment of mild to moderate hypertension. Clin Autonom Res 3:385–390, 1993.
116. Maibach HI. Clonidine transdermal delivery system: cutaneous toxicity studies. In: Weber MA, Drayer JIM, Kolloch R, eds. Low Dose Oral and Transdermal Clonidine Therapy of Hypertension. Darmstadt: Steinkopff, pp 55–59, 1985.
117. Gordon CR, Shupak A, Doweck I, Spitzer O. Allergic contact dermatitis caused by transdermal hyoscine. Br Med J 298:122–131, 1989.
118. van der Willigen AH, de Graaf YP, van Joost T. Periocular dermatitis from atropine. Contact Dermatitis 17:56, 1987.
119. Muller P, Imhof PR, Burkart F, Chu LC, Gerardin A. Human pharmacological studies of a new transdermal system containing nitroglycerin. Eur J Clin Pharmacol 22:473–480, 1982.
120. Letendre PW, Barr C, Wilkens K. Adverse dermatologic reaction to transdermal nitroglycerin. Drug Int Clin Pharm 18:69–70, 1984.
121. Fischer RG, Tyler M. Severe contact dermatitis due to nitroglycerin patches. South Med J 78:1523–1524, 1985.
122. Hendricks AA, Dec GW. Contact dermatitis due to nitroglycerin ointment. Arch Dermatol 115:853–855, 1979.
123. Fisher AA. Dermatitis due to transdermal therapeutic systems. Cutis 34:526, 1984.
124. Russell MAH, Sutherland G, Costello J. Nicotine replacement: the role of blood nicotine levels, their rate of change, and nicotine tolerance, In: Pomerleau OF, Pomerleau CS, eds. Nicotine Replacement: A Critical Evaluation. New York: Liss, p 63, 1988.
125. Smith EW, Smith KA, Maibach HI, Andersson P–O, Cleary G, Wilson D. The local side effects of transdermally absorbed nicotine. Skin Pharmacol 5:69–76, 1992.
126. Abelin T, Buchler A, Muller P, Besanen K, Imhof PR. Controlled trial of transdermal nicotine patch in tobacco withdrawal. Lancet 1:7, 1989.
127. Russell MAH, Stapleton JA, Feyerabend C, Wiseman SM, Gustavson G. Targeting heavy smokers in general practice: randomised controlled trial of transdermal nicotine patches. Br Med J 306:1308, 1993.
128. Martin PD, Robinson GM. The safety, tolerability and efficacy of transdermal nicotine (Nicotinell TTS) in initially hospitalised patients. NZ Med J 108:6, 1995.
129. Bals–Pratsch M, Yoon YD, Knuth UA, Meschlag E. Transdermal testosterone substitution therapy for male hypogonadism. Lancet 2:943–946, 1986.
130. SmithKline Beecham. Androderm Package Insert.

131. Miser AW, Narang PK, Dothage JA, Young RC, Sindelar W, Miser JS. Transdermal fentanyl for pain control in patients with cancer. Pain 37:15–21, 1989.

132. Gourley GK, Kowalsky SR, Plummer JL. The efficacy of transdermal fentanyl in the treatment of postoperative pain: a double blind comparison of fentanyl and placebo systems. Pain 40:21–28, 1990.

133. Calis KA, Kohler DR, Corso DM. Transdermally administered fentanyl for pain management. Clin Pharm 11:22–36, 1992.

134. Singh J, Roberts MS. Transdermal delivery of drugs by iontophoresis: a review. Drug Des Deliv 4:1–12, 1989.

135. Singh J. Effect of ionization on in-vitro iontophoretic and passive transport of thyrotropin-releasing hormone through human epidermis. Pharm Sci 1:111, 1995.

136. Singh J, Roberts MS. Iontophoretic transport of amphoteric solutes through human epidermis: p-aminobenzoic acid and amphotericin. Pharm Sci 1:223, 1995.

137. Singh J, Bhatia KS. Topical iontophoretic drug delivery: pathways, principles, factors, and skin irritation. Med Res Rev 16:285–296, 1996.

138. Singh J, Maibach HI. Topical iontophoretic drug delivery in vivo: historical development, devices and future perspectives. Dermatology 187:235, 1993.

139. Camel E, O'Connell M, Sage B, Gross M, Maibach H. The effect of saline iontophoresis on skin integrity in human volunteers. Fundam Appl Toxicol 32:168–178, 1996.

140. Nannmark U, Buch F, Albrektsson T. Vascular reactions during electrical stimulation. Vital microscopy of the hamster cheek pouch and the rabbit tibia. Acta Orthop Scand 56:52–56, 1985.

141. Shelley WB, Horvath PN, Weidman FD, Pillsbury DM. Experimental miliaria in man, I. Production of sweat retention anidrosis and vesicles by means of iontophoresis. J Invest Dermatol 11:275, 1949.

142. Singh J, Gross M, O'Connell M, Sage B, Maibach HI. Effect of iontophoresis in different ethnic groups' skin function. Proc Int Symp Control Release Bioact Mater 21:365–366, 1994.

143. Berardesca E, Maibach HI. Racial differences in skin pathophysiology. J Am Acad Dermatol 34:667, 1996.

144. Singh J, Gross M, O'Connell M, Sage B, Maibach HI. Effect of skin preparation and iontophoresis on skin functions and irritation in human volunteers. Pharm Res 11:S101, 1994.

145. Bishop GF. Responses to electrical stimulation of single sensory units of skin. J Neurophysiol 6:361, 1963.

146. Grimnes S. Electrovibration, cutaneous sensation of microampere current. Acta Physiol Scand 118:19, 1983.

147. Laitenen LV, Eriksson AT. Electrical stimulation in the measurement of cutaneous sensibility. Pain 22:139, 1985.

148. Mason JL, Mackay NAM. Pain sensations associated with electrocutaneous stimulation. IEE Trans Biomed Eng 23:405–409, 1976.

149. Reilly JP, Larkin WD. Electrocutaneous stimulation with high voltage capacitive discharges. IEE Trans Biomed Eng 30:631, 1983.

150. Saunders FA. Information transmission across the skin: high resolution tactile sensory aids for the deaf and the blind. Int J Neurosci 19:21–28, 1983.

151. Sigel H. Price threshold stimulation with square-wave current: a new measure of skin sensibility. Yale J Biol Med 26:145, 1953.

152. Tuckett RP. Itch evoked by electrical stimulation of the skin. J Invest Dermatol 79:368–373, 1982.

153. Gransworthy RK, Gully RL, Kenins P, Westerman RA. Transcutaneous electrical stimulation and the sensation of prickle. J Neurophysiol 59:1116–1127, 1988.

154. Collins WF, Nulsen FE, Randt CT. Relation of peripheral fiber size and sensation in man. Arch Neurol 3:381, 1973.
155. Westerman RA, Margerl W, Handwercker HO, Szolcsányi J. Itch sensation evoked by percutaneous microintophoresis of histamine on human hairy skin. Proc Aust Physiol Pharmacol Soc 16:79, 1985.
156. Gibson RH. Electrical stimulation of pain and touch. In: Kenshalo DR, ed. The Skin Senses. Springfield, IL: Charles C Thomas, pp 223–261, 1968.
157. Saunders FA. Recommended procedures for electrocutaneous displays. In: Hambrecht FT, Reswick JB, eds. Functional Electrical Stimulation. New York: Marcel Dekker, p 303, 1977.
158. Sachs RM, Miller JD, Grant KW. Perceived magnitude of multiple electrocutaneous pulses. Percept Psychophys 28:255, 1980.
159. Hahn JF. Cutaneous vibratory thresholds for square wave electrical pulses. Science 127:879–880, 1958.
160. Notermans SLH. Measurement of the pain threshold determined by electrical stimulation and its clinical application. I. Method and factors possibly influencing the pain threshold. Neurology 16:1071–1086, 1966.
161. Pfeiffer EA. Electrical stimulation of sensory nerves with skin electrodes for research, diagnosis, communication and behavioral conditioning: a survey. Med Biol Eng 6:637–651, 1968.
162. Yosipovitch G, Szolar C, Hui X, Maibach HI. Effect of topically applied menthol on thermal, pain and itch sensation and biophysical properties of the skin. Arch Dermatol Res 288:245, 1996.
163. Fisher AA. Dermatitis associated with transcutaneous electrical nerve stimulation. Cutis 21:24–47, 1978.
164. Shealy CN, Maurer D. Transcutaneous nerve stimulation for control of pain. Surg Neurol 2:45, 1974.
165. Burton C. Transcutaneous electrical nerve stimulation to relieve pain. Postgrad Med 59:105, 1976.
166. Plutchik R, Bender H. Electrocutaneous pain thresholds in humans to low frequency square-wave pulses. J Psychol 62:151–154, 1966.
167. Potts RO. Transdermal peptide delivery using electroporation. In: Proceedings of the Third TDS Technology Symposium: Polymers and Peptides in Transdermal Delivery. Tokyo: Nichon Toshi Center, pp 47–67, 1993.

Index

For Product Safety Concerns and Information please contact our EU
representative GPSR@taylorandfrancis.com
Taylor & Francis Verlag GmbH, Kaufingerstraße 24, 80331 München, Germany